SYSTEM IDENTIFICATION

Advances and Cases Studies

ACADEMIC PRESS RAPID MANUSCRIPT REPRODUCTION

This is Volume 126 in
MATHEMATICS IN SCIENCE AND ENGINEERING
A Series of Monographs and Textbooks
Edited by RICHARD BELLMAN, *University of Southern California*

The complete listing of books in this series is available from the Publisher upon request.

SYSTEM IDENTIFICATION

Advances and Case Studies

edited by

Raman K. Mehra
Division of Engineering and Applied Physics
Harvard University
and
Scientific Systems, Inc.
Cambridge, Massachusetts

Dimitri G. Lainiotis
Department of Electrical Engineering
State University of New York
Buffalo, New York

Academic Press New York San Francisco London 1976
A Subsidiary of Harcourt Brace Jovanovich, Publishers

ACADEMIC PRESS, INC.
111 Fifth Avenue, New York, New York 10003

United Kingdom Edition published by
ACADEMIC PRESS, INC. (LONDON) LTD.
24/28 Oval Road, London NW1

Library of Congress Cataloging in Publication Data

Main entry under title:

System identification.

(Mathematics in science and engineering ;)
Includes bibliographical references.
1. System analysis. 2. Time-series analysis.
3. Estimation theory. I. Mehra, Raman K.
II. Lainiotis, Demetrios G. III. Series.
QA402.S956 003 76-46267
ISBN 0-12-487950-0

PRINTED IN THE UNITED STATES OF AMERICA

CONTENTS

LIST OF CONTRIBUTORS

Numbers in parentheses indicate the pages on which the authors' contributions begin.

Hirotugu Akaike (27), The Institute of Statistical Mathematics, Tokyo, Japan

Samer Attasi (289), Iria-Laboria, Rocquencourt, France

T. Bohlin (441), Royal Institute of Technology, Stockholm, Sweden

P. E. Caines (349), Systems Control Group, Department of Electrical Engineering, University of Toronto, Toronto, Canada

C. W. Chan (349), Systems Engineering Section, Unilever Research Laboratories, Port Sunlight, England

Pierre L. Faurre (1), Iria-Laboria and Sagem, Rocquencourt, France

Lennart Ljung (121), Lund Institute of Technology, Lund, Sweden

Raman K. Mehra (211), Division of Engineering and Applied Physics, Harvard University, and Scientific Systems, Inc., Cambridge, Massachusetts

Kumpati S. Narendra (165), Yale University, New Haven, Connecticut

G. C. Goodwin (251), Department of Electrical Engineering, University of Newcastle, New South Wales, Australia

Gustaf Olsson (519), Department of Automatic Control, Lund Institute of Technology, Lund, Sweden

R. L. Payne (251), Department of Systems and Control, University of New South Wales, New South Wales, Australia

J. Rissanen (97), IBM Research Laboratory, San Jose, California

P. M. Robinson (407), Harvard University, Cambridge, Massachusetts

PREFACE

The field of system identification and time series analysis is currently in a state of rapid development. Significant contributions have been made in the past few years by researchers from such diverse fields as statistics, control theory, system theory, econometrics, and information theory. The specialized jargon of each field, geographic isolation of researchers, and the difficulty of working on what Wiener called "cracks between disciplines" has hampered a rich cross fertilization of ideas among different specialties. The purpose of this book is to promote this activity by presenting in one volume promising new approaches and results in the field of system identification, approaches and results that are not easily available elsewhere.

The idea of putting together the current volume originated from this editor's experience with a special issue of the IEEE Transactions on Automatic Control (December 1974).* The limitations on the length of the journal papers made it very difficult for authors to expand fully on their ideas. Furthermore, significant new developments took place, which deserved widespread exposure. The effort turned out to be truly international in character with contributions from seven different countries. The authors were invited to write chapters on their current fields of interest, making their presentations self-contained and summarizing the state-of-art in their subject areas. To achieve depth and completeness in their presentations, the authors have assumed on the part of readers a basic background in statistical estimation and time series analysis, equivalent to that contained in texts such as Jenkins and Watts,[1] Box and Jenkins,[2] Graupe,[3] Sage and Melsa,[4] Eykhoff,[5] Schweppe,[6] and Åström.[7]

Following Box and Jenkins,[2] the four steps in system identification are shown schematically in Fig. 1. The chapters in this book are organized accordingly under the following headings: (1) model structure determination, (2) parameter estimation, (3) experimental design, (4) special topics, and (5) case studies.

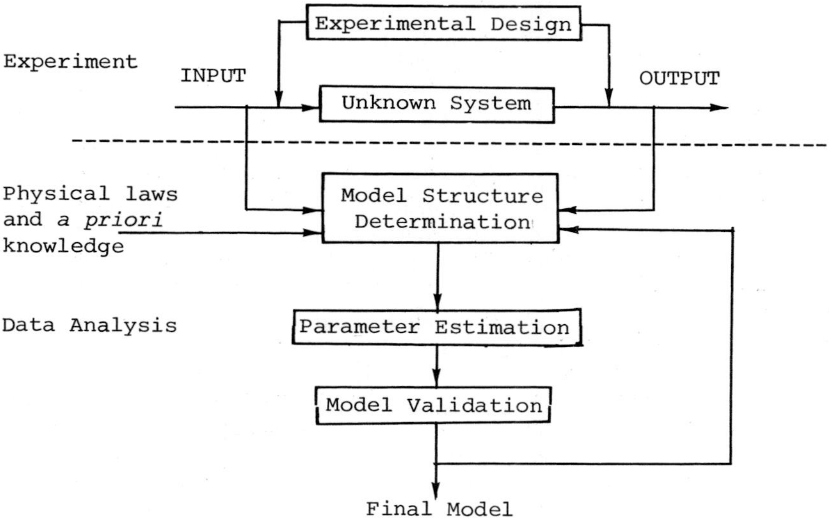

Fig. 1. Steps in System Identification

A brief description of each chapter is given below.

In Chapter I, Faurre introduces the linear Markovian representation of a time series and discusses the problem of obtaining a whole class of representations from the covariance function. He points out the importance of two special Markovian representations, one of which corresponds to the minimum variance Kalman filter for the process. Akaike, in Chapter II, further expands on this representation and discusses in detail his elegant method for determining the structure of this representation from noisy input–output data. For model order determination, Akaike uses an information criterion and illustrates his method with a number of interesting examples. Akaike's procedure is easy to implement and constitutes a major contribution to the analysis of multiple time series. It is interesting to note that a solution to this long-standing problem in time series analysis requires use of concepts from modern control and system theory, such as canonical forms and state vector models. Chapter III by Rissanen develops a new criterion for model structure determination based on the information-theoretic concept of entropy. These concepts are likely to play an increasingly important role in future developments of the system identification.

Chapters IV and V by Ljung and Narendra respectively consider the problem of consistent and stable estimation of parameters in adaptive closed-loop systems. Ljung presents new methods for proving consistency and shows that the prediction error minimization method is consistent under very weak conditions. Narendra discusses on-line estimation of parameters using a model reference approach and Lyapunov's direct method. The effectiveness of this method is demonstrated by numerous examples and extensive simulation results.

Chapters VI and VII by Mehra and Goodwin and Payne respectively present new results on the choice of inputs and sampling rates. In practice, the success of system identification is often dependent on these two factors, which are generally chosen on an ad hoc basis for convenience in experimentation. A study of these two chapters reveals that methods are now available for computing both optimal and good suboptimal experimental designs for system identification.

The special topics discussed in Chapters VIII, IX, and X by Attasi, Caines and Chan, and Robinson respectively pertain to the identification and estimation of doubly indexed time series (or random fields), feedback systems, and continuous-time systems. Attasi presents a new state vector model for discrete random fields, such as those encountered in image processing and gravity modeling, and develops a complete theory of stochastic realization and recursive estimation for these models. The parallels between his theory and that discussed by Faurre in Chapter I are remarkable considering the fact that causality does not hold in the case considered by Attasi. A special feature of Attasi's model for random fields is that vector white noise inputs are used to obtain a recursive structure for the model, and for the statistical smoother. In Chapter IX, Caines and Chan present a thorough rigorous discussion of feedback and the identification of closed-loop systems. They also present results from applications in the areas of economics, power systems, and physiology. In Chapter X, Robinson discusses the important problem of identifying a continuous-time model using discrete or sampled data. He considers the effect of "aliasing" on the cross-spectral method for obtaining both parametrized and unparametrized models for multiple time series. Robinson's chapter provides a very good balance to the rest of the book in that it contains clear exposition of the spectral methods, which do not receive their full share of attention in the other chapters.

The last two chapters of the book are devoted to case studies. Bohlin (Chapter XI) presents four case studies relating to dryer control in a paper mill, EEG signals with changing spectra, machine failure forecasting, and load forecasting in power systems. A unified procedure based on Gauss–Markov models for changing system parameters, Kalman filtering, and maximum likelihood estimation is used successfully in all four applications. The chapter contains important insights that the author has gained over the years through extensive experience with real data. In Chapter XII, Olsson presents a detailed and careful study relating to the modeling and identification of a nuclear reactor, a problem that is of great current interest for safety reasons. The chapter serves as a good example of the way a practical system has to be studied using different methods. The application of different techniques for system identification is not a luxury but a necessity when one is dealing with complex real-life systems that never fit neatly into any standard theoretical mold. Each technique properly applied gives some insight into the system and helps to reinforce the results obtained from other techniques.

The references at the end of each chapter constitute an extensive bibliography on the subject of system identification.

This volume would not have been possible without the full and dedicated participation of all the authors, to whom the editors are highly indebted. Special thanks are due to Mrs. Renate D'Arcangelo for typing most of the book in such a short period of time with partial help from Karin Young. The international scale of the effort required special coordination skills for which thanks are due to Marie Cedrone.

Finally, I would like to thank my wife, Anjoo, for her patience and understanding during long hours of work in preparing this volume.

Raman K. Mehra

References

1. G. M. Jenkins and D. G. Watts, *Spectral Analysis and its Applications.* Holden Day, San Francisco, 1968.
2. G. E. P. Box and G. M. Jenkins, *Time Series Analysis, Forecasting and Control.* Holden Day, San Francisco, 1976 (revised edition).
3. D. Graupe, *Identification of Systems.* Robert E. Krieger Pub. Co., Huntington, New York, 1976.
4. A. P. Sage and J. L. Melsa, *System Identification.* Academic Press, New York, 1971.
5. P. Eykhoff, *System Identification, Parameter and State Estimation.* Wiley, New York, 1974.
6. F. Schweppe, *Uncertain Dynamic Systems.* Prentice Hall, Englewood Cliffs, New Jersey, 1973.
7. K. J. Åström, *Introduction to Stochastic Control Theory.* Academic Press, New York, 1970.

*Special Issue; System Identification and Time Series Analysis, *IEEE Trans. on Automatic Control,* Dec. 1974.

STOCHASTIC REALIZATION ALGORITHMS

Pierre L. Faurre
Iria-Laboria and Sagem
France

1. INTRODUCTION

The progress of mathematical methods for signal processing and the simultaneous progress of digital data processing hardware has generated new interest in Markovian models which have been known for a long time [9, 22, 18, 19, 6].

However although very many papers have appeared on filtering, detection or control using such models, very little attention has been given to

— the study of the properties of such models

— the set of all models which can represent a given stochastic process and

— the design of efficient algorithms to compute such models.

The stochastic realization problem could be viewed as one of building two blocks of a stochastic identification procedure starting from raw data and giving a Markovian model for the data, as is shown on the following diagram:

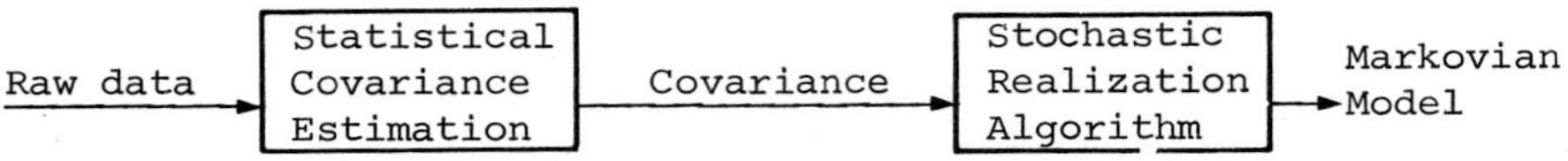

Statistical Identification

We shall here deal with the theoretical problem of studying *all* Markovian models corresponding to a given stochastic process. This study will appear to have interest from a practical point of view by sorting out peculiar Markovian models, such as the statistical filter of the process. Moreover this study and the related constructing proofs will lead us to the design of efficient algorithms.

2. RATIONAL TIME-SERIES AND MARKOVIAN REPRESENTATIONS

A. *RATIONAL TIME SERIES AND MARKOVIAN REPRESENTATIONS*

We shall be concerned with a vectorial stochastic process (time scale $T = \mathbb{R}$) or a vectorial time-series (time scale $T = \mathbb{Z}$) of dimension m:

$$\{y(t), \; t \in T\}, \quad y(t) = \begin{bmatrix} y_1(t) \\ \vdots \\ y_m(t) \end{bmatrix} \tag{1}$$

We shall assume in this chapter that $y(t)$ is zero mean ($E\{y(t)\} = 0$) and Gaussian.

We define:

— the covariance of $y(t)$

$$\Lambda(t,s) = E\{y(t)y'(s)\} \;^{(*)} \tag{2}$$

(*) y' stands for the transpose of y.

— the following Hilbert spaces generated by the random variables which are the components of the random vectors included in the brackets

$$\begin{aligned} \mathcal{Y} &= [y(t),\ t \in T] \\ \mathcal{Y}_t &= [y(t)] \\ \mathcal{Y}_t^+ &= [y(t+\tau),\ \tau \geq 0] \\ \mathcal{Y}_t^- &= [y(t+\tau),\ \tau < 0] \end{aligned} \tag{3}$$

We recall that orthogonal projections (denoted by ./.) and conditional expectations (denoted by $E\{.|.\}$) are the same, because all random variables considered are zero mean Gaussian. For example

$$y(t)/\mathcal{Y}_t^- = E\{y(t)|y(t-1),\ y(t-2),\ldots\} \ .$$

We define the *innovation* $\mathcal{V}(t)$ of the time-series $y(t)$ as the time-series

$$\mathcal{V}(t) = y(t) - y(t)/\mathcal{Y}_t^- \tag{4}$$

which is clearly a (Gaussian) white noise. We will write for the corresponding space:

$$\mathcal{V}_t = [\mathcal{V}_i(t),\quad i = 1,\ldots,m] \ . \tag{5}$$

We define now the *state-space* of the time-series as the Hilbert space

$$\mathcal{X}_t = \mathcal{Y}_t^+/\mathcal{Y}_t^- \ . \tag{6}$$

The dimension of this Hilbert space is in general infinite. However the case of finite dimension is of great interest.

DEFINITION 1: *The time-series* $\{y(t),\ t \in \mathcal{Z}\}$ *is said to be rational if and only if* $\dim\ \mathcal{X}_t = n_t < \infty$ *for all* t.

Let $x(t)$ be a random n_t - vector which generates $\mathcal{X}_t$ (i.e. a basis for $\mathcal{X}_t$).

Then using basic properties of orthogonal projection over the space $\mathcal{Y}^-_{t+1}$ which appears as the direct sum of two orthogonal subspaces:

$$\mathcal{Y}^-_{t+1} = \mathcal{V}_t \oplus \mathcal{Y}^-_t \tag{7}$$

we can write directly

$$\mathcal{Y}^+_{t+1}/\mathcal{Y}^-_{t+1} = \mathcal{Y}^+_{t+1}/\mathcal{Y}^-_t \oplus \mathcal{Y}^+_{t+1}/\mathcal{V}_t \tag{8}$$

Moreover $\mathcal{Y}^+_{t+1}/\mathcal{Y}^-_t$ is clearly a subspace of $\mathcal{Y}^+_t/\mathcal{Y}^-_t$:

$$\mathcal{Y}^+_{t+1}/\mathcal{Y}^-_t \subset \mathcal{Y}^+_t/\mathcal{Y}^-_t$$

and $y(t)/\mathcal{Y}^-_t$ belongs to the space $\mathcal{Y}^+_t/\mathcal{Y}^-_t = \mathcal{X}_t$.

So using the basis $x(t+1)$ and $x(t)$, one can find matrices of (deterministic) coefficients $H(t)$, $F(t)$ and $T(t)$ such that expressions (8) and (4) become

$$\begin{cases} x(t+1) = F(t)x(t) + T(t)\mathcal{V}(t) & (9) \\ y(t) = H(t)x(t) + \mathcal{V}(t) & (10) \end{cases}$$

The relations (9), (10) above are called a *Markovian representation* of $y(t)$ (the time-series $x(t)$ is clearly Markovian, because $\mathcal{V}(t)$ is white-noise).

So it appears that a time-series $y(t)$ is Markovian if and only if

$$\mathcal{X}_t = \mathcal{Y}_{t-1} \tag{11}$$

and then $y(t)$ obeys the following model:

$$y(t) = F(t)y(t-1) + \mathcal{V}(t) \tag{12}$$

Conversely suppose that the time-series $y(t)$ has a Markovian representation in the sense of the definition below.

DEFINITION 2: *A Markovian representation for the time-series* $y(t)$ *is a model of the form*

$$\begin{cases} x(t+1) = F(t)x(t) + v(t) & (13) \\ y(t) = H(t)x(t) + w(t) & (14) \end{cases}$$

where

$$\begin{bmatrix} v(t) \\ w(t) \end{bmatrix}$$

is white noise (so $x(t)$ *is a Markovian time-series of dimension* n_t'*).*

Then, we see that the space

$$X_t = Y_t^+/Y_t^- \subset [x_i(t),\ i = 1,\ldots,n_t']$$

is of finite dimension. So $y(t)$ *is rational in the sense of Definition 1. We conclude by stating that rational time-series and time-series which admit Markovian representation are the same.*

B. STATIONARY RATIONAL TIME-SERIES

We are going to investigate in more detail the rational time-series which are *stationary*.

Let us then define the covariance function

$$\Lambda(k) = E\{y(t+k)y'(t)\} \tag{15}$$

and consider the infinite vectors (matrices) corresponding to the space Y_t^+ (future) and Y_t^- (past):

$$Y^+(t) = \begin{bmatrix} y(t) \\ y(t+1) \\ y(t+2) \\ \vdots \end{bmatrix} \qquad Y^-(t) = \begin{bmatrix} y(t-1) \\ y(t-2) \\ y(t-3) \\ \vdots \end{bmatrix} \tag{16}$$

We see immediately that

$$E\{Y^+Y^{-\prime}\} = \begin{bmatrix} \Lambda(1) & \Lambda(2) & \Lambda(3) & . \\ \Lambda(2) & \Lambda(3) & . & \\ \Lambda(3) & . & \cdot\cdot & \\ . & & . & \end{bmatrix} = \mathcal{H} \tag{17}$$

$$E\{Y^-Y^{+\prime}\} = \mathcal{H}' \tag{18}$$

$$E\{Y^-Y^{-\prime}\} = \begin{bmatrix} \Lambda(0) & \Lambda(1) & . \\ \Lambda'(1) & \Lambda(0) & \\ . & & . \end{bmatrix} = \mathbb{\Lambda} \tag{19}$$

$$E\{Y^+Y^{+\prime}\} = \begin{bmatrix} \Lambda(0) & \Lambda'(1) & . \\ \Lambda(1) & \Lambda(0) & \\ . & & . \end{bmatrix} = \tilde{\mathbb{\Lambda}} \tag{20}$$

where $\mathcal{H}$ is a Hankel matrix and $\mathbb{\Lambda}, \tilde{\mathbb{\Lambda}}$ are symmetrical Toeplitz matrices.

From Hilbert space geometry it follows

THEOREM 1: $y(t)$ *is rational and has a Markovian representation of dimension* n *(dimension of* $x(t) = n$*) if and only if*

$$\text{rank } \mathcal{H} = n \quad . \tag{21}$$

Then one knows from the *deterministic realization problem* as solved by [16, 25, 26] that such is the case if and only if there exists three matrices $m \times n$, $n \times n$ and $n \times m$, H, F, G such that

$$\begin{aligned} (H,F) &= \text{completely observable pair} \\ (F,G) &= \text{completely controllable pair} \\ \Lambda(k) &= HF^{k-1}G, \; k = 1,2,... \end{aligned} \tag{22}$$

Such matrices are unique modulo a change of basis, i.e. if T is any regular matrix, HT, $T^{-1}FT$, $T^{-1}G$ "realize" also $\Lambda(k)$ in the sense of formula (22).

We shall assume also that the time-series $y(t)$ is purely

nondeterministic (following the WOLD decomposition) in the sense that

$$\Lambda(k) \to 0 \qquad \text{if} \qquad k \to \infty \qquad . \tag{23}$$

This is equivalent to

$$F \;=\; \text{asymptotically stable matrix} \tag{24}$$

(i.e. all eigenvalues of F have negative real parts).*

It is useful to note that (22) implies that

$$\mathcal{H} \;=\; \mathcal{O}\mathcal{C} \;=\; \begin{bmatrix} H \\ HF \\ \vdots \end{bmatrix} \; [G,\ FG,\ ..] \tag{25}$$

where $\mathcal{O}$ and $\mathcal{C}$ are the so-called observability and controllability matrices associated with $\{H,F,G\}$:

$$\mathcal{O} \;=\; \begin{bmatrix} H \\ HF \\ HF^2 \\ \vdots \end{bmatrix}$$

$$\mathcal{C} \;=\; [G,\ FG,\ F^2G,\ldots] \qquad . \tag{26}$$

There exist efficient algorithms which could be used to compute n, H, F, G by factorizing the Hankel matrix $\mathcal{H}$ (of rank n) built from the covariance sequence $\Lambda(k)$. See [16, 25, 26].

In [26, 2], it is shown that use of canonical forms in such algorithms gives good results.

We conclude that the covariance $\Lambda(k)$ of a rational stationary time series can be written

$$\Lambda(k) \;=\; HF^{k-1}G\ 1_k + G'F'^{-k-1}H'\ 1_{-k} + \Lambda_0\ \delta_{k0} \tag{27}$$

---

*For a complete treatment when y is not necessarily purely nondeterministic see [13].

where

$$1_k = \begin{cases} 1 & \text{if } k > 0 \\ 0 & \text{otherwise} \end{cases}$$

and δ_{k0} is the Kronecker index.

The *spectrum* of $y(t)$, Z-transform of the covariance $\Lambda(k)$ is then

$$S(z) = \sum_k \Lambda(k) z^{-k} = H(zI-F)^{-1}G + G'(z^{-1}I-F')^{-1}H' + \Lambda_0 \tag{28}$$

which appears as a *rational* function of z. The terminology "*rational* time series" we are using is thus explained.

C. *STATIONARY MARKOVIAN REPRESENTATIONS*

A stationary Markovian representation--in the sense of Definition 2--for a stationary rational time-series is a model of the following kind

$$x(t+1) = Fx(t) + v(t) \tag{29}$$

$$y(t) = Hx(t) + w(t) \tag{30}$$

where

$$\begin{bmatrix} v(t) \\ w(t) \end{bmatrix}$$

is stationary white noise:

$$E\left\{\begin{bmatrix} v(t) \\ w(t) \end{bmatrix} [v'(s)w'(s)]\right\} = \begin{bmatrix} Q & S \\ S' & R \end{bmatrix} \delta_{t,s} \tag{31}$$

with of course

$$\begin{bmatrix} Q & S \\ S' & R \end{bmatrix} \geq 0.^{*} \tag{32}$$

---

* $A \geq 0$ [$A > 0$] means that the symmetrical matrix A is non-negative semidefinite [positive definite].

Stationarity implies an equilibrium value for the covariance of x:

$$E\{x(t)x'(t)\} = P = \text{constant}$$

But if follows easily from (29) that

$$E\{x(t+1)x'(t+1)\} = FE\{x(t)x'(t)\}F' + Q$$

and also

$$x(t) = v(t) + Fv(t-1) + F^2v(t-2) + \ldots$$

So that necessarily

$$P - FPF' = Q \qquad\qquad (33)$$
$$P = Q + FQF' + F^2QF'^2 + \ldots$$

The first relation--which is a Liapunov equation--implies some stability conditions on F. Here we shall assume as in the preceding paragraph

$$F = \text{asymptotically stable matrix} \qquad\qquad (34)$$

and

$$P > 0 \quad .$$

Then one can compute the covariance of $y(t)$ which appears as

$$\Lambda(k) = E\{y(t+k)y'(t)\}$$
$$= HF^{k-1}(FPH' + S)\, 1_k + (HPF' + S')F^{-k-1}H'1_{-k}$$
$$+ (HPH' + R)\delta_{k0} \quad , \qquad\qquad (35)$$

which looks like expression (27) when one has set

$$FPH' + S = C \qquad\qquad (36$$

$$HPH' + R = \Lambda_0 \quad . \qquad\qquad (37)$$

We are now in a position to state the main problem of this

chapter.

STOCHASTIC REALIZATION PROBLEM

Given the covariance $\Lambda(k)$ *of a rational stationary time-series* $y(t)$, *find a Markovian representation for it.*

As stated above, since $y(t)$ is rational, we can apply a deterministic realization algorithm to find the size n of the realization (rank of $\mathcal{H}$) and three matrices H, F, G such that $\Lambda(k)$ could be expressed by formula (27).

In this basis, any Markovian representation (29)-(30) will correspond to unknown matrices P, Q, S and R satisfying (33), (36) and (37):

$$\left\{\begin{array}{ll} P - FPF' = Q & (38) \\ G - FPH' = S & (39) \\ \Lambda_0 - HPH' = R & (40) \\ \begin{bmatrix} Q & S \\ S' & R \end{bmatrix} \geq 0, \quad P > 0 & (41) \end{array}\right.$$

We have to solve these linear equations (38)-(40) with the highly nonlinear constraints (41), where

- H, F, G, Λ_0 are given
- P, Q, S, R are unknown.

From equations (38)-(40) one sees that Q, S and R are uniquely determined from P. So to any P matrix corresponds a Markovian representation and conversely. We shall then identify P and its corresponding Markovian representation. With this convention, we can give the definition.

DEFINITION 3: *We define the set of all Markovian representations associated with a given covariance* $\Lambda(k)$ *as expressed by* (27)*--or equivalently with* $\{H, F, G, \Lambda_0\}$*--as the set* $\mathcal{P}$ *of all symmetrical matrices* P *verifying* (38)-(41).

D. CONTINUOUS TIME CASE

In the continuous time case, a *rational stochastic process* $y(t)$ is a process which can be expressed by a Markovian representation

$$\dot{x}(t) = F(t)x + v(t) \tag{42}$$

$$y(t) = H(t)x(t) + w(t) \tag{43}$$

where

$$\begin{bmatrix} v(t) \\ w(t) \end{bmatrix}$$

is continuous-time white noise, i.e.

$$E\left\{\begin{bmatrix} v(t) \\ w(t) \end{bmatrix} [v'(s)w'(s)]\right\} = \begin{bmatrix} Q & S \\ S' & R \end{bmatrix} \delta(t-s) \tag{44}$$

In the stationary case, H, F, Q, R and S are constant matrices and

$$\begin{aligned} \Lambda(\tau) &= E\{y(t+\tau)y'(t)\} \\ &= He^{F\tau}G\,1(\tau) + G'e^{-F'\tau}H'1(-\tau) + R\delta(\tau) \end{aligned} \tag{45}$$

where

$$1(\tau) = \begin{cases} 1 & \text{if} \quad \tau > 0 \\ 1/2 & \text{if} \quad \tau = 0 \\ 0 & \text{if} \quad \tau < 0 \end{cases}$$

$\delta(\tau)$ is the Dirac impulse and one has set

$$G = PH' + S \tag{46}$$

where P is the covariance of $x(t)$

$$P = E\{x(t)x'(t)\} \tag{47}$$

solution of the Liapunov equation

$$FP + PF' = -Q \tag{48}$$

also expressed by

$$P = \int_0^{\infty} e^{F\tau} Q e^{F'\tau} \, d\tau \tag{49}$$

As in the discrete time case H, F and G can be computed from $\Lambda(\tau)$ by an algorithm of the Ho-Kalman type [16], and the set $\mathcal{P}$ of all Markovian representations is the set of all P matrices verifying

$$\begin{cases} FP + PF' = -Q & (50) \\ T - PH' = S & (51) \\ P > 0, \quad \begin{bmatrix} Q & S \\ S' & R \end{bmatrix} \geq 0 & (52) \end{cases}$$

where H, F, G and R are obtained from $\Lambda(\tau)$ and P, Q and S are unknown.

3. SET OF ALL MARKOVIAN REPRESENTATIONS

This part will be devoted to the study of the set $\mathcal{P}$, subset of symmetrical $n \times n$ matrices as defined by (38)-(41) in the discrete time case, or by (50)-(52) in the continuous time case. As already stated this set is an image of all Markovian representations of the initial time-series.

It is easy to see that $\mathcal{P}$ is *a closed bounded convex set.*

Finer results on its structure will be given after we have proved an important result, known as the positive real lemma [29, 24, 20]. We give an original proof which can be found in [10, 13].

A. *POSITIVE REAL LEMMA*

This lemma gives a characterization of positive realness for operators as expressed by (27) or (45). It is closely related to our subject because one knows from Kolmogorov's theorem [22] that a symmetrical operator is a covariance if and only if it is positive real.

DEFINITION 4: *An operator* $\Lambda(k)$ *is positive real* (p.r.) *if and only if for any sequence* $u(i)$

$$U'\Lambda U = \sum_{i,j} u'(i)\Lambda(j-i)u(j) \tag{53}$$

is nonnegative.

A criterion to recognize if $\Lambda(k)$ as given by (27) is p.r. is the following.

THEOREM 2 (Positive Real Lemma): $\Lambda(k)$ *is p.r. if and only if the associated* $\mathcal{P}$ *set is nonvoid.*

We now give the proof. First we need a lemma that is easy to prove.

LEMMA: *If* P, Q, R *and* S *verify* (38)-(40) *then*

$$U'\Lambda U = x'(0)Px(0) + \sum_{-\infty}^{-1} (x'Qx + u'Ru + 2x'Su) \tag{54}$$

where the sequence $x(i)$ *is given by*

$$\begin{cases} x(-\infty) = 0 \\ x(i+1) = F'x(i) + H'u(i) \end{cases} \tag{55}$$

i) Proof of Theorem 2. *Sufficiency.* If P, Q, R, and S verify (38)-(40) and the nonnegativeness conditions (41), it is clear that expression (54) is nonnegative and so Λ is p.r.

ii) Proof of Theorem 2. *Necessity.* Let us assume that Λ is p.r. Then define the nonnegative definite matrix P^* by

$$x'P^*x = \inf_{U \in \mathcal{E}(x)} U'\Lambda U \tag{56}$$

where $\mathcal{E}(x)$ denotes the set of U sequences:

$$\mathcal{E}(x) = \left\{ u(\cdot) \,\middle|\, x = \sum_{-\infty}^{-1} F'^{-i-1}H'u(i) \right\}. \tag{57}$$

Define now

$$Q^* = P^* - FP^*F'$$

$$S^* = G - FP^*H'$$

$$R^* = \Lambda_0 - HP^*H'$$

We are going to prove that

$$\begin{bmatrix} Q^* & S^* \\ S^{*\prime} & R^* \end{bmatrix} \geq 0$$

which will complete the proof ($\mathcal{P} \neq \emptyset$, because $P^* \in \mathcal{P}$).

Let us consider the sequence $U = \{u(i),\ i \leq -1\}$ which drives the system (55) to state ξ at time 0.

Then consider $V = \{v(i),\ i \leq -1\}$ with $v(i) = u(i+1)$ which drives the system (55) to state ξ at time -1 and state ζ at time 0.

Using the lemma, we can write

$$V'\Lambda V - U'\Lambda U = \zeta'P^*\xi - \xi'P^*\xi + [\xi' \ v'(-1)] \begin{bmatrix} Q^* & S^* \\ S^{*\prime} & R^* \end{bmatrix} \begin{bmatrix} \xi \\ v(-1) \end{bmatrix}$$

or

$$[\xi' \ v'(-1)] \begin{bmatrix} Q^* & S^* \\ S^{*\prime} & R^* \end{bmatrix} \begin{bmatrix} \xi \\ v(-1) \end{bmatrix} = (V'\Lambda V - \zeta'P^*\zeta) - (U'\Lambda U - \xi'P^*\xi)$$

From the way we define P^*, it is clear that

i) $V'\Lambda V - \zeta'P^*\zeta \geq 0$

ii) $U'\Lambda U - \xi'P^*\xi\ (\geq 0)$ can be set arbitrarily small by choice of U.

Thus we have shown that

$$[\xi' \ v'(-1)] \begin{bmatrix} Q^* & S^* \\ S^{*\prime} & R^* \end{bmatrix} \begin{bmatrix} \xi \\ v(-1) \end{bmatrix} \geq 0$$

for all ξ and $v(-1)$. Therefore, it follows that

$$\begin{bmatrix} Q^* & S^* \\ S^{*\prime} & R^* \end{bmatrix} \geq 0 \quad .$$

B. *STRUCTURE OF THE SET* $\mathcal{P}$

We have exhibited a particular point P^* of the set $\mathcal{P}$. One of its properties is the following

THEOREM 3: P^* *is a maximal point of* $\mathcal{P}$, *i.e.*

$$P \leq P^* \qquad \textit{for all} \qquad P \in \mathcal{P} \tag{*}$$

This follows from the definition of P^* and the lemma which allows us to write

$$x'P^*x = \inf_{u \in E(x)} \{x'Px + \Sigma(x'Qx + u'Ru + 2x'Su)\}$$

It is clear that for any $P \in \mathcal{P}$

$$x'P^*x - x'Px = \inf_{u \in E(x)} \Sigma(x'Qx + u'Ru + 2x'Su) \geq 0 \quad .$$

and so

$$P^* \geq P \quad .$$

Moreover one can introduce the dual time-series $\tilde{y}(t)$ by inverting time: $\tilde{y}(t) = y(-t)$. Its covariance is $\tilde{\Lambda}$, where we define

$$\tilde{\Lambda}(k) = \Lambda(-k) \tag{58}$$

and we write $\tilde{\mathcal{P}}$ for the associated set of Markovian representations.

Then one can prove (see [10]) that $\mathcal{P}$ and $\tilde{\mathcal{P}}$ correspond to each other by matrix inversion, i.e.

$$P \in \mathcal{P} \Leftrightarrow P^{-1} \in \tilde{\mathcal{P}} \tag{59}$$

---

* We write $A \leq B$ if the matrix $B - A$ is nonnegative definite.

We can immediately deduce that $\mathcal{P}$ has a minimal point $P_* = (\tilde{P}^*)^{-1}$ such that

$$P \geq P_* \quad \textit{for all} \quad P \in \mathcal{P}$$

So when it is not void, $\mathcal{P}$ *appears as a closed bounded convex set with two interesting extreme points* P^* *and* P_* *such that*

$$\text{for all} \quad P \in \mathcal{P}, \quad P_* \leq P \leq P^* \quad . \tag{60}$$

Further properties of $\mathcal{P}$ can be found in [15]. For example, if not void, $\mathcal{P}$ consists of positive definite matrices. An important result lies also in the equivalence of the three properties below:

i) the operator $\mathbb{A}$ is coercive, i.e. there exists a scalar $\rho > 0$ such that

$$U'\mathbb{A}U \geq \rho U'U = \rho \sum_i \|u(i)\|^2$$

ii) R_* and $P^* - P_*$ are positive definite

iii) F_* is an asymptotically stable matrix where we write

$$F_* = F - S_* R_*^{-1} H$$

(F_* could be shown as the dynamical matrix of the statistical filter associated with time-series $y(t)$, see below part 5).

C. CONTINUOUS TIME CASE

For a rational stationary stochastic process $y(t)$, the set $\mathcal{P}$ of all its Markovian representations is defined as the set of matrices P such that

$$\begin{cases} FP + PF' = -Q & (61) \\ G - PH' = S & (62) \\ P > 0, \begin{bmatrix} Q & S \\ S' & R \end{bmatrix} \geq 0 & (63) \end{cases}$$

We have similar results for the continuous-time case. The operator $\Lambda(\tau)$

$$\Lambda(\tau) = He^{F\tau}G1(\tau) + G'e^{-F'\tau}H'1(-\tau) + R\delta(\tau) \tag{64}$$

is positive real if and only if $\mathcal{P}$ is nonvoid.

Moreover $\mathcal{P}$ is a bounded closed convex set with a maximal point P^* and a minimal point P_* where P^* is defined as

$$x'P^*x = \inf_{u(\cdot)\in \mathcal{E}(x)} U'\mathbb{\Lambda}U \tag{65}$$

with

$$\mathcal{E}(x) = \left\{ u(\cdot) \,\middle|\, x = \int_{-\infty}^{0} e^{-F'\alpha}H'u(\alpha)\ d\alpha \right\} \tag{66}$$

and

$$U'\mathbb{\Lambda}U = \int_{-\infty}^{0}\int_{-\infty}^{0} u'(\alpha)\Lambda(\beta-\alpha)u(\beta)\ d\alpha\ d\beta \quad . \tag{67}$$

And P_* is defined as $P_* = (\tilde{P}^*)^{-1}$ where $\tilde{P}^*$ is the maximal point of the set $\tilde{\mathcal{P}}$ associated with the dual stochastic process $\tilde{y}(t) = y(-t)$.

In the regular case $(R > 0)$, we have also the equivalence between the three propositions:

i) $\mathbb{\Lambda}$ is coercive

ii) $P^* - P_*$ is positive definite

iii) $F_* = F - S_*R^{-1}H$ is asymptotically stable (F_* is also the dynamical matrix of the statistical filter associated with time-series $y(t)$).

4. ALGORITHMS

Now we are going to describe some algorithms which allow us to compute the two extreme points P^* and P_* of the set $\mathcal{P}$. Thus we generate the complete segment $[P_*, P^*]$ of realizations.

These algorithms involve integrating a Riccati type equation and start from a description of the covariance function Λ in terms of H, F, G matrices derived from a Ho-Kalman-type algorithm [16].

A. DISCRETE-TIME CASE

We compute P^* as defined by (56) from the limit

$$P^* = \lim_{n\to\infty} P_n \tag{68}$$

where

$$x'P_n x = \inf_{U \in E_n(x)} U'\Lambda U \tag{69}$$

and $E_n(x)$ is the intersection of $E(x)$ and the set of U sequences of length n:

$$E_n(x) = \{u(\cdot)\,|\,x = \mathcal{O}_n'U\} \tag{70}$$

with

$$\mathcal{O}_n' = [H', F'H', \ldots, F'^{n-1}H'] \tag{71}$$

It is easy to show that

$$P_n = [\mathcal{O}_n'\Lambda_n^{-1}\mathcal{O}_n]^{-1} \tag{72}$$

where Λ_n is the $n \times n$ blocks covariance matrix (19).

Now if we note that

$$\begin{aligned} P_{N=1}^{-1} &= \mathcal{O}_{n+1}' \Lambda_{n+1}^{-1} \mathcal{O}_{n+1} \\ &= [H', F'\mathcal{O}_n']\Lambda_{n+1}^{-1} \begin{bmatrix} H \\ \mathcal{O}_n'F \end{bmatrix} \end{aligned}$$

and that

$$\begin{aligned} \Lambda_{n+1}^{-1} &= \begin{bmatrix} \Lambda_0 & G', \mathcal{O}_n' \\ \mathcal{O}_n G, & \Lambda_n \end{bmatrix}^{-1} \\ &= \begin{bmatrix} E_n & -E_n G'\mathcal{O}_n'\Lambda_n^{-1} \\ -\Lambda_n^{-1}\mathcal{O}_n G E_n & \Lambda_n^{-1}+\Lambda_n^{-1}\mathcal{O}_n G E_n G'\mathcal{O}_n\Lambda_n^{-1} \end{bmatrix} \end{aligned}$$

with $E_n = (\Lambda_0 - G'P_n^{-1}G)^{-1}$ we see that

$$P_{n+1}^{-1} = F'P_n^{-1}F + (H'-F'P_n^{-1}G)(\Lambda_0-G'P_n^{-1}G)^{-1}(H-G'P_n^{-1}F) \quad . (73)$$

We deduce from (73) the first algorithm

ALGORITHM 1: *The max* P^* *of* $\mathcal{P}$ *in the discrete time case can be computed by generating the sequence*

$$\Sigma_0 = 0$$

$$\Sigma_{n+1} = F'\Sigma_n F + (H'-F'\Sigma_n G)(\Lambda_0-G'\Sigma_n G)^{-1}(H-G'\Sigma_n F) \quad (74)$$

until it converges to Σ_∞. *Then*

$$P^* = \Sigma_\infty^{-1}$$

$$Q^* = P^* - FP^*F'$$

$$S^* = G - FP^*H'$$

$$R^* = \Lambda_0 - HP^*H' \quad .$$

Now if we use this algorithm for the dual problem we get immediately

ALGORITHM 2: *The min* P_* *of* $\mathcal{P}$ *in the discrete time case can be computed by generating the sequence*

$$\Omega_0 = 0$$

$$\Omega_{n+1} = F\Omega_n F' + (G-F\Omega_n H')(\Lambda_0 - H\Omega_n H')^{-1}(G'-H\Omega_n F') \quad (75)$$

until it converges to Ω_∞. *Then*

$$P_* = \Omega_\infty$$

$$Q_* = P_* - FP_*F'$$

$$S_* = G - FP_*H'$$

$$R_* = \Lambda_0 - HP_*H' \quad .$$

B. CONTINUOUS TIME CASE

In the continuous time case a similar approach will allow us to compute P^* and P_* as limits of solution of Riccati equations. All details can be found in [13] and [15].

Here we have to distinguish between the *regular* case, i.e. the case when $R > 0$, the *mixed* case, i.e. $R \geq 0$ but $R \not> 0$ and $R \neq 0$, and the *singular* case, i.e. $R = 0$.

In the regular case we get algorithms similar to Algorithms 1 and 2:

ALGORITHM 3: *The max* P^* *of* $\mathcal{P}$ *is given in the regular case by integrating*

$$\begin{aligned} \Sigma(0) &= 0 \\ \dot{\Sigma} &= F'\Sigma + \Sigma F + (H'-\Sigma G)R^{-1}(H-G'\Sigma) \end{aligned} \tag{76}$$

until the equilibrium value Σ_∞. *Then*

$$P^* = \Sigma_\infty^{-1}$$

ALGORITHM 4: *The min* P_* *of* $\mathcal{P}$ *is given in the regular case by integrating*

$$\begin{aligned} \Omega(0) &= 0 \\ \dot{\Omega} &= F\Omega + \Omega F' + (G-\Omega H')R^{-1}(G'-H\Omega) \end{aligned} \tag{77}$$

until the equilibrium value Ω_∞. *Then*

$$P_* = \Omega_\infty \quad .$$

Germain [15] has generalized these algorithms by extending them to the singular case $(R = 0)$ or the mixed ease (R positive semidefinite) by introducing *equivalent problems*: problems $\{H_1, F_1, G_1, R_1\}$ and $\{H_2, F_2, G_2, R_2\}$ are said *eqivalent* if they have the same set $\mathcal{P}$.

For example, singular problem $\{H,F,G,0\}$ is equivalent to the problem $\{HF,F,-FG,-(HFG + G'F'H')\}$. So Algorithms 3 and 4 allow one to compute P^* and P_* in any case.

Stability of these algorithms can be proved by using stability results for Riccati equations.

For example, Algorithm 4 converges towards P_* for any initial value $\Omega(0) = B$ such that

$$B \leq P^*$$

In particular, if we choose $B = 0$ we can insure convergence (see [15] for a proof).

These algorithms are efficient for computing Markovian realizations. Previous methods were based on spectral factorization ideas; although both approaches are comparable in case of a scalar process, in the multivariable case our algorithms seem to be much more efficient.

Generalizations have been worked out recently by Attasi for processes depending of two indices (images), both for models and realization algorithms. See Attasi [4, 5].

5. MINIMAL REALIZATION AND FILTER

An important application of statistical modeling is for statistical filtering.

An optimal statistical filter is an algorithm or a system which generates at time t the best estimate of a signal $z(t)$ from past observations $\{y(t-1), y(t-2), \ldots\}$.

When using a Markovian representation as (29)-(30), one has to compute

$$\hat{x}(t) = E\{x(t)\,|\,y(t-1),\ y(t-2), \ldots\} \tag{78}$$

Let us recall the result due to Kalman that under the preceding hypothesis, $\hat{x}(t)$ is generated by the so-called Kalman filter:

$$\hat{x}(t+1) = F\hat{x}(t) + T\mathcal{V}(t) \tag{79}$$

$$y(t) = H\hat{x}(t) + \mathcal{V}(t) \tag{80}$$

where the optimal gain T is given by

$$T = (F\Sigma H' + S)(H\Sigma H' + R)^{-1} \tag{81}$$

Σ being the unique positive definite solution of

$$\Sigma = F\Sigma F' - (F\Sigma H' + S)(H\Sigma H' + R)^{-1}(H\Sigma F' + S') + Q \qquad (82)$$

and also the variance of the "error" $\tilde{x} = x - \hat{x}$:

$$\Sigma = E\{\tilde{x}(t)\tilde{x}'(t)\} \quad . \qquad (83)$$

From (79)-(80) we see that the optimal filter is itself a Markovian representation of $y(t)$. In fact, it is the particular Markovian representation we introduced in Section 2 (see expressions (9)-(10)).

From the fact that $\tilde{x}$ and $\hat{x}$ are independent it follows easily that

$$P = \hat{P} + \Sigma \qquad (84)$$

where P is the variance of the initial Markovian representation and $\hat{P}$ the variance of $\hat{x}$. So

$$P \geq \hat{P}$$

and we see immediately that

$$\hat{P} = P_* \quad . \qquad (85)$$

We conclude

THEOREM: *The minimal Markovian representation* P_* *is nothing else but the statistical filter, i.e.*

$$x_*(t) = E\{x_*(t) \mid y(t-1), y(t-2), \ldots\} \qquad (86)$$

The same result applies also in the continuous time case.

Thus Algorithms 2 and 4 given above not only solve the sto stochastic realization problem, but also, at the same time, the filtering problem.

REFERENCES

1. Akaike, H., "Markovian Representation of Stochastic Processes by Canonical Variables," *SIAM J. of Control,* 1974.

2. Akaike, H., Chapter in this book

3. Anderson, B. D. O., "The Inverse Problem of Stationary Co-Variance Generation," *J. of Statistical Physics, Vol. 1, No. 1,* 1969, pp. 133-147.

4. Attasi, S., "A Generalization of Kalman Statistical Technique to Image Processing," to appear in *Applied Mathematics and Optimization* (1975).

5. Attasi, S., Chapter in this book.

6. Bryson, A. E., M. Frazier, "Smoothing for Linear and Non-linear Dynamic Systems," Proc. Optimum System Synthesis Conf., U.S. Air Force Tech. Dept. ASD-TDR, Feb. 1963, pp. 63-119.

7. Bryson, A. E., Y. C. Ho, *Applied Optimal Control,* Blaisdell, 1969.

8. Clerget, M., "Systèmes linéaires positifs non stationnaires," Rapport Laboria, No. 65, April 1974.

9. Doob, J. L., "The Elementary Gaussian Processes," *Ann. Math. Stat. 15,* 1944, pp. 229-282.

10. Faurre, P., J. P. Marmorat, "Un algorithme de réalisation stochastique," *C. R. Acad. Sci. Paris,* T.268, Série A, April 1969, pp. 978-981.

11. Faurre, P., "Identification par minimisation d'une représentation markovienne de processus aléatoire," Symp. on Optimization, Nice, June 1969, published by Springer-Verlag, Lecture Notes in Mathematics 132.

12. Faurre, P., P. Chataignier, "Identification en temps réel et en temps différé par factorisation de matrices de Hankel," Colloque Franco-Suédois sur la Conduite des Procédés, Iria, October 1971.

13. Faurre, P., "Realisations markoviennes de processus stationnaires," Rapport Laboria No. 13, March 1973 (Iria, Voluceau, 78-Rocquencourt, France).

14. Faurre, P., M. Clerget, F. Germain, *Opérateurs Rationnels Positifs et Applications*, (to appear).

15. Germain, F., "Algorithmes continus de calcul de réalisations markoviennes, Cas singulier et stabilité," Rapport Laboria No. 66, April 1974.

16. Ho, B. L., R. E. Kalman, "Effective Construction of Linear State Variable Models from Input/Output Data," Proc. 3rd Allerton Conference, 1965, pp. 449-459.

17. Kailath, T., "An Innovation Approach to Least Square, Estimation, Part I-IV," *IEEE Trans. AC 13-16*.

18. Kalman, R. E., "A New Approach to Linear Filtering and Prediction Problems," *J. of Basic Engineering*, March 1960, pp. 35-45.

19. Kalman, R. E., R. S. Bucy, "New Results in Linear Filtering and Prediction Theory," *J. of Basic Engineering*, March 1961, pp. 95-108.

20. Kalman, R. E., "Liapunov Functions for the Problem of Lur'e in Automatic Control," Proc. N.A.S., 49, 1963, pp. 201-205.

21. Kalman, R. E., "Linear Stochastic Filtering - Reappraisal and Outlook," Symposium on System Theory, Polytechnic Institute of Brooklyn, April 1965, pp. 197-205.

22. Kolmogorov, A. N., "Sur l'interpolation et l'extrapolations des suites stationnaires, *C.R. Acad. Sc. Paris, 208*, 1939, pp. 2043-45.

23. Mehra, R. K., "On the Identification of Variances and Adaptive Kalman Filtering," *IEEE Trans. on AC, Vol. AC-15*, April 1970, pp. 175-184.

24. Popov, V. M., *L'hyperstabilité des Systèmes Automatiques*, Dunod, 1973.

25. Rissanen, J., "Recursive Identification of Linear Systems," *J. SIAM Control, Vol. 9, No. 3*, August 1971, pp. 420-430.

26. Rissanen, J., Chapter in this book.

27. Wiener, N., *Extrapolation, Interpolation and Smoothing of Stationary Time Seires*, The M.I.T. Press, 1949.

28. Wold, H., *A Study in the Analysis of Stationary Time-Series*, Uppsala, 1938.

29. Yakubovich, V. A., "Absolute Stability of Nonlinear Control Systems in Critical Cases; I: *Avt. i. Telem. 24, 3,* March 1963, pp. 293-303; II: *Avt. i. Telem., 24, 6,* June 1963, pp. 717-731.

CANONICAL CORRELATION ANALYSIS OF TIME SERIES AND THE USE OF AN INFORMATION CRITERION

Hirotugu Akaike
The Institute of Statistical Mathematics
Tokyo, Japan

1. INTRODUCTION

We start this chapter with a brief introductory review of some of the recent developments of time series analysis. One of the most established procedure of time series analysis is the method of estimation of power spectrum through windowed sample covariance sequence. This method of power spectrum estimation, of which systematic account is given in the book by Blackmann and Tukey [1], found various fields of application in science and engineering. The importance of the method was further increased when it was extended to the estimation of the frequency response function through the estimates of the cross and power spectra. An example of application of this method to the estimation of the power spectra of the input and output and the frequency response function of a noisy linear system is illustrated in Fig. 1. Although the method is generally considered to be quite useful, the secret of the success of the method lies entirely in the application of the windowing procedure which transforms the original sample covariance sequence into windowed covariance sequence by multiplying the sample covariances with appropriately chosen numerical factors which constitute the lag-window. This basic idea is schematically illustrated in Fig. 2.

Since the power spectral density function is defined as the Fourier transform of the theoretical covariance sequence the effectiveness of the windowing procedure is merely due to the fact that an appropriately windowed sample covariance sequence provides a better estimate of the covariance sequence than the original sample covariance sequence. When it is viewed as a method of estimation of a covariance sequence this windowing procedure is obviously a very naive one, unless it is supplied with some rules for the adaptive selection of the lag-window using the information supplied by the observed sample covariance sequence. Not much attention was paid to this aspect and the selection of the window remained as the most subjective and empirical part of the procedure in practical applications [2, 3].

In contrast to the so-called nonparametric approach which is based on the pointwise local estimates of the spectra, the parametric approach which uses global parametric models has a long history in time series analysis. The analysis of the sunspot number series by G. U. Yule in 1927 [4] by using an autoregressive model is considered to be a pioneering work in this parametric approach. In the autoregressive model fitting of a scalar stationary time series $y(n)$ with zero mean it is assumed that $y(n)$ is generated by the relation

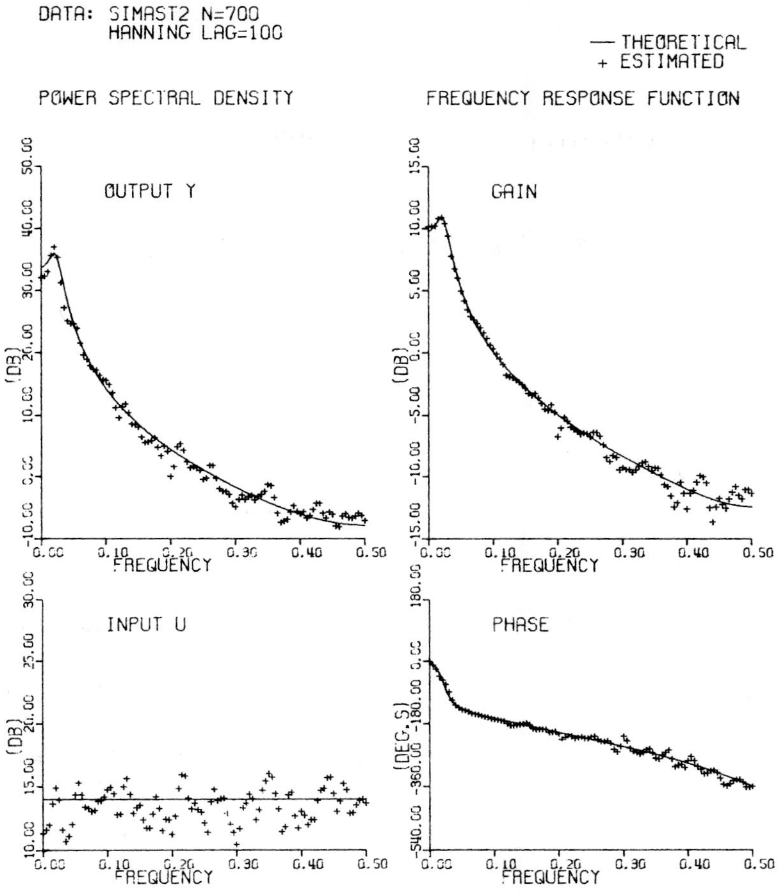

Fig. 1. Spectrum analysis through windowed covariance sequence.

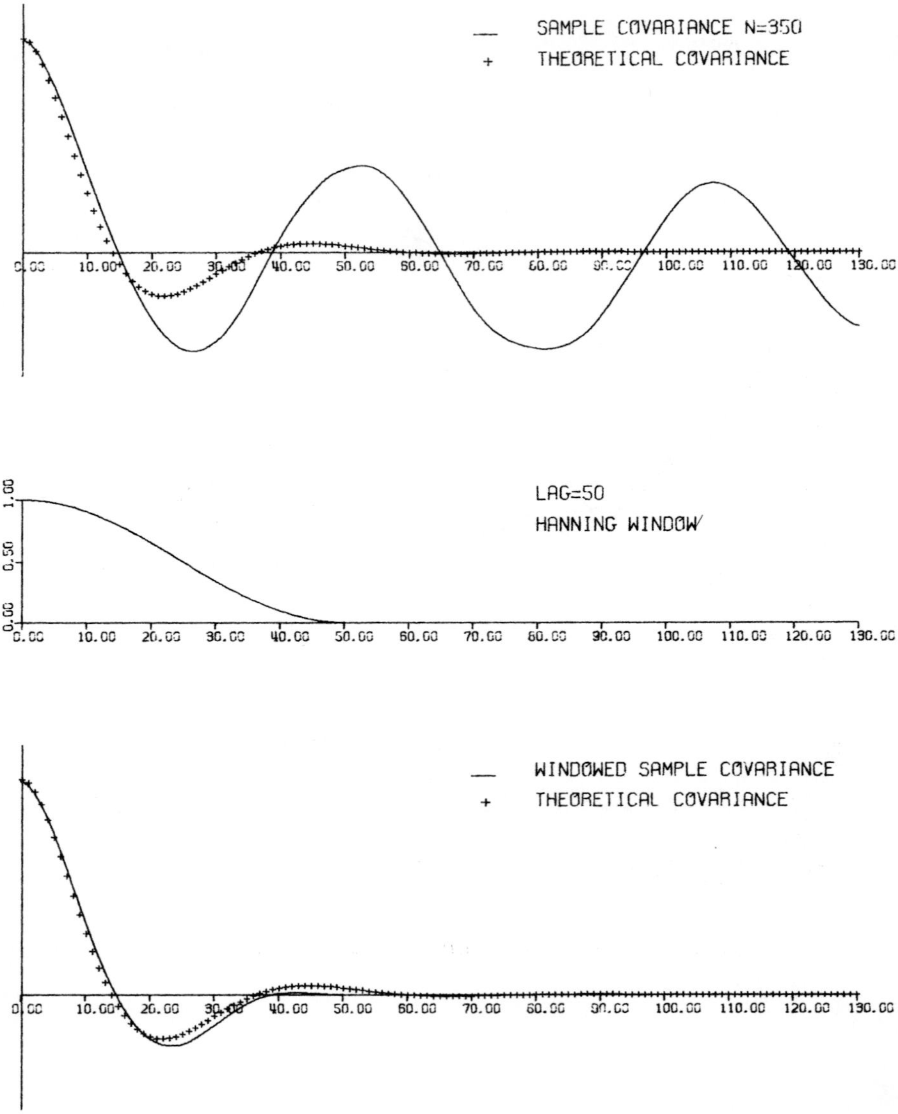

Fig. 2. Effect of windowing a sample covariance sequence.

$$y(n) + a_1 y(n-1) + \cdots + a_M y(n-M) = x(n),$$

where $x(n)$'s are the innovations of $y(n)$ and they are independently and identically distributed random variables.

When a record of a finite number of observations $y(n)$ $(n = 1,2,...,N)$ is given the m-shifted series $y_m(n)$ $(m = 0,1,...,M)$ are defined by $y_m(n) = y(n-m)$ for $n = m + 1, m + 2, ..., m + N$ and 0 otherwise. By minimizing the sum of squares

$$\frac{1}{N} \sum_{n=-\infty}^{\infty} (y_0(n) + a_1 y_1(n) + \ldots + a_M y_M(n))^2 ,$$

one can get the least squares estimates of the autoregressive coefficients a_m $(m = 1,2,...,M)$ and the minimized sum of squares provides an estimate of the variance of the innovation $x(n)$. The model of the stationary time series defined by these estimates is an autoregressive process such that its first $M + 1$ covariances are identical to the corresponding sample covariances $C(m)$ $(m = 0,1,...,M)$, where by definition

$$C(m) = \frac{1}{N} \sum_{n=1}^{N-m} y(n + m) y(n) .$$

Thus the autoregressive model fitting procedure can again be viewed as a method of estimation of the covariance sequence of a time series. Since the estimates are completely specified by the values of the first $M+1$ sample covariances $C(m)$ $(m = 0,1,...,M)$, it is obvious that the performance of the procedure as an estimation procedure of the covariance sequence is entirely dependent on the choice of the order M. Thus, except for the special situations where the orders are specified in advance, any method of the autoregressive model fitting is not well defined as a method of estimation of the covariance sequence without the description of the rules for the determination of the order. Since the covariance sequence determines the power spectrum, once an appropriate rule for the order determination is given the autoregressive model fitting procedure automatically provides an estimate of the power spectrum.

A solution to the problem of order determination of an autoregressive model was obtained in 1969 by using the concept of final prediction error (FPE) [5,6]. FPE is defined as the one-step ahead prediction error variance when the least squares estimates of the autoregressive coefficients are used for prediction. An appropriate estimate of FPE is computed for each value of M within some limited range and the value of M which gives the minimum of the estimates is chosen as the order of the model. This procedure, called the minimum FPE procedure, was found to be quite useful as a method of autoregressive model fitting and produced an almost automatic procedure of estimation of power spectra [7,8,9].

In spite of its intuitive appeal the concept of the one-step ahead prediction error variance had a difficulty in extending it to the multivariate situation. This was due to the non-uniqueness of the measure of variance in the multivariate situation and a solution was found with the aid of the Gaussian model and the concept of maximum likelihood which suggested the use of the generalized variance, the determinant of the variance matrix, of the one-step ahead prediction error [10]. The multivariate autoregressive model fitting procedure based on this extended FPE was found quite useful for the analysis and control of multivariate stochastic systems [11, 12]. Fig. 3 shows the results obtained by fitting an autoregressive model by the minimum FPE procedure to the data treated in Fig. 1. The improvements of the accuracies of the estimates attained by the new procedure are quite obvious. The development of the parametric approach to time series analysis after this stage constitutes the subject of this chapter.

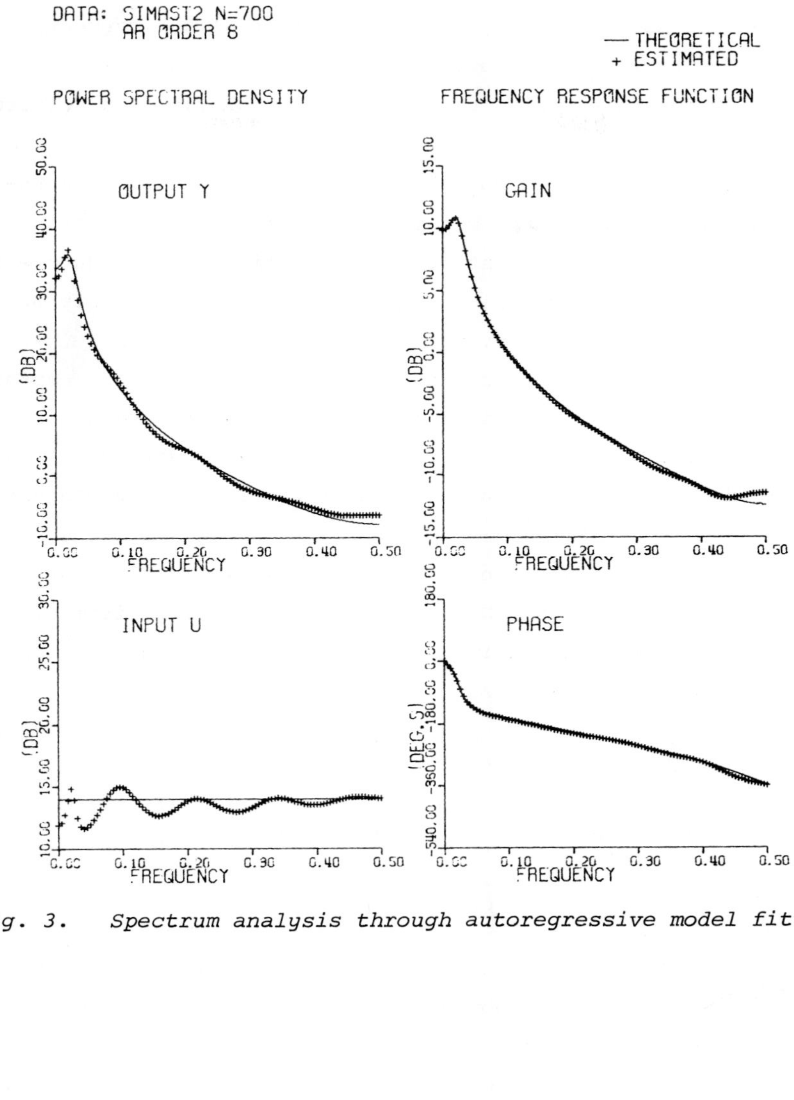

Fig. 3. Spectrum analysis through autoregressive model fitting.

The success obtained by the introduction of the concept of FPE in the autoregressive model fitting clearly suggests the importance of the role played by the criterion of fit in a statistical model fitting procedure. Traditionally the method of maximum likelihood was considered to be a method of estimation of parameters within a statistical model, and the fact that the method could be considered as a statistical model fitting procedure with respect to a natural criterion of fit was almost completely overlooked. For a set of observations $x(n)$ $(n = 1,2,\ldots,N)$ the value of the likelihood function for the statistical model specified by a vector $\underline{\Theta}$ of parameters is defined as the value $f(x(1), x(2), \ldots, x(N)|\underline{\Theta})$ of the probability density function determined by the model. In the case of the simple independent observations the likelihood function for the model with the probability density function $f(x|\underline{\Theta})$ of $x(n)$ is given by $f(x(1)|\underline{\Theta})f(x(2)|\underline{\Theta}) \ldots f(x(N)|\underline{\Theta})$. By taking the logarithm of this likelihood function and taking the average with respect to the number of observations we get

$$\frac{1}{N} \sum_{n=1}^{N} \log f(x(n)|\underline{\Theta}) \quad , \tag{1.1}$$

which is a natural estimate of a quantity defined by

$$N(g;f(\cdot|\underline{\Theta})) = \int \{\log f(x|\underline{\Theta})\} \, g(x) \, dx ,$$

where $g(x)$ denotes the true probability density of the observed values. It is now almost obvious that the method of maximum likelihood, which defines the maximum likelihood estimate $\hat{\underline{\Theta}}$ of $\underline{\Theta}$ as the value of $\underline{\Theta}$ which maximizes (1.1), is implicitly aimed at finding a value of $\underline{\Theta}$ which maximizes the above criterion or, equivalently, minimizes the following information criterion:

$$I(g;f(\cdot|\underline{\Theta})) = \int \log \left(\frac{g(x)}{f(x|\underline{\Theta})} \right) g(x) \, dx \quad .$$

This last quantity is known as Kullback's information quantity and is positive unless $f(x|\underline{\Theta})$ is equal to $g(x)$ almost everywhere. This quantity can be viewed as a measure of improbability of getting a sample distribution $g(x)$ when the true structure is given by $f(x|\underline{\Theta})$. Taking into account the limiting distribution of the maximum likelihood estimates under the regularity conditions an information criterion for the statistical model fitting was introduced in 1971 [13,14]. This information criterion is denoted by AIC and is defined by

AIC = -2 (maximum log likelihood) + 2 (number of independently adjusted parameters).

When several competing models are being fitted by the method of maximum likelihood the one with the smallest value of AIC is chosen as the best model. This procedure is called the minimum AIC procedure and the final estimate is called the minimum AIC estimate (MAICE). The definition of MAICE gives a clear formulation of the principle of parsimony in statistical model building. When two or more models fit equally well to a given set of data in terms of the likelihood the MAICE is the one with the smallest number of free parameters.

In extending the concept of AIC to time series modelling it is found that it is more natural first to develop an estimate of an information theoretic criterion which measures the deviation of a parametric stochastic model from the true structure and then to optimize the parameter values with respect to the estimate of the criterion. This estimate of the criterion is an approximation to the log likelihood and will be called the modified log likelihood. The definition of AIC of a stationary time series model is obtained by replacing the log likelihood in the original definition of AIC by this modified log likelihood. It can be shown that the minimum FPE procedure developed for the autoregressive model fitting is actually a minimum AIC procedure. The concept of AIC is discussed in detail in Section 2.

Although the general advantage of the autoregressive model fitting through the minimum FPE or the minimum AIC procedures over the conventional windowing procedure is quite obvious from the comparison of Figs. 1 and 3, the concept of MAICE suggests the necessity of introduction of some ultimately parsimonious model for the estimation of the covariance sequence of a stationary time series. A solution to this problem is obtained by the introduction of the autoregressive moving average model. The use of the autoregressive moving average model in time series analysis was extensively discussed in a book by Box and Jenkins [15]. Yet the procedure of fitting the model was found to be rather time consuming. This is partly due to the need of the subjective judgements in choosing the orders of the model and partly due to the necessary maximum likelihood computation. Although it was not treated in the book the extension of the procedure to the multivariate situation contains another serious difficulty. This difficulty is concerned with the problem of identifiability of the model and has a deep connection with the concept of canonical representation of a multivariate linear system.

A very simple stochastic interpretation of the state vector of a stochastic linear system is given as the vector composed of a basis of the linear space spanned by the components of the predictors of the present and future outputs of the system based on the present and past inputs to the system [16]. With the aid of this interpretation the Markovian representation of a stationary time series $\underline{y}(n)$ is obtained in the form

$$\underline{v}(n) = \underline{F}\,\underline{v}(n-1) + \underline{G}\,\underline{w}(n)$$

$$\underline{y}(n) = \underline{H}\,\underline{v}(n) \quad ,$$

where $\underline{w}(n)$ is the innovation of $\underline{y}(n)$ at time n and is a white noise. It can be shown that for a Markovian representation there is an equivalent autoregressive moving average representation

$$\underline{y}(n) + \underline{B}_1\underline{y}(n-1) + \ldots + \underline{B}_M\underline{y}(n-M) = \underline{w}(n) + \underline{A}_1\underline{w}(n-1) + \ldots + \underline{A}_L\underline{w}(n-L) \quad .$$

Thus the problem of fitting an autoregressive moving average model is essentially reduced to the problem of fitting a Markovian model.

Since the state vector $\underline{v}(n)$ can be constructed by using the predictors of $\underline{y}(n)$, $\underline{y}(n+1), \ldots$ at time n, a canonical representation is obtained by specifying $\underline{v}(n)$ as the first maximum set of linearly independent predictors of $y_1(n), y_2(n), \ldots, y_r(n), y_1(n+1), y_2(n+1), \ldots,$ where r is the dimension of $\underline{y}(n)$ and $y_i(n)$ denotes the i-th component of $\underline{y}(n)$. For this choice of $\underline{v}(n)$ the matrix $\underline{F}$ takes a simple form which is usually very sparse. This canonical representation constitutes the theoretical basis for the identification of a Markovian model and its derivation is discussed in Section 3.

For the application of the above stated canonical Markovian representation to the identification of a real time series the most significant problem is the identification of the structure of the state vector. It was suggested in [17] that by doing the canonical correlation analysis between the set of the present and past observations and the set of the present and future observations a reasonable estimate of the structure of the state vector could be obtained. In Section 4 of this chapter, the practicability of this procedure is demonstrated by using the results of application to various artificial and real time series. The organization of the necessary computational procedure is described in detail and the use of an AIC type criterion for the decision on the structure of the state vector is discussed. The discussion reveals the essential role played by the criterion of fit in a real statistical model building situation, where any theoretical assumptions such as stationarity would never hold exactly.

With the aid of the autoregressive model fitting by the minimum AIC procedure, the canonical correlation analysis of time series can provide an estimate of the parameters within the Markovian representation of a stationary time series. Yet the accuracies of the estimates are not necessarily high and it is expected that the maximum likelihood estimates of the parameters would be more useful. In Section 5, the organizations of the computational procedures of the gradient and an approximate Hessian of the modified log likelihood are described for the Gaussian Markovian model. With these procedures it is possible to develop a numerical procedure for the maximization of the log likelihood, using the results of the canonical correlation analysis as the initial guesses of the parameters. Some numerical results are given to show the improvement of accuracies attained by the application of the maximum likelihood procedure.

To give a feeling how the procedures to be described in this chapter works, the results of application of the procedures to the data treated in Fig. 1 are illustrated in Fig. 4. By comparing Figs. 1, 3 and 4 we can very clearly see the advances of the techniques of time series analysis achieved within the last decade.

In concluding this introduction it must be reiterated that the success of our procedure is very much dependent on the introduction of the concept of MAICE. Taking into account the fact that the choice of the orders of an autoregressive moving average model is essentially a choice of the functional form of the spectrum, we can clearly see that the problems being treated by MAICE are certainly beyond the scope of the original theory of estimation developed by R. A. Fisher, where the functional form of the population is assumed to be given [35, p. 250].

2. AN INFORMATION CRITERION FOR TIME SERIES MODEL FITTING

The procedure of statistical model identification may be defined as a procedure of finding a model which fits best to a set of observed data with respect to a criterion. Thus the performance of an identification procedure depends very much on the choice of the criterion of fit. It is sometimes argued that the choice of the criterion also depends on the purpose of the identification. By this argument, in the case of a controller design of a stochastic system, the criterion must be chosen so as to make the result of identification useful for designing a controller.

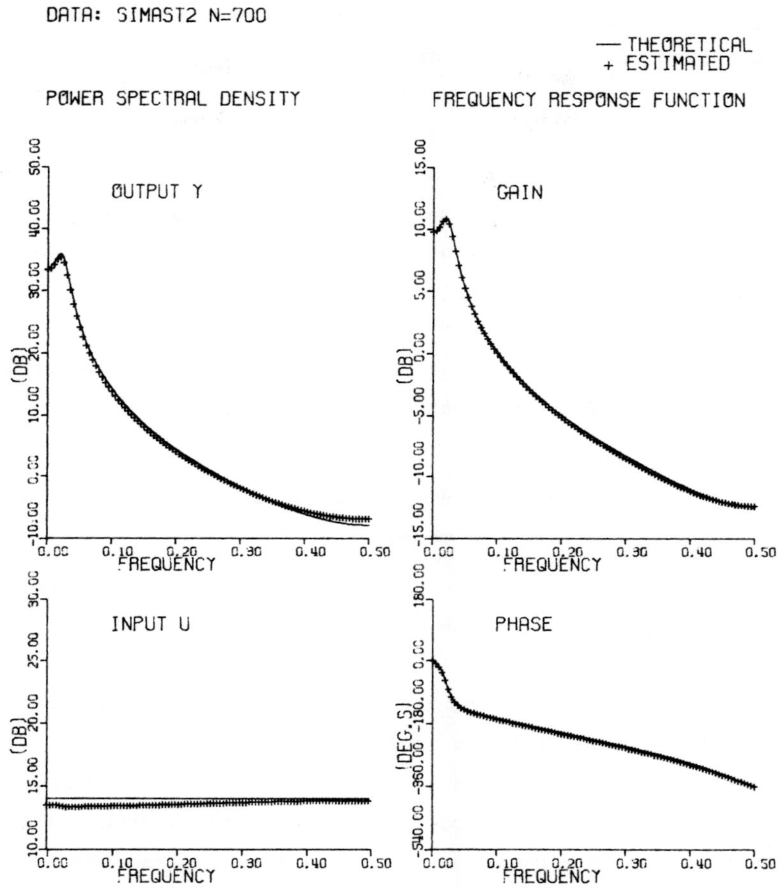

Fig. 4. Spectrum analysis through Markovian model fitting.

Although this idea sounds quite natural, in many practical situations the type of the controller to be introduced can be determined only as a result of some preliminary analysis of the related time series. This suggests the necessity of introducing a criterion which is independent of the specific use of the results of identification. At first it may seem that the introduction of this type of criterion is impossible, yet the successful results obtained by applying the simple method of least squares to various problems show that this is not the case.

The criterion on which we are going to base our identification procedure is an information criterion (AIC) defined by

$$\begin{aligned} \text{AIC} \;=\; & -\,2(\text{maximum log likelihood}) \\ & +\,2(\text{number of independently adjusted parameters} \\ & \qquad \text{within the model}). \end{aligned}$$

This criterion was first introduced in 1971. It was obtained with the aid of an information theoretic interpretation of the method of maximum likelihood [13, 14].

In the simplest case of the observation of a sequence of independently and identically distributed random variables $x(n)$ the method of maximum likelihood is applied in the following manner: First we assume a model which specifies a probability density function $f(x|\underline{\theta})$ of $x(n)$ with a free parameter vector $\underline{\theta}$ and then find the maximum likelihood estimate (MLE) $\hat{\underline{\theta}}$ of $\underline{\theta}$ by maximizing the likelihood function

$$L(x(1),\, x(2),\ldots,x(N)\,|\,\underline{\theta}) \;=\; \prod_{n=1}^{N} f(x(n)\,|\,\underline{\theta})$$

with respect to $\underline{\theta}$. Here $(x(1),\, x(2),\ldots,x(N))$ denotes the set of the observed data. By taking the natural logarithm of the likelihood function and dividing it by the sample size N we get

$$\frac{1}{N}\sum_{n=1}^{N} \log f(x(n)\,|\,\underline{\theta}) \qquad . \tag{2.1}$$

For a fixed value of $\underline{\theta}$, as N is increased indefinitely, this quantity will converge with probability one to

$$B(g;f(\cdot|\underline{\theta})) = \int \log f(x|\underline{\theta})\ g(x)\ dx \quad ,$$

where $g(x)$ denotes the true probability density function of $x(n)$. By taking the difference of $B(g;g)$ and $B(g;f(\cdot|\underline{\theta}))$ we get

$$I(g;f(\cdot|\underline{\theta})) = \int \log \left(\frac{g(x)}{f(x|\underline{\theta})}\right) g(x)\ dx \quad ,$$

which is known as Kullback's information quantity [18]. The nature of Kullback's information quantity can be viewed most clearly when it is related to the quantity

$$S(g;f(\cdot|\underline{\theta})) = -\int \left(\frac{g(x)}{f(x|\underline{\theta})}\right) \log \left(\frac{g(x)}{f(x|\underline{\theta})}\right) f(x|\underline{\theta})\ dx \ .$$

Obviously $S(g;f(\cdot|\underline{\theta})) = -I(g;f(\cdot|\underline{\theta}))$ and $S(g;f(\cdot|\underline{\theta}))$ is directly connected with the probabilistic interpretation of entropy introduced by Boltzmann [19]. This interpretation is based on the fact that when the related distributions are concentrated on discrete points the logarithm of the probability of getting a sample distribution $g(x)$, when N independent samples are drawn from the population with the distribution specified by $f(x|\underline{\theta})$, is asymptotically equal to $NS(g;f(\cdot|\underline{\theta}))$. It can be shown that an analogous interpretation holds generally [20, 21]. Thus we can see that $S(g;f(\cdot|\underline{\theta}))$ is a natural measure of similarity between $g(x)$ and $f(x|\underline{\theta})$. $S(g;f(\cdot|\underline{\theta}))$ is called the *generalized entropy* of $g(x)$ with respect to $f(x|\theta)$ and $I(g;f(\cdot\ \underline{\theta}))$ the *neg-entropy* of $g(x)$ with respect to $f(x|\underline{\theta})$.

Now the method of maximum likelihood can be viewed as an identification procedure of a model represented by $f(x|\underline{\theta})$ through the criterion (2.1) which is a consistent estimate of the neg-entropy of $g(x)$ with respect to $f(x|\underline{\theta})$. If we follow this interpretation of the maximum likelihood method it becomes obvious that there is no need for $g(x)$ to be a member within the family of $f(x|\underline{\theta})$. Only the analysis of the asymptotic properties of the maximum likelihood estimates is usually done under the assumption that there is a value $\underline{\theta}_0$ of $\underline{\theta}$ such that $g(x) = f(x|\underline{\theta}_0)$. The fact that the maximum likelihood procedure is a robust procedure of model fitting which is useful even without the above stated restrictive assumption of $g(x) = f(x|\underline{\theta}_0)$ has not been fully recognized. The assumption was introduced only to provide estimates of the accuracies of the maximum likelihood estimates. When the assumption holds, the first term in the definition of AIC asymptotically shows, as an estimate of $-2NB(f(\cdot|\underline{\theta}_0);f(\cdot|\underline{\theta}_0))$, a downward bias of the amount equal to the number of the free parameters within the model. The first half of the second term in the definition of AIC is used to compensate for this bias. Now the fact that the maximum likelihood estimate introduces this amount of bias means that the fitted model $f(x|\hat{\underline{\theta}})$ shows on the average a deviation from $f(x|\underline{\theta}_0)$ equal to this amount as measured in terms of the criterion which is defined as $2N$ times the negentropy. The second half of the second term in the definition of AIC is thus introduced to make AIC an estimate of $E(-2)N\int \log f(x|\hat{\underline{\theta}})\, f(x|\underline{\theta}_0)\, dx$ with the bias corrected to the order $O(1)$. When the assumption $g(x) = f(x|\underline{\theta}_0)$ does not hold, the first term in the definition of AIC provides an estimate of $-2NB(g;f(\cdot|\underline{\theta}^*))$, where $f(\cdot|\underline{\theta}^*)$ is such that $I(g;f(\cdot|\underline{\theta}^*)$ is equal to the minimum of $I(g;f(\cdot|\underline{\theta}))$ as a function of $\underline{\theta}$. When $f(x|\underline{\theta}^*)$ is sufficiently close to $g(x)$, the above corrections by the second term are still useful. When $f(x|\underline{\theta}^*)$ is very far from $g(x)$ this will be reflected in the value of the first term and can be detected by comparison with other models. Thus when there are several models the parameters within the models are estimated by the maximum likelihood procedure and the values of the AIC's are computed and compared to find a model with the minimum value of AIC. This procedure is called the minimum AIC procedure and the model with the minimum AIC is called the minimum AIC estimate (MAICE) [14].

The source of the main difficulty in applying the minimum AIC procedure to time series analysis is the analytical complexity of the likelihood function. In the case of a stationary time series $\underline{y}(1), \underline{y}(2),\ldots,\underline{y}(N)$ the limit of the time average of the expected log likelihood,

$$K(g;f(\cdot|\underline{\theta}))$$
$$= \lim_{N\to\infty} \frac{1}{N} \int \log f(\underline{y}(1),\underline{y}(2),\ldots,\underline{y}(N)|\underline{\theta})g(\underline{y}(1),\underline{y}(2),\ldots,\underline{y}(N)) \cdot d\underline{y}(1)\, d\underline{y}(2)\ldots d\underline{y}(N) \quad ,$$

will define a natural criterion of fit of the model, which is specified by the finite dimensional distributions $f(\underline{y}(1), \underline{y}(2), \ldots,\underline{y}(N))$ $(N = 1,2,\ldots)$. The criterion $K(g;f(\cdot|\underline{\theta}))$ reduces to the Kullback's information quantity when the observations are assumed to be mutually independent. $K(g;f(\cdot|\underline{\theta}))$ will be called the average information criterion. In general it is expected that, under some regularity conditions, the average log likelihood

$$\frac{1}{N} \log f(\underline{y}(1), \underline{y}(2),\ldots,\underline{y}(N)|\underline{\theta})$$

will converge with probability one to the above defined criterion as N tends to infinity. The average log likelihood may then be viewed as an estimate of the average information criterion $K(g;f(\cdot|\underline{\theta}))$. Thus instead of directly maximizing the log likelihood, $\log f(\underline{y}(1), \underline{y}(2),\ldots,\underline{y}(N)|\underline{\theta})$, with respect to $\underline{\theta}$ we would rather develop an estimate of the criterion $K(g;f(\cdot|\underline{\theta}))$ first and then find an optimum value $\hat{\underline{\theta}}$ of $\underline{\theta}$ with respect to this estimate. This $\hat{\underline{\theta}}$ will show the characteristics of the ordinary maximum likelihood estimates. For simplicity, we will call $\hat{\underline{\theta}}$ the maximum likelihood estimate of $\underline{\theta}$.

Now let us consider the case of a scalar autoregressive model fitting. Assume that $y(n)$ is an M-th order autoregressive process generated by the relation

$$y(n) + A_1\, y(n-1) + \cdots + A_M\, y(n-M) = x(n) \quad , \qquad (2.2)$$

where $x(n)$ is a sequence of mutually independent Gaussian random variables with mean zero and variance C and $x(n)$ is assumed to be independent of $y(n-1)$, $y(n-2), \ldots$. From the assumption the likelihood function admits the following factorization

$$f(y(1), y(2), \ldots, y(N) | \theta)$$
$$= f(y(1), \ldots, y(M) | \underline{\theta}) \prod_{n=M+1}^{N} f(y(n)\ y(n-1), \ldots, y(n-M), \underline{\theta}),$$

where $\underline{\theta}$ denotes the vector of the parameters $A_1, A_2, \ldots, A_M$ and C and

$$f(y(n)\ y(n-1), \ldots, y(n-M), \underline{\theta})$$
$$= \frac{1}{\sqrt{2\pi C}} \exp \left\{ -\frac{1}{2C} (y(n) + A_1 y(n-1) + \cdots + A_M y(n-M))^2 \right\}$$

The criterion $K(g; f(\cdot | \underline{\theta}))$ for this case is obtained as

$$K(g; f(\cdot | \underline{\theta})) = -\frac{1}{2} \log (2\pi C) - \frac{1}{2C} \sum_{j=0}^{M} \sum_{k=0}^{M} A_j A_k R(k-j) ,$$

where by definition $A_0 = 1$ and $R(k) = Ey(n+k)y(n)$ is the k-lag covariance of the actual process and the expectation E is taken with respect to the distribution specified by $g(y(n), y(n+1), \ldots, y(n+k))$. Since the values of $R(k)$ are unknown, for the purpose of evaluation of the goodness of fit of a model specified by $A_1, A_2, \ldots, A_M$ and C, we have to replace $R(k)$ by an appropriate estimate. A natural choice is the sample covariance $C(k)$ defined by

$$C(k) = \frac{1}{N} \sum_{n=1}^{N-k} y(n+k)y(n) \quad \text{for} \quad k = 0, 1, \ldots, M ,$$

and

$$C(-k) = C(k) .$$

The sequence of $R(k)$'s must be a positive definite sequence and the present choice of $C(k)$ satisfies this condition. Thus our estimate of $K(g;f(\cdot|\underline{\theta}))$ for this case is given by

$$\hat{K}(g;f(\cdot|\underline{\theta})) = -\frac{1}{2}\log(2\pi C) - \frac{1}{2C}\sum_{j=0}^{M}\sum_{k=0}^{M} A_j A_k\, C(j-k)\ .$$

We will call $N\hat{K}(g;f(\cdot|\underline{\theta}))$ the modified log likelihood. The values of the parameters which maximizes this modified log likelihood are obtained by the relations

$$C(j) = -\sum_{k=1}^{M} \hat{A}_k\, C(j-k) \qquad j = 1,2,\ldots,M \tag{2.3a}$$

and

$$\hat{C} = C(0) + \sum_{k=1}^{M} \hat{A}_k\, C(-k)\ . \tag{2.3b}$$

From the asymptotic distribution of $C(k)$'s we can derive the asymptotic distribution of the maximum likelihood estimates $\hat{A}_1$, $\hat{A}_2,\ldots,\hat{A}_M$ and $\hat{C}$ [6]. From this asymptotic distribution we can see that the asymptotic behavior of an information criterion defined by

$$AIC = (-2)\, N\hat{K}(g;f(\cdot|\hat{\underline{\theta}})) + 2(M+1) \tag{2.4}$$

is identical to the case of independent observations. When the problem is only the comparison of autoregressive models of different orders, the common additive constants do not affect the result of comparison and the present definition of AIC can be replaced by

$$AIC = N\log(\hat{C}) + 2M\ .$$

The performance of the minimum AIC procedure is illustrated by an example in Fig. 5.

When $\underline{y}(n)$ is an r-dimensional vector the Gaussian autoregressive model of order M is defined by (2.2) with the matrices $\underline{A}_m$ and $\underline{C}$ of dimension $r \times r$, where $\underline{C}$ is the covariance of the r-dimensional white Gaussian noise $x(n)$. For this case the definition of $C(k)$ is replaced by

$$\underline{C}(k) = \frac{1}{N}\sum_{n=1}^{N-k} \underline{y}(n+k)\,\underline{y}(n)' \qquad \text{for} \quad k = 0,1,\ldots,M\ ,$$

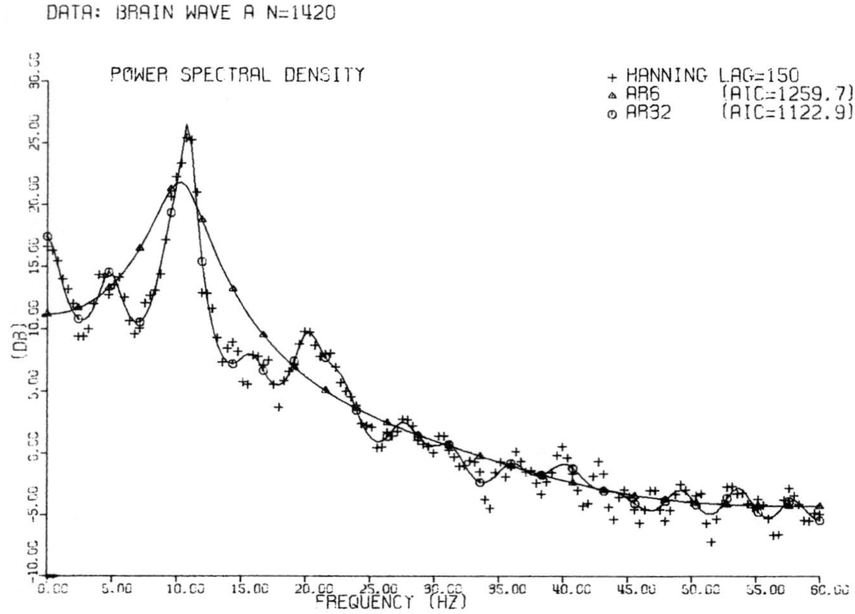

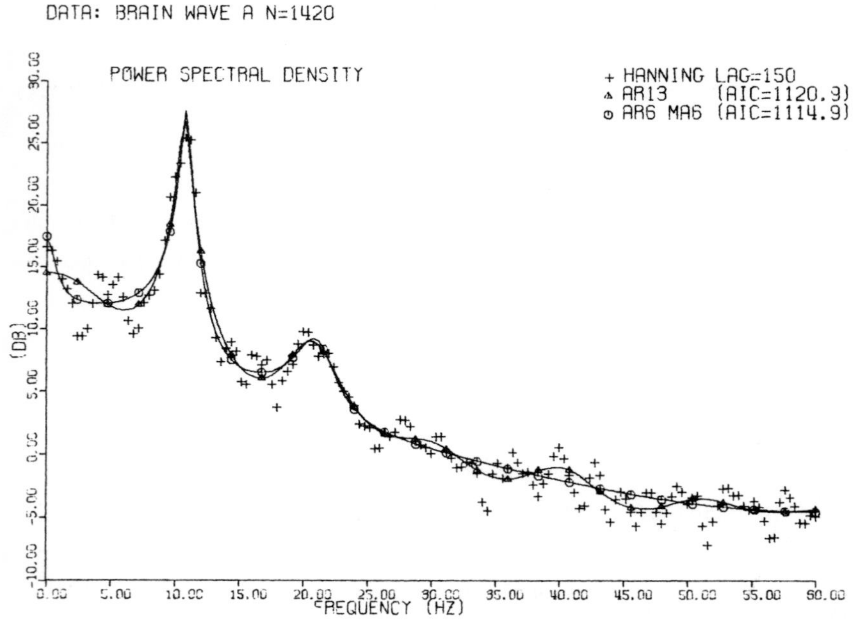

Fig. 5. Use of the minimum AIC procedure in the autoregressive model fitting. Fig. 5(a) shows the examples of under and over fitting. The minimum AIC estimates of the autoregressive and autoregressive moving average models are shown in Fig. 5(b).

and

$$\underline{C}(-k) = \underline{C}(k)' .$$

Here ' denotes the transposition. With this modification the estimates $\hat{\underline{A}}_1, \hat{\underline{A}}_2, \ldots, \hat{\underline{A}}_M$ and $\hat{\underline{C}}$, which are optimum with respect to the information criterion are obtained from the relations (2.3a) and (2.3b). The definition (2.4) of AIC is replaced by

$$\text{AIC} = N \log (|\hat{\underline{C}}|) + 2Mr^2 , \tag{2.5}$$

where $|\underline{C}|$ denotes the determinant of $\underline{C}$. This definition of AIC was used to get the results illustrated in Fig. 3. The results illustrated in Figs. 3 and 5 show that the autoregressive model fitting by the minimum AIC procedure can provide a very reasonable estimate of the covariance sequence of a scalar or vector stationary time series.

What has been clarified by the introduction of AIC is the importance of the concept of parsimony in statistical model building. By explicitly taking into account the penalty of introducing the free parameters into a model the definition of AIC clearly shows why we have to be parsimonious in choosing a model. Now this concept of parsimony suggests that the choice of the autoregressive models as the basic family of models to be used for the purpose of identification is by no means optimal. An ultimate finite parameter linear model of a stationary time series $\underline{y}(n)$ is an autoregressive moving average model given by

$$\underline{y}(n) + \underline{B}_1\underline{y}(n-1) + \cdots + \underline{B}_M\underline{y}(n-M)$$
$$= \underline{x}(n) + \underline{A}_1\underline{x}(n-1) + \cdots + \underline{A}_L\underline{x}(n-L) ,$$

where $\underline{x}(n)$ is the innovation of $\underline{y}(n)$ at time n. The use of this autoregressive moving average model in the modelling of a scalar time series was extensively discussed by Box and Jenkins in their book [15]. In the field of engineering or science the most interesting part of time series analysis is usually the analysis of dependence between various simultaneous time series.

The extension of the application of autoregressive moving average models to the analysis of multivariate time series is not so simple as in the case of autoregressive models. Since it can be shown that an autoregressive moving average model has an equivalent Markovian representation which is to be discussed in the next section, the following sections of this chapter which are concerned with the statistical identification of Markovian models of multivariate stationary time series are essentially concerned with this extension. The extension is made possible through a process of fusion of the results on the system realization obtained by the mathematical system theory and the statistical concept of information.

3. CANONICAL REPRESENTATIONS OF LINEAR STOCHASTIC SYSTEMS

It is well recognized that an essential difficulty in the identification of a multivariate linear system lies in the choice of a suitable canonical form [22]. The concept of canonical form or representation is also related with the concept of parsimony of a model, the statistical implication of which is already made clear by the introduction of AIC. It can be shown by following the discussion of this section that the autoregressive moving average models with the lowest possible orders of the autoregression and moving average parts constitute a system of canonical forms for the representation of a scalar stationary time series. This fact forms a theoretical basis of the use of autoregressive moving average models in time series analysis. In spite of its popularity in the scalar case, no significant practical example of application of the model is ever known in the multivariate case. It is fairly safe to say that this was caused by the lack of a proper canonical representation of a multivariate stationary time series. In this section we will introduce a useful canonical representation of a general multivariate linear stochastic system. The representation is obtained by using the concept of the predictor space which was first developed in [16].

Consider a stochastic system with the input $\underline{u}(n)$ and output $\underline{y}(n)$. It is assumed that $\underline{u}(n)$ and $\underline{y}(n)$ are multivariate stationary time series of dimension q and r, respectively. It is further assumed that both $\underline{u}(n)$ and $\underline{y}(n)$ are with zero mean, and finite second order moments and that they are stationarily correlated. For the sake of simplicity we will further assume that $(\underline{u}(n), \underline{y}(n))$ is a (r+q)-dimensional stationary Gaussian process. Hereafter we will be concerned with various linear spaces spanned by the components of the vectors $\underline{u}(n)$ and $\underline{y}(n)$. It is assumed throughout the chapter that the metric in these spaces is given by the root mean square and the spaces are closed with this metric. If we adopt the definition of the state space of a system at time n as the totality of the information from the present and past input to be transmitted to the present and future output, a natural mathematical representation of the state space in this case will be given by the linear space spanned by the components of the projections $\underline{y}(n+k|n)$ (k = 0,1,...), where $\underline{y}(n+k|n)$ denotes the projection of $\underline{y}(n+k)$ onto the linear space $R_u(n-)$ spanned by the components of the present and past inputs $\underline{u}(n), \underline{u}(n-1), \ldots$. The projection $\underline{y}(n+k|n)$ is composed of the elements of $R_u(n-)$ and is characterized by the relation

$$E\underline{y}(n+k|n)\underline{u}(n-m)' = E\underline{y}(n+k)\underline{u}(n-m)' \quad \text{for} \quad m = 0,1,\ldots \tag{3.1}$$

The very basic assumption to be made in this section is that the dimension of the linear space spanned by the components of $y(n+k|n)$ (k = 0,1,...) is finite. This space which is spanned by the components of the predictors is called the *predictor space* and is denoted by $R(n+|n-)$. The dimension of the space will be denoted by p. Obviously $R(n+|n-)$ is a subspace of $R_u(n-)$. (3.1) shows that the whole information of $\underline{y}(n+k)$ within $R_u(n-)$ is condensed in $\underline{y}(n+k|n)$, in the sense that the dependence of $\underline{y}(n+k)$ on $R_u(n-)$ can be explained by $\underline{y}(n+k|n)$. Take an arbitrary basis $v_1(n), v_2(n), \ldots, v_p(n)$ of the predictor space

$R(n+|n-)$ and denote by $\underline{v}(n)$ the vector $(v_1(n), v_2(n), \ldots, v_p(n))'$. Since $\underline{y}(n|n)$ is composed of the elements of $R(n+|n-)$, there is an $r \times p$ matrix $\underline{H}$ such that

$$\underline{y}(n|n) = \underline{H}\,\underline{v}(n) \quad .$$

Thus $\underline{y}(n)$ has a representation

$$\underline{y}(n) = \underline{H}\,\underline{v}(n) + \underline{z}(n), \tag{3.2}$$

where $\underline{z}(n)$ is independent of the elements of $R_u(n-)$, i.e., $\underline{z}(n)$ is independent of the present and past inputs. Denote by $\underline{v}(n+1)$ the random vector obtained by replacing $\underline{u}(n), \underline{u}(n-1), \ldots$ in the definition of $\underline{v}(n)$ by $\underline{u}(n+1), \underline{u}(n), \ldots$. From the stationarity the components of $\underline{v}(n+1)$ form a basis of the space $R((n+1)+|(n+1)-)$. By an analogous reasoning as in the case of $\underline{y}(n)$, it can be seen that $\underline{v}(n+1)$ admits a representation

$$\underline{v}(n+1) = \underline{F}\,\underline{v}(n) + \underline{w}(n+1) \quad , \tag{3.3}$$

where $\underline{w}(n+1)$ is independent of $\underline{u}(n), \underline{u}(n-1), \ldots$. From (3.3), since the components of $\underline{v}(n+1)$ and $\underline{v}(n)$ are elements of $R_u((n+1)-)$, $\underline{w}(n+1)$ is composed of the elements of $R_u((n+1)-)$ but its projection onto $R_u(n-)$ is equal to zero. Thus $\underline{w}(n+1)$ can be represented in the form

$$\underline{w}(n+1) = \underline{G}\,\underline{x}(n+1) \quad , \tag{3.4}$$

where $\underline{G}$ is a $p \times q$ matrix and

$$\underline{x}(n+1) = \underline{u}(n+1) - \underline{u}(n+1|n) \ . \tag{3.5}$$

$\underline{u}(n+1|n)$ denotes the projection of $\underline{u}(n+1)$ onto $R_u(n-)$ and $\underline{x}(n+1)$ is the innovation of the input $u(n)$ at time $n+1$. The covariance matrix of the innovation $\underline{x}(n)$ is denoted by $\underline{C}$, i.e., $\underline{C} = E\underline{x}(n)\underline{x}(n)'$. (3.2), (3.3), (3.4) and (3.5) give a representation of a general stationary linear stochastic system. In this representation $\underline{z}(n)$ denotes the part of $\underline{y}(n)$ which is independent of the present and past inputs. $\underline{x}(n)$'s with different values of n are mutually independent but $\underline{z}(n)$'s are not necessarily so. $\underline{v}(n)$ provides a specification of the predictor space $R(n+|n-)$ and thus may be considered to be a representation of the state of the stochastic system at time n. $\underline{v}(n)$ is called the *state vector*.

When there are feedbacks from the outputs to the inputs the distinction of the variables in terms of the inputs and outputs becomes meaningless. To get a representation of this most general system we have only to take $\underline{u}(n) = \underline{y}(n)$. For this case, since $\underline{y}(n)$ is composed of the elements of $R_u(n-)$, we have $\underline{y}(n|n) = \underline{y}(n)$ and $\underline{z}(n)$ in (3.2) vanishes. Thus we get a representation

$$\begin{aligned} \underline{v}(n+1) &= \underline{F}\,\underline{v}(n) + \underline{G}\,\underline{x}(n+1) , \\ \underline{y}(n) &= \underline{H}\,\underline{v}(n) , \end{aligned} \tag{3.6}$$

where the innovation $\underline{x}(n+1)$ of $\underline{y}(n)$ at time n+1 is defined by

$$\underline{x}(n+1) = \underline{y}(n+1) - \underline{y}(n+1|n) . \tag{3.7}$$

This representation is called a *Markovian representation* of $\underline{y}(n)$.

Since any basis of the predictor space $R(n+|n-)$ can play the role of the state vector $\underline{v}(n)$, the matrices $\underline{F}$, $\underline{G}$ and $\underline{H}$ of the above representation are not unique. A proper specification of the structure of $\underline{v}(n)$ defines a canonical representation of a stochastic linear system.

Our canonical representation is obtained by choosing the first maximum set of linearly independent elements within the sequence of the predictors $y_1(n|n)$, $y_2(n|n)$, ..., $y_r(n|n)$, $y_1(n+1|n)$, $y_2(n+1|n)$, ..., $y_r(n+1|n)$, ..., $y_1(n+k|n)$, $y_2(n+k|n)$, ... as the state vector, where $y_h(n+k|n)$ denotes the h-th component of $\underline{y}(n+k|n)$. Conceptually the state vector $\underline{v}(n)$ can be constructed by the following procedure:

1) Define the vector $\underline{s}$ by

$$\underline{s} = \begin{bmatrix} \underline{y}(n|n) \\ \underline{y}(n+1|n) \\ \vdots \end{bmatrix} . \tag{3.8}$$

the i-th component s_i of $\underline{s}$ is given by

$$s_i = y_j(n+k|n) ,$$

where $i = kr + j$ and r is the dimension of $\underline{y}(n)$.

2) Successively test the linear dependence of s_i on its antecedents

$$s_{i-1}, s_{i-2}, \ldots, s_1 \qquad \text{for} \quad i = 1,2,\ldots$$

3) $\underline{v}(n)$ is defined as the vector obtained by discarding from $\underline{s}$ those components which are dependent on its antecedents. The dimension of $\underline{v}(n)$ is denoted by p.

$\underline{v}(n)$ is completely characterized by a vector $\underline{h} = (h_1, h_2, \ldots, h_p)'$ which is defined by the relation

$$h_j = i \qquad \text{when} \qquad v_j(n) = s_i .$$

h_j gives the specification of the j-th component of $\underline{v}(n)$ as an element within the sequence of the predictors. This vector $\underline{h}$ will be called the *structural characteristic vector* of $\underline{v}(n)$. Once the specification of $\underline{v}(n)$ is given the matrices $\underline{F}$, $\underline{G}$ and $\underline{H}$ of the corresponding canonical representation can be obtained very easily.

When there is a representation of $s_i = y_j(n+k|n)$ as a linear combination of its antecedents, by replacing n by $n + m$ $(m = 1,2,\ldots)$ in the definition of the elements in the representation, one can see that $s_{i+mr} = y_j(n+m+k|n)$ can also be expressed as a linear combination of its antecedents. Thus for each component $y_j(n)$ of $\underline{y}(n)$ there is a smallest value k_j of k such that $y_j(n+k|n)$ is linearly dependent on its antecedents in $\underline{s}$. The r-dimensional vector $\underline{k} = (k_1, k_2, \ldots, k_r)'$ can also specify the structure of the state vector. The state vector $\underline{v}(n)$ is obtained by arranging $y_j(n+k|n)$'s with $k = 0,1,\ldots,k_j-1$ and $j = 1,2,\ldots,r$ in the increasing order of $j + kr$. Those $v_i(n)$'s which correspond to $y_j(n+k_j-1|n)$'s are characterized by the fact that (h_i+r)'s do not belong to the set of the components of the structural characteristic vector $\underline{h}$. Now for $v_i(n) = y_j(n+k_j-1|n)$ we have $v_i(n+1) = y_j(n+k_j|n+1)$ and its projection on $R_y(n-)$, denoted by $v_i(n+1|n)$, is equal to $y_j(n+k_j|n)$ which can be expressed as a linear combination of its antecedents in $\underline{s}$ of (3.8). Thus for this case there is a unique representation

$$v_i(n+1|n) = \sum_{r=1}^{p} F(i,r)\, v_r(n) \quad , \tag{3.9}$$

where $F(i,r) = 0$ for $r = r(i) + 1, r(i) + 2,\ldots,p$. where $r(i)$ is the number of h_m's which are smaller than $h_i + r$. $F(i,r)$ defines the (i,r)-th element of the matrix of $\underline{F}$. When $v_i(n)$ is defined to be equal to $y_j(n+k-1|n)$ with $k < k_j$ then $y_j(n+k|n)$ is another component of $\underline{v}(n)$ which will be denoted by $v_{r(i)}(n)$. The corresponding component of the structural characteristic vector $\underline{h}$ is given by $h_{r(i)} = h_i + r$. $r(i)$ is equal to the number of h_m's which are smaller than or equal to $h_i + r$. For this case, since $v_i(n+1)$ is given by $y_j(n+k|n+1)$, $v_i(n+1|n)$ is equal to $y_j(n+k|n)$ which is the $r(i)$-th component of $\underline{v}(n)$. Thus we get a unique representation

$$v_i(n+1|n) = \sum_{r=1}^{p} F(i,r)\, v_r(n) \quad , \tag{3.10}$$

where $F(i,r) = 1$ for $r = r(i)$ and 0 otherwise. (3.9) and (3.10) give a complete specification of the $p \times p$ matrix $\underline{F}$, the transition matrix. The (i,s)-th element $G(i,s)$ of the input matrix $\underline{G}$, which is $p \times r$ is characterized as the impulse response of $v_i(n)$ to the input $x_s(n)$, the s-th component of the innovation. If $v_i(n)$ is identical to $y_j(n+k|n)$ then $G(i,s)$ can be obtained as the impulse response of $y_j(n)$ at time lag k to the input $x_s(n)$. The output matrix $\underline{H}$, which is $r \times p$, can be obtained as the matrix of regression coefficients of $\underline{y}(n)$ on $\underline{v}(n)$.

Under the assumption of nonsingularity of the covariance matrix of $\underline{y}(n|n)$ $\underline{H}$ takes the form

$$\underline{H} = [\underline{I}\ \underline{0}] \quad ,$$

where $\underline{I}$ is an $r \times r$ identity matrix and $\underline{0}$ is an $r\times(p-r)$ zero matrix. This is the case when $\underline{u}(n) = \underline{y}(n)$ and the covariance matrix $\underline{C}$ of the innovation $\underline{x}(n)$ is nonsingular. When $\underline{u}(n) = \underline{y}(n)$, under the same assumption as above, the uppermost $r \times r$ submatrix of the input matrix $\underline{G}$ is an identity matrix. Hereafter we will only be concerned with the case where $\underline{u}(n) = \underline{y}(n)$ and the matrix $\underline{C}$ is nonsingular.

It is obvious that in the above canonical Markovian representation of $\underline{y}(n)$ the matrices $\underline{F}$, $\underline{G}$, $\underline{H}$ and $\underline{C}$ are uniquely determined. Any modification of the elements of these matrices, which are not explicitly specified as equal to 1 or 0, produces a representation of a time series which is different from the original one. This shows that under the assumptions of nonsingularity of $\underline{C}$ and the finiteness of the dimension of the predictor space the correspondence between the covariance structure of $\underline{y}(n)$ and the set of $\underline{h}$, $\underline{F}$, $\underline{G}$, $\underline{H}$, and $\underline{C}$ is one to one. Here $\underline{F}$, $\underline{G}$, $\underline{H}$ and $\underline{C}$ are such that they are in the forms given by the above canonical representation of $\underline{y}(n)$ and define a covariance structure with the linear dependence characteristics of the predictors specified by $\underline{h}$.

The totality of the sets $\{\underline{h}, \underline{F}, \underline{G}, \underline{H}, \underline{C}\}$ provides a basic framework for the identification of a stationary time series. For a scalar time series $\underline{h}$ has a very simple structure given by $h_i = i$ $(i = 1,2,\ldots,p)$ and $\underline{k} = (k_1) = (p)$. k_1 is called the order of the scalar stochastic linear system. The fundamental difficulty in identifying a multivariate stochastic linear system is caused by the complex structure of $\underline{h}$ which is also reflected in the fact that $\underline{k}$ is a vector with the same dimension as that of $\underline{y}(n)$. We will call k_i the *order of the* i-*th component* of $\underline{y}(n)$. It should be remembered that the value of k_i is dependent on the ordering of the components within the vector $\underline{y}(n)$. The quantity $K = \max(k_1,k_2,...,k_r)$ is independent of this ordering and will be called the *order* of $\underline{y}(n)$ or of the linear stochastic system.

The relation between the Markovian and the autoregressive moving average models of a stationary time series can be analyzed as follows. From the relation (3.9) we get, for $j = 1,2,...,r$,

$$y_j(n+k_j|n) = \sum_{m=0}^{k_j} \underline{B}_m^-(j,\cdot)\, \underline{y}(n+k_j-m|n), \tag{3.11}$$

where $\underline{B}_m^-(j,\cdot)$ is $1 \times r$ and $B_0^-(j,h)$, the h-th component of $\underline{B}_0^-(j,\cdot)$, is zero for $h \geqq j$. Since $\underline{y}\,(n+k_j-m)$'s $(m = 0,1,...,k_j)$ are composed of the elements of $R_y((n+k_j)-)$, from the above relations (3.11) follow the representations $(j = 1,2,...,r)$

$$y_j(n+k_j) - \sum_{m=0}^{k_j} \underline{B}_m^-(j,\cdot)\underline{y}(n+k_j-m) = x_j(n+k_j) + \sum_{\ell=0}^{k_j-1} \underline{A}_\ell^+(j,\cdot)\underline{x}(n+k_j-\ell)$$

where $\underline{A}_\ell^+(j,\cdot)$ is $1 \times r$ and $\underline{A}_0^+(j,\cdot) = -\underline{B}_0^-(j,\cdot)$. The assumption of nonsingularity of the varaince matrix $\underline{C}$ of the innovation $\underline{x}(n)$ assures the uniqueness of the coefficients within the right hand sides of the above equations. Summarizing these equations we get

$$\underline{y}(n+K) + \sum_{m=0}^{K} \underline{B}_m^+\, \underline{y}(n+K-m) = \underline{x}(n+K) + \sum_{\ell=0}^{K-1} \underline{A}_\ell^+\underline{x}(n+K-\ell),$$

where the matrix $\underline{B}_m^+$ is $r \times r$ and its j-th row is equal to

$-\underline{B}_m^-(j,\cdot)$, for $m < k_j$, and zero, otherwise, and the matrix $\underline{A}_\ell^+$ is defined from $\underline{A}_\ell^+(j,\cdot)$ analogously. Since $\underline{I} + \underline{B}_0^+$ is lower triangular we can define the representation

$$\underline{y}(n+K) + \sum_{m=1}^{K} \underline{B}_m \underline{y}(n+K-m) = \underline{x}(n+K) + \sum_{\ell=1}^{K-1} \underline{A}_\ell \underline{x}(n+k-\ell), \quad (3.12)$$

where $\underline{B}_m + (\underline{I} + \underline{B}_0^+)^{-1}\underline{B}_m^+$ and $\underline{A}_\ell = (\underline{I} + \underline{B}_0^-)^{-1}\underline{A}_\ell^+$. Obviously this autoregressive moving average representaiton of $\underline{y}(n)$ is uniqely determined by the original canonical Markovian representation. When $\underline{y}(n)$ admits an autoregressive moving average representation

$$\underline{y}(n+K) + \sum_{m=1}^{K} \underline{D}_m \underline{y}(n+K-m) = \underline{x}(n+K) + \sum_{\ell=1}^{K-1} \underline{C}_\ell \underline{x}(n+K- \ ) ,$$

we have

$$\underline{y}(n+K|n) + \sum_{m=1}^{K} \underline{D}_m \underline{y}(n+K-m|n) = 0 .$$

This means that the dimension of the predictor space of $\underline{y}(n)$ is finite and thus $\underline{y}(n)$ has a canonical Markovian representation which determines the corresponding autoregressive moving average representation (3.12) uniquely. Thus we can see that the problem of identification of a multivariate autoregressive moving average model is essentially solved by the use of the present canonical Markovian representation. Hereafter we will limit our attention to the identification of the canonical Markovian model.

4. CANONICAL CORRELATION ANALYSIS OF TIME SERIES

For a stationary time series we now have a class of the Gaussian models defined by the sets $\{\underline{h}, \underline{F}, \underline{G}, \underline{H}, \underline{C}\}$, which is obtained as the result of the discussions of Section 3. Since we already have a criterion AIC of the fit of a statistical model the statistical identification of a Gaussian model of a stationary time series can be relaized by a direct application of the minimum AIC procedure to this class of Gaussian models. In this procedure the maximum likelihood estimates of the free parameters within the models are computed and the model with the minimum value of AIC is chosen as the best estimate. Although the procedure is simple in principle, the maximum likelihood computation usually requires a time-consuming iterative procedure and the success of the computation is strongly dependent on the choice of the initial guesses of the parameters. Furthermore, for a multivariate time series, even if the highest possible value of the orders k_i $(i = 1,2,\ldots r)$ is limited in advance, there are usually enormous numbers of possible choices of the characteristic vector $\underline{h}$. Thus the practicability of application of the minimum AIC procedure to the identification of the Markovian model of a multivariate stationary time series is very much dependent on whether we have a procedure to get reasonable initial guesses of the vector $\underline{h}$. The purpose of this chapter is to show that the desired initial guesses of the vector $\underline{h}$ and the parameters within the matrices $\underline{F}$ and $\underline{G}$ can be obtained by the analysis of canonical correlations between the set of the present and past values and the set of the present and future values of the time series. This result may be viewed as a significant recent advance in time series analysis.

For the sake of simplicity of explanation let us assume that the original time series $\underline{y}(n)$ is Gaussian. We will further assume that the predictor space of $\underline{y}(n)$ is of finite dimension. Our problem here is the analysis of the dependence relations within the components of the predictors $\underline{y}(n|n)$, $\underline{y}(n+1|n),\ldots$, which are the elements of $R_y(n-)$, the linear space spanned by the components of $\underline{y}(n)$, $\underline{y}(n-1),\ldots$. Denote by $\underline{y}(n+k|n,n-M)$ the projections of $\underline{y}(n+k)$ on the space $R_y(n,n-M)$ which is a linear space spanned by the components of $\underline{y}(n)$, $\underline{y}(n-1),\ldots,\underline{y}(n-M)$. Now any element of $R_y(n-)$ can be approximated arbitrarily closely by a finite linear combination of the components of $\underline{y}(n)$, $\underline{y}(n-1),\ldots$. Thus if a fixed finite linear combination of the components of the projections $\underline{y}(n|n,n-M)$, $\underline{y}(n+1|n,n-M),\ldots$ is identically equal to zero for any choice of M the same linear combination of the components of the predictors $\underline{y}(n|n)$, $\underline{y}(n+1|n),\ldots$ must be equal to zero.

This means that the dependence relations among a finite number of components of the predictors can be determined by analyzing the dependence relations of the corresponding components of the projections $\underline{y}(n|n, n-M)$, $\underline{y}(n+1|n, n-M),\ldots$, for a sufficiently large value of M. Hereafter we will assume that M is chosen large enough so that we can replace the analysis of dependence within the components of the predictors by the analysis of dependence within the components of $\underline{y}(n|n, n-M)$, $\underline{y}(n+1|n, n-M),\ldots$. Now $y_j(n+k)$ admits a representation

$$y_j(n+k) = y_j(n+k|n, n-M) + w_j(n+k, n-M) ,$$

where $w_j(n+k, n-M)$ is independent of the elements of $R_y(n,n-M)$ and $y_j(n+k|n, n-M)$ can be represented in the form

$$y_j(n+k|n, n-M) = \sum_{m=0}^{M} \sum_{\ell=1}^{r} a(j,k;\ell,m)\, y_\ell(n-m),$$

where $a(j,k;\ell,m)$'s are the coefficients of regression of $y_j(n+k)$ on $y_\ell(n-m)$ ($\ell = 1,2,\ldots,r$, $m = 0,1,\ldots,M$). Under the assumption of nonsingularity of the covaraince matrix of the innovation these $y_\ell(n-m)$'s are always linearly independent and the coefficients of regression are uniquely determined. Accordingly the number of the linearly independent elements within the set of the components of a vector which is composed of a finite number of $y_j(n+k|n,n-M)$'s can be determined as the rank of the matrix of the coefficients of regression of the vector on $y_\ell(n-m)$ ($\ell = 1,2,\ldots,r$, $m = 0,1,\ldots,M$). It is a well known fact that the multivariate regression analysis with constraints on the rank of the matrix of the regression coefficients leads to the analysis of the canonical correlation coefficients [23].

The canonical correlation analysis is concerned with a canonical representation of the correlation between the two sets of random variables. Consider a pair of vectors of random variables $\underline{u} = (u_1,u_2,\ldots,u_s)'$ and $\underline{v} = (v_1,v_2,\ldots,v_t)'$. It is assumed here that s is greater than or equal to t and the random variables have zero means and finite variances. For the sake of simplicity it is also assumed that the covariance matrices $E\underline{u}\underline{u}'$ and $E\underline{v}\underline{v}'$ are nonsingular. Then there are matrices $\underline{S}$ and $\underline{T}$ such that $E\underline{S}\underline{u}(\underline{S}\underline{u})'$ and $E\underline{T}\underline{v}(\underline{T}\underline{v})'$ are identity matrices of dimension $s \times s$ and $t \times t$, respectively. The matrices $\underline{S}$ and $\underline{T}$ are not unique. Now for the $t \times s$ matrix $E\underline{T}\underline{v}(\underline{S}\underline{u})'$ there are orthogonal matrices $\underline{U}$ and $\underline{V}$ which satisfy the relation

$$\underline{V}E\underline{T}\underline{v}(\underline{S}\underline{u})'\underline{U}' = \begin{bmatrix} c_1 & 0 & \cdots & 0 & 0 & \cdots & 0 \\ 0 & c_2 & \cdots & 0 & 0 & \cdots & 0 \\ \vdots & & & \vdots & \vdots & & \vdots \\ 0 & 0 & \cdots & c_t & 0 & \cdots & 0 \end{bmatrix}, \tag{4.1}$$

where on the right hand side of the equation $c_1 \geqq c_2 \geqq \cdots \geqq c_t \geqq 0$ and all the other elements are zero [24]. $\underline{V}'\,\mathrm{diag}(c_1, c_2, \ldots, c_t)\underline{U}$ defines the singular value decomposition of the matrix $E\underline{T}\underline{v}(\underline{S}\underline{u})'$, where $\mathrm{diag}(c_1, c_2, \ldots, c_t)$ stands for the right hand side of (4.1) and c_i's are the singular values of the matrix. In the present situation c_i is the coefficient of correlation between the i-th components $(i \leqq t)$ of the vectors $\underline{U}\underline{S}\underline{u}$ and $\underline{V}\underline{T}\underline{v}$ and is called the i-th canonical correlation coefficient between $\underline{u}$ and $\underline{v}$. The i-th components of $\underline{U}\underline{S}\underline{u}$ and $\underline{V}\underline{T}\underline{v}$ are called the i-th canonical variables. The i-th rows of $\underline{U}\underline{S}$ and $\underline{V}\underline{T}$ are called the canonical weights of the i-th canonical variables. Obviously c_i is not greater than 1. The concept of canonical correlation is a classical one in statistics and is concerned with the extraction of the most useful information of a vector of Gaussian random variables from another vector of Gaussian random variables [23]. The number of linearly independent components of the projection of $\underline{v}$ onto the linear space spanned by the components of $\underline{u}$ is identical to the number of nonzero canonical correlation coefficients between $\underline{u}$ and $\underline{v}$. By combining this result with the observation of the preceding paragraph it is now obvious that by putting $\underline{u} = (y(n)', y(n-1)', \ldots, y(n-M)')'$ and $\underline{v}$ = (a vector composed of some finite number of components $y_j(n+k)$ $(j = 1,2,\ldots,r,\ k = 0,1,\ldots)$) the number of linearly independent elements within the corresponding projections $y_j(n+k|n,n-M)$ can be determined as the number of the nonzero canonical correlation coefficients between $\underline{u}$ and $\underline{v}$. Thus at least theoretically we can determine the structural characteristic vector $\underline{h}$ through the analysis of the canonical correlation coefficients between the set of the present and past values and the set of the present and future values of the time series. This result shows the inherent relation among the canonical correlation analysis, the singular value decomposition and the canonical representation of a linear stochastic system.

It might be of interest to note that in the Ho-Kalman algorithm of the minimal realization of time invariant linear systems [25] the Hankel matrix which is composed of the impulse response matrix sequence of the system can be interpreted as the covariance matrix between the present and future outputs and the present and past inputs of the system when the system is driven by a white noise with unit covariance matrix [16]. The singular value decomposition of a finite portion of the Hankel matrix constitutes the basic part of the algorithm, which is equivalent to the canonical correlation analysis of the present and past inputs and the present and future outputs of the corresponding stochastic system [26].

The importance of the stochastic interpretation in terms of the concepts of canonical correlation is that it directly leads to the realization of a statistical procedure for identification. When only a record of observations of finite length of the related variables is available the theoretical covariances are replaced by the corresponding sample covariances and the numerical procedure developed for the singular value decomposition of a matrix [24] can be used for the computation of the sample canonical correlation coefficients. The problem of the determination of the rank of the matrix of the regression coefficients is thus reduced to the problem of the statistical decision on the number of non-zero canonical correlation coefficients [17].

To see how the canonical correlation analysis really works, let us consider a system with a scalar input $y_1(n)$ and a scalar output $y_2(n)$ satisfying the relation

$$y_2(n) - 2.184\, y_2(n-1) + 1.493\, y_2(n-2) - 0.294\, y_2(n-3)$$
$$= 0.142\, y_1(n-1) + 0.214\, y_1(n-2) - 0.212\, y_1(n-3) + d(n), \quad (4.2)$$

where $d(n)$ is a disturbance which is independent of the input process $y_1(n)$ and is generated by the relation

$$d(n) = x_2(n) - 1.080\, x_2(n-1) + 0.288\, x_2(n-2),$$

where $x_2(n)$ is a Gaussian white noise with mean zero and variance 0.36. This system is a close approximation to the ship model obtained by Åström and Källström [27] to describe the yawing response of a ship to the rudder input under stochastic environment. For the purpose of simulation study, we assume that the input $y_1(n)$ is a Gaussian white noise with mean zero and variance 25. The innovation of $\underline{y}(n) = (y_1(n), y_2(n))'$ at time n is given by $\underline{x}(n) = (x_1(n), x_2(n))'$, where $x_1(n) = y_1(n)$. Hereafter we will simply call this model the ship model. In this model the feedback from the output to the input is absent, but we simply assume that we only know that $y_1(n)$ and $y_2(n)$ constitute a two-dimensional stationary time series $y(n)$. The state vector of our canonical representation of $\underline{y}(n)$ is obtained by successively testing the linear dependences within the components of the vector $\underline{s} = (y_1(n|n), y_2(n|n), y_1(n+1|n), y_2(n+1|n), \ldots)'$. Obviously $y_1(n|n) = y_1(n)$ and $y_2(n|n) = y_2(n)$ and they are linearly independent. Thus in the notation of Section 3 we have $v_1(n) = s_1 = y_1(n)$ and $v_2(n) = s_2 = y_2(n)$. Accordingly we have $h_1 = 1$ and $h_2 = 2$. As $y_1(n+1)$ is independent of the past input and output we have $y_1(n+1|n) = 0$, which means that $y_1(n+1|n)$ admits a representation

$$y_1(n+1|n) = 0\, y_1(n|n) + 0\, y_2(n|n) .$$

Thus the order of the first component of $\underline{y}(n)$ is equal to one and we have $k_1 = 1$ and in the vector $\underline{s}$ those $y_1(n+k|n)$'s with $k = 1,2,\ldots$ are crossed out and the search for the basis of the predictor space is limited to the remaining sequence $y_1(n)$, $y_2(n)$, $y_2(n+1|n)$, $y_2(n+2|n),\ldots$. From the relation (3.2) we have

$$\begin{aligned} y_2(n+3|n) - 2.184\, y_2(n+2|n) + 1.493\, y_2(n+1|n) - 0.294\, y_2(n|n) \\ = -\,0.212\, y_1(n|n) \quad . \end{aligned} \tag{4.3}$$

Thus we know that k_2, the order of the second component, is not greater than 3. If k_2 is equal to 3, i.e., if $y_1(n|n)$, $y_2(n|n)$, $y_2(n+1|n)$ and $y_2(n+2|n)$ are linearly independent, the coefficients of the linear relation (4.3) are unique and the state vector $\underline{v}(n)$ is defined by

$$\underline{v}(n) = \begin{bmatrix} y_1(n|n) \\ y_2(n|n) \\ y_2(n+1|n) \\ y_2(n+2|n) \end{bmatrix} .$$

It can be shown that k_2 is equal to 3 in this case. Without going into the detailed discussion of the proof we will assume here that we know this fact. A simple idea of the proof may be obtained from the fact that the covariances between $y_2(n+k)$ and $y_1(n+m)$ remain unchanged even when $d(n)$ in (4.2) is completely suppressed. It is easy to check that for the present choice of $y_1(n)$, when $d(n)$ is suppressed in (4.2), $y_2(n)$ defines a third order stochastic linear system and thus k_2 cannot be smaller than 3.

By using a sequence of random numbers generated from a physical noise source a realization of $\underline{y}(n)$ of length 1500 was generated with zero initial conditions and the first 100 points were discarded to eliminate the effect of the initial transient. The resultant sequence of length 1400 is denoted by $\underline{y}(n)$ $(n = 1, 2,\ldots,1400)$. To get the feeling of the statistical behavior of the related statistics the sample canonical correlation coefficients were computed by using the first 700 points of $\underline{y}(n)$ and then the whole set of data of length 1400. The results are illustrated in Table 1. They are designated by $N=700$ and $N=1400$, respectively. The canonical correlation coefficients were computed for the vector $\underline{v}$ of successively increasing number of present and future values of $\underline{y}(n)$ and the fixed vector $\underline{u} = (\underline{y}(n)', \underline{y}(n-1)',\ldots,\underline{y}(n-M)')'$ with $M = 8$. This value of M is the order of the autoregressive model chosen by the minimum AIC procedure for the whole set of data $\underline{y}(n)$ $(n = 1,2,\ldots,1400)$. From Table 1 we can see that the sample canonical correlation coefficients corresponding to the theoretical values which are equal to zero generally decrease as the length of data is increased. Obviously this is due to the reduction of the sampling fluctuations by the increase of the data length. In contrast to this, those sample canonical correlation coefficients corresponding to nonzero canonical correlation coefficients are showing rather steady increase of the values. By a numerical analysis it was confirmed that this phenomenon is due to the reduction of the bias by the increase of the data length used for the computation of the sample covariances. The result is a clear separation of the sample canonical correlation coefficients into the two groups corresponding to zero and nonzero canonical correlation coefficients. This suggests the feasibility of the determination of the structural characteristic vector $\underline{h}$ by observing the behavior of the sample canonical correlation coefficients. The fact that this is not so simple in the case of a real time series is shown by the results illustrated in Table 2. The results were obtained by the canonical correlation analysis of a record of the rudder movement $(y_1(n))$ and the yaw angle $(y_2(n))$ of a real ship. We cannot see very

TABLE 1

Behavior of Sample Canonical Correlation Coefficients of an Artificial Time Series of a Ship Model

		N=700 $\underset{\sim}{u}=(y(n)',y(n-1)',\ldots y(n-8)')'$		N=1400 $\underset{\sim}{u}=(y(n)',y(n-1)',\ldots y(n-8)')'$	
$\underset{\sim}{v}$	q	CANONICAL CORRELATION COEFFICIENT	DIC(q)	CANONICAL CORRELATION COEFFICIENT	DIC(q)
$y_1(n)$	0	1.0000	∞	1.0000	∞
$y_2(n)$	1	1.0000	∞	1.0000	∞
$y_1(n+1)$	2	0.1646*	-12.78**	0.1135*	-13.85**
$y_1(n)$	0	1.0000	∞	1.0000	∞
$y_2(n)$	1	1.0000	∞	1.0000	∞
$y_1(n+1)$	2	0.9570	1687.41	0.9662	3751.69
$y_2(n+1)$	3	0.1645*	-10.79**	0.1135*	-11.86**
$y_1(n)$	0	1.0000	∞	1.0000	∞
$y_2(n)$	1	1.0000	∞	1.0000	∞
$y_1(n+1)$	2	0.9571	1677.12	0.9663	3738.41
$y_2(n+1)$	3	0.1819*	-20.39**	0.1209*	-23.59**
$y_1(n+2)$	4	0.1506*	-11.95	0.1059*	-12.21
$y_1(n)$	0	1.0000	∞	1.0000	∞
$y_2(n)$	1	1.0000	∞	1.0000	∞
$y_1(n+1)$	2	0.9656	1890.36	0.9730	4254.20
$y_2(n+1)$	3	0.3520	41.85	0.3994	184.81
$y_1(n+2)$	4	0.1816*	-16.73**	0.1084*	-24.49**
$y_2(n+2)$	5	0.1494*	-10.21	0.1031*	-11.03
$y_1(n)$	0	1.0000	∞	1.0000	∞
$y_2(n)$	1	1.0000	∞	1.0000	∞
$y_1(n+1)$	2	0.9656	1879.66	0.9730	4238.96
$y_2(n+1)$	3	0.3598	33.03	0.4005	171.47
$y_1(n+2)$	4	0.1880*	-28.04**	0.1310*	-37.23**
$y_2(n+2)$	5	0.1496*	-21.24	0.1033*	-29.47
$y_1(n+3)$	6	0.1452*	-9.09	0.0731*	-16.50
$y_1(n)$	0	1.0000	∞	1.0000	∞
$y_2(n)$	1	1.0000	∞	1.0000	∞
$y_1(n+1)$	2	0.9674	1965.18	0.9746	4501.12
$y_2(n+1)$	3	0.4662	85.24	0.5234	351.57
$y_1(n+2)$	4	0.1899*	-48.33**	0.1342*	-58.66**
$y_2(n+2)$	5	0.1546*	-40.04	0.1038*	-50.11
$y_1(n+3)$	6	0.1489*	-26.97	0.0779*	-35.29
$y_2(n+3)$	7	0.0872*	-16.66	0.0546*	-17.82

* denotes the sample canonical correlation coefficients of which theoretical values are equal to zero.

** denotes the minimum of DIC(p).

TABLE 2

Behavior of Sample Canonical Correlation Coefficients of a Time Series of a Real Ship

		N=400 $\underset{\sim}{u}=(y(n)',y(n-1)',\ldots y(n-10)')'$		N=800 $\underset{\sim}{u}=(y(n)',y(n-1)',\ldots y(n-10)')'$	
$\underset{\sim}{v}$	q	CANONICAL CORRELATION COEFFICIENT	DIC(q)	CANONICAL CORRELATION COEFFICIENT	DIC(q)
$y_1(n)$	0	1.0000	∞	1.0000	∞
$y_2(n)$	1	1.0000	∞	1.0000	∞
$y_1(n+1)$	2	0.4560	53.24*	0.4061*	104.15**
$y_1(n)$	0	1.0000	∞	1.0000	∞
$y_2(n)$	1	1.0000	∞	1.0000	∞
$y_1(n+1)$	2	0.4863	111.60	0.5094	284.03
$y_2(n+1)$	3	0.4344	45.64**	0.3785*	85.67**
$y_1(n)$	0	1.0000	∞	1.0000	∞
$y_2(n)$	1	1.0000	∞	1.0000	∞
$y_1(n+1)$	2	0.4979	103.93	0.5142	301.77
$y_2(n+1)$	3	0.4412	33.99	0.3817*	100.06
$y_1(n+2)$	4	0.2383	-12.61**	0.2464	14.11**
$y_1(n)$	0	1.0000	∞	1.0000	∞
$y_2(n)$	1	1.0000	∞	1.0000	∞
$y_1(n+1)$	2	0.5393	134.17	0.6022	424.05
$y_2(n+1)$	3	0.4780	42.69	0.3879*	109.66
$y_1(n+2)$	4	0.2950	-19.09**	0.2754*	21.19
$y_2(n+2)$	5	0.2010	-17.51	0.1921*	-3.93**
$y_1(n)$	0	1.0000	∞	1.0000	∞
$y_2(n)$	1	1.0000	∞	1.0000	∞
$y_1(n+1)$	2	0.5396	121.34	0.6027	445.02
$y_2(n+1)$	3	0.4840	31.68	0.4076*	131.92
$y_1(n+2)$	4	0.3028	-31.07**	0.2807*	30.58
$y_2(n+2)$	5	0.2431	-29.53	0.2319*	4.90
$y_1(n+3)$	6	0.1861	-17.90	0.1876	-3.33**
$y_1(n)$	0	1.0000	∞	1.0000	∞
$y_2(n)$	1	1.0000	∞	1.0000	∞
$y_1(n+1)$	2	0.5899	158.00	0.6504	564.77
$y_2(n+1)$	3	0.5163	36.90	0.4435*	174.87
$y_1(n+2)$	4	0.3159	-41.10	0.3189	45.65
$y_2(n+2)$	5	0.2610	-41.14**	0.2435*	1.67
$y_1(n+3)$	6	0.2335	-31.36	0.2249*	-9.05
$y_2(n+3)$	7	0.1587	-19.79	0.1290*	-16.57**

* denotes that the sample canonical correlation coefficient decreased its value when N is increased from 400 to 800.

** denotes the minimum of DIC(p).

systematic behavior of the sample canonical correlation coefficients when the data length is increased. This result may be explained as partly due to the sampling fluctuations or the possible nonstationarity of the data and partly due to the fact that the finite order model is only an approximation to the real structure which will be of infinite order. This observation suggests that we have again to resort to the introduction of some criterion of fit.

In the case of ordinary multivariate analysis where the data are taken from the sequence of independent observations of a pair of multivariate Gaussian random vectors, the canonical correlation analysis can be considered to be the maximum likelihood estimation procedure of a linear model which defines the covariance structure between the two random vectors [23]. The number of the free parameters within the model is controlled by the rank of the matrix of the regression coefficients of the components of one vector on those of the other. For the two random vectors $\underline{u}$ and $\underline{v}$ with dimensions s and t, respectively, the model is defined by the representation

$$\underline{v} = \underline{A}\,\underline{u} + \underline{w} \qquad ,$$

where $\underline{A}$ is the matrix of the regression coefficients of the components of $\underline{v}$ on the components of $\underline{u}$ and the components of $\underline{w}$ are uncorrelated with those of $\underline{u}$. The number of the free parameters within the model can be obtained as the sum of the numbers of the free parameters within the covariance matrices of $\underline{u}$ and $\underline{w}$ and within the matrix $\underline{A}$. It is assumed that t is not greater than s. Under the assumption that the rank of $\underline{A}$ is $q(q \leqq t \leqq s)$ these numbers are respectively $s(s+1)/2$, $t(t+1)/2$ and $q(s+t-q)$. This last quantity is equal to ts when q is equal to t. The number of the free parameters is the sum of the above three numbers and will be denoted by $F(q)$. When N independent observations were made with the two Gaussian random vectors $\underline{u} = (u_1,u_2,\ldots,u_s)'$ and $\underline{v} = (v_1,v_2,\ldots,v_t)'$ $(s \geqq t)$

and it is assumed that the number of non-zero canonical correlation coefficients is equal to $q (q \leqq t)$, AIC for the corresponding model can be defined by

$$\mathrm{AIC}(q) = N \log \prod_{i=1}^{q} (1 - c_i^2) + 2F(q) \quad ,$$

where c_i is the i-th largest sample canonical correlation coefficient. In Table 1, the statistic DIC(q) for the model with the matrix of the regression coefficients of rank q is defined by

$$\mathrm{DIC}(q) = \mathrm{AIC}(q) - \mathrm{AIC}(t) \quad ,$$

where AIC(t) is the value of AIC when there is no constraint on the matrix of the regression coefficients. In the present case of time series the statistical behavior of AIC(q) is not identical to the case of ordinary independent observations. From the definition of AIC(q) we have

$$\mathrm{DIC}(q) = -N \log \prod_{i=q+1}^{t} (1-c_i^2) - 2(t-q)(s-q) \; . \qquad (4.4)$$

When the true values of the canonical correlation coefficients are equal to zero except for the first q largest ones it can be expected that under fairly general conditions the expectation of the first term in the right hand side of (4.4) will be approximated by $(t-q)$ times some positive constant. The constant is equal to 1 in the case of independent observations. Thus when $s-q$ is large, DIC(q) will take negative values when q is larger or equal to q_0, the number of nonzero canonical correlation coefficients. Hopefully, within the range $q_0 \leqq q \leqq t$, DIC(q) will often take the minimum value at $q = q_0$, if only the length of the data is sufficient to detect the drop of the values of the canonical correlation coefficients to zero. When q is smaller than q_0 the first term in the right hand side of (4.4) will grow indefinitely as N is increased. Thus it is certain that if we choose the value of q which gives the minimum of DIC(q) it will not remain below q_0 as N is increased indefinitely. There will remain the possibility of q being larger than q_0, but the probability will be small when $s-q$ is kept large. Certainly this probability will be made arbitrarily small if DIC(q) is defined by (4.4) with $2(t-q)(s-q)$ replaced by $2N^b(t-q)(s-q)$ with b satisfying the relation $0 < b < 1$, but some subjective judgement is required in choosing b. As can be seen from Table 1, the performance of the statistics DIC(q) in the decision on the number of nonzero canonical correlation coefficients is quite satisfactory in the case of our simulation experiment. In the case of the real data treated in Table 2 the value of q which gives the minimum of the statistics DIC(q) is showing consistent increase when the length of observations is increased from $N = 400$ to $N = 800$. This result can be interpreted as an indication of the possible nonstationarity of the original data or the infinite dimensional structure of the predictor space. Even for this type of data, with the aid of the criterion DIC(q), we can define a stationary finite dimensional Markovian model which fits best to the data. Obviously this is very convenient for many practical applications, yet we must always remember that when the procedure is applied carelessly there is also a danger of fitting a definitely inadequate model to a real time series.

Based on the results of experimental applications to simulated and real data the following procedure of canonical correlation analysis of a stationary time series $\underline{y}(n)$ $(n = 1,2,\ldots,N)$ is suggested:

1. Define the sample autocovariance matrices $\underline{C}(k)$ $(k = 0,1,2,\ldots)$ by

$$\begin{aligned} \underline{C}(k) &= \frac{1}{N}\sum_{n=1}^{N-k} (\underline{y}(n+k) - \bar{\underline{y}})(\underline{y}(n) - \bar{\underline{y}})' && \text{for} \quad 0 \leqq k \leqq N-1 \\ &= 0 && \text{for} \quad N \leqq k \end{aligned} \tag{4.5}$$

and

$$\underline{C}(k) = \underline{C}(-k)' \qquad \text{for} \quad k < 0,$$

where

$$\bar{\underline{y}} = \frac{1}{N} \sum_{n=1}^{N} \underline{y}(n) \quad .$$

2. Fit an autoregressive model $\underline{y}(n) + \underline{A}_1\underline{y}(n-1) + \underline{A}_2\underline{y}(n-2) + \cdots + \underline{A}_M\underline{y}(n-M) = \underline{x}(n)$ by the minimum AIC procedure. M denotes the MAICE of the order of the autoregressive model. For the computational procedure of autoregressive model fitting, see [10,28, 29].

3. Define the vector $\underline{u}$ by $\underline{u} = (\underline{y}(n)', \underline{y}(n-1),',\ldots,\underline{y}(n-M)')'$. The dimension s of $\underline{u}$ is given by $s = (M+1)r$, where r is the dimension of $\underline{y}(n)$. Also define the vector $\underline{s}$ by $\underline{s} = (\underline{y}(n)', \underline{y}(n+1)',\ldots,\underline{y}(n+M)')'$.

4. The t-dimensional vector $\underline{v}$ is defined by some t components of $\underline{s}$ and its i-th component is denoted by v_i. It is assumed that the canonical correlation coefficients between $\underline{v}^- = (v_1,v_2,\ldots,v_{t-1})$ and $\underline{u}$ are all positive and the first $t-1$ components of the state vector $\underline{v}(n)$ are defined by $v_i(n) = v_i$ $(i = 1,2,\ldots,t-1)$.

At the very beginning of the present procedure t is set equal to $r+1$ and $\underline{v} = (y_1(n), y_2(n),\ldots,y_r(n), y_1(n+1))$.

5. Do the canonical correlation analysis of $\underline{u}$ and $\underline{v}$ assuming the structure of vocariance defined by the sample covariance matrices $\underline{C}(k)$.

6. If $DIC(t-1)$ is negative, $v_t|n$, the projection of the last component of $\underline{v}$ onto $R_y(n-)$, is judged to be linearly dependent on its antecedents in the sequence of predictors, i.e., the minimum canonical correlation coefficient between $\underline{v}$ and $\underline{u}$ is considered to be zero. Assume that the t-th canonical variable is defined by

$$b_1v_1 + b_2v_2 + \cdots + b_tv_t \quad ,$$

where b_i's are the canonical weights of the t-th canonical variable which is judged to have zero canonical correlation coefficient. From the assumption on $\underline{v}^-$, b_t cannot be equal to zero and we have the relation

$$v_t|n = -\frac{b_1}{b_t} v_1(n) - \frac{b_2}{b_t} v_2(n) - \dots - \frac{b_{t-1}}{b_t} v_{t-1}(n) .$$

If $v_t = v_i(n+1) (1 \leqq i \leqq t-1)$ this last equation determines the i-th row of the transition matrix $\underline{F}$ by (3.9) of Section 3, i.e.,

$$\begin{aligned} F(i,k) &= -\frac{b_k}{b_t} \qquad k = 1,2,\dots,t-1 \\ &= 0 \qquad \text{otherwise.} \end{aligned}$$

If $v_i(n+1) = y_j(n+k_j)$, discard the variables $y_j(n+k_j)$, $y_j(n+k_j+1),\dots$ from the vector $\underline{s}$ to define the updated version of $\underline{s}$. If some components of s are still left for further test, return to stage 4. Otherwise put $\underline{v}(n) = (v_1(n), v_2(n), \dots, v_{t-1}(n))$.

7. If DIC(t-1) is positive, $v_t|n$ is judged to be linearly independent of its antecedents and is accepted as the t-th component $v_t(n)$ of the state vector $\underline{v}(n)$. Increase the value of the variable t by one and return to the stage 4.

By applying the above procedure to a set of ten simulated series of the ship model the results illustrated in Table 3 were obtained. The results denoted by $N = 350$ and 700 were obtained by using the first 350 and 700 data points of the ten series with $N = 1400$. Remember that the i-th component h_i of the structural characteristic vector $\underline{h}$ denotes the position of $v_i(n)$ within the sequence of the predictors $y_1(n|n)$, $y_2(n|n),\dots,y_r(n|n)$, $y_1(n+1|n)$, $y_2(n+1|n),\dots$. The results of Table 3 are quite encouraging. In every case with the data length $N = 1400$ the estimate of the structural characteristic vector was exact. Since it may be argued that the above results may be critically dependent on the assumption of stationarity and whiteness of the simulated rudder input another experiment was made with $y_1(n)$ generated by the non-stationary first order autoregression

TABLE 3

Estimation of the Structural Characteristic Vector of a Ship Model Using DIC(p) *Statistics*

N = 350	
estimated structural characteristic vector	frequency
(1,2,4)'	4
(1,2,4,6)'	6

N = 700	
estimated structural characteristic vector	frequency
(1,2,4,6)'	9
(1,2,3,4,5,6)'	1

N = 1400	
estimated structural characteristic vector	frequency
(1,2,4,6)'	10

$$y_1(n) = a(n)y_1(n) + 3x_0(n) ,$$

where $x_0(n)$ is a Gaussian white noise with mean zero and variance 1 and $a(n)$ is defined by

$$a(n) = \begin{cases} 0.7 & n \leqq 450 \\ 0.8 & 450 < n \leqq 800 \quad \text{and} \quad 1150 < n \leqq 1400 \\ 0.9 & 800 < n \leqq 1150. \end{cases}$$

Two sets of records, each of length 1500, were generated starting at $n = -99$ and the first 100 points of each record were discarded. The estimates of the structural characteristic vector produced by the canonical correlation analysis procedure were both identical to $(1,2,4,6)'$ and the maximum likelihood estimates of the transition matrix, which were obtained by using the procedure to be described in the next section, were very close to the time average of the transition matrices corresponding to the four stationary periods, each of length 350. This result shows that the present procedure of canonical correlation analysis is a fairly robust procedure of identification of the structural characteristic vector.

By the above stated canonical correlation analysis procedure we can get an estimate of the transition matrix $\underline{F}$. An estimate of the covariance matrix $\underline{C}$ of the innovation is already obtained by fitting the autoregressive model. Also by using the autoregressive model the impulse response matrices of the system to the innovation input can be estimated and an estimate of the input matrix $\underline{G}$ is obtained. This process can be organized as follows:

1. Assume the autoregressive model

$$\underline{y}(n) + \underline{A}_1\underline{y}(n-1) + \underline{A}_2\underline{y}(n-2) + \cdots + \underline{A}_M\underline{y}(n-M) = \underline{x}(n) .$$

2. Compute the impulse response matrices $\underline{W}_k$ successively by the relation

$$\underline{W}_k + \underline{A}_1\underline{W}_{k-1} + \underline{A}_2\underline{W}_{k-2} + \cdots + \underline{A}_M\underline{W}_{k-M} = \underline{0} ,$$

where $\underline{W}_0 = \underline{I}$, an $r \times r$ identity matrix, and $\underline{W}_k = \underline{0}$, a zero matrix, for $k < 0$.

3. When $v_i(n) = y_j(n+k|n)$, i.e., if $h_i = j + rk$, put

$$G(i,s) = W_k(j,s), \qquad s = 1,2,\ldots,r ,$$

where $G(i,s)$ is the (i,s)-th element of G and $W_k(j,s)$ denotes the (j,s)-th element of $\underline{W}_k$.

A natural question is how good these estimates are. Our experience suggests that even when the structural characteristic vectors are identified correctly the accuracies of the estimates of the parameters within the matrices $\underline{F}$ and $\underline{G}$ are often rather low compared with those of the maximum likelihood estimates which were obtained by using the results of canonical correlation analysis as the initial values to start the maximum likelihood computation.

Also there is a possibility of getting an estimate of $\underline{F}$ which defines an unstable system and thus cannot be used as the initial values for the maximum likelihood computation. Further, even the decision on the structural characteristic vector is not necessarily always satisfactory and the final MAICE obtained by using the maximum likelihood estimates of various models with different structural characteristic vectors may prove to be with different structural characteristic vector. In spite of these limitations the canonical correlation analysis procedure produced very reasonable initial guesses in many applications, including the tests by various simulated data. Taking into account the complexity of the decision on the structural characteristic vectors, especially when the system is multivariate, the procedure constitutes a significant step towards the practical use of Markovian or autoregressive moving average models in time series analysis. The computation for the canonical correlation analysis of a stationary time series $\underline{y}(n)$ can be organized as follows:

1. Define the covariance matrices $\underline{R}_{vv}$, $\underline{R}_{vu}$ and $\underline{R}_{uu}$ by

$$\underline{R}_{vv} = E^S \underline{v}\underline{v}'$$

$$\underline{R}_{vu} = E^S \underline{v}\underline{u}'$$

$$\underline{R}_{uu} = E^S \underline{u}\underline{u}'$$

where E^S symbolically denotes the expectation operator with respect to the probability distribution of $\underline{y}(n)$ which is assumed to be Gaussian and with mean zero and the covariance matrices

$\underline{C}(k)$ defined by (4.5).

2. Factorize $\underline{R}_{uu}$ and $\underline{R}_{vv}$ into

$$\underline{R}_{uu} = \underline{S}^{-1}(\underline{S}^{-1})'$$

and

$$\underline{R}_{vv} = \underline{T}^{-1}(\underline{T}^{-1})' ,$$

where $\underline{S}$ and $\underline{T}$ are lower triangular matrices.

3. Apply the singular value decomposition computation procedure to $\underline{T}\underline{R}_{vu}\underline{S}'$ to get

$$\underline{V}\underline{T}\underline{R}_{vu}\underline{S}'\underline{U}' = \begin{bmatrix} c_1 & 0 & \cdots & 0 & 0 & \cdots & 0 \\ 0 & c_2 & \cdots & 0 & 0 & \cdots & 0 \\ \vdots & \vdots & \ddots & \vdots & \vdots & & \vdots \\ 0 & 0 & & c_t & 0 & \cdots & 0 \end{bmatrix}$$

where $1 \geqq c_1 \geqq c_2 \geqq \cdots \geqq c_t \geqq 0$ and $\underline{U}$ and $\underline{V}$ are orthogonal matrices. The i-th row of $\underline{V}\underline{T}$ gives the canonical weights for the i-th canonical variable. DIC(t-1) is computed by the formula

$$\mathrm{DIC}(t-1) = -N \log(1-c_t^2) - 2(s-t+1) .$$

4. For the computation of the matrix $\underline{T}\underline{R}_{vu}\underline{S}'$ use the following iterative procedure:

The triangular matrix $\underline{S}$ is obtained once for all in the form

$$\underline{S} = \begin{bmatrix} \underline{L}_0 & & & & \\ \underline{L}_1\underline{B}_1^1 & \underline{L}_1 & & & \\ \underline{L}_2\underline{B}_2^2 & \underline{L}_2\underline{B}_1^2 & \underline{L}_2 & & \\ \vdots & \vdots & \vdots & \ddots & \\ \underline{L}_M\underline{B}_M^M & \underline{L}_M\underline{B}_{M-1}^M & \underline{L}_M\underline{B}_{M-2}^M & \cdots & \underline{L}_M \end{bmatrix} ,$$

where $\underline{B}_i^m$ $(i = 1,2,\ldots,m)$ denote the matrices of the coefficients of the m-th order "backward" autoregressive model and are defined by the relation

$$E^S(\underline{y}(n-m) + \underline{B}_1^m \underline{y}(n-m+1) + \cdots + \underline{B}_m^m \underline{y}(n))\underline{y}(n-k)' = 0$$
$$k = 0,1,\ldots,m-1 ,$$

and $\underline{L}_m$ is the lower triangular matrix defined by the relation

$$\underline{L}_m \, \underline{D}_m \, \underline{L}_m' = \underline{I} ,$$

where $\underline{D}_m$ is the covariance matrix of the residuals and is defined by

$$\underline{D}_m = E^S(\underline{y}(n-m) + \underline{B}_1^m \underline{y}(n-m+1) + \cdots + \underline{B}_m^m \underline{y}(n))\underline{y}(n-m)' .$$

By definition $E^S \underline{y}(n-m)\underline{y}(n-k)' = \underline{C}(k-m)$ for $k \geqq m$ and $\underline{C}(m-k)'$ for $m > k$. The matrices $\underline{B}_i^m$ and $\underline{D}_m$ are obtained during the computation for the "forward" autoregressive model fitting by the Levinson-Whittle type iterative procedure [12, 28, 29].

At the start of the computation put $\underline{v} = (y_1(n), y_2(n), \ldots, y_r(n))'$ and $\underline{T} = \underline{L}_0$. $\underline{R}_{vu}$ is given by

$$\underline{R}_{vu} = (\underline{C}(0), \underline{C}(1), \ldots, \underline{C}(M) .$$

Also put $t = r$.

At an intermediate stage of the computation, when the t-dimensional vector $\underline{v}$ is augmented with a new component v_{t+1}, denote the augmented vector by $\underline{v}^+$, i.e.,

$$\underline{v}^+ = \begin{bmatrix} \underline{v} \\ v_{t+1} \end{bmatrix} .$$

The matrix which corresponds to $\underline{T}$ of $\underline{v}$ is denoted by $\underline{T}^+$ and is obtained in the form

$$\underline{T}^+ = \begin{bmatrix} \underline{T} & 0 \\ \underline{f}' & g \end{bmatrix} ,$$

where f and g are determined by the relation

$$\begin{bmatrix} \underline{T} & 0 \\ \underline{f}' & g \end{bmatrix} \begin{bmatrix} \underline{R}_{vv} & \underline{r} \\ \underline{r}' & s \end{bmatrix} \begin{bmatrix} \underline{T}' & \underline{f} \\ 0 & g \end{bmatrix} = \underline{I} ,$$

where r and s are defined by the relation

$$E^s \underline{v}^+ \underline{v}^{+'} = \begin{bmatrix} \underline{R}_{vv} & \underline{r} \\ \underline{r}' & s \end{bmatrix} .$$

$\underline{f}$ and g are obtained as

$$\underline{f} = - \underline{T}'\underline{T}\underline{r}g$$

and

$$g = \left(\frac{1}{s-(\underline{T}\underline{r})'(\underline{T}\underline{r})} \right)^{1/2} .$$

The matrix $\underline{T}^+ E^s \underline{v}^+ \underline{u}' \underline{S}'$ which is to be used in the singular value decomposition computation in the next stage is obtained in the form

$$\underline{T}^+ E^s \underline{v}^+ \underline{u}' \underline{S}' = \begin{bmatrix} \underline{T}\underline{R}_{vu}\underline{S}' \\ \underline{f}'\underline{R}_{vu}\underline{S}' + g\underline{w}'\underline{S}' \end{bmatrix} ,$$

where $\underline{w}$ is the vector of the covariances between the new component v_t of $\underline{v}^+$ and the components of $\underline{u}$ and is defined by the relation

$$E^s \underline{v}^+ \underline{u}' = \begin{bmatrix} \underline{R}_{vu}' \\ \underline{w}' \end{bmatrix} .$$

When $v_{t+1}|n$ is adopted as the $(t+1)$-st component $v_{t+1}(n)$ of the state vector $\underline{v}(n)$, replace $\underline{v}$, $\underline{T}$, $\underline{T}\underline{R}_{vu}\underline{S}'$ and t by $\underline{v}^+$, $\underline{T}^+$, $\underline{T}^+ E^s \underline{v}^+ \underline{u}' \underline{S}'$, and $t+1$, respectively. Otherwise retain the original $\underline{v}$, $\underline{T}$, $\underline{T}\underline{R}_{vu}\underline{S}'$ and t.

5. MAXIMUM LIKELIHOOD COMPUTATION OF MARKOVIAN MODELS

The computational aspects of the maximum likelihood estimation is discussed extensively by Gupta and Mehra [30]. It is pointed out by Mehra [31] that the direct maximization of the exact Gaussian likelihood function of a stochastic system is a formidable problem and a useful approximation for a constant system is developed in [32]. Our approach here is very close to using this approximation and is realized by maximizing a modified log likelihood function. It is different from Mehra's approximation in that its definition is free from the initial condition of the system. This point is important for the application of AIC which must be defined unambiguously up to the order $O(1)$ for the purpose of comparison of various structures.

A. MODIFIED LOG LIKELIHOOD AND ITS FOURIER REPRESENTATION

The definition of our modified log likelihood for the Markovian model is given by following the idea described in Section 2 for the case of an autoregressive model. The basic idea is to calculate the average information criterion $K(g; f(\cdot|\underline{\theta}))$ for the discrimination of the fitted model $f(\cdot|\underline{\theta})$ from the true structure defined by $g(\cdot)$ and then replace the theoretical moments required for the computation of the criterion by the appropriate sample values. In the case of a Gaussian model of a stationary time series $\underline{y}(n)$ the only required moments are the first and second order moments, and the sample mean $\bar{\underline{y}}$ and the sample covariance matrices $\underline{C}(k)$ are used to define the modified log likelihood. Usually we replace $\underline{y}(n)$ by $\underline{y}(n) - \bar{\underline{y}}$ and assume that the mean of $\underline{y}(n)$ is zero.

For the Gaussian Markovian model

$$\begin{aligned} \underline{v}(n+1) &= \underline{F}\underline{v}(n) + \underline{G}\underline{x}(n+1) , \\ \underline{y}(n) &= \underline{H}\underline{v}(n) , \end{aligned}$$

with the covariance matrix of the innovation $x(n)$ equal to $\underline{C}$, it can be shown that if the original process is stationary and ergodic we have

$$\begin{aligned} &\lim_{N\to\infty} \frac{1}{N} \log f(\underline{y}(1), \underline{y}(2), \ldots, \underline{y}(N)|\underline{\theta}) \\ &= -\frac{r}{2} \log 2\pi - \frac{1}{2} \log |\underline{C}| - \frac{1}{2} E\mathrm{tr}\underline{C}^{-1}\underline{x}(n)\underline{x}(n)' , \end{aligned}$$

where $f(\underline{y}(1), \underline{y}(2), \ldots, \underline{y}(N)|\underline{\theta})$ denotes the likelihood of the model, $\underline{\theta}$ stands for the free parameters within the model, tr $\underline{A}$

denotes the trace of a matrix $\underline{A}$ and $\underline{x}(n)$ is defined by the relations

$$\begin{aligned} \underline{x}(n) &= \underline{y}(n) - \underline{H}\underline{F}\underline{v}(n-1) \quad , \\ \underline{v}(n) &= \underline{F}\underline{v}(n-1) + \underline{G}\underline{x}(n) \quad , \end{aligned} \tag{5.1}$$

and the expectation is taken with respect to the distribution of the original process. Our modified log likelihood for the present Gaussian Markovian model is then given by

$$- \frac{rN}{2} \log 2\pi - \frac{N}{2} \log |\underline{C}| - \frac{N}{2} \operatorname{tr} \underline{C}^{-1} E^S \underline{x}(n)\underline{x}(n)' \quad ,$$

where E^S stands for the expectation taken with respect to the Gaussian distribution of $\underline{y}(n)$ which is assumed to be zero mean and with the covariance sequence equal to the sample covariance sequence defined by (4.5).

If we denote by $\underline{K}(f)$ the frequency response function of the linear system which transforms $\underline{x}(n)$ into the original process $\underline{y}(n)$, $E^S\underline{x}(n)\underline{x}(n)'$ can be expressed in the form

$$E^S\underline{x}(n)\underline{x}(n)' = \int_{-1/2}^{1/2} \underline{K}(f)^{-1}\underline{P}_N(f)\underline{K}(f)^{*-1}\, df \quad , \tag{5.2}$$

where $\underline{K}(f)^*$ denotes the conjugate transpose of $\underline{K}(f)$ and $\underline{P}_N(f)$ denotes the Fourier transform of the sample covariance matrix sequence. Hereafter the limits of integrations are always 1/2 and -1/2 and they are omitted. $\underline{K}(f)$ is given by

$$\underline{K}(f) = \underline{H}\,\{\underline{I} - \exp(-i2\pi f)\underline{F}\}^{-1}\underline{G} \quad ,$$

where i within the exponential function denotes a purely imaginary number. When the elements of the covariance matrix of the innovation, $\underline{C}$, are within the set of the free parameters it can be shown (see, for example, [33]) that, for a given set of $\underline{F}$, $\underline{G}$ and $\underline{H}$, the $\underline{C}$ that maximizes the modified log likelihood is given by $\underline{C}_0 = E^S\underline{x}(n)\underline{x}(n)'$. Thus for this case the maximum likelihood computation reduces to the minimization of $\log|\underline{C}_0|$, where $\underline{C}_0$

is defined by the right hand side of (5.2) as a function of the free parameters within the matrices $\underline{F}$, $\underline{G}$ and $\underline{H}$. Under the assumption of nonsingularity of $\underline{C}$ the matrix $\underline{H}$ takes the form $\underline{H} = [\underline{I}\ \underline{0}]$ and thus can be left out of our consideration.

$\underline{P}_N(f)$ admits a representation

$$\underline{P}_N(f) = \underline{Y}(f)\underline{Y}(f)^* \quad ,$$

where $Y(f)$ is defined by

$$\underline{Y}(f) = \frac{1}{\sqrt{N}} \sum_{n=1}^{N} \exp(-i2\pi fn)\, \underline{y}(n) \quad . \tag{5.3}$$

It is assumed here that the sample mean is already deleted from $\underline{y}(n)$. By replacing $\underline{P}_N(f)$ of (5.2) by the present representation $\underline{C}_0$ is expressed in the form

$$\underline{C}_0 = \int \underline{X}(f)\underline{X}(f)^*\ df \quad ,$$

where

$$\underline{X}(f) = \underline{K}(f)^{-1}\underline{Y}(f) \quad .$$

If we define the *Fourier transform*, or the n-th *Fourier coefficient*, of a function $\underline{u}(f)$ $(-1/2 \leqq f \leqq 1/2)$ by

$$\underline{u}(n) = \int \exp(i2\pi fn)\underline{u}(f)\ df \qquad n = 0, \pm 1, \ldots,$$

it holds that

$$\int \underline{u}(f)\underline{v}(f)^*\ df = \sum_{n=-\infty}^{\infty} \underline{u}(n)\underline{v}(n)' \quad ,$$

where $\underline{v}(n)$ is the Fourier transform of $\underline{v}(f)$. It is assumed that the integral and the infinite sum take finite values. As an example of application of the above relation we have

$$\int \underline{Y}(f)\underline{Y}(f)^*\ df = \frac{1}{N} \sum_{n=1}^{N} \underline{y}(n)\underline{y}(n)' \ .$$

From the obvious analogy we will call $\sum_{n=-\infty}^{\infty} \underline{u}(n)\underline{v}(n)'$ the *covariance*, or the covariance matrix, between the two time series $\underline{u}(n)$ and $\underline{v}(n)$. The distinction between the present definition of the covariance between the two deterministic sequences, also called time series here, and the stochastic definition should be clear enough to avoid any confusion, yet the analogy is quite useful to develop the understanding of the meaning of some analytical operations applied to the Gaussian likelihood function.

B. FOURIER REPRESENTATIONS OF GRADIENTS AND HESSIAN

From the definitions of $\underline{Y}(f)$ and $\underline{K}(f)$ it is obvious that the Fourier transform of $\underline{X}(f) = \underline{K}(f)^{-1}\underline{Y}(f)$ is an estimate of the sequence of innovations, scaled by the factor $1/\sqrt{N}$. Accordingly $\partial\underline{X}(f)/\partial\theta_i$ defines the sensitivity sequence of the estimated sequence of innovations to the variation of a parameter θ_i, where θ_i stands for one of the free parameters within $\underline{F}$ and $\underline{G}$. We have

$$\frac{\partial \underline{C}_0}{\partial \theta_i} = 2\int \underline{X}(f) \frac{\partial \underline{X}(f)^*}{\partial \theta_i} df \quad , \tag{5.4}$$

which shows that $\partial\underline{C}_0/\partial\theta_i$ is obtained as twice the covariance of the estimated sequence of innovations and its sensitivity sequence. We will denote the free parameters within $\underline{F}$ and $\underline{G}$ by f_{ij} and g_{ik} when the parameter are the (i,j)-th and the (i,k)-th elements of $\underline{F}$ and $\underline{G}$, respectively. From the definitions of $\underline{X}(f)$ and $\underline{K}(f)$ we have

$$\frac{\partial \underline{X}(f)}{\partial f_{ij}} = -\underline{K}(f)^{-1}\underline{\underline{HA}}(f)^{-1}\underline{F}_{ij}\underline{V}(f) \tag{5.5}$$

and

$$\frac{\partial \underline{X}(f)}{\partial g_{ik}} = -\underline{K}(f)^{-1}\underline{\underline{HA}}(f)^{-1}\underline{G}_{ik}\underline{X}(f) \quad , \tag{5.6}$$

where $\underline{A}(f) = \underline{I} - \exp(-i2\pi f)\underline{F}$, $\underline{V}(f) = \exp(-i2\pi f)\underline{A}(f)^{-1}\underline{\underline{GX}}(f)$, $\underline{F}_{ij} = \partial\underline{F}/\partial f_{ij}$ and $\underline{G}_{ik} = \partial\underline{G}/\partial g_{ik}$. $\underline{F}_{ij}$ and $\underline{G}_{ik}$ are the matrices with the (i,j)-th and the (i,k)-th elements equal to 1 and others equal to zero, respectively. The Fourier transform of $\underline{V}(f)$ is a sequence of the estimates of the state vectors, shifted backward by one unit of time and scaled by the factor $1/\sqrt{N}$, i.e., the estimates of $\underline{v}(n-1)/\sqrt{N}$'s.

For the calculation of the first and second order derivatives of $\log|C_0|$ the following general relations are used:

$$\frac{\partial}{\partial\theta_i}\log|\underline{A}| = \mathrm{tr}\,\frac{\partial\underline{A}}{\partial\theta_i}\underline{A}^{-1}, \tag{5.7}$$

and

$$\frac{\partial}{\partial\theta_i}\underline{A}^{-1} = -\underline{A}^{-1}\frac{\partial\underline{A}}{\partial\theta_i}\underline{A}^{-1}, \tag{5.8}$$

where the matrix $\underline{A}$ is assumed to be nonsingular and its elements are functions of a set of parameters θ_i and the derivative of the matrix is defined as the matrix of the derivatives of the elements.

From (5.4), (5.5) and (5.7) and by rotating the factors under the trace sign we get

$$\frac{\partial}{\partial f_{ij}}\log|\underline{C}_0| = -2\int\underline{O}(f)^*\underline{X}(f)\underline{V}(f)^*\,df(i,j), \tag{5.9}$$

where $\underline{O}(f) = \underline{C}_0^{-1}\underline{K}(f)^{-1}\underline{H}\underline{A}(f)^{-1}$ and (i,j) denotes the (i,j)-th element of the preceding matrix. Analogously we get

$$\frac{\partial}{\partial b_{jk}}\log|\underline{C}_0| = -2\int\underline{O}(f)^*\underline{X}(f)\underline{X}(f)^*\,df(j,k). \tag{5.10}$$

The Hessian of $\log|C_0|$ is given by

$$\frac{\partial^2}{\partial\theta_i\partial\theta_j}\log|\underline{C}_0| = \mathrm{tr}\left[\frac{\partial^2\underline{C}_0}{\partial\theta_i\partial\theta_j}\underline{C}_0^{-1} - \frac{\partial\underline{C}_0}{\partial\theta_i}\underline{C}_0^{-1}\frac{\partial\underline{C}_0}{\partial\theta_j}\underline{C}_0^{-1}\right]. \tag{5.11}$$

When the model is exact and the process is ergodic we have with probability one

$$\int g(f)\underline{X}(f)\underline{X}(f)^*\,df \sim \int \underline{g}(f)\underline{C}_0\,df, \tag{5.12}$$

where $g(f)$ is an arbitrary bounded continuous function and $\sim$ denotes the asymptotic equality in the sense that the difference of the both hand sides tends to zero as the length N of the data is increased indefinitely. From (5.5) we have $\partial \underline{X}(f)/\partial f_{ij} = \exp(-i2\pi f)\underline{M}(f)\underline{X}(f)$, where $\underline{M}(f) = \underline{K}^{-1}(f)\underline{HA}(f)^{-1}\underline{F}_{ij}\underline{A}(f)^{-1}\underline{G}$. Since the filter with the frequency response function $\underline{M}(f)$ is physically realizable we have

$$\int \exp(-i2\pi f)\underline{M}(f)\, df = 0 \quad .$$

Also for $\partial \underline{X}(f)/\partial g_{ik} = -\underline{K}(f)^{-1}\underline{HA}(f)^{-1}\underline{G}_{ik}\underline{X}(f)$ we have

$$\int \underline{K}(f)^{-1}\underline{HA}(f)^{-1}\underline{G}_{ik}\, df = 0 \quad .$$

This last relation is due to the fact that the linear system defined by $\underline{H}$, $\underline{F}$, $\underline{G}$ introduces a delay of at least one unit of time to the input specified by $\underline{G}_{ik}$. These results and (5.12) show that at the true values of the parameters θ_i we have

$$\begin{aligned} \frac{\partial \underline{C}_0}{\partial \theta_i} &= 2\int \frac{\partial \underline{X}(f)}{\partial \theta_i}\, \underline{X}(f)^*\, df \\ &\sim 0 \quad . \end{aligned}$$

This observation suggests that when θ_i's are close to the theoretical values we may get a reasonable approximation to the Hessian by ignoring the second term in the right hand side of (5.11).

We have

$$\frac{\partial^2 \underline{C}_0}{\partial \theta_i \partial \theta_j} = 2\int \left(\frac{\partial \underline{X}(f)}{\partial \theta_i}\, \frac{\partial \underline{X}(f)^*}{\partial \theta_j} + \frac{\partial^2 \underline{X}(f)}{\partial \theta_i \partial \theta_j}\, \underline{X}(f)^* \right) df \quad .$$

By following the same type of reasoning as in the case of the evaluation of $\partial \underline{C}_0/\partial \theta_i$ we have

$$\int \frac{\partial^2 \underline{X}(f)}{\partial \theta_i \partial \theta_j}\, \underline{X}(f)^*\, df \sim 0 \quad .$$

Our approximate Hessian is thus defined as a matrix $\underline{L}$ of which (i,j)-th element is given by

$$L(\theta_i,\theta_j) = 2\,\mathrm{tr}\left(C_0^{-1}\int \frac{\partial \underline{X}(f)}{\partial\theta_i}\;\frac{\partial \underline{X}(f)^*}{\partial\theta_j}\,df\right) \qquad (5.13)$$

C. COMPUTATION ORGANIZATION IN TIME DOMAIN

For the computation of the gradients of $\log|\underline{C}_0|$ and the approximate Hessian $\underline{L}$ we use the time domain representation of the functions given in the frequency domain. From (5.9) the time domain representation of $\partial\log|\underline{C}_0|/\partial f_{ij}$ is given by

$$\frac{\partial}{\partial f_{ij}}\log|\underline{C}_0| = -2\left(\sum_{m=-\infty}^{\infty} \underline{O}^*_{-m}(\underline{XV}^*)_{-m}\right)(i,j) \quad ,$$

where $\underline{O}^*_{-m}$ and $(\underline{XV}^*)_m$ are the Fourier transforms of $\underline{O}(f)^*$ and $\underline{X}(f)\underline{V}(f)^*$ and are respectively given by

$$\underline{O}^*_{-m} = \int \exp(i2\pi fm)\underline{O}(f)^*\,df$$

and

$$(\underline{XV}^*)_m = \int \exp(i2\pi fm)\underline{X}(f)\underline{V}(f)^*\,df \quad .$$

Here $(\underline{XV}^*)_m$ is the m-lag covariance between the two time series corresponding to $\underline{X}(f)$ and $\underline{V}(f)$. If we denote by $\underline{O}_{-m}$ the Fourier transform of $\underline{O}(f)$ we have $\underline{O}^*_{-m} = \underline{O}'_{-m}$. Since $\underline{O}(f)$ is the frequency response function of a physically realizable system we have

$$\frac{\partial}{\partial f_{ij}}\log|\underline{C}_0| = -2\left(\sum_{n=0}^{\infty} \underline{O}'_{-n}(\underline{XV}^*)_n\right)(i,j) \quad .$$

Computationally when $\underline{O}_{-n}$'s with n greater than M are very close to zero we truncate $\underline{O}_{-n}$ at $n = M$ and get an approximation

$$\frac{\partial}{\partial f_{ij}}\log|\underline{C}_0| = -2\left(\sum_{m=0}^{M} \underline{O}'_{-m}(\underline{XV}^*)_m\right)(i,j) \quad .$$

Analogously we get an approximation

$$\frac{\partial}{\partial g_{ik}} \log |\underline{C}_0| = -2\left(\sum_{m=0}^{M} \underline{O}'_m (\underline{XX}^*)_m\right)(i,j) .$$

The covariance matrices $(\underline{XV}^*)_m$ and $(\underline{XX}^*)_m$ are obtained by passing the original covariance matrices $(\underline{YY}^*)_m$ through various linear filters, where $(\underline{YY}^*)_m$ is by definition

$$(\underline{YY}^*)_m = \frac{1}{N}\sum_{n=1}^{N-m} \underline{y}(n+m)\underline{y}(n)' \qquad m = 0,1,2,\ldots,N-1$$

$$= 0 \qquad m = N, N+1, \ldots,$$

and

$$(\underline{YY}^*)_m = (\underline{YY}^*)'_{-m} \qquad m = -1,-2,\ldots .$$

By using an appropriately truncated impulse response matrices we get

$$(\underline{YX}^*)_m = \sum_{k=0}^{K} (\underline{YY}^*)_{m+k} \underline{X}'_k \tag{5.14}$$

and

$$(\underline{XX}^*)_m = \sum_{k=0}^{K} \underline{X}_k (\underline{YX}^*)_{m-k} , \tag{5.15}$$

where $\underline{X}_m$ is the Fourier transform of $\underline{K}(f)^{-1}$ and is assumed to be close to zero for m greater than K. Analogously we get

$$(\underline{XV}^*)_m = \sum_{k=1}^{L+1} (\underline{XY}^*)_{m+k} \underline{U}'_{k-1} , \tag{5.16}$$

where $\underline{U}_k$ is the Fourier transform of $\underline{A}(f)^{-1}\underline{GK}(f)^{-1}$. The required impulse responses are generated recursively by the following relations:

$$\underline{U}_k = \underline{FU}_{k-1} + \underline{GZ}_k \tag{5.17a}$$

and

$$\underline{Z}_{k+1} = -\underline{HFU}_k . \tag{5.17b}$$

where the initial conditions are $\underline{Z}_0 = \underline{I}$, an $r \times r$ identity matrix, and $\underline{U}_{-1} = \underline{0}$, a $p \times r$ zero matrix, where r and p are the dimensions of $\underline{y}(n)$ and the state vector, respectively.

From (5.5), (5.6) and (5.13) we can see that we need the covariance matrices $(XX^*)_m$, $(VV^*)_m$ and $(XV^*)_m$ for the computation of the approximate Hessian. $(XX^*)_m$ and $(XV^*)_m$ are obtained by (5.15) and (5.16) and $(VV^*)_m$ is computed by the formulas

$$(\underline{YV}^*)_m = \sum_{k=1}^{L+1} (\underline{YY}^*)_{m+k} \underline{U}'_{k-1}$$

and

$$(\underline{VV}^*)_m = \sum_{k=1}^{L+1} \underline{U}_{k-1} (\underline{YV}^*)_{m-k} \quad .$$

We use the relation

$$\mathrm{tr}\, \underline{AE}_{ij} \underline{BE}_{nm} = \underline{A}(m,i) \underline{B}(j,n) \quad ,$$

where $\underline{E}_{mn}$ denotes a matrix with an appropriate dimension and with the (m,n)-th element equal to 1 and others equal to zero. By substituting (5.5) and (5.6) into (5.13) and using the above relation after an appropriate rotation of the factors under the trace sign we get the following formulas for the computation of the approximate Hessian:

$$L(f_{ij}, f_{mn}) = 2 \left\{ C_0(m,i) \cdot (VV^*)_0(j,n) + \sum_{k=1}^{M} C_k(m,i) \cdot (VV^*)_k(n,j) + \sum_{k=1}^{M} C_k(i,m) \cdot (VV^*)_k(j,n) \right\} \quad ,$$

$$L(f_{ij}, g_{mn}) = 2 \left\{ C_0(m,i) \cdot (XV^*)_0(n,j) + \sum_{k=1}^{M} C_k(m,i) \cdot (XV^*)_k(n,j) + \sum_{k=1}^{M} C_k(i,m) \cdot (XV^*)_{-k}(n,j) \right\} \quad ,$$

and

$$L(g_{ij}, g_{mn}) = 2\left\{C_0(m,i)\cdot(XX^*)_0(j,n) + \sum_{k=1}^{M} C_k(m,i)\cdot(XX^*)_k(n,j) + \sum_{k=1}^{M} C_k(i,m)\cdot(XX^*)_k(j,n)\right\}$$

where $C_k(m,i)$ denotes the (m,i)-th element of the matrix $\underline{C}_k$ which is the k-th Fourier coefficient of the function $(\underline{K}(f)^{-1}\underline{\underline{HA}}(f)^{-1})^* \underline{C}_0^{-1}(\underline{K}(f)^{-1}\underline{\underline{HA}}(f)^{-1})$ and can be obtained by the formula

$$\underline{C}_k = \sum_{n=0}^{K} \underline{Z}_n' \underline{C}_0^{-1} \underline{Z}_{k+n} ,$$

where $\underline{Z}_n$ is given by (5.17). $(VV^*)_k(n,j)$, $(XV^*)_k(n,j)$ and $(XX^*)_k(n,j)$ denote the (n,j)-th elements of $(\underline{\underline{VV}}^*)_k$, $(\underline{\underline{XV}}^*)_k$ and $(\underline{\underline{XX}}^*)_k$.

The inverse of the approximate Hessian can be used for the implementation of the Newton-Raphson type procedure or for the starting of the Davidon's variance algorithm type procedure for the minimization of $\log |C_0|$.

Since the transition matrix $\underline{F}$ is a rather sparse matrix the multiplication of a p-dimensional vector $\underline{v} = (v(1), v(2),\ldots,v(p))$ by $\underline{F}$, i.e. $\underline{\underline{Fv}}$, can be realized economically. We store the free parameters within $\underline{F}$ in the form of a vector $\underline{f}$ of which components are defined by the relation

$$f\left(\sum_{i=1}^{s-1} q_i + h\right) = F(j_s, h) \qquad \begin{array}{l} h = 1,2,\ldots,q_s \\ s = 1,2,\ldots,r \end{array} ,$$

where $f(i)$ denotes the i-th component of $\underline{f}$ and j_s and q_s are defined as follows:

1. Given the structural characteristic vector $\underline{h}$, construct the vectors $\underline{d} = (d(1), d(2),\ldots,d(p))'$ and $\underline{r} = (r(1), r(2),\ldots, r(p))'$ by the following procedure:

For the successive values $1,2,\ldots,p$ of i, if $h_i + r < h_p$, count the number R_i of h_j's $(j = i+1, i+2,\ldots)$ which satisfy the relation $h_j \leqq h_i + r$. When the last and the largest h_j which satisfies the inequality is equal to $h_i + r$, put $d_i = 0$ and $r(i) = i + R_i$. When the last h_j is less than $h_i + r$ put $d(i) = 1$ and $r(i) = i + R_i$. For those values of i for which $h_i + r > h_p$ holds, put $d(i) = 1$ and $r_i = p$. $d(i) = 1$ means that the i-th row of the transition matrix $\underline{F}$ contains the free parameters and $r(i)$ denotes the column of the last non-zero element within the i-th row of $\underline{F}$.

2. Given the vectors $\underline{d}$ and $\underline{r}$ construct the r-dimensional vectors $\underline{j}$ and $\underline{q}$ by the following procedure:

First put $s = 0$. Successively scan $d(i)$ for $i = 1, 2, \ldots, p$. When $d(i) = 1$, increase the value of s by 1, and put $j_s = i$ and $q_s = r(i)$. s takes the values $1,2,\ldots,r$. The j_s-th row of the transition matrix $\underline{F}$ is the s-th row containing the free parameters and q_s is the number of the free parameters within the row.

By using the vector $\underline{r}$ the product $\underline{v}^+ = \underline{F}\underline{v}$ can be obtained by the following procedure:

1. First put $t = 0$.

2. Successively examine the value of $d(i)$ for $i = 1,2,\ldots,p$. If $d(i) = 0$ put $v^+(i) = v(r(i))$, where $v^+(i)$ and $v(r(i))$ denote the i-th and the $r(i)$-th components of $\underline{v}^+$ and $\underline{v}$, respectively. If $d(i) = 1$ compute $v^+(i)$ by $v^+(i) = \sum_{j=1}^{r(i)} f(t+j)v(j)$ and increase the value of t by the amount $r(i)$.

The computation of the impulse response matrices $\underline{U}_k$ and $\underline{Z}_k$ by (5.17) can be organized quite efficiently by using the above procedure. Also it should be remembered that now the upper most $r \times r$ submatrix of $\underline{G}$ is an identity matrix and that $\underline{H} = [\underline{I}\ \underline{0}]$.

D. NUMERICAL EXAMPLES

1. *The Ship Model*

The maximum likelihood estimates of the parameters were obtained for the data treated in Table 1. Since the identified structural characteristic vectors were identical to the true structural characteristic vector of the case N = 1400, the sample means and the sample standard deviations of the estimates of the free parameters within the matrix $\underline{F}$ were computed. They are given in Table 4 along with the corresponding statistics for the estimates obtained by the canonical correlation analysis. The superiority of the maximum likelihood estimates is quite obvious.

TABLE 4

Comparison of the Accuracies of the Maximum Likelihood Estimates of the Free Parameters Within $\underline{F}$ and the Estimates Obtained by the Canonical Correlation Analysis. The Mean and Standard Deviation Denote the Sample Mean and Standard Deviation of Ten Cases

True Values	Maximum Likelihood Estimates		Estimates by the Canonical Correlation Analysis	
	mean	(standard deviation)	mean	(standard deviation)
0	-0.0042	(0.0200)	-0.0041	(0.0201)
0	-0.0014	(0.0080)	-0.0013	(0.0080)
-0.212	-0.2121	(0.0059)	-0.2304	(0.0319)
0.294	0.2955	(0.0135)	0.3287	(0.0616)
-1.493	-1.4971	(0.0287)	-1.5880	(0.1594)
2.184	2.1862	(0.0161)	2.2460	(0.1011)

2. *Comparison of Different Structures*

For the cases N = 350 and 700, the identified structural characteristic vectors were not always identical to the true structural characteristic vector. Thus it is impossible to compare the accuracies directly as in the case of N = 1400. By a slight reflection we realize that the goodness of fit of an estimated model should be measured by the information criterion. The information criterion takes the form

$$\frac{1}{2}\left\{r \log 2\pi + \log |\underline{C}| + \operatorname{tr} \underline{C}^{-1} E\underline{x}(n)\underline{x}(n)'\right\} ,$$

where E is taken with respect to the true structure and $\underline{x}(n)$ is defined by (5.1) for the assumed model specified by $\underline{F}$, $\underline{G}$, $\underline{H}$ and $\underline{C}$. The criterion takes its minimum value at the true structure which is assumed to be specified by the matrices $\underline{F}_0$, $\underline{G}_0$, $\underline{H}_0$ and $\underline{C}_0$. We are interested in the difference of the criterion from its minimum, which, after multiplication by the factor 2, is given by

$$\text{TIC} = \left\{\log \frac{|\underline{C}|}{|\underline{C}_0|} + \operatorname{tr} \underline{C}^{-1} E\underline{x}(n)\underline{x}(n)' - r\right\} .$$

When this value is multiplied by N, the length of the data used for the identification, it gives a measure of the lack of fit of the identified model in the scaling unit of the criterion AIC. The computation of $E\underline{x}(n)\underline{x}(n)'$, the covariance of the estimated innovation, can be organized as follows:

1. Compute $\underline{W}_m$ (m = 0, 1, 2, ...) by using the relations

$$\underline{V}_m^0 = \underline{F}_0\underline{V}_{m-1}^0 + \underline{G}\underline{X}_m$$

$$\underline{W}_m^0 = \underline{H}_0\underline{V}_m^0$$

$$\underline{W}_m = \underline{W}_m^0 - \underline{H}\underline{F}\underline{V}_{m-1}$$

$$\underline{V}_m = \underline{F}\underline{V}_{m-1} + \underline{G}\underline{W}_m ,$$

where $\underline{V}_{-1}^0 = \underline{V}_{-1} = \underline{0}$, a $p \times r$ zero matrix, $\underline{X}_m = \underline{I}$, an $r \times r$ identity matrix, for m = 0, and $\underline{0}$, an $r \times r$ zero matrix, for $m \neq 0$.

2. Compute $E\underline{x}(n)\underline{x}(n)'$ by the formula

$$E\underline{x}(n)\underline{x}(n)' = \sum_{m=0}^{L} \underline{W}_m \underline{C}_0 \underline{W}'_m ,$$

where L is such that $\underline{W}_m$ is close to $\underline{0}$ for m greater than L.

Unfortunately, in the case of the ship model, the estimates obtained by the canonical correlation analysis and autoregressive model fitting did not always produce an invertible model with $\underline{W}_m$ convergent to $\underline{0}$ and the direct comparison of the estimates was impossible. Although this result is disappointing, it should be remembered that at least theoretically a direct comparison of various models with different structures is possible by the present approach. From the stand point of the statistical model fitting, it is the goodness of the fitted model in explaining the stochastic structure under observation and not the accuracy of an individual parameter that matters. Thus the evaluation of the performance of an identification procedure should always be based on the distribution of some criterion of fit of the identified model.

3. *Applications to Real Data*

To test the practical utility of the Markovian model fitting procedure described in this chapter, the procedure was applied to a record of four-dimensional vector time series of a cement rotary kiln process. The record was composed of $y_1(n)$ = cooler grate speed, $y_2(n)$ = fuel rate, $y_3(n)$ = under cooler grate pressure, $y_4(n)$ exit gas temperature. The length of the data is $N = 741$. The physical characteristics of these variables are described in [11]. By fitting the multivariate autoregressive model with the orders up to 12 it was found that AIC attains the minimum at $M = 6$. This means that $4 \times 4 \times 6 = 96$ parameters are required for the autoregressive coefficients. When the canonical correlation analysis procedure was applied the identified structural characteristic vector was $\underline{h} = (1,2,3,4,5,6,7,8)'$. The number of free parameters within the matrices $\underline{F}$ and $\underline{G}$ is computed as 48 which is a half of the number of the parameters within the matrices of the coefficients of autoregression. The values of AIC for these two models are as follows:

AIC for AR6 (autoregressive model of order 6) = 12110

AIC for the Markovian model = 12076.

This result suggests that by using the procedures described in this chapter there are possibilities of getting a parsimonious model with a better fit than the multivariate autoregressive model in analyzing a real multivariate stochastic system. This reduction of the number of parameters also implies the simplification of the controller to be designed on the basis of the identified model of the process. The only disadvantage of the Markovian model fitting is the complexity of the required maximum likelihood computation. The lists of the computer programs in a Fortran IV type language for the canonical correlation analysis and the maximum likelihood computation are available in the form of a monograph [34]. This monograph also contains a list of an automatic autoregressive moving average model fitting procedure for scalar time series which automatically searches for a best combination of orders with the aid of AIC. The program produced the autoregressive moving average model of Fig. 5(b).

As the final example of application, Fig. 6 shows the estimates of the frequency response characteristic of a power generator. The generator was under a feedback control and the record of the command signal to this feedback system formed the first component $y_1(n)$ of a two dimensional time series $\underline{y}(n)$. The record of the generator output was taken as $y_2(n)$. Since the generator was only a part of a larger automatic frequency control system, the feedback from $y_2(n)$ to $y_1(n)$ was assumed to be negligible and the estimates of the frequency response function were obtained

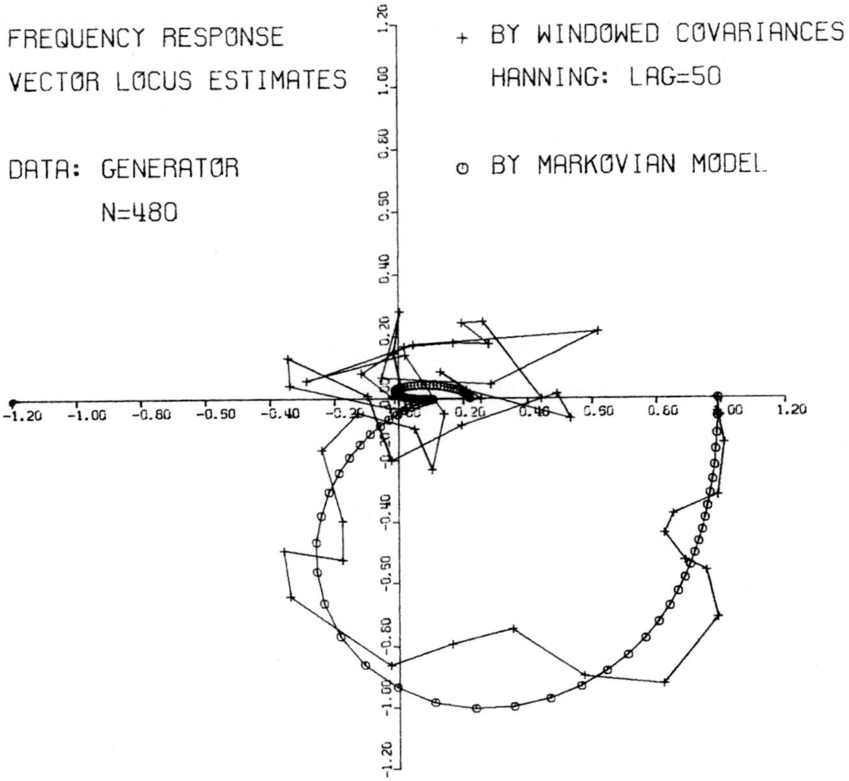

Fig. 6. Estimates of the frequency response characteristic of a power generator.

as the ratios of the estimated cross and power spectra, which were obtained either by the windowed covariance sequences or by the Markovian model of $\underline{y}(n)$. The actual feedback system contained some nonlinearity but it was found that the coherence at the lower frequency band was very close to 1.0. The estimate obtained through the Markovian model was considered to be in much better agreement with the engineer's concept of a power generator characteristic.

Acknowledgements

The author would like to express his thanks to Professor K. Sato, Nagasaki University, for the record of EEG and to Mr. H. Nakamura, Kyushu Electric Company, for the record of the power generator. Thanks are also due to Mr. K. Ohtsu, Tokyo University of Marcantile Marine, for the record of the ship.

REFERENCES

1. Blackmann, R. B. and J. W. Tukey, *The Measurement of Power Spectra,* 2nd ed., Dover, New York, 1959.

2. Jenkins, G. M. and D. G. Watts, *Spectral Analysis and Its Applications,* Holden-Day, San Francisco, 1968.

3. Akaike, H., "On the Use of an Index of Bias in the Estimation of Power Spectra," *Ann. Inst. Statist. Math., 20,* 1968, pp. 55-69.

4. Yule, G. U., "On a Method of Investigating Periodicities in Disturbed Series, with Special Reference to Wolfer's Sunspot Numbers," *Phil. Trans. A226,* 1927, pp. 267-298.

5. Akaike, H., "Fitting Autoregressive Models for Prediction," *Ann. Inst. Statist. Math., 21,* 1969, pp. 243-247.

6. Akaike, H., "Statistical Predictor Identification," *Ann. Inst. Statist. Math., 22,* 1970, pp. 203-217.

7. Akaike, H., "On a Semiautomatic Power Spectrum Estimation Procedure," in Proc. 3rd Hawaii Int. Conf. System Sciences, 1970, pp. 974-977.

8. Gersch, W. and D. R. Sharpe, "Estimation of Power Spectra with Finite-Order Autoregressive Models," *IEEE Trans. Automat. Contr., AC-18,* 1973, pp. 367-379.

9. Jones, R. H., "Identification and Autoregressive Spectrum Estimation," *IEEE Trans. Automat. Contr., AC-19,* 1974, pp. 894-898.

10. Akaike, H., "Autoregressive Model Fitting for Control," *Ann. Inst. Statist. Math., 23,* 1971, pp. 163-180.

11. Otomo, T., T. Nakagawa and H. Akaike, "Statistical Approach to Computer Control of Cement Rotary Kilns," *Automatica, 8,* 1972, pp. 35-48.

12. Akaike, H. and T. Nakagawa, *Statistical Analysis and Control of Dynamic Systems,* Saiensu-sha, Tokyo, 1972. (In Japanese, with a list of a computer program package TIMSAC for time series analysis and control written in a Fortran IV type language and with English comments.)

13. Akaike, H., "Information Theory and an Extension of the Maximum Likelihood Principle," in 2nd International Symposium on Information Theory, B. N. Petrov and F. Csaki, eds., Akademiai Kiado, Budapest, 1973, pp. 267-281.

14. Akaike, H., "A New Look at the Statistical Model Identification," *IEEE Trans. Automat. Contr., AC-19,* 1974, pp. 716-723.

15. Box, G. E. P. and G. M. Jenkins, *Time Series Analysis, Forecasting and Control,* Holden-Day, California, 1970.

16. Akaike, H., "Stochastic Theory of Minimal Realization," *IEEE Trans. Automat. Contr., AC-19,* 1974, pp. 667-674.

17. Akaike, H., "Markovian Representation of Stochastic Processes and its Application to the Analysis of Autoregressive Moving Average Processes," *Ann. Inst. Statist. Math., 26,* 1974, pp. 363-387.

18. Kullback, S., *Information Theory and Statistics,* Wiley, New York, 1959.

19. Boltzmann, L., "Über die Beziehung zwischen dem zweiten Hauptsatze der mechanischen Wärmetheorie und der Wahrscheinlichkeitsrechnung respektive den Sätzen über das Wärmegleichgewicht," *Wiener Berichte, 76,* 1877, pp. 373-435.

20. Chernoff, H., "Large Sample Theory - Parametric Case," *Ann. Math. Stat., 27,* 1956, pp. 1-22.

21. Rao, C. R., "Efficient Estimates and Optimum Inference Procedure in Large Sample," *J. Roy. Statist. Soc., B., 24,* 1962, pp. 46-72.

22. Åström, K. J. and P. E. Eykhoff, "System Identification - A Survey," *Automatica, 7,* 1971, pp. 123-162.

23. Anderson, T. W., *Introduction to Multivariate Statistical Analysis,* Wiley, New York, 1958.

24. Goloub, G. H., "Matrix Decompositions and Statistical Calculations," in *Statistical Computation,* R. C. Milton and J. A. Nelder, eds., Academic Press, New York, 1969, pp. 365-397.

25. Ho, B. L. and R. E. Kalman, "Effective Construction of Linear State-Variable Models from Input/Output Functions," *Regelungestechnik, 14,* 1966, pp. 545-548.

26. Akaike, H., "Markovian Representation of Stochastic Processes by Canonical Variables," *SIAM J. Control, 13,* 1975, pp. 162-173.

27. Åström, K. J. and C. G. Källström, "Application of System Identification Techniques to the Determination of Ship Dynamics," in *Identification and System Parameter Estimation,* P. Eykhoff, ed., North-Holland Publishing Co., Amsterdam, 1973, pp. 415-424.

28. Whittle, P., "On the Fitting of Multivariate Autoregressions, and the Approximate Factorization of a Spectral Density Matrix," *Biometrika, 50,* 1963, pp. 129-134.

29. Akaike, H., "Block Toeplitz Matrix Inversion," *SIAM J. Appl. Math., 24,* 1973, pp. 234-241.

30. Gupta, N. K. and R. K. Mehra, "Computational Aspects of Maximum Likelihood Estimation and Reduction in Sensitivity Function Calculation," *IEEE Trans. Automat. Contr., AC-19,* 1974, pp. 774-783.

31. Mehra, R. K., "Maximum Likelihood Identification of Aircraft Parameters," in 1970 Joint Automatic Control Conf., Preprints, Atlanta, Georgia, June 1970.

32. Mehra, R. K., "Identification of Stochastic Linear Dynamic Systems," *AIAA Journal, 9,* 1971, pp. 28-31.

33. Akaike, H., "Maximum Likelihood Identification of Gaussian Autoregressive Moving Average Models," *Biometrika, 60,* 1973, pp. 255-265.

34. Akaike, H., E. Arahata and T. Ozaki, "TIMSAC-74 - A Time Series Analysis and Control Program Package - (1)," *Computer Science Monographs, No. 5,* The Institute of Statistical Mathematics, Tokyo, March 1975.

35. Fisher, R. A., "Uncertain Inference," Proceedings of the American Academy of Arts and Sciences, 71, 1936, pp. 245-258.

MINMAX ENTROPY ESTIMATION OF MODELS FOR VECTOR PROCESSES

J. Rissanen
IBM Research Laboratory
San Jose, California

1. INTRODUCTION

A widely studied problem in estimation of models for vector stochastic processes may be described as follows: For an observed sequence of, say, p-component real valued vectors $y(0),\ldots,y(N)$ find the model of either one of the following two types,

$$y(t) + A_1 y(t-1) + \cdots + A_n y(t-n) = e(t) + B_1 e(t-1) + \cdots + B_n e(t-n) \qquad (1)$$

$$y(t) = 0 \qquad \text{for} \quad t < n ,$$

or

$$x(t+1) = Fx(t) + Ge(t)$$

$$y(t) = Hx(t) + e(t) , \quad x(0) = 0 , \qquad (2)$$

which minimizes a suitable criterion of the also p-component error vectors $e(t)$. Often, there is an additional term of observed inputs, $C_1 u(t-1) + \cdots + C_n u(t-n)$, in the right hand side of (1), and an analogous term in (2), but since these terms are treated as deterministic we leave them out. Their inclusion is straightforward.

A perfect fit, $e(t) = 0$ for $t > 0$, is achieved only in the special case where the sequence $y(t)$ is defined by $y(0)$ and the impulse response matrix of a system of type (1) or (2) as: $y(t) = H_t y(0)$, where

$$H(z) = H_0 + H_1 z^{-1} + H_2 z^{-2} + \cdots \qquad (3)$$

is the transfer function matrix (z-transform of H_t). This so-called realization problem is by no means trivial, and, in fact, its algorithmic solutions have to a degree contributed to a better understanding of linear systems of the given two types. The first solutions to the problem were given by Kalman [28], Ho [1], and Silverman [2]; the best algorithms subsequently by us [3], [4], the latter being an order of a magnitude faster than the others. A deep analysis of realization algorithms has been made by De Jong, [5], who also developed superior variants to the algorithm described in [3].

In all cases of practical significance the perfect fit for any meaningful values for n cannot be attained, and the question of choosing the error criterion becomes a subtle one. In an, again, idealized situation the sequence $y(t)$ results from observed samples of a stationary process with a rational spectrum, i.e., a process generated by a system of one of the two types above. Then for correct parameters in (1) or (2) the error process $e(t)$ will be uncorrelated defining the least squares predictions $y(t/t-1) = y(t) - e(t)$ of $y(t)$ given all the past $y(t-1), y(t-2), \ldots$. In this case clearly we have again a well defined attainable goal for the optimum model: to minimize a criterion of the prediction errors. This appears to be an intuitively appealing criterion even in the general case where nothing is known of the process y beyond its sample.

The aforestated goal of minimizing the prediction errors in a certain sense would be achieved by the maximum likelihood criterion. However, there is a serious difficulty in applying this technique because of the fact that several parameters in the models are capable of producing the same maximum. A way to overcome this is to represent systems by canonical forms, but since there are no universal canonical forms for the models of the considered type we are faced with the problem of estimating which of the several possible ones best fits with the data. Here one visualizes all the models of a given order n as being partitioned into a finite number of classes, and to each class there corresponds a canonical form with a number of parameters. Each value assignment for these parameters then defines a particular model having that canonical form.

Now, the likelihood function is not suitable at all for selecting a correct canonical form, even if the data actually was generated by some such true system. This is because the likelihood function is determined by the impulse response of the model, and the "true" impulse response can be arbitrarily closely approximated by a model even in a "wrong" canonical form. The estimated parameters in a "wrong" canonical form cannot converge, however, as easily shown.

An important step towards improving the maximum likelihood criterion has recently been made by Akaike, [6], [7]. Reasoning by information theoretic arguments he arrived at the criterion,

$$\text{AIC} = -\log(\text{maximum likelihood}) + (\text{number of independent parameters in the model}), \tag{4}$$

to be minimized. This criterion, as argued by Akaike, incorporates the sound idea of parsimony in model estimation, "Occam's razor", in that redundant parameters will be eliminated by the second term. Hence, no hypothesis testing is needed to settle the question of how many parameters should be included in the model.

But in view of the above outlined description of the models (1) and (2) not even Akaike's criterion can make a distinction between canonical forms with the same number of parameters. And besides, the structure of a system is not determined by the number of its parameters. This then means that while Akaike's criterion clearly improves the maximum likelihood criterion and, to be sure, seems to be adequate for estimating the input-output behavior of the models, it still falls short for the estimation of the internal description of the models; i.e., models of type (1) or (2).

In the following sections we shall introduce a criterion based on an entropy, which is aimed at supplying the missing structure dependent term. Whether or not it turns out to be entirely satisfactory should be decided on experiments. Our aim here is to introduce the problem, or at any rate, what we think is a problem and suggest a remedy in the form of a criterion which itself is based on a very broad and intuitively attractive principle. Another related approach to the estimation problem was done by Ljung and Rissanen in [8], in which the seeds to the present study were sown. Other studies in the same general vein have been made by Parzen [9], and Tse and Weinert [10]. In all the studies known to us, the structure is being estimated separately from the other parameters in contrast with the approach taken here. See however, [29], where in the scalar case the order estimation can be done together with the other parameters. We also refer to [30], where the entropy estimators are shown to be consistent, and to [31], where the connection to shortest description of the observed data is made.

2. ENTROPY

In [11] Jaynes proposed the Principle of Minimum Prejudice. This very general principle can in vague terms be stated thus:

"the minimally prejudiced assignment of probabilities is that which maximizes the entropy subject to the given information about the situation". The principle, also discussed in [12], can be extended to the estimation problems by addition of the second clause stating that *"the parameters in a model which determine the value of the maximum entropy should be assigned values which minimize the maximum entropy"*. So complemented Jaynes' principle was applied to certain estimation problems in [13] and [14].

Simple and sound as these principles are it is by no means obvious how they can be applied in specific estimation problems. They do not spell out what the entropy to be maximized and minimized really should be and under what constraints it should be determined. As a case in point, Akaike's criterion results from a related principle, and yet it is rather different from ours.

The following discussion is somewhat informal and vague, especially as it relates to the different independent parametrizations of models (1) or (2). Roughly speaking, we think of a set of such parametrizations to consist of a set of integer-valued structure parameters together with another set of real-valued system parameters. The former set assigns certain elements in the matrices of the models to be either 0 or 1, while the remaining elements constitute the real-valued system parameters. Hence both their number and their location in the matrices are determined by the structure parameters. Examples of this type of parametrizations are the usual canonical forms [15], [16], and [17].

Let the symbol s stand for a structure either in models (1) or (2) ranging over a tacitly understood set, and let θ denote the vector of, say, k real-valued system parameters. The number $k = k_s$ depends on s and remains constant within one and the same structure. It is quite important that these parameters form a complete and independent set in the sense of [18] so that no two impulse responses get mapped into one set of parameters, and that any values in the parameters with the exception of a certain "thin" set define canonical representations of some impulse response; i.e., system. In Section 4 we shall describe a fairly large family of structures where these requirements are fulfilled.

For a parameter (s,θ) the corresponding model, say (1) to be specific, defines the prediction errors $e(t) = y(t) - y^*(t)$ by the well known formula:

$$e(t) + B_1 e(t-1) + \cdots + B_n e(t-n) = A_1 y(t-1) + \cdots + A_n y(t-n)$$

$$e(t) = 0 \quad \text{for} \quad t < n \quad . \tag{5}$$

This then defines a function:

$$e : (t,s,\theta,y^t) \longmapsto e(t,s,\theta,y^t) \quad , \tag{6}$$

where $y^t = \text{col}(y(0),\ldots,y(t))$.

Let now for each structure s,

$$\phi_s : y^N \longmapsto \phi_s(y^N) \triangleq \hat{\theta} \ , \tag{7}$$

be an estimator; i.e. a function which takes the observations y^N to a system parameter vector. The images $\hat{\theta}$, as y^N runs through all the possible observations with their density function, define a random variable with an induced density function $h_s(\hat{\theta})$.

If we insert these images $\phi_s(y^N)$ in place of the fixed parameter θ in (6) we obtain the random variables:

$$e(t,s,\phi_s(y^N),y^t) \triangleq e(t),\ e^t = \text{col}(e(0),\ldots,e(t)) \ ,$$

where we now regard y^t and y^N as random variables rather than samples.

To summarize, $e(t)$ is obtained by first estimating the parameter θ, Eq. (7), with a chosen structure s, from the observations y^N by some estimation method. Then the prediction error $e(t)$ is calculated from Eqs. (5). Needless to say, this prediction error may have no optimal properties whatsoever; in fact, it is not uncorrelated nor normally distributed even if the process y is that.

We shall now consider the joint random variables $\hat{\theta}$ and e^t, which we shall treat on an equal basis. Indeed, they both represent natural estimation errors about their mean values.

The density function f_s of the joint random variables $\hat{\theta}$ and e^t may be factored as follows:

$$f_s(\hat{\theta}, e^t) = g_s(e^t/\hat{\theta}) \cdot h_s(\hat{\theta}) \quad . \tag{8}$$

The meaning of the conditional density g_s is clear: With a fixed parameter values s and θ Eqs. (5) define a density function for e^t from the distribution for y^t, which is $g_s(-/\theta)$. Although we shall take advantage of the Bayes' rule (8), our approach has nothing whatsoever to do with Bayesian viewpoint in estimation, which the reader should carefully keep in mind. In particular, we do not assume $h_s(-)$ to be an a priori given density function.

Suppose next, rather importantly, that we are willing to regard the sample estimation of the covariance of $e(t)$ corresponding to a parameter value θ as a relevant statistic about $e(t)$;

$$R_s(\theta) = \frac{1}{N-1} \sum_{t=1}^{N} e(t)e'(t) \quad . \tag{9}$$

In addition, we shall need a second statistic about $\hat{\theta}$. For this, too, we select a suitable estimate of the covariance of $\hat{\theta}-m$, where $m = E\hat{\theta}$. We shall postpone with the exact definition of this estimate until a bit later. For the time being we shall just denote it by $Q_s(\theta)$ to indicate that $E(\hat{\theta}-m)(\hat{\theta}-m)'$ has been estimated with a function evaluated at the parameters s and θ.

Returning to (8) we may write the entropy defined by f_s as follows:

$$H_s(\hat{\theta}, e^t) = H_s(e^t/\hat{\theta}) + H_s(\hat{\theta}) \quad . \tag{10}$$

This entropy depends clearly on the densities g_s and h_s, which are not known to us. However, if we agree to regard $R_s(\theta)$ and $Q_s(\theta)$ as the only relevant statistics about the random variables $\hat{\theta}$ and e^t we may ask for the density functions g_s and h_s which maximize the entropy $H_s(\hat{\theta}-m, e^N)$ subject to the natural requirements that the covariances of $e(t)$ and $\hat{\theta}-m$ equal the sample estimates. Such density functions are by a theorem due to Shannon [19], normal and independent, and the corresponding entropies are given by:

$$\max H_s(\hat{\theta}-m, e^N) = \frac{N}{2}(\log \det R_s(\theta) + p(1+\log 2\pi))$$
$$+ \frac{1}{2}\log \det Q_s(\theta) + \frac{1}{2}k(1+\log 2\pi) \quad . \tag{11}$$

Observe that the maximization under the described constraints automatically treats the $e(t)$-process as uncorrelated. By the same token, we could also have maximized $H_s(\hat{\theta})$ so as to regard the components of $\hat{\theta}$ as uncorrelated. This results in:

$$\max H_s(\hat{\theta}-m, e^N) = \frac{N}{2}(\log \det R_s(\theta) + p(1+\log 2\pi))$$
$$+ \frac{1}{2}\sum_{i=1}^{k} \log q_{ii}^{s}(\theta) + \frac{1}{2}k(1+\log 2\pi), \tag{12}$$

where $q_{ii}^{s}(\theta)$ denote the diagonal elements of $Q_s(\theta)$. Eq. (12) should be used if some pair (s,θ) exists for which the components of $\hat{\theta}$ are uncorrelated; otherwise, Eq. (11) is appropriate.

We shall finish this section by a very neat interpretation: The maximum of the entropy $H_s(e^N/\hat{\theta})$ is precisely the negative of the logarithm of the likelihood function $g_s(e^N/\theta)$ corresponding to the entropy-maximizing distribution for e^N:

$$\max_{\text{cov } e(t)=R_s(\theta)} H_s(e^N/\hat{\theta}) = -\log g_s(e^N/\theta)$$
$$= \frac{N}{2}(\log \det R_s(\theta) + p(1+\log 2\pi)). \tag{13}$$

This follows at once by using the identity, trace $[e(t)e'(t)R_s^{-1}(\theta)] = e'(t)R_s^{-1}(\theta)e(t)$. This should not be confused with the closely related result due to Eaton [20], where the last equality was shown to hold when "$-\log g_s$" is replaced by its maximized value over θ.

As a concluding remark observe how the usual normality and independence assumptions in deriving the maximum likelihood criterion are elegantly avoided by the introduction of the relevant statistics and applying Shannon's theorem. The subtle point is that we do not assume normality and independence. The processes e and $\hat{\theta}$ would, however, have an entropy given by (13) only when these two properties hold for some value θ. The real test of the meaningfulness of the so obtained criterion is a consistency proof which we shall present in a future communication, [30].

3. ESTIMATION CRITERION

The maximum entropy (11) is a function of the model parameters s and θ; the number k which gives the number of the adjustable real-valued components in θ is determined by s and is therefore not a free variable in the expression for the entropy. A minimization, then, leads to the following characterization of the optimum estimations:

$$\min_{s,\theta} \max \frac{1}{N} H_s(\hat{\theta}-m, e^N) \quad ,$$

or, equivalently,

$$\min_{s,\theta} \left[V_N(s,\theta) \triangleq \log \det R_s(\theta) + \frac{1}{N} \sum_{i=1}^{k} \log q_{ii}^s(\theta) + \frac{k}{N}(1 + \log 2\pi) \right] , \tag{14}$$

which is meaningful only when the determinants do not vanish. It is a counter intuitive peculiarity of continuous distributions that their entropy which actually is a relative one, [19], may be negative, but this does not really cause any problems in (14). By a suitable discretization the entropies in (14) will always be positive giving in fact a lower bound for the description of the data y^N; see [31].

A. DERIVATION OF $Q_s(\theta)$

Our aim is to derive an expression for the statistic $Q_s(\theta)$. This term is to give an estimate of the covariance of the estimation error $\hat{\theta}-m$, which clearly depends on the statistic $Q_s(\theta)$ itself. The derived expression for $Q_s(\theta)$ will therefore have the intended interpretation only when it is evaluated at the minimizing point θ^* of $V_N(s,\theta)$.

For any analytic function $Q_s(\theta)$ the function $V_N(s,-)$ is itself analytic by (5) and (9). It therefore admits the expansion

$$V_N(s,\theta) = V_N(s,m^*) + \text{grad}_\theta V_N(s,m^*)'(\theta-m) + \frac{1}{2}(\theta-m^*)' P(\alpha)(\theta-m^*) \tag{15}$$

about the point $E\theta^* = m^*$, where the $k\times k$ - Hessian matrix,

$$P(\theta) = \left\{\frac{\partial^2 V_N(s,\theta)}{\partial\theta_i\ \partial\theta_j}\right\}, \tag{16}$$

is evaluated at a neighboring point α of m^*.

From (15) we then obtain:

$$\text{grad}_\theta V_N(s,\theta^*) = 0 = \text{grad}_\theta V_N(s,m^*) + P(\alpha)(\theta^*-m^*). \tag{17}$$

In view of the facts that the maximum likelihood estimator is strongly consistent, and that $V_N(s,\theta)$, as a function of θ, differs from the likelihood function only by a term of size $O\left(\frac{1}{N}\right)$, it is plausible that also the estimations θ^* converge a.s. as $N\to\infty$. Therefore, for a large N the points θ^*, α, and m^* are not too far from each other a.s., and if we regard $P(\alpha) \cong P(\theta^*)$ as constants we get:

$$\mathrm{cov}(\theta^*-m^*) \cong P^{-1}(\theta^*)\, E\left\{\frac{\partial V_N(s,\theta^*)}{\partial\theta_i} \cdot \frac{\partial V_N(s,\theta^*)}{\partial\theta_j}\right\} \cdot P^{-1}(\theta^*) \tag{18}$$

Next, from (14):

$$\frac{\partial V_N(s,\theta)}{\partial\theta_i} = \frac{\partial \log\det R_s(\theta)}{\partial\theta_i} + O\left(\frac{1}{N}\right) . \tag{19}$$

Further, from (13) with the identity in θ, [21]:

$$-E\,\frac{\partial^2 \log g_s(e^N/\theta)}{\partial\theta_i\,\partial\theta_j} = E\,\frac{\partial \log g_s(e^N/\theta)}{\partial\theta_i} \cdot \frac{\partial \log g_s(e^N/\theta)}{\partial\theta_j} ,$$

we establish the desired equality:

$$E\,\frac{\partial^2 \log\det R_s(\theta^*)}{\partial\theta_i\,\partial\theta_j} = \frac{N}{2} \cdot E\,\frac{\partial \log\det R_s(\theta^*)}{\partial\theta_i} \cdot \frac{\partial \log\det R_s(\theta^*)}{\partial\theta_j} . \tag{20}$$

By the above stated a.s. convergence of θ^* we put here $EP(\theta^*) \cong P(\theta^*)$, and with (18)-(19) we write the estimate $Q_s(\theta^*)$ of $\mathrm{cov}(\theta^*-m^*)$ in the final form:

$$Q_s(\theta^*) = \frac{2}{N}\left\{\frac{\partial^2 \log\det R_s(\theta^*)}{\partial\theta_i\,\partial\theta_j}\right\}^{-1} . \tag{21}$$

The minimization (14) with the expression (9) for $R_s(\theta)$ and (21) provides a valid characterization of the estimators - albeit an implicit one - since the minimizing parameter θ^* appears in the third term. Except for small values for N the dependence of $V_N(s,\theta)$ on θ is dominated by the first term. Therefore,

$Q_s(\theta^*)$ may be replaced by $Q_s(\theta)$ and θ^* can be found iteratively.

B. DISCUSSION

It was already stated above that within a "true" structure s the criterion (14) is asymptotically equivalent to the maximum likelihood criterion, and hence it gives consistent and quite likely also asymptotically efficient estimates θ^*. This last part for the maximum likelihood estimator was proved in the scalar case in [22].

The first term in (14) is by (5) and (9) determined by the impulse response of the model. Therefore, models in equivalent canonical forms, i.e., structures, which have a common impulse response, produce the same value for the first term and cannot be distinguished by it. Instead, they will be compared by the second and the third terms. That these terms are weighted by $\frac{1}{N}$ is, of course, now immaterial.

Among the equivalent structures which moreover have the same number of parameters k the second term selects the one or, perhaps, ones whose parameters can be estimated with the smallest covariance. And this, to us seems as quite reasonable. However, to fully accept this we must consider the parameters in the various canonical forms to represent meaningful statistics about the process y in the similar way as the mean or the higher order moments. This is because, otherwise, we could define $\bar{\theta} = 10^{-3}\theta$, and find out that $\bar{\theta}$ describes our model with a much greater accuracy than θ! A more satisfactory way to deal with such scale changes is described in [31].

It remains to discuss the choice among the non-equivalent canonical forms, of which, perhaps, one represents the true system. The first term being dominant favors strongly the models which are capable of producing small values for the prediction errors as measured by log det $R_s(\theta)$. But what happens if the first term is nearly minimized by a model in a wrong canonical form? As shown in [30] such a wrong canonical form gives a high value for the second term. This basically follows from the fact that the parameters in a wrong canonical form cannot be consistently estimated, hence, large entropy $H(\theta)$ and the second term.

Observe that (14) does not reduce to Akaike's criterion in the scalar case because of the existence of the second term. This term reinforces the third term, and because of this, we feel that Akaike's criterion does not adequately capture the intended effect of the number of parameters on the criterion. In contrast, the first two terms are remarkably analogous to the criterion in [29], when again, we consider the degenerate scalar case. Nevertheless, it was Akaike's work that was the inspirational source to us and that led us to systematically study estimation problems with information theoretic means.

Observe that both Akaike's and our criterion incorporate in a formal way the intuitive idea that in estimation one should first fit the model to a set of data and then check the result by calculating the prediction error from another set of data. This way one avoids "tailoring" the model for, perhaps, a special set of observations by taking too many parameters. This sensible idea is so important that we wish to amplify it by a simple example. Suppose that the observations $y(0),\ldots,y(N)$ are generated by the single-output system:

$$y(t+1) + 0.8\,y(t) = \varepsilon(t+1) + 0.5\,\varepsilon(t) .$$

A fit of a second order model to these observations with the maximum likelihood technique by picking the first term in $V_N(s,\theta)$ only might have given the estimate:

$$y(t+1) + 0.75\,y(t) + 0.1\,y(t-1) = e(t+1) + 0.6\,e(t) + 0.2\,e(t-1) .$$

This means that the predictions of the set $y(0),\ldots,y(N)$ with this model are better than with the true system. However, if we take another set of observations and yet another etc., we will find that the found model would perform worse on the average than the true system.

4. FAMILY OF STRUCTURES

The minimization in (14) calls for a listing of structures in a set for the models of type (1) or (2). To make this practicable we should select a suitably large family of structures and find the minimum among those. But even this may be too laborious. A closer examination of the problem seems to suggest that the minimization task can virtually be carried out in two stages: First one examines the family of structures and, in fact, determines a near optimum one without even determining the parameters θ. Then the parameters are found for this structure by minimizing the first term in (14).

The reason why this seems to be at least approximately true is that a good structure tends to allow both for a good fit; i.e., small prediction error, and for a small covariance in the parameter estimates. Both of these desired properties appear to be closely related to the problem of finding a well conditioned submatrix in the block Hankel matrix defined by the impulse response of the models (1) or (2).

In the following two subsections we shall give a brief discussion of the relevant matters; a more detailed account is given in [8]. The proposed two-stages turn out to form the first step in a quite interesting iterative minimization process, to be studied in another context.

A. *STRUCTURES AND BASES*

We shall begin by recalling the stochastic realization problem, which particularly neatly has been described by Akaike in [23]. From (1) or (2):

$$y(t) = \sum_{j=0}^{\infty} H_j \, e(t-j) \quad , \tag{22}$$

where the $p \times p$-matrices $H_0, H_1, \ldots$ define the impulse response of these systems. Suppose for a moment that $\{e(t)\}$ is an orthogonal process and that (22) is a one-sided moving average representation of $y(t)$. Then the best least squares prediction $y(t/t-r)$ of $y(t)$ based on $y(t-r)$, $y(t-r-1), \ldots$ is given by:

$$y(t/t-r) = \sum_{j=r}^{\infty} H_j e(t-j) \quad . \tag{23}$$

If H^t denotes the block Hankel matrix:

$$H^t = \begin{pmatrix} H_1 & H_2 & \cdots \\ H_2 & H_3 & \cdots \\ \vdots & & \\ H_t & H_{t+1} & \cdots \end{pmatrix} \tag{24}$$

we can write from (23)

$$u^n \triangleq \begin{pmatrix} y(t+1/t) \\ y(t+2/t) \\ \vdots \\ y(t+n/t) \end{pmatrix} = H^n \begin{pmatrix} e(t) \\ e(t-1) \\ \vdots \end{pmatrix} \quad . \tag{25}$$

Now, it is well known that the rank of the matrix H^n, as n increases, will only grow up to a point, which is the order, say, q of the system (1) or (2), [24]. Therefore, among the random variables in the list u^n on the left-hand side of (25) no more than q are linearly independent, and for a large enough value n,

$$y(t+n/t) + A_1\, y(t+n-1/t) + \cdots + A_n\, y(t+1/t) = 0 \ , \tag{26}$$

for certain matrices A_i. Here we, moreover, have picked a least value for n for which (26) holds for all t.

In the matrices A_i we clearly may put the r'th column as zero if the r'th component of $y(t+n-i/t)$ is not one of the q basis elements picked among the elements in the list u^n. The remaining elements in the matrices A_i are then uniquely determined by this choice of the basis. Observe in (25) that instead of the random variables in u^n we may consider the rows in H^n.

The so-determined matrices A_i together with H^n determine in turn n+1 matrices $B_0,\ldots,B_n$ such that,

$$\sum_{i=0}^{\infty} H_i z^{-i} = (I+A_1 z + \ldots + A_n z^n)^{-1}(B_0+B_1 z + \cdots + B_n z^n),$$

and hence a "canonical" model of type (1) for the system defined by H^n. In an analogous manner we would have obtained a canonical model of the other type (2) for the same system. (A closer examination shows that these are not quite canonical, but this does not matter here.)

The result of this brief study of parametrizations of the models (1) indicates that we may define a set of structures for these models to be in 1-1 correspondence with the set of indices for bases that one can pick among the rows of H^n for a large enough n. For each such structure the elements in the A_i's described above and all the elements in the B_i's form the parameter vector θ.

The so-obtained structures define workable "canonical" forms for the models (1). Their parameters are complete and independent for each structure in the sense of [18]. Hence, they can be estimated; i.e., they are what could be called parameter identifiable structures, [8], and distinct forms in each structure define distinct impulse responses or systems. Observe, though, that if in a given q-element basis the variable elements of the matrices A_i and B_j run through all their possible values, then some such values may define a system whose Hankel matrix H^n has rank less than q. In other words, such values do not define a valid canonical form. There is, fortunately, no harm in this for the purpose of model estimation, since such exceptional parameter values cannot minimize the criterion $V_N(s,\theta)$, not even the first term. This will become apparent from the discussion in the next subsection.

B. ESTIMATION PROCEDURE

In the preceding section we identified a structure for the models (1) and (2) with the index set of a basis selected among the rows of H^n. The parameters in the matrices A_i are then determined by expressing the other rows as linear combinations of the basis elements. It is therefore clear that the parameters get determined with a greater numerical accuracy if the basis elements, normalized to the length one, are nearly orthogonal than if they make small angles between themselves. This immediately implies that if the basis elements themselves are determined from statistical data and become thus random variables, then the estimation of the parameters can be carried out with smaller error covariance in the former type of basis than in the latter.

The reasoning above suggests that we could select promising candidates for the structure of the model in the following way: Form preliminary estimates of $y_i(t+1/t)$, $y_i(t+2/t),\ldots,i = 1,\ldots,p$, for instance, by determining these as orthogonal projections of $y_i(t+1)$, $y_i(t+2),\ldots$ on the linear space spanned by the observations $y(t),\ldots,y(0)$. This gives a matrix $\bar{H}^n$ as in (25) except that the number of the columns is finite, say, Np, $N > n$. Among the np rows of this matrix pick those q rows for $q = 1,2,\ldots,$ which are "most linearly independent". What this means is explained in a moment.

Why would such a structure also allow for a good fit, i.e., a small prediction error? The explanation lies in the way we interpret this notion of q "most linearly independent" rows in $\bar{H}^n$ or, equivalently, the q "most linearly independent" elements in the set $y_i(t+1/t)$, $y_i(t+2/t),\ldots$. We regard this problem as one of finding the q vectors in this set which has as much of the information in the full set as possible. This amounts to finding that $q \times Np$-submatrix of $\bar{H}^n$ which has the least complexity in the sense of van Emden [13].

The vanEmden-complexity $C(A)$ of an $n\times m$-matrix A is defined as follows:

$$C(A) = -\frac{1}{2}\sum_{i=1}^{n}\log(n\lambda_i) , \tag{27}$$

where λ_i is an eigenvalue of $R = (\mathrm{tr}AA')^{-1}AA'$. By expanding the logarithm around $\lambda_i = \frac{1}{n}$ up to second order terms, $C(A)$ can be approximated by:

$$C(A) \cong \frac{1}{n}\sum_{i,j=1}^{n} r_{ij}^2 - \frac{1}{n^2} , \tag{28}$$

where $R = \{r_{ij}\}$; observe that $\mathrm{tr}R = 1$.

This notion of complexity is closely related to the problem of Hotelling's, [25], namely, to find the q orthogonal linear combinations of the np random variables,

$$\bar{u}^n = \begin{pmatrix} y_1(t+1/t) \\ \vdots \\ y_p(t+1/t) \\ y_1(t+2/t) \\ \vdots \end{pmatrix} = \bar{H}^n \begin{pmatrix} e(t) \\ e(t-1) \\ \vdots \\ \\ \\ \end{pmatrix} \tag{29}$$

such that the projections of the components of $\bar{u}^n$ on the subspace spanned by the q linear combinations are nearest to the components of $\bar{u}^n$ in the least squares sense. (Actually, Hotelling's problem was formulated somewhat differently.) The solution consists of the q largest eigenvectors of the covariance matrix of $\bar{u}^n$.

In much the same manner the q rows of $\bar{H}^n$ with indices $i_1,\ldots,i_q$ defined by the least complex $q\times Np$ submatrix of $\bar{H}^n$ also give the components of $\bar{u}^n$ with the property that when all the components are projected orthogonally onto the subspace spanned by these the differences are minimized in a constrained least squares sense. This implies that a model with a structure defined by these q components is capable of giving about as small a prediction error as any other comparable structure, when the parameters are determined by minimizing the first term (14). This is the justification for the proposed two-stage estimation method. It also follows that no $q-1$ element basis exists with as small prediction error, which explains why minimization with the given models is "safe"; i.e., that the minimizing parameters indeed describe workable "canonical" forms.

We add to this end that closely related numerical procedures to those discussed here have also been described by Akaike in his contribution to the present volume.

5. NUMERICAL COMPUTATIONS

In this last section we discuss briefly the numerical computations required for performing the minimization in (14) with respect to the parameter vectors θ. These calculations are the same as those required in the maximum likelihood method. The important point is that the partial derivatives of $\log \det R_s$ needed both for the "steepest gradient" type of minimization and for the expression for $Q_s(\theta)$, Eq. (21), can be calculated from certain formulas, rather than from laborous approximations by differences. The derivation of these formulas is quite straightforward, and, of course, well-known in various versions; e.g., [22]. We shall carry the derivations only to the first partial derivatives; the second ones are obtained entirely analogously.

We shall need the derivatives $\partial V_N(s,\theta)/\partial\theta_i$, which by (19) are approximately given as:

$$\frac{\partial \log \det R(\theta)}{\partial\theta_i} = \frac{\partial \det R(\theta)}{\det R(\theta)\,\partial\theta_i}, \tag{30}$$

where we dropped the subindex s in $R_s(\theta)$. The derivative of a determinant is by a well-known formula as follows:

$$\frac{\partial \det R(\theta)}{\partial \theta_q} = \sum_{i,j=1}^{k} \frac{\partial \det R(\theta)}{\partial r_{ij}} \cdot \frac{\partial r_{ij}}{\partial \theta_q}$$

$$= \det R(\theta) \sum_{i,j=1}^{k} c_{ij} \cdot \frac{\partial r_{ij}}{\partial \theta_q}, \quad (31)$$

where r_{ij} and c_{ij} are the (i,j)'th elements of $R(\theta)$ and $R^{-1}(\theta)$, respectively.

Further, by (9):

$$r_{ij} = \frac{1}{N-1} \sum_{t=1}^{N} e_i(t) e_j(t) \quad ,$$

so that

$$\frac{\partial r_{ij}}{\partial \theta_q} = \frac{1}{N-1} \sum_{t=1}^{N} \left(\frac{\partial e_i(t)}{\partial \theta_q} \cdot e_j(t) + e_i(t) \frac{\partial e_j(t)}{\partial \theta_q} \right). \quad (32)$$

Finally, for the partial derivatives $\partial e_i(t)/\partial \theta_q$ we obtain from (5) the following recurrence equations:

$$\frac{\partial e(t)}{\partial B_r(i,j)} + B_1 \frac{\partial e(t-1)}{\partial B_r(i,j)} + \cdots + B_n \frac{\partial e(t-1)}{\partial B_r(i,j)} + \delta_{ij} e(t-r) = 0 \quad (33)$$

for $\theta_q = B_r(i,j)$, the (i,j)'th element of B_r, and

$$\frac{\partial e(t)}{\partial A_r(i,j)} + B_1 \frac{\partial e(t-1)}{\partial A_r(i,j)} + \cdots + B_n \frac{\partial e(t-n)}{\partial A_r(i,j)} = \delta_{ij} y(t-r) , \quad (34)$$

for $\theta_q = A_r(i,j)$, the (i,j)'th element of A_r. Here,

$$\delta_{ij}x = \begin{pmatrix} 0 \\ \vdots \\ 0 \\ x_j \\ 0 \\ \vdots \\ 0 \end{pmatrix}, \quad x_j \text{ in the i'th position.} \qquad (35)$$

We have given formulas for all the parts of (30). In [26] R. L. Kashyap proposed simpler recurrence relations for the first partial derivatives above. We do not understand his derivation, in particular, his Eqs. (30), although it seems to give correct expressions for the first partial derivatives. The numerical maximization of the likelihood function is also discussed by Gupta and Mehra, [27].

REFERENCES

1. B. L. Ho and R. E. Kalman, "Effective Construction of Linear State-Variable Models from Input/Output Functions," Proc. Third Allerton Conf., Urbana, Ill., 1966, pp. 449-459.

2. L. Silverman, "Representation and Realization of Time-Variable Linear Systems," Tech. Rep. 94, Dept. of EE, Columbia Univ., New York, 1966.

3. J. Rissanen, "Recursive Identification of Linear Systems," *SIAM J. Control, Vol., 9, No. 3,* August 1971, pp. 420-430.

4. J. Rissanen, "Realization of Matrix Sequences," IBM Res. Rep. RJ1032, May 15, 1972.

5. L. S. De Jong, Numerical Aspects of Realization Algorithms in Linear Systems Theory, (Doctoral Thesis), Department of Mathematics, Technological University, Eindhoven, The Netherlands, 1975.

6. H. Akaike, "Use of an Information Theoretic Quantity for Statistical Model Identification," Proc. 5'th Hawaii Intern. Conf. on System Sciences, *Western Periodicals Co.,* 1972, pp. 249-250.

7. H. Akaike, "A New Look at the Statistical Model Identification," *IEEE Trans., Vol. AC-19, No. 6,* Dec. 1974, pp. 716-723.

8. L. Ljung, J. Rissanen, "On Canonical Forms, Parameter Identifiability and the Concept of Complexity," to appear in the 4'th IFAC Symposium on Identification and System Parameter Estimation, Sept. 1976, Tbilisi.

9. E. Parzen, "Some Recent Advances in Time Series Modeling," *IEEE Trans. Vol. AC-19, No. 6,* Dec. 1974, pp. 723-730.

10. E. Tse and H. L. Weinert, "Structure Determination and Parameter Identification for Multivariable Stochastic Linear Systems," Proc. of Joint Automatic Control Conf., Columbus, Ohio, 1973, pp. 604-610.

11. E. T. Jaynes, "Information Theory and Statistical Mechanics," *Phys. Rev., Vol. 106,* 1957, pp. 620-630.

12. M. Tribus, *Recent Developments in Information and Decision Processes,* Machol-Gray (ed.), MacMillan Co., 1962, pp. 102-140.

13. M. H. vanEmden, "An Analysis of Complexity," Mathematical Centre Tracts *35,* Mathematisch Centrum, Amsterdam, 1971.

14. I. J. Good, *The Estimation of Probabilities,* MIT Press, Cambridge, Mass., 1965.

15. D. G. Luenberger, "Canonical Forms for Linear Multivariable Systems," *IEEE Trans. Vol. AC-12,* 1967, pp. 290-293.

16. V. M. Popov, "Invariant Description of Linear, Time-Invariant Controllable Systems," *SIAM J. Control 10,* 1972, 254-264.

17. M. J. Denham, "Canonical Forms for the Identification of Multivariable Linear Systems," *IEEE Trans., Vol. AC-19, No. 6,* Dec. 1974, pp. 646-656.

18. J. Rissanen, "Basis of Invariants and Canonical Forms for Linear Dynamic Systems," *Automatica, Vol. 10,* 1973, pp. 175-182.

19. C. E. Shannon, "A Mathematical Theory of Communication," *Bell System Tech. J. 27,* 1948, pp. 379-423, 623-656.

20. J. Eaton, "Identification for Control Purposes," IEEE 1967 International Convention Record, N.Y., Part 3, AC, 1967, pp. 38-52.

ON THE CONSISTENCY OF PREDICTION ERROR IDENTIFICATION METHODS

Lennart Ljung
Department of Automatic Control
Lund Institute of Technology
S-220 07 Lund, Sweden

1. INTRODUCTION

The problem of identification is to determine a model that describes input-output data obtained from a certain system.

The choice of model is as a rule made using some criterion of closeness to the data, see e.g. Åström-Eykhoff (1971), Soudack *et al.* (1971). In *output error methods* the discrepancy between the model output and the measured output is minimized. The common model-reference methods are of this type, see e.g. Lüders-Narendra (1974). *Equation error methods* minimize the discrepancy in the input-output equation describing the model. Mendel (1973) has given a quite detailed treatment of these methods.

Output- and equation error methods are originally designed for noiseless data and deterministic systems. If they are applied to noisy data or stochastic systems they will give biased parameter estimates, unless the noise characteristics either are known or are of a very special kind.

A natural extension of these methods is to take the noise characteristics into account and compare the predicted output of the model with the output signal of the system. Minimization of criteria based on this discrepancy leads to the class of *prediction error identification methods*. This class contains under suitable conditions the maximum likelihood method.

The maximum likelihood method (ML method) was first introduced by Fisher (1912) as a general method for statistical parameter estimation. The problem of consistency for this method has been investigated by e.g. Wald (1949) and Cramér (1946) under the assumption that the obtained observations are independent. The first application of the ML method to system identification is due to Åström-Bohlin (1965), who considered single input-single output systems of difference equation form. In this case the mentioned consistency results are not applicable. Åström-Bohlin (1965) showed one possibility to relax the assumption on independent observations.

ML identification using state space models have been considered by e.g. Caines (1970), Woo (1970), Aoki-Yue (1970), and Spain (1971). Caines-Rissanen (1974) have discussed vector difference equations. All these authors consider consistency with probability one (strong consistency). Tse-Anton (1972) have proved consistency in probability for more general models.

Balakrishnan has treated ML identification in a number of papers, see e.g. Balakrishnan (1968).

In the papers dealing with strong consistency, one main tool usually is an ergodic theorem. To be able to apply such a result, significant idealization of the identification experiment conditions must be introduced. The possibilities to treat input signals that are partly determined as feedback are limited, and an indispensable condition is that the likelihood function must converge w.p.1. To achieve this usually strict stationarity of the output is assumed. These conditions exclude many practical identification situations. For example, to identify unstable systems some kind of stabilizing feedback must be used. Other examples are processes that inherently are under time-varying feedback, like many economic systems.

In this paper strong consistency for general prediction error methods, including the ML method is considered. The results are valid for general process models, linear as well as nonlinear. Also quite general feedback is allowed.

A general model for stochastic dynamic systems is discussed in Section 2. There also the identification method is described.

Different identifiability concepts are introduced in Section 3, where a procedure to prove consistency is outlined. In Section 4 consistency is shown for a general system structure as well as for linear systems. The application of the results to linear time-invariant systems is discussed in Section 5.

2. SYSTEMS, MODELS AND PREDICTION ERROR IDENTIFICATION METHODS

2.1 SYSTEM DESCRIPTION

A causal discrete time, deterministic system, denoted by $\mathcal{S}$, can be described by a rule to compute future outputs of the systems from inputs and previous outputs:

$$y(t+1) = f_S(t;y(t),y(t-1),\ldots,y(1);u(t),\ldots,u(1);\mathcal{Y}_0) \tag{2.1}$$

where $\mathcal{Y}_0$, "the initial conditions", represents the necessary information to compute $y(1)$.

Often $y(t+1)$ is not expressed as an explicit function of old variables, but some recursive way to calculate $y(t+1)$ is preferred. Linear difference equations and state space models are well-known examples. The advantage with such a description is that only a finite and fixed number of old values are involved in each step.

For a stochastic system future outputs cannot be exactly determined by previous data as in (2.1). Instead the conditional probability distribution of $y(t+1)$ given all previous data should be considered. It turns out to be convenient to subtract out the conditional mean and consider an innovations representation of the form

$$y(t+1) = E[y(t+1)|\mathcal{Y}_t,S] + \varepsilon(t+1,\mathcal{Y}_t,S) \tag{2.2}$$

where $E[y(t+1)|\mathcal{Y}_t,S)]$ is the conditional mean given all previous outputs and inputs,

$$E[y(t+1)|\mathcal{Y}_t,S] = g_S(t,y(t),\ldots,y(1),u(t),\ldots,u(1);\mathcal{Y}_0) \tag{2.3}$$

Here $\mathcal{Y}_t$ denotes the σ-algebra generated by $\{y(t),\ldots,y(1); u(t),\ldots,u(1);\mathcal{Y}_0\}$, and $\mathcal{Y}_0$, "the initial condition", represents the information available at time $t = 0$ about the previous behavior of the system.

The sequence $\{\varepsilon(t+1,\mathcal{Y}_t,S)\}$ is a sequence of random variables for which holds

$$E[\varepsilon(t+1,\mathcal{Y}_t,S)|\mathcal{Y}_t,S] = 0 \quad .$$

It consists of the *innovations*, see Kailath (1970).

The conditional mean $E[y(t+1)|\mathcal{Y}_t, \mathcal{S}]$ will also be called the *prediction* of $y(t+1)$ based on $\mathcal{Y}_t$. Since it will frequently occur in this report a simpler notation

$$\hat{y}(t+1|\mathcal{S}) = E[y(t+1)|\mathcal{Y}_t, \mathcal{S}]$$

will be used.

REMARK. It should be remarked that the description (2.2) to some extent depends on $\mathcal{Y}_0$. Two cases of choice of $\mathcal{Y}_0$ will be discussed here. The most natural choice is of course $\mathcal{Y}_0$ = the actual *a priori* information about previous behavior known to the "model-builder". A disadvantage with this choice is that in general $E(y(t+1)|\mathcal{Y}_t, \mathcal{S})$ will be time varying even if the system allows a time-invariant description. This point is further clarified below. A second choice is $\mathcal{Y}_0 = \bar{\mathcal{Y}}_0$ = the information equivalent (from the viewpoint of prediction) to knowing *all* previous $y(t)$, $u(t)$, $t < 0$. This choice gives simpler representations $E(y(t+1)|\mathcal{Y}_t, \mathcal{S})$, but has the disadvantage that $\bar{\mathcal{Y}}_0$ is often not known to the user. Both choices will be discussed in more detail for linear systems below.

It should also be clear that the function $g_{\mathcal{S}}$ in (2.3) can be taken as independent of any feedback relationships between u and y such that $u(t)$ is independent of future $\varepsilon(k, \mathcal{Y}_{k-1}, \mathcal{S})$ $k > t$. General stochastic systems can be described by (2.2), just as (2.1) is a general description of deterministic systems. The main results of this paper will be formulated for this general system description (2.2).

For practical reasons, in the usual system descriptions the output is not given explicitly as in (2.2). Various recursive ways to calculate $y(t+1)$ are used instead. Examples are given below.

EXAMPLE 2.1. Linear Systems in State Space Representation

State space representations are a common and convenient way of describing linear, time-varying systems. The input-output relation for the system S is then defined by

$$\begin{aligned} x(t+1) &= A_S x(t) + B_S u(t) + e(t) \\ y(t) &= C_S x(t) + v(t) \end{aligned} \tag{2.4}$$

where $\{e(t)\}$ and $\{v(t)\}$ are sequences of independent Gaussian random vectors with zero mean values and $Ee(t)e^T(t) = R_S(t)$, $Ee(t)v(t)^T = R_S^c(t)$ and $Ev(t)v(t)^T = Q_S(t)$. The system matrices may very well be time-varying but the time argument is suppressed.

The function

$$E(y(t+1)|Y_t, S) = \hat{y}(t+1|S)$$

where Y_t is the σ-algebra generated by $\{y(t),\ldots,y(1), u(t),\ldots,u(1), Y_0\}$ is obtained as follows:

$$\hat{y}(t+1|S) = C_S \hat{x}(t+1|S) \tag{2.5}$$

where the state estimate $\hat{x}$ is obtained from standard Kalman filtering:

$$\hat{x}(t+1|S) = A_S \hat{x}(t|S) + B_S u(t) + K_S(t)\{y(t) - C_S \hat{x}(t|S)\} \tag{2.6a}$$

$K_S(t)$ is the Kalman gain matrix, determined from A_S, B_S, C_S, R_S, R_S^c, and Q_S as

$$\begin{aligned} K_S(t) &= (A_S P_S(t) C_S^T + R_S^c)(C_S P_S(t) C_S^T + Q_S)^{-1} \\ P_S(t+1) &= A_S P_S(t) A_S^T - K_S(t)(C_S P_S(t) C_S^T + Q_S)^{-1} K_S^T(t) + R_S \end{aligned} \tag{2.6b}$$

The information Y_0 is translated into an estimate of the initial value $\hat{x}(0|S)$ with corresponding covariance $P_S(0)$. Then (2.6) can be solved recursively from $t = 0$. The representation (2.6) then holds for any Y_0, and in this case it is convenient to let Y_0 be the actual *a priori* knowledge about the previous behavior of the system. Notice that if the system matrices and covariance matrices all are time invariant and $\mathcal{Y}_0 = \bar{\mathcal{Y}}_0$, then also $K_{\mathcal{S}}$ will be time invariant.

A continuous time state representation can be chosen instead of (2.4). In e.g. Åström-Källström (1973) and Mehra-Tyler (1973) it is shown how $E[y(t+1)|\mathcal{Y}_t, \mathcal{S}]$, where $\mathcal{Y}_t$ is as before, can be calculated. The procedure is analogous to the one described above. □

EXAMPLE 2.2. General Linear, Time-Invariant Systems

A linear time-invariant system can be described as

$$y(t+1) = G_{\mathcal{S}}(q^{-1})u(t) + H_{\mathcal{S}}(q^{-1})e(t+1) \tag{2.7}$$

where q^{-1} is the backward shift operator: $q^{-1}u(t) = u(t-1)$ and $G_{\mathcal{S}}(z)$ and $H_{\mathcal{S}}(z)$ are matrix functions of z (z replaces q^{-1}). The variables $\{e(t)\}$ form a sequence of independent random variables with zero mean values and covariance matrices $Ee(t)e(t)^T = \Lambda_{\mathcal{S}}$ (which actually may be time-varying). It will be assumed that $G_{\mathcal{S}}(z)$ and $H_{\mathcal{S}}(z)$ are matrices with rational functions of z as entries and that $H_{\mathcal{S}}(0) = I$. The latter assumption implies that $e(t)$ has the same dimension as $y(t)$, but this is no loss of generality. Furthermore, assume that $\det H_{\mathcal{S}}(z)$ has no zeroes on or inside the unit circle. This is no serious restriction, cf. the spectral factorization theorem. Then $H_{\mathcal{S}}^{-1}(q^{-1})$ is a well defined exponentially stable linear filter, which is straightforwardly obtained by inverting $H_{\mathcal{S}}(z)$.

To rewrite (2.7) on the form (2.2) requires some caution regarding the initial values. If $\mathcal{Y}_0$ does not contain enough information the representation (2.2) will be time varying, even though (2.7) is time-invariant. In such a case a state space representation can be used. A simpler approach is to assume that $\mathcal{Y}_0 = \bar{\mathcal{Y}}_0$ = the information equivalent to knowing all previous $y(t)$, $u(t)$, $t < 0$. It will follow from the analysis in the following sections that this assumption is quite relevant for identification problems.

From (2.7) we have

$$H_S^{-1}(q^{-1})y(t+1) = H_S^{-1}(q^{-1})G_S(q^{-1})u(t) + e(t+1)$$

and

$$y(t+1) = \{I - H_S^{-1}(q^{-1})\}y(t+1) + H_S^{-1}(q^{-1})G_S(q^{-1})u(t) + e(t+1) \tag{2.8}$$

Since $H_S^{-1}(0) = I$, the right hand side of (2.8) contains $y(s)$ and $u(s)$ only up to time t. The term $e(t+1)$ is independent of these variables, also in the case u is determined from output feedback. Hence, if $\mathcal{Y}_0 = \bar{\mathcal{Y}}_0$,

$$E(y(t+1)|\mathcal{Y}_t,S) = \{I - H_S^{-1}(q^{-1})\}y(t+1) + H_S^{-1}(q^{-1})G_S(q^{-1})u(t) \tag{2.9}$$

Now, linear systems are often not modelled directly in terms of the impulse response functions G_S and H_S. A frequently used representation is the vector difference equation (VDE):

$$\bar{A}_S(q^{-1})y(t) = \bar{B}_S(q^{-1})u(t) + \bar{C}_S(q^{-1})e(t) \tag{2.10}$$

Another common representation is the state space form (in the time-invariant innovations representation form):

$$\begin{aligned} x(t+1) &= A_S x(t) + B_S u(t) + K_S e(t) \\ y(t) &= C_S x(t) + e(t) \end{aligned} \tag{2.11}$$

It is easily seen that these two representations correspond to

$$G_S(z) = \bar{A}_S^{-1}(z)\bar{B}_S(z) \qquad H_S(z) = \bar{A}_S^{-1}(z)\bar{C}_S(z) \tag{2.12}$$

and

$$G_S(z) = C_S(I-zA_S)^{-1}B_S \qquad H_S(z) = zC_S(I-zA_S)^{-1}K_S + I \tag{2.13}$$

respectively. In these two cases $G_S(z)$ and $H_S(z)$ will be matrices with rational functions as entries.

Inserting (2.12) into (2.9) it is seen that $E(y(t+1)|Y_t,S) = \hat{y}(t+1|S)$ is found as the solution of

$$C_S(q^{-1})\hat{y}(t+1|S) = \{C_S(q^{-1})-A_S(q^{-1})\}y(t+1) + B_S(q^{-1})u(t) \tag{2.14}$$

for the case of a VDE model. Solving (2.14) requires knowledge of $y(0),\ldots,y(-n)$, $u(0),\ldots,u(-n)$, $\hat{y}(0|S),\ldots,\hat{y}(-n|S)$. This information is contained in the information $\bar{Y}_0$.

For the state space model (2.11) $\hat{y}(t+1|S)$ is found from

$$\begin{aligned} \hat{x}(t+1) &= A_S\hat{x}(t) + B_Su(t) + K_S\{y(t) - C_S\hat{x}(t)\} \\ \hat{y}(t+1|S) &= C_S\hat{x}(t+1) \end{aligned} \tag{2.14b}$$

where the initial condition $\hat{x}(0)$ is obtained from $\bar{Y}_0$.

Notice that there is a parameter redundancy in the representations (2.10) and (2.11). All matrix polynomials $\bar{A}_S$, $\bar{B}_S$, and $\bar{C}_S$ and all matrices A_S, B_S, C_S, K_S that satisfy (2.12) and (2.13) respectively, correspond to the same system (2.7). □

These examples cover linear, possibly time varying systems. Clearly, also non-linear systems can be represented by (2.3). A simple example is

$$y(t+1) = f(y(t),u(t)) + \sigma(y(t))e(t+1)$$

It should, however, be remarked that it is in general no easy problem to transform a nonlinear system to the form (2.2). This is, in fact, equivalent to solving the nonlinear filter problem. It is therefore advantageous to directly model the nonlinear system on the form (2.2), if possible.

2.2 MODELS

In many cases the system characteristics, i.e. the function g_S in (2.2) and the properties of $\{\varepsilon(t+1,Y_t,S)\}$ are not known *a priori*. One possibility to obtain a model of the system is to use input-output data to determine the characteristics. In this paper we shall concentrate on the problem how the function g_S can be found.

Naturally, it is impossible to find a general function $g_S(t;y(t),\ldots,y(1);u(t),\ldots,u(1);\mathcal{Y}_0)$. Therefore the class of functions among which g is sought must be restricted. We will call this set of functions the *model set* or the *model structure*. Let it be denoted by $\mathcal{M}$ and let the elements of the model set be indexed by a parameter vector θ. The set over which θ varies will be denoted by $D_\mathcal{M}$. A certain element of $\mathcal{M}$ will be called a *model* and be denoted by $\mathcal{M}(\theta)$ or written as

$$E[y(t+1)|\mathcal{Y}_t,\mathcal{M}(\theta)] = g_{\mathcal{M}(\theta)}(t;y(t),\ldots,y(1);\ u(t),\ldots,u(1);\ \mathcal{Y}_0)$$

Hence

$$\mathcal{M} = \{\mathcal{M}(\theta)|\theta \in D_\mathcal{M}\}$$

A complete model of the system also models the sequence $\{\varepsilon(t+1,\mathcal{Y}_t,S)\}$ so that it is described by

$$y(t+1) = E[y(t+1)|\mathcal{Y}_t,\mathcal{M}(\theta)] + \varepsilon(t+1,\mathcal{Y}_t,\mathcal{M}(\theta))$$

where $\{\varepsilon(t+1,\mathcal{Y}_t,\mathcal{M}(\theta))\}$ is a sequence of random variables with conditional distribution that depends on $\mathcal{M}(\theta)$.

For brevity, the notation

$$\hat{y}(t+1|\theta) = E[y(t+1)|\mathcal{Y}_t,\mathcal{M}(\theta)]$$

is also used for the prediction.

REMARK. The notions $E[y(t+1)|\mathcal{Y}_t,\mathcal{M}(\theta)]$ and $\hat{y}(t+1|\theta)$ are used to indicate that certain probabilistic arguments underlie the construction of the rules how to calculate a good estimate of the value $y(t+1)$ from previous data. It must however be stressed that the notions are purely formal and should be understood just as replacements for the function $g_{\mathcal{M}(\theta)}$. There is no underlying "model-probability space" in the discussion, and all random variables, including "$\hat{y}(t+1|\theta)$" belong to the probability space of the true system. (The author is indebted to Prof. L. E. Zachrisson, who suggested this clarification.)

The model structures can be chosen in a completely arbitrary way. For example, g can be expanded into orthogonal function systems:

$$g_{M(\theta)} = \sum_{i=1}^{N} \theta_i f_i$$

Such choices are discussed by e.g. Lampard (1955). If there is no natural parametrization of the model, such an expansion may be advantageous. Tsypkin (1973) has discussed models of this type in connection with identification of nonlinear systems. However, the usual choice is to take one of the models in Example 2.1 or 2.2 and introduce unknown elements θ_i into the system matrices.

A vector difference equation model, e.g., is then described by

$$A_{M(\theta)}(q^{-1})y(t) = B_{M(\theta)}(q^{-1})u(t) + C_{M(\theta)}(q^{-1})\varepsilon(t;M(\theta))$$

where

$$A_{M(\theta)}(z) = I + A_{1,M(\theta)}z + \ldots + A_{n(\theta),M(\theta)}z^{n(\theta)} \qquad \text{etc.}$$

$\{\varepsilon(t;M(\theta))\}$ is a sequence of independent random variables with zero mean values and $E\varepsilon(t,M(\theta))\varepsilon(t,M(\theta))^T = \Lambda_{M(\theta)}$. The unknown elements may enter quite arbitrarily in the matrices $A_{i,M(\theta)}$. Some elements may be known from basic physical laws, or *a priori* fixed. Other elements may be related to each other, etc. Generally speaking, M can be described by the way the parameter vector θ enters in the matrices: the model parameterization.

Thus, for time-invariant linear systems the choice of model type, (vector difference equation, state space representation, etc.) and parameters can be understood as a way of parametrizing G and H in (2.8): $G_{M(\theta)}$ and $H_{M(\theta)}$ via (2.12) or (2.13).

REMARK. Like for the system description, also the model description depends on the initial conditions $\mathcal{Y}_0$. It would be most sensible to choose $\mathcal{Y}_0$ as the actual *a priori* knowledge, but as remarked previously, this gives more complex algorithms for computing the prediction. For time-invariant systems it will therefore be assumed in the sequel that $\mathcal{Y}_0 = \bar{\mathcal{Y}}_0$ = knowledge of all previous history. Since $\bar{\mathcal{Y}}_0$ is in general not known it has to be included in the model: $\bar{\mathcal{Y}}_0 = \bar{\mathcal{Y}}_0(\theta)$. Often it is sufficient to take $\bar{\mathcal{Y}}_0(\theta) = 0$ all θ, i.e. $u(t) = y(t) = 0, t < 0$, corresponding to zero initial conditions in (2.14) and (2.14b).

2.3 *IDENTIFICATION CRITERIA*

The purpose of the identification is to find a model $\mathcal{M}(\theta)$ that in some sense suitably describes the measured input and output data.

The prediction of $y(t+1)$ plays an important role for control. In, e.g., linear quadratic control theory, the optimal input shall be chosen so that $E[y(t+1)|\mathcal{Y}_t, \mathcal{S}]$ has desired behavior. This is the separation theorem, see e.g. Åström (1970).

Therefore, it is very natural to choose a model that gives the best possible prediction. That is, some function of the prediction error

$$y(t+1) - E(y(t+1)|\mathcal{Y}_t, \mathcal{M}(\theta))$$

should be minimized with respect to θ.

We will consider the following class of criteria. Introduce the matrix

$$Q_N(\mathcal{M}(\theta)) = \sum_{t=1}^{N} \sqrt{R(t)}(y(t) - \hat{y}(t|\theta))\{\sqrt{R(t)}(y(t) - \hat{y}(t|\theta))\}^T \tag{2.15}$$

Its dimension is $n_y \times n_y$, where n_y is the number of outputs. $\{R(t)\}$ is a sequence of positive definite matrices. It is assumed that $\{|R(t)|\}$ is bounded. The selection of the matrices naturally effects the relative importance given to the components of the prediction. A special choice of weighting matrices is discussed in Section 2.4.

A scalar function, $h[Q_N(\mathcal{M}(\theta))]$, of the matrix of prediction errors will be minimized with respect to θ. For the minimization to make sense, some simple properties of the function h must be introduced.

PROPERTIES OF h. Let h be a continuous function with $n_y \times n_y$, symmetric matrices as domain. Assume that

$$h(\lambda A) = g(\lambda)h(A), \ \lambda, \ g(\lambda) \ \text{ scalars and } \ g(\lambda) > 0 \ \text{ for } \ \lambda > 0 \tag{2.16a}$$

Let $\delta I < A < 1/\delta I$ be a symmetric positive definite matrix, and let B be symmetric, positive semidefinite and nonzero. Assume that then

$$h(A+B+C_\varepsilon) \geq h(A) + p(\delta) \text{ tr } B \quad \text{where} \quad p(\delta) > 0 \tag{2.16b}$$

for $\text{tr } C_\varepsilon C_\varepsilon^T < \varepsilon_0$, where ε_0 depends only on δ and $\text{tr } B$. □

If h satisfies (2.16), it defines a well posed identification criterion by

$$\tilde{V}_N(\theta) = h[Q_N(\mathcal{M}(\theta))]$$

or

$$V_N(\theta) = h\left[\frac{1}{N} Q_N(\mathcal{M}(\theta))\right] \tag{2.17}$$

In particular, $h(A)$ will be taken as $\text{tr } A$, which clearly satisfies (2.16). This criterion is probably the easiest one to handle, theoretically as well as computationally. Then

$$\text{tr } Q_N(\mathcal{M}(\theta)) = \sum_1^N |y(t) - \hat{y}(t|\theta)|^2_{R(t)}$$

where

$$|x|^2_{R(t)} = x^T R(t) x \quad .$$

Another possible choice is $h(A) = \det(A)$, which is of interest because of its relation to the likelihood function, cf. Section 2.4.

LEMMA 2.1: $h(A) = \det(A)$ *satisfies* (2.16).

<u>Proof</u>. Condition (2.16a) is trivially satisfied.

$$\det(A+B+C_\varepsilon) = \det A^{1/2} \det(I+A^{-1/2}(B+C_\varepsilon)A^{-1/2}) \det A^{1/2}$$
$$= \det A \prod_{i=1}^{n_y} (1+d_i)$$

where d_i are the eigenvalues of $A^{-1/2}(B+C_\varepsilon)A^{-1/2}$.

Let λ be the largest eigenvalue of B. Then $\lambda \geq \operatorname{tr} B/n_y$. Also, $A^{-1/2}BA^{-1/2}$ has one eigenvalue that is larger or equal to $\lambda\delta$. (Consider $A^{-1/2}BA^{-1/2}x$, where $A^{-1/2}x$ is an eigenvector to B with eigenvalue λ.) Now, adding C_ε to B can distort the eigenvalues at most ε/δ where $\varepsilon = \|C_\varepsilon\|$, the operator norm of C_ε, and

$$\prod_{i=1}^{n_y} (1+d_i) \geq \left[\prod_{i=1}^{n_y-1} (1-\varepsilon/\delta)\right] (1+\delta\lambda-\varepsilon/\delta)$$
$$\geq \left(1 - \frac{n_y\varepsilon}{\delta}\right)\left(1 + \delta\frac{\operatorname{tr} B}{n_y} - \varepsilon/\delta\right) \geq 1 + \frac{\delta}{2n_y}\operatorname{tr} B$$

$$\text{for} \quad \varepsilon < \frac{\delta \operatorname{tr} B}{2n_y\left(\frac{n_y}{\delta} + \operatorname{tr} B\right)} = \sqrt{\varepsilon_0}$$

which concludes the proof. ∎

In this chapter we will consider the limiting properties of the estimate θ that minimizes (2.17) as N tends to infinity. Of particular interest is of course whether the limiting values of θ gives models $\mathcal{M}(\theta)$ that properly describe $\mathcal{S}$. This is the problem of consistency of prediction error identification methods.

So far we have only discussed how the function $E[y(t+1)|\mathcal{Y}_t,\mathcal{S}]$ can be estimated. The properties of $\{\varepsilon(t+1,\mathcal{Y}_t,\mathcal{S})\}$ can then be estimated from the residuals

$$y(t+1) - E[y(t+1)|\mathcal{Y}_t, \mathcal{M}(\theta^*)] = \varepsilon(t+1;\mathcal{Y}_t, \mathcal{M}(\theta^*))$$

where θ^* is the minimizing value. In particular, if $\{\varepsilon(t+1,\mathcal{Y}_t, \mathcal{S}) = \{e(t+1)\}$ is a stationary sequence of independent random variables with zero mean values and we are only interested in the second order moment properties then $\Lambda = Ee(t)e^T(t)$ can be estimated as $1/N\ Q_N(\mathcal{M}(\theta^*))$ where Q_N is defined by (2.15) with $R(t) = I$.

2.4 *CONNECTION WITH MAXIMUM LIKELIHOOD ESTIMATION*

It is well known that prediction error criteria are intimately connected with maximum likelihood estimates. This section contains a brief discussion of how the formal relations can be established.

Consider the model

$$y(t+1) = E(y(t+1)|\mathcal{Y}_t, \mathcal{M}(\theta)) + \varepsilon(t+1;\mathcal{M}(\theta)) \tag{2.18}$$

with

$$E\,\varepsilon(t;\mathcal{M}(\theta))\varepsilon^T(t;\mathcal{M}(\theta)) = \hat{\Lambda}(t)$$

Let the innovations $\{\varepsilon(t,\mathcal{M}(\theta))\}$ be assumed to be independent and normally distributed. The probability density of $y(t+1)$ given $\mathcal{Y}_t$ and given that (2.18) is true then is

$$f(x_{t+1}|\mathcal{Y}_t) = \frac{1}{\sqrt{2\pi \det \hat{\Lambda}(t+1)}} \cdot$$

$$\cdot\, e^{-1/2[x_{t+1}-\hat{y}(t+1|\theta)]^T\hat{\Lambda}^{-1}(t+1)[x_{t+1}-\hat{y}(t+1|\theta)]}$$

Here $f(x|\mathcal{Y}_t) = F'(x|\mathcal{Y}_t)$ where $F(x|\mathcal{Y}_t) = P(y(t+1) \leq x|\mathcal{Y}_t)$.

Using Bayes' rule the joint probability density of $y(t+1)$ and $y(t)$ given $\mathcal{Y}_{t-1}$ can be expressed as

$$f(x_{t+1}, x_t | \mathcal{Y}_{t-1}) = f(x_{t+1} | y(t) = x_t, \mathcal{Y}_{t-1}) f(x_t | \mathcal{Y}_{t-1})$$

$$= f(x_{t+1} | \mathcal{Y}_t) f(x_t | \mathcal{Y}_{t-1})$$

$$= [2\pi \det \hat{\Lambda}(t+1) \det \hat{\Lambda}(t)]^{-1/2}$$

$$\cdot \exp\left\{- 1/2 \ [x_{t+1} - \hat{y}(t+1|\theta)]^T \hat{\Lambda}^{-1}(t+1)\right.$$

$$\left.\cdot [x_{t+1} - \hat{y}(t+1|\theta)]\right\}$$

$$\cdot \exp\left\{- 1/2 \ [x_t - \hat{y}(t|\theta)]^T \hat{\Lambda}^{-1}(t) [x_t - \hat{y}(t|\theta)]\right\}$$

where $y(t)$ in $\hat{y}(t+1|\theta)$ should be replaced by x_t. In case $E\{y(t+1|\mathcal{Y}_t, \mathcal{M}(\theta))\}$ does not depend linearly on $y(t)$, the distribution of $(y(t+1), y(t))$ is not jointly normal. Iteration directly gives the joint probability density of $y(t+1), y(t), \ldots,$ $y(1)$ given $\mathcal{Y}_0$. The logarithm of the likelihood function, given $\mathcal{Y}_0$, then is obtained as

$$\log f(y(t+1), \ldots, y(1) | \mathcal{Y}_0) = - 1/2 \sum_{s=0}^{t} [y(s+1) - \hat{y}(s+1|\theta)]^T \hat{\Lambda}^{-1}$$

$$(s+1) [y(s+1) - \hat{y}(s+1|\theta)] - \frac{t}{2} \log 2\pi - \frac{1}{2} \sum_{s=1}^{t} \log \det \hat{\Lambda}(s+1)$$

The maximum likelihood estimate (MLE) of θ therefore is obtained as the element that minimizes

$$\sum_{s=1}^{t} [y(s+1) - \hat{y}(s+1|\theta)]^T \hat{\Lambda}^{-1}(s+1) [y(s+1) - \hat{y}(s+1|\theta)] + \sum_{s=1}^{t} \log \det \hat{\Lambda}(s+1)$$

If the matrices $\hat{\Lambda}(t)$ are known, the MLE is consequently obtained as the minimizing point of the loss function (2.17) with $h(A) =$ $\mathrm{tr}(A)$ and $R(t) = \hat{\Lambda}^{-1}(t)$. When $\hat{\Lambda}(t)$ are unknown, the minimization should be performed also with respect to $\{\hat{\Lambda}(s)\}$. In case $\hat{\Lambda}(t)$ does not depend on t, the minimization with respect to can be performed analytically, Eaton (1967), yielding the problem to minimize $\det[Q_N(M(\theta))]$ giving $\theta(N)$ [where $R(t) = I$ in $Q_N(M(\theta))$] and then take

$$\hat{\Lambda} = \frac{1}{N} Q_N(M(\theta(N)))$$

Summing up, the loss function (identification criterion) (2.17) which *per se* has good physical interpretation, also corresponds to the log likelihood function in the case of independent and normally distributed innovations. In the analysis, however, this will not be exploited. The results are therefore valid for general distributions of the innovations.

3. CONSISTENCY AND IDENTIFIABILITY

The question of identifiability concerns the possibility to determine the characteristics of a system using input output data. This question is obviously closely related to the problem of consistency of the parameter estimate θ. A way to connect the two concepts is introduced in this section. The definitions given here are consistent with those of Ljung-Gustavsson-Söderström (1974). The consistency of the parameter estimate θ depends on a variety of conditions, such as noise structure, choice of input signal, model parametrization etc. One specific problem is that there usually is parameter redundancy in the models. It was demonstrated in Examples 2.1 and 2.2 that several sets of matrices give the same input output relationships, and hence cannot be distinguished from each other from measurements of inputs and outputs.

Introduce the set

$$D_T(S,M) = \Big\{ \theta \,\Big|\, \theta \in D_M,$$

$$\lim_{N\to\infty} \frac{1}{N} \sum_1^N \left| E(y(t+1)\,|\,Y_t, Y_0, S) - E(y(t+1)\,|\,Y_t, Y_0(\theta), M(\theta))\right|^2 = 0 \quad \text{all } Y_t \Big\} \tag{3.1}$$

The set $D_T(\mathcal{S},\mathcal{M})$ consists of all parameters in $D_{\mathcal{M}}$ which give models that describe the system without error in the mean square sense. There might be differences between $\mathcal{S}$ and $\mathcal{M}(\theta)$, $\theta \in D_T(\mathcal{S},\mathcal{M})$ due to initial conditions and discrepancies at certain time instances, but on the whole they are indistinguishable from input-output data only.

For the case of linear, time-invariant systems it is easy to see that $D_T(\mathcal{S},\mathcal{M})$ can be described as

$$D_T(\mathcal{S},\mathcal{M}) = \{\theta | \theta \in D_{\mathcal{M}},\ G_{\mathcal{M}(\theta)}(z) = G_{\mathcal{S}}(z) \ ; \ H_{\mathcal{M}(\theta)}(z) = H_{\mathcal{S}}(z) \text{ a.e.} z\} \tag{3.2}$$

Clearly, it is not meaningful to consider consistency if $D_T(\mathcal{S},\mathcal{M})$ is empty. Therefore, unless otherwise stated it will be assumed that $\mathcal{M}$ is such that $D_T(\mathcal{S},\mathcal{M})$ is nonempty. Naturally, this is a very strong assumption in practice, since it implies that the actual process can be modelled exactly. However, the theory of consistency does not concern approximation of systems, but convergence to "true" values. In Ljung (1975) general convergence results are given, which are valid also for the case when $D_T(\mathcal{S},\mathcal{M})$ is empty.

The estimate based on N data, $\theta(N)$, naturally depends on $\mathcal{S}$ and $\mathcal{M}$ and on the identification method used, $\mathcal{I}$. It also depends on the experimental conditions, like the choice of input signals, possible feedback structures etc. The experimental conditions will be denoted by $\mathcal{X}$. When needed, these dependencies will be given as arguments.

Suppose now that

$$\theta(N) \to D_T(\mathcal{S},\mathcal{M}) \qquad \text{w.p.1 as } N \to \infty \tag{3.2}$$

REMARK. By this is meant that $\inf_{\theta' \in D_T} |\theta(N) - \theta'| \to 0$ with probability one as $N \to \infty$. It does not imply that the estimate converges. □

Then the models that are obtained from the identification all give the same input-output characteristics as the true system. If we understand a system basically as an input-output relation, it is natural to say that we have identified the system if (3.2) holds:

DEFINITION 3.1: *A system* S *is said to be System Identifiable* (SI(M,I,X)) *under given* M, I, *and* X, *if* $\theta(N) \to D_T(S,M)$ w.p. 1 *as* $N \to \infty$.

If the objective of the identification is to obtain a model that can be used to design control laws, the concept of SI is quite adequate. Since all elements in $D_T(S,M)$ give the same input-output relation, they also give equivalent feedback laws.

When minimizing the criterion function, it may however lead to numerical difficulties if there is a non-unique minimum. If the objective is to determine some parameters that have physical significance another conept is more natural.

DEFINITION 3.2: *A system* S *is said to be Parameter Identifiable* (PI(M,I,X)) *under given* M, I, *and* X, *if it is* SI(M,I,X) *and* $D_T(S,M)$ *consists of only one point.*

REMARK. Parameter identifiability is the normal identifiability concept, and it has been used by several authors, see e.g. Åström-Bohlin (1965), Balakrishnan (1968), Bellman-Åström (1970), Tse-Anton (1972) and Glover-Willems (1973). Usually the system matrices are assumed to correspond to a certain parameter value θ^0 for the given model parametrization. In such a case the parameter θ^0 is said to be identifiable w.p. 1 (or in probability) if there exists a sequence of estimates that tends to θ^0 w.p.1 (or in probability). Now, the sequence of estimates converges to θ^0 w.p.1 if and only if it is PI(M,I,X) according to Def. 3.2 and $D_T(S,M) = \{\theta^0\}$. Therefore the definition just cited is a special case of the Definition 3.2 above.

Clearly, a system S can be $\mathrm{PI}(M,I,X)$ only if $D_T(S,M) = \{\theta^0\}$. This means that there exists a one to one correspondence between the transfer function and the parameter vector θ^0. This one to one correspondence can hold globally or locally around a given value. The terms global and local identifiability have been used for the two cases, see e.g. Bellman and Åström (1970). Definition 3.2 clearly corresponds to global parameter identifiability.

The problem to obtain such a one to one correspondence for linear systems is related to canonical representation of transfer functions. This is a field that has received much attention. The special questions related to canonical forms for identification have been treated by e.g. Åström-Eykhoff (1971), Caines (1971), Mayne (1972) and Rissanen (1974).

From the above discussion we conclude that the problem of consistency and identifiability can be treated as three different problems:

1. First determine a set $D_I(S,M,I,X)$ such that

$$\theta(N) \to D_I(S,M,I,X) \text{ w.p.1 as } N\to\infty$$

This is a statistical problem. To find such a set, certain conditions, mainly on the noise structure of the system, must be imposed.

2. Then demand that

$$D_T(S,M) \supset D_I(S,M,I,X)$$

i.e. that S is $\mathrm{SI}(M,I,X)$. This introduces requirements on the experimental conditions, X, choice of input signal, feedback structures etc.

3. If so desired, require that

$$D_T(S,M) = \{\theta^0\}$$

This is a condition on the model structure only, and for linear systems it is of algebraic nature.

In Lemma 4.1 and in Theorems 4.1 and 4.2 of the following section the set D_I is determined for general model structures (2.18), and linear systems respectively. Problem 2 is discussed in Section 5 for linear time-invariant systems. In Gustavsson-Ljung-Söderström (1974) problem 2 is extensively treated for vector difference equations. Problem 3 is, as mentioned, the problem of canonical representation and can be treated separately from the identification problem. It will not be discussed in this paper.

REMARK. In the following, the arguments S, M, I, X in D_I, D_T, SI and PI will be suppressed when there is no risk of ambiguity.

4. CONSISTENCY FOR GENERAL MODEL STRUCTURES

The problem to determine a set D_I is, as mentioned above, a statistical problem. The approach used in most works is to apply an ergodicity result to the criterion function (2.17) and then show that D_I is the set of global minima of the limit of the criterion function. However, to assume ergodicity of the involved processes introduces rather limiting conditions on the system, possible feedback structure and on the input signal. Furthermore, uniform (in θ) inequalities for the loss functions must be established. This is a fairly difficult problem, which in fact has been overlooked in many of the previous works.

The set into which the estimates converge will here be shown to be

$$D_I = \left\{ \theta \mid \theta \in D_M \, , \liminf_{N\to\infty} \frac{1}{N} \sum_{1}^{N} \left| \hat{y}(t+1|S) - \hat{y}(t+1|\theta) \right|^2_{R(t)} = 0 \right\} \tag{4.1}$$

The reason for using limit inferior is that, under some circumstances, the limit may fail to exist.

It should also be noted that D_I may depend on the realization ω, $D_I(\omega)$, although in most applications it does not (a.e.), see Section 5. For adaptive regulators it is, however, sometimes useful to consider D_I as a function of ω.

If convergence into a set that does not depend on ω is desired, this can be achieved by showing that

$$D_I(\omega) = \bar{D}_I \quad \text{w.p.1} \quad \text{or} \quad D_I(\omega) \subset \bar{D}_I \quad \text{w.p.1} \tag{4.2}$$

Then $\theta(N) \to \bar{D}_I$ w.p.1 as $N \to \infty$.

4.1 *MAIN RESULT*

LEMMA 4.1. *Consider the system*

$$y(t+1) = E(y(t+1)|\mathcal{Y}_t, \mathcal{S}) + \varepsilon(t+1, \mathcal{Y}_t, \mathcal{S})$$

where

$$E(|\varepsilon(t+1, \mathcal{Y}_t, \mathcal{S})|^4 | \mathcal{Y}_t) < C \quad .$$

Consider a set of models, $\mathcal{M}$, such that $D_T(\mathcal{S}, \mathcal{M})$ is non-empty. Let $\theta(N)$ minimize the identification criterion (2.17), $V_N(\theta) = h[1/N \; Q_N(\mathcal{M}(\theta))]$, where h satisfies (2.16), over a compact set $D_{\mathcal{M}}$. Let $D_I(\omega)$ be defined by (4.1). Suppose that

$$z(t) = \sup_{\theta \in D'_{\mathcal{M}}} \max_{1 \le i \le n_y} \left| \frac{\partial}{\partial \theta} \hat{y}^{(i)}(t|\theta) \right| \quad \text{((i) denotes i-th row)}$$

where $D'_{\mathcal{M}}$ is an open set containing $D_{\mathcal{M}}$, satisfies the following condition

$$\lim_{N\to\infty} \sup \frac{1}{N} \sum_1^N z(t)^2 < \infty \qquad \text{w.p.1} \tag{4.3}$$

Assume further that

$$\limsup_{N\to\infty} \operatorname{tr} \frac{1}{N} Q_N(M(\theta)) < \infty \quad \text{w.p.1 for any fixed} \quad \theta \in D_M \tag{4.4}$$

and that

$$E(\varepsilon(t+1,Y_t,S)\,\varepsilon(t+1,Y_t,S)^T | Y_t) \geq \delta I \quad \text{all} \quad t \tag{4.5}$$

(for $h = \operatorname{tr}$ this assumption (4.5) is not necessary).

Then the estimate

$$\theta(N,\omega) \to D_I(\omega) \qquad \text{a.e. as} \quad N \to \infty$$

The proof of the lemma is given in the Appendix.

To apply Lemma 4.1 conditions (4.3) and (4.4) have to be checked. This requires some analysis of the model structures.

REMARK. If the search is restricted to a finite number of models, conditions (4.3) and (4.4) can be removed and as the set D_I the smaller set

$$\left\{\theta | \theta \in D_M \sum_{1}^{\infty} |\hat{y}(t+1|S) - \hat{y}(t+1|\theta)|^2_{R(t)} < \infty\right\}$$

can be chosen, see Ljung (1974).

4.2 LINEAR SYSTEMS

For the linear, time-invariant model described in Example 2.2 it is relatively easy to find the variable $z(t)$ defined in Lemma 4.1. If the search for models is restricted to those corresponding to stable predictors, (4.3) will be satisfied if the input and output are subject to similar conditions. This is shown in Theorem 4.1.

THEOREM 4.1: *(LINEAR, TIME-INVARIANT SYSTEMS). Consider the system* (2.7)

$$y(t+1) = G_S(q^{-1})\,u(t) + H_S(q^{-1})\,e(t+1)$$

where $E|e(t)|^4 < C$ *(and, if the general criterion* (2.17) *is used* $Ee(t)e(t)^T > \delta I$*), and let the model set be described by*

$$y(t+1) = G_{M(\theta)}(q^{-1})u(t) + H_{M(\theta)}(q^{-1})\varepsilon(t+1); \quad \theta \in D_M$$

where D_M *is compact and* $D_{M(\theta)}(z)$ *and* $H_{M(\theta)}(z)$ *are matrices with rational functions as entries. Assume*

- *for VDE-parametrizations* (2.12), *that* $\det \bar{C}_{M(\theta)}(z)$ *has all zeroes outside the unit circle for* $\theta \in D_M$.
- *for state space realizations* (2.13), *that* $A_{M(\theta)} - K_{M(\theta)}C_{M(\theta)}$ *has all eigenvalues inside the unit circle for* $\theta \in D_M$.
- *for the general case, that* $\det H_{M(\theta)}(z)$ *as well as the least common denominator to the denominator polynomials in* $G_{M(\theta)}(z)$ *and* $H_{M(\theta)}(z)$ *have all zeroes outside the unit circle for* $\theta \in D_M$.

Suppose also that $D_T(S,M)$ *(defined in* (3.2)*) is non-empty. Any feedback relationships between* u *and* y *may exist, but assume that*

$$\limsup_{N\to\infty} \frac{1}{N}\sum_{1}^{N} (y(t)^T y(t) + u(t)^T u(t)) < \infty \quad \text{w.p.1} \tag{4.7}$$

Then the identification estimate $\theta(N)$ *converges into* D_I w.p.1 *as* N *tends to infinity. In this case* D_I *can be expressed as*

$$D_I = \left\{\theta \,\middle|\, \theta \in D_M, \quad \liminf_{N\to\infty} \frac{1}{N}\sum_{1}^{N} |(H^{-1}_{M(\theta)} - H^{-1}_S)y(t+1) + (H^{-1}_S G_S - H^{-1}_{M(\theta)}G_{M(\theta)})u(t)|^2 = 0\right\}$$

■

The proof is given in the Appendix. It uses the fact that the linear filters that determine $\hat{y}(t|\theta)$ and $d/d\theta\, \hat{y}(t|\theta)$ from u and y are exponentially stable.

For the time-varying model described in Example 2.1 it is known that the Kalman filter (see e.g. Jazwinski (1970),

Theorem 7.4) is exponentially stable if the pair $(A(t), \sqrt{R(t)})$ is completely uniformly controllable. Furthermore, the basis for the exponential depends only on the bounds on the observability and controllability Gramians. Hence we have the following theorem:

THEOREM 4.2: *(TIME-VARYING LINEAR SYSTEMS IN STATE SPACE FORM). Consider the linear system described in Example 2.1 and suppose that the covariance matrices are uniformly bounded from above. (For the general criterion (2.17) assume also that* $C_S(t)P_S(t)C_S(t)^T + Q(t) > \delta I$.) *Let the set of models be defined by*

$$\begin{aligned} x(t+1) &= A_{M(\theta)}(t)x(t) + B_{M(\theta)}(t)u(t) + \varepsilon(t) \\ y(t) &= C_{M(\theta)}(t)x(t) + w(t) \end{aligned} \qquad \theta \in D_M$$

where $\{\varepsilon(t)\}$ *and* $\{w(t)\}$ *are sequences of independent Gaussian random variables with zero mean values and* $E\varepsilon(t)\varepsilon(t)^T = R_{M(\theta)}(t)$, $E\varepsilon(t)w(t)^T = R^C_{M(\theta)}(t)$ *and* $Ew(t)w(t)^T = Q_{M(\theta)}(t)$. D_M *is a compact set such that* $D_T(S,M)$ *defined by* (3.1) *is nonempty and such that* (A,C) *is uniformly (in* t *and in* $\theta \in D_M$*) completely observable and* $(A, \sqrt{R})$ *is uniformly (in* t *and in* $\theta \in D_M$*) completely controllable.*

Any feedback relationships between u *and* y *may exist but assume that* (4.7) *is satisfied. Then the identification estimate* $\theta(N)$ *converges into* D_I *with probability one as* N *tends to infinity.* ■

Theorems 4.1 and 4.2 determine the set D_I under quite general and weak conditions. Actually, the imposed conditions: bounded fourth moments of the innovations, model search over stable predictors and the condition on the overall system behavior (4.7) are believed to be satisfied in almost all conceivable applications. For actual applications it is of interest to study D_I more closely: When is it possible to find a set D_I satisfying (4.2) and what additional assumptions on the input generation must be imposed in order to assure $D_I \subset D_T$, i.e. system identifiability. These questions are discussed in the next section.

5. IDENTIFIABILITY RESULTS

As outlined in Section 3, the identifiability and consistency questions can be separated into three problems. The first problem to determine a set D_I was solved in the previous section. The second problem, investigation of the set D_I and in particular the relation $D_I \subset D_T$ will be the subject of the present section.

5.1 *A DETERMINISTIC SET* D_I

D_I as defined in (4.1) is a random variable. However, in most applications

$$D_I = \bar{D}_I \quad \text{w.p.1} \tag{5.1}$$

where

$$\bar{D}_I = \left\{ \theta \,\middle|\, \theta \in D \,,\quad \liminf_{N\to\infty} \frac{1}{N} \sum_{1}^{N} E\left|\hat{y}(t+1|S) - \hat{y}(t+1|\theta)\right|^2_{R(t)} = 0 \right\} \tag{5.2}$$

where the expectation is with respect to the sequence of innovations. This deterministic set $\bar{D}_I$ may be easier to handle.

For linear systems the relation (5.1) will hold if the system is in open loop and is stable, or contains linear feedback which makes the closed loop system exponentially stable. To include also nonlinear feedback terms, which makes the closed loop system nonlinear, the concept of exponential stability has to be extended to stochastic, nonlinear systems.

DEFINITION 5.1: *Consider the linear system*

$$y(t+1) = E(y(t+1)|\mathcal{Y}_t, \mathcal{S}) + e(t+1)$$

where $e(t)$ *are independent random variables, and where part of the input* $u(t)$ *is determined as (nonlinear) output feedback. Let the system and regulator be started up at time* $t-N$, *with zero initial conditions, yielding at time* t *the outputs and inputs,* $y_N^0(t)$ *and* $u_N^0(t)$ *respectively. Suppose that*

$$|y(t) - y_N^0(t)| \leq C(\mathcal{Y}_{t-N})\lambda^N, \quad |u(t) - u_N^0(t)| \leq C(\mathcal{Y}_{t-N})\lambda^N \quad ;$$

some $\lambda < 1$, *where* $C(\mathcal{Y}_{t-N})$ *is a scalar function of* $\mathcal{Y}_{t-N}$, *such that* $EC(\mathcal{Y}_{t-N})^4 < C$. *Then the closed loop system is said to be exponentially stable.*

For linear feedback this definition is consistent with that the closed loop poles be inside the unit circle.

It turns out that exponential stability assures not only (5.1) but also (4.7). Hence we have the following lemma:

LEMMA 5.1: *Consider the linear systems of Example 2.1 or Example 2.2. Let the input have the general form*

$$u(t) = f_t(y(t),\ldots,y(0);\ u(t-1),\ldots,u(0)) + u_r(t) + w(t)$$

where $u_r(t)$ *is a signal that is independent of* $y(s)$, $u(s)$, $s < t$ *and such that*

$$\limsup_{N\to\infty} \frac{1}{N}\sum |u_r(t)|^2 < \infty \quad .$$

$\{w(t)\}$ *is a sequence of disturbances of a filtered white noise character, say, which is independent of* $\{e(t)\}$ *and such that* $E|w(t)|^4 < C$. *The function* f_t *may be unknown to the experiment designer. Assume that the input is such that the closed loop systems is exponentially stable (Def. 5.1) and that* D_M *satisfies the assumptions of Theorem* 4.2 *or* 4.1 *respectively. Suppose that* e(t), y(t) *and* u(t) *have uniformly bounded fourth moments. Then* (4.7) *and* (5.1) *hold.*

Proof. The proof is based on the following theorem due to Cramer and Leadbetter (1967): If

$$|\mathrm{Cov}(\xi(s),\ \xi(t))| \leq K \frac{s^p + t^p}{1 + |t-s|^q} \qquad 0 \leq 2p < q < 1 \tag{5.3}$$

then

$$\lim_{N\to\infty} \frac{1}{N} \sum_{s=1}^{N} (\xi(s) - E\xi(s)) = 0$$

with probability one.

It follows by straightforward calculations from the assumptions on exponential stability and on $D_{\mathcal{M}}$ that

$$\xi(t) = |\hat{y}(t|\mathcal{S}) - \hat{y}(t|\theta)|^2_{R(t)}$$

and

$$\eta(t) = |y(t)|^2 + |u(t)|^2$$

satisfy (5.3). (For details, see Ljung (1974), Lemma 5.2.) This proves the lemma.

5.2 *LINEAR TIME-INVARIANT SYSTEMS*

Let us now study in more detail linear time-invariant systems as treated in Example 2.2 and Theorem 4.1. Since this class includes any parametrization of vector difference equations or state space realizations or any other parametrization of a linear time-invariant system, it is believed that such analysis is sufficient for most applications.

From Theorem 4.1 and Lemma 5.1 it follows that the estimates tend to the set

$$\bar{D}_I = \left\{ \theta \,\middle|\, \liminf_{N\to\infty} \frac{1}{N} \sum_{1}^{N} E|(H^{-1}_{M(\theta)} - H^{-1}_S)y(t+1) + (H^{-1}_S G_S - H^{-1}_{M(\theta)} G_{M(\theta)})u(t)|^2 = 0 \right\}$$

This set clearly depends on the input signal. If the input is not sufficiently general, the set may contain parameters corresponding to models that describe the system well for the used input, but fail to describe it for other inputs. This is the case if the input contains too few frequencies or if it has certain relationships with the output. Then $\bar{D}_I$ is not contained in D_T and the system is not System Identifiable for this input (experiment condition).

The set $\bar{D}_I$ has been analysed in Ljung-Gustavsson-Söderström (1974) in detail for the case of time-varying feedback. Here we will consider a case with linear feedback and an extra input signal (or noise). Let the input be

$$u(t) = F(q^{-1})y(t) + u_R(t)$$

where $\{u_R(t)\}$ is a sequence that does not depend on $\{e(t)\}$. Suppose that $F(z)$ is a matrix with rational functions as entries, such that the closed loop system is stable. The closed loop system is

$$y(t+1) = (I - q^{-1}G_S(q^{-1})F(q^{-1}))^{-1}H_S(q^{-1})e(t+1) + (I - q^{-1}G_S(q^{-1})F(q^{-1}))^{-1}G_S(q^{-1})u_R(t)$$

Introduce

$$\tilde{e}(t+1) = (I - q^{-1}G_S(q^{-1})F(q^{-1}))^{-1}H_S(q^{-1})e(t+1)$$

$$\tilde{u}_R(t) = (I - q^{-1}G_S(q^{-1})F(q^{-1}))^{-1}G_S(q^{-1})u_R(t)$$

$$K_\theta(q^{-1}) = H_{M(\theta)}^{-1}(q^{-1}) - H_S^{-1}(q^{-1})$$

$$L_\theta(q^{-1}) = H_S^{-1}(q^{-1})G_S(q^{-1}) - H_{M(\theta)}^{-1}(q^{-1})G_{M(\theta)}(q^{-1})$$

Then

$$\bar{D}_I = \left\{\theta \middle| \liminf_{N\to\infty} \frac{1}{N}\sum_{1}^{N} E\left|K_\theta(q^{-1})(\tilde{e}(t+1) + \tilde{u}_R(t)) - L_\theta(q^{-1})(F(q^{-1})\tilde{e}(t+1) + F(q^{-1})\tilde{u}_R(t) + u_R(t))\right|^2 = 0\right\}$$

Since $\tilde{e}$ is independent of $\tilde{u}_R$ and u_R, the expectation can be written

$$E\left|(K_\theta(q^{-1}) - L_\theta(q^{-1})F(q^{-1}))\tilde{e}(t+1)\right|^2 + E\left|(K_\theta(q^{-1}) - L_\theta(q^{-1})F(q^{-1}))\tilde{u}_R(t) + L_\theta(q^{-1})u_R(t)\right|^2$$

If $Ee(t)e(t)^T > \delta I$, then it follows that

$$K_\theta(q^{-1}) - L_\theta(q^{-1})F(q^{-1}) = 0 \qquad \text{for} \quad \theta \in \bar{D}_I \tag{5.4}$$

since the first term has to be arbitrarily close to zero infinetely often for $\theta \in \bar{D}_I$. This in turn implies that

$$\liminf_{N\to\infty} \frac{1}{N}\sum_{1}^{N} |L_\theta(q^{-1})u_R(t)|^2 = 0 \quad \text{for} \quad \theta \in \bar{D}_I .$$

If u_R is persistently exciting (see e.g. Mayne (1972)) of sufficiently high order then this implies that

$$L_\theta(q^{-1}) \equiv 0 \qquad \text{for} \quad \theta \in \bar{D}_I$$

which, via (5.4) implies that

$$K_\theta(q^{-1}) \equiv 0 \qquad \text{for} \quad \theta \in \bar{D}_I \quad .$$

That is, $\bar{D}_I = D_T$.

REMARK. Let $U_M(t) = \text{col}(u_R(t),\ldots,u_R(t-M))$. Then it is sufficient to assume that

$$\delta I < \frac{1}{N}\sum_{1}^{N} U_M(t)U_M(t)^T < \frac{1}{\delta} I \;; \qquad N > N_0 \tag{5.5}$$

The limit of the sum does not have to exist, as in the definition of persistent excitation in Mayne (1972).

The number M for which (5.5) has to be satisfied depends on $\mathcal{S}$ and on the parametrization of $\mathcal{M}$. For state space representations M can be related to the orders of the system and model, see e.g. Mayne (1972). For the unspecified models, which we deal with here, we can require that (5.5) holds for any M.

Summing up this discussion, Lemma 5.1 and Theorem 4.1, we have the following theorem.

THEOREM 5.1: *Consider the system* (2.7), $\mathcal{S}$,

$$y(t+1) = G_{\mathcal{S}}(q^{-1})u(t) + H_{\mathcal{S}}(q^{-1})e(t+1)$$

where $\{e(t)\}$ *is a sequence of independent random variables such that* $E|e(t)|^4 < C$ *and* $Ee(t)e(t)^T > \delta I$. *The input is*

$$u(t) = F(q^{-1})y(t) + u_R(t)$$

where $\{u_R(t)\}$ *is independent of* $\{e(t)\}$ *and satisfies* (5.5) *for any* M. *Assume that* F *is such that the closed loop system is exponentially stable. Let the model set,* $\mathcal{M}$, *be described by*

$$y(t+1) = G_{\mathcal{M}(\theta)}(q^{-1})u(t) + H_{\mathcal{M}(\theta)}(q^{-1})e(t+1) \;; \quad \theta \in D_{\mathcal{M}}$$

where $D_{\mathcal{M}}$ *is compact and such that* $H_{\mathcal{M}(\theta)}(z)$ *satisfies the same conditions as in Theorem* 4.1 *for* $\theta \in D_{\mathcal{M}}$. *Assume that* $D_T(\mathcal{S},\mathcal{M})$ *is nonempty. Let* $\theta(N)$ *be the estimate of* θ *based on* N *data points, obtained by minimizing the general criterion* (2.17). *Then*

$$\theta(N) \to D_T(\mathcal{S},\mathcal{M}) \quad \text{with probability one as} \quad N \to \infty$$

where

$$D_T(S,M) = \left\{\theta \middle| G_S(z) = G_{M(\theta)}(z);\ H_S(z) = H_{M(\theta)}(z) \quad \text{a.e.z.}\right\}$$

That is, S is System Identifiable. ∎

REMARK. Notice that, when evaluating the criterion (2.17), the predictor $\hat{y}(t|\theta)$ does not have to be based on the true initial data. As remarked several times above, it is most suitably chosen as the time-invariant, steady state predictor (2.9) initialized with zero initial state.

The chapter by Caines and Chan in this volume contains several interesting, related results on identifiability of feedback systems.

6. CONCLUSIONS

In this contribution consistency and identifiability properties of prediction error identification methods have been investigated. A separation of the problem into three different tasks has been introduced, and results of varying generality and on varying levels have been given. The results can be used as a kit for "doing-your-own consistency theorem" in a specific application. They also solve the identifiability and consistency problem for linear, time-invariant systems under quite general assumptions, as shown in Theorem 5.1.

The hard part in consistency theorems is believed to be to determine the set into which the estimates converge (Problem 1 in the formulation of Section 3). This has been solved for quite general (Lemma 4.1) as well as for linear systems (Theorems 4.1 and 4.2). Due to the very weak conditions imposed, these results are applicable to adaptive systems of almost any kind, in addition to the more straightforward cases treated in Section 5.

The difficult and vital problem of choosing a parametrization (model set) that gives identifiable parameters and consistent parameter estimates has not been considered here. However, this problem is most conveniently treated as a problem of its own, and it does not depend on the identification method chosen.

APPENDIX

A.1 PROOF OF LEMMA 4.1

The idea of the proof is to show that

$$\inf V_N(\theta) > V_N(\tilde{\theta}) \qquad \text{for} \quad N > N_0(\theta^x, \rho, \omega)$$

where the infimum is taken over an open sphere around θ^x with radius ρ, and where $\tilde{\theta} \in D_T$. Then the minimizing point $\theta(N)$ cannot belong to this sphere for $N > N_0(\theta^x, \rho, \omega)$. This result is then extended to hold for the complement of any open region containing D_I, by applying the Heine-Borel theorem.

Let without loss of generality $R(t) = I$. Introduce, for short,

$$e(t) = \varepsilon(t, \mathcal{Y}_{t-1}, \mathcal{S}) = y(t) - \hat{y}(t|\mathcal{S})$$

and consider

$$Q_N(\mathcal{S}) = \sum_1^N e(t)e(t)^T$$

Let $E[e(t)e(t)^T|\mathcal{Y}_{t-1}] = S_t$. According to the assumptions $S_t > \delta I$ for all t. Each element of the matrix

$$Z(t) = \sum_1^t [e(k)e(k)^T - S_k]/k$$

clearly is a martingale with bounded variance, from which follows that

$$\frac{1}{N} Q_N(\mathcal{S}) - \frac{1}{N}\sum_1^N S_t \to 0 \qquad \text{w.p.1 as } N \to \infty$$

and

$$2/\delta' \geq \frac{1}{N} Q_N(\mathcal{S}) \geq \frac{\delta'}{2} I \quad \text{for} \quad n > N_I(\omega), \quad \omega \in \Omega_1$$

$$\text{where} \quad P(\Omega_1) = 1$$

where $\delta' = \min(\delta, 1/C)$. (The argument ω will as a rule be suppressed in the variables y, e, $\hat{y}$ etc., but used explicitly in bounds.)

Introduce also

$$\beta(t) = \hat{y}(t|S) - \hat{y}(t|\theta^x)$$

Then it follows from (4.4) and (A.1) that

$$\frac{1}{N}\sum_1^N |\beta(t)|^2 < C_1(\omega,\theta^x) \ ; \ \omega \in \Omega_2(\theta^x),\ P(\Omega_2(\theta^x)) = 1 \qquad (A.2)$$

Now take a fixed element $\tilde{\theta} \in D_T$ and consider

$$Q_N(\tilde{\theta}) = \sum_1^N (y(t) - \hat{y}(t|\tilde{\theta}))(y(t) - \hat{y}(t|\tilde{\theta}))^T$$

Introduce

$$\alpha(t) = \hat{y}(t|S) - \hat{y}(t|\tilde{\theta})$$

Then

$$Q_N(\tilde{\theta}) = Q_N(S) + \sum_1^N (e(t)\alpha(t)^T + \alpha(t)e(t)^T + \alpha(t)\alpha(t)^T)$$

Since $\tilde{\theta} \in D_T$, by definition

$$\frac{1}{N}\sum_1^N \alpha(t)^T\alpha(t) \to 0 \qquad \text{w.p.1 as} \quad N \to \infty$$

and

$$\left| \frac{1}{N}\sum_1^N e(t)\alpha(t)^T + \alpha(t)e(t)^T \right|^2 \leq 4\left(\frac{1}{N}\sum_1^N |e(t)|^2 \cdot \frac{1}{N}\sum_1^N |\alpha(t)|^2\right)$$

But from (A.1)

$$\frac{1}{N}\sum_1^N |e(t)|^2 < 2n_y/\delta \quad \text{for} \quad N > N_1(\omega) \quad (n_y = \text{number of outputs})$$

Hence

$$Q_N(\tilde{\theta}) - Q_N(S) \to 0 \qquad \text{w.p.1 as} \quad N \to \infty$$

and, since h is continuous,

$$V_N(\tilde{\theta}) < V_N(S) + \varepsilon \quad \text{for} \quad N > N_2(\omega,\varepsilon) \quad \text{and} \quad \omega \in \Omega_3, \; P(\Omega_3) = 1 \tag{A.3}$$

Now consider $Q_N(\theta)$ and decompose

$$y(t) - \hat{y}(t|\theta) = y(t) - \hat{y}(t|S) + \hat{y}(t|S) - \hat{y}(t|\theta^x) + \hat{y}(t|\theta^x) - y(t|\theta)$$

where $\theta^x \in D_M$ is a fixed point and $\theta \in B(\theta^x,\rho) = \{\theta|\ |\theta - \theta^x|<\rho\}$. Introduce for short

$$\gamma(t) = \hat{y}(t|\theta^x) - \hat{y}(t|\theta)$$

From the mean value theorem

$$|\gamma(t)| \leq \rho z(t) \tag{A.4}$$

if ρ is sufficiently small. Then

$$Q_N(\theta) = Q_N(S) + \sum_1^N \beta(t)\beta(t)^T + \sum_1^N \gamma(t)\gamma(t)^T + \sum_1^N (e(t)\beta(t)^T + \beta(t)e(t)^T) + \sum_1^N (e(t)\gamma(t)^T + \gamma(t)e(t)^T) + \sum_1^N (\beta(t)\gamma(t)^T + \gamma(t)\beta(t)^T) \tag{A.5}$$

We will first show that each element of the matrix

$$\frac{1}{N} Q_N^{(1)}(\theta^x) = \frac{1}{N} \sum_1^N (e(t)\beta(t)^T + \beta(t)e(t)^T)$$

tends to zero w.p.1 as N tends to infinity: Let

$$r(t) = \max\left(t, \sum_1^t \beta_i^2(k)\right) \qquad (\beta_i = \text{i-th component of } \beta)$$

It is easy to see that

$$m_N = \sum_1^N e_j(t)\beta_i(t)/r(t)$$

is a martingale with respect to $\{\mathcal{Y}_N\}$.

Furthermore,

$$Em_N^2 = E\sum_1^N E(e_j^2(t)|\mathcal{Y}_{t-1}) \cdot \beta_i^2(t)/r^2(t)$$

$$\leq E\sum_1^N \frac{C\ \beta_i^2(t)}{\sum_1^t \beta_i^2(k)} \leq C$$

Hence m_N converges with probability one (for $\omega \in \Omega_4(\theta^x)$, $P(\Omega_4(\theta^x)) = 1)$ according to the martingale convergence theorem, Doob (1953). It now follows from Kronecker's lemma (see e.g. Chung (1968)) that, since $r(N) \to \infty$,

$$\frac{1}{r(N)}\sum_1^N e_j(t)\beta_i(t) \to 0 \quad \text{as} \quad N \to \infty \quad \text{for} \quad \omega \in \Omega_4(\theta^x)$$

But $1 \leq r(N)/N \leq C_1(\omega,\theta^x)$ for $\omega \in \Omega_2$ according to (A.2). Hence

$$\frac{1}{N}\sum_1^N e_j(t)\beta_i(t) \to 0 \quad \text{as} \quad N \to \infty \quad \text{for} \quad \omega \in \Omega_4(\theta^x) \cap \Omega_2$$

and so

$$\text{tr}\ \frac{1}{N}Q_N^{(1)}(\theta^x)\ \frac{1}{N}Q_N^{(1)}(\theta^x)^T < \varepsilon \quad \text{for} \quad N > N_3(\varepsilon,\omega,\theta^x)$$

$$\text{and} \qquad \omega \in \Omega_2 \cap \Omega_4(\theta^x) \qquad \text{(A.6)}$$

From (A.4), (A.1), (A.2) and (4.3) it follows that

$$\text{tr}\ \frac{1}{N}Q_N^{(2)}(\theta,\theta^x)\ \frac{1}{N}Q_N^{(2)}(\theta,\theta^x)^T < C_2(\omega,\theta^x) \cdot \rho \qquad \text{(A.7)}$$

where

$$Q_N^{(2)}(\theta,\theta^x) = \sum_1^N \gamma(t)\gamma(t)^T + e(t)\gamma(t)^T + \gamma(t)e(t)^T$$

$$+ \beta(t)\gamma(t)^T + \gamma(t)\beta(t)^T$$

Property (2.16) can now be applied to (A.5) with

$$A = \frac{1}{N} Q_N(S), \qquad B = \frac{1}{N}\sum_1^N \beta(t)\beta(t)^T \qquad \text{and}$$

$$C_\varepsilon = \frac{1}{N} Q_N^{(1)}(\theta^x) + \frac{1}{N} Q_N^{(2)}(\theta,\theta^x)$$

First introduce a countable subset $\tilde{D}_M$ of D_M that is dense in D_M. Also introduce a subset Ω^x of the sample space

$$\Omega^x = \Omega_1 \cap \prod_{\theta^x \in \tilde{D}_M} \Omega_2(\theta^x) \cap \Omega_3 \cap \prod_{\theta^x \in \tilde{D}_M} \Omega_4(\theta^x)$$

Clearly, $P(\Omega^x) = 1$. Consider from now on a fixed realization $\omega \in \Omega^x$ and introduce the set

$$D_M^x(\varepsilon,\omega) = \left\{ \theta \,|\, \theta \in D_M, \inf_{\theta' \in D_I(\omega)} |\theta - \theta'| \geq \varepsilon \right\}$$

Choose $\theta^x \in D_M^x(\varepsilon,\omega) \cap \tilde{D}_M$. (If this set is empty for any $\varepsilon > 0$, the assertion of the theorem is trivially true.)

Since $\theta^x \notin D_I$,

$$\text{tr } B = \frac{1}{N} \sum \beta(t)^T\beta(t) > \delta_2(\theta^x) \qquad \text{for } N > N_4(\omega)$$

According to (2.16) there is an $\varepsilon_0 = \varepsilon_0(\text{tr } B,\delta) = \varepsilon_0(\delta_2(\theta^x),\delta)$. Choose $N > N_0(\theta^x,\omega) = \max(N_4(\omega), N_3(\varepsilon_0/2,\omega,\theta^x), N_2(\omega, p(\delta)\delta_2(\theta^x)/2))$ and choose

$$\rho < \rho^x(\theta^x) = \varepsilon_0/2C_2(\omega,\theta^x)$$

Then

$$V_N(\theta) = h\left(\frac{1}{N} Q_N(S) + \frac{1}{N}\sum_{1}^{N} \beta(t)\beta(t)^T + \frac{1}{N} Q_N^{(1)}(\theta^x)\right.$$

$$\left. + \frac{1}{N} Q_N^{(2)}(\theta,\theta^x)\right) \geq h\left(\frac{1}{N} Q_N(S)\right) + p(\delta)\delta_2(\theta^x)$$

$$= V_N(S) + p(\delta)\delta_2(\theta^x) > V_N(\tilde{\theta}) + p(\delta)\delta_2(\theta^x)/2$$

for $N > N_0(\theta^x,\omega)$ and $\theta \in B(\theta^x,\rho^x(\theta^x))$. Hence

$$\inf_{\theta \in B(\theta^x,\rho^x)} V_N(\theta) \geq V_N(\tilde{\theta}) + p(\delta)\delta_2(\theta^x)/2 \quad \text{for} \quad N > N_0(\theta^x,\omega) \tag{A.8}$$

The family of open sets

$$\left\{B(\theta^x,\rho^x(\theta^x)), \quad \theta^x \in D_M^x(\varepsilon,\omega) \cap \tilde{D}_M\right\}$$

clearly covers the compact set $D_M^x(\varepsilon,\omega)$. According to the Heine Borel theorem there exists a finite set

$$\left\{B(\theta_i,\rho^*(\theta_i)), \quad i = 1,\ldots,K\right\}$$

that covers $D_M^x(\varepsilon,\omega)$. Let

$$N_0(\omega,\varepsilon) = \max_{1\leq i\leq K} N_0(\theta_i,\rho^*(\theta_i),\omega)$$

It then follows from (A.8) that

$$\inf_{\theta \in D_M^x(\varepsilon,\omega)} V_N(\theta) > V_N(\theta^0) \qquad \text{for} \quad N > N_0(\omega,\varepsilon)$$

which means that the minimizing element $\theta(N)$ cannot belong to $D_M^x(\varepsilon,\omega)$ for $N > N_0(\omega,\varepsilon)$, i.e.

$$|\theta(N) - D_I| < \varepsilon \qquad \text{for} \quad N > N_0(\omega,\varepsilon)$$

which, since ε is an arbitrary small number, is the assertion of the theorem.

A.2 *PROOF OF THEOREM* 4.1

The theorem follows from Lemma 4.1 if (4.3) and (4.4) can be shown to be satisfied.

We will consider the general case, and let $\alpha_{M(\theta)}(z)$ can be the least common denominator to the denominator polynomials in $G_{M(\theta)}(z)$ and $H_{M(\theta)}(z)$. Introduce the matrix polynomials

$$C_{M(\theta)}(z) = \alpha_{M(\theta)}(z)\, H_{M(\theta)}(z)$$

$$B_{M(\theta)}(z) = \alpha_{M(\theta)}(z)\, G_{M(\theta)}(z)$$

Due to the assumptions, det $C_{M(\theta)}(z)$ has all zeroes outside the unit circle for $\theta \in D_M$, i.e. $C^{-1}_{M(\theta)}(z)$ is an exponentially stable filter. (There are several pole-zero cancellations between α and det H. Therefore the requirement that α has all zeroes outside the unit circle is sufficient, but not necessary. Stability of C_M^{-1} is the only thing that matters.) From (2.14)

$$C_{M(\theta)}(q^{-1})\hat{y}(t|\theta) = \Big((C_{M(\theta)}(q^{-1}) - \alpha_{M(\theta)}(q^{-1})I)y(t)\Big) + B_{M(\theta)}(q^{-1})u(t-1) \tag{A.9}$$

and, with

$$C'(q^{-1}) = \frac{d}{d\theta} C_{M(\theta)}(q^{-1})$$

etc., and suppressed q^{-1},

$$C_{M(\theta)} \frac{d}{d\theta} \hat{y}(t|\theta) = (C' - \alpha' I)y(t) + B'u(t-1) - C'\hat{y}(t|\theta) \tag{A.10}$$

From (A.9) and (A.10) it follows that

$$\hat{y}(t|\theta) = \sum_{k=0}^{t} \Big(h^{(1)}_{t,k}(\theta)y(t-k) + h^{(2)}_{t,k}(\theta)u(t-k)\Big)$$

and

$$\frac{d}{d\theta}\hat{y}(t|\theta) = \sum_{k=0}^{t}\left(h_{t,k}^{(3)}(\theta)y(t-1k) + h_{t,k}^{(4)}(\theta)u(t-k) + h_{t,k}^{(5)}(\theta)\hat{y}(t-k|\theta)\right)$$

Since $C_{M(\theta)}^{-1}(q^{-1})$ is exponentially stable for all $\theta \in D_M$, and D_M is compact

$$\sup_{\substack{\theta \in D_M \\ 1 \leq i \leq 5}} |h_{t,k}^{(i)}(\theta)| < C_1 \cdot \lambda_1^k \quad ; \quad \lambda_1 < 1 \quad . \qquad \text{(A.11)}$$

and hence

$$\sup_{\theta \in D_M} |\hat{y}(t|\theta)| < C_1 \sum_{k=0}^{t} \lambda_1^k \{|y(t-k)| + |u(t-k)|\} \qquad \text{(A.11)}$$

and

$$\sup_{\theta \in D_M} \left|\frac{d}{d\theta}\hat{y}(t|\theta)\right| < C_1 \sum_{k=0}^{t} (\lambda_1^k + k\lambda_1^k)\{|y(t-k)| + |u(t-k)|\} \qquad \text{(A.12)}$$

Let C and λ ; $\lambda < 1$, be such that

$$C_1(\lambda_1^k + k\lambda_1^k) < C\lambda^k$$

Introduce for brevity the notation

$$\eta(t) = C[|y(t)| + |u(t)|]$$

and define

$$\tilde{z}(t) = \sum_{1}^{t} \lambda^k \eta(t-k)$$

Then $z(t) \leq \tilde{z}(t)$ and

$$\tilde{z}(t+1) = \lambda\tilde{z}(t) + \eta(t) \qquad \tilde{z}(0) = 0$$

or

$$\tilde{z}(t+1)^2 = \lambda^2\tilde{z}(t)^2 + \eta(t)^2 + 2\lambda\tilde{z}(t)\eta(t)$$

Sum over $t = 0,\ldots,N$ and divide by N:

$$\frac{1}{N}\sum_{1}^{N} \tilde{z}(t)^2 \leq \lambda^2 \frac{1}{N}\sum_{1}^{N} \tilde{z}(t)^2 + \frac{1}{N}\sum_{0}^{N} \eta(t)^2 + 2\lambda \frac{1}{N}\sum_{0}^{N} z(t)\eta(t)$$

or

$$(1-\lambda^2)\frac{1}{N}\sum_{1}^{N} \tilde{z}(t)^2 \leq \frac{1}{N}\sum_{0}^{N} \eta(t)^2 + 2\lambda \left[\frac{1}{N}\sum_{0}^{N}\eta(t)^2 \cdot \frac{1}{N}\sum_{0}^{N}\tilde{z}(t)^2\right]^{\frac{1}{2}}$$

According to the assumptions of the lemma,

$$\frac{1}{N}\sum_{0}^{N} \eta(t)^2$$

is bounded w.p.1, from which directly follows that

$$\frac{1}{N}\sum_{1}^{N} \tilde{z}(t)^2$$

is bounded w.p.1. Since $z(t) \leq \tilde{z}(t)$ this concludes the proof of (4.3). Condition (4.4) follows from (A.12) and (4.7), in the same way since

$$\operatorname{tr}\frac{1}{N} Q_N = \frac{1}{N}\sum_{1}^{N} |y(t) - \hat{y}(t|\theta)|^2_{R(t)} \leq \frac{2}{N}\sum_{1}^{N} |y(t)|^2_{R(t)} + \frac{2}{N}\sum_{1}^{N} |\hat{y}(t|\theta)|^2_{R(t)} \quad .$$

Acknowledgements

I would like to thank Professor Karl Johan Åström for important discussions on the subject and Torsten Söderström and Daniel Thorburn for many valuable comments on the manuscript. The support of the Swedish Board for Technical Development under Contract No. 733546 is also gratefully acknowledged.

REFERENCES

Aoki, M., and Yue, P. C. (1970), "On Certain Convergence Questions in System Identification," *SIAM J. Control, Vol. 8, No. 2.*

Åström, K. J., and Bohlin, T. (1965), "Numerical Identification of Linear Dynamic Systems from Normal Operating Records," IFAC (Teddington) Symposium.

Åström, K. J. (1970), *Introduction to Stochastic Control Theory,* Academic Press, New York.

Åström, K. J., and Eykhoff, P. (1971), "System Identification - A Survey," *Automatica, 7,* 123-162.

Åström, K. J., and Källström, C. (1973), "Application of System Identification Techniques to the Determination of Ship Dynamics," Preprints of 3rd IFAC Symposium on Identification and System Parameter Estimation, The Hague, pp. 415-425.

Balakrishnan, A. V. (1968), "Stochastic System Identification Techniques" In: *Stochastic Optimization and Control,* M. F. Karreman, Ed., New York, Wiley, pp. 65-89.

Bellman, R., and Åström, K. J. (1970), "On Structural Identifiability," *Math. Biosc., Vol. 7,* pp. 329-339.

Caines, P. E. (1970), "The Parameter Estimation of State Variable Models of Multivariable Linear Systems," Ph.D. dissertation, Imperial College, London.

Caines, P. E. (1971), "The Parameter Estimation of State Variable Models of Multivariable Linear Systems," Proceedings of the U.K.A.C. Conference on Multivariable Systems, Manchester, England.

Caines, P. E., and Rissanen, J. (1974), "Maximum Likelihood Estimation of Parameters in Multivariable Gaussian Stochastic Processes," *IEEE Trans. IT-20, No. 1.*

Chung, K. L. (1968), *A Course in Probability Theory,* Harcourt, Brace & World, Inc.

Cramér, H. (1946), *Mathematical Methods of Statistics,* Princeton University Press, Princeton.

Cramér, H., and Leadbetter, M. R. (1967), *Stationary and Related Stochastic Processes,* Wiley, New York.

Doob, J. L. (1953), *Stochastic Processes,* Wiley, New York.

Eaton, J. (1967), "Identification for Control Purposes," *IEEE Winter meeting,* New York.

Fisher, R. A. (1912), "On an Absolute Criterion for Fitting Frequency Curves," *Mess. of Math., Vol. 41,* p. 155.

Glover, K., and Willems, J. C. (1973), "On the Identifiability of Linear Dynamic Systems," Preprints of the 3rd IFAC Symposium on Identification and System Parameter Estimation, The Hague, pp. 867-871.

Gustavsson, I., Ljung, L., and Söderström, T. (1974), "Identification of Linear, Multivariable Process Dynamics Using Closed Loop Experiments," Report 7401, Division of Automatic Control Lund Inst. of Techn.

Jazwinski, A. H. (1970), *Stochastic Processes and Filtering Theory,* Academic Press, New York.

Kailath, T. (1970), "The Innovations Approach to Detection and Estimation Theory," *Proc. IEEE, Vol. 58, No. 5.*

Lampard, D. G. (1955), "A New Method of Determining Correlation Functions of Stationary Time Series, *Proc. IEE, 102C,* pp. 35-41

Ljung, L. (1974), "On Consistency for Prediction Error Identification Methods," Report 7405, Division of Automatic Control, Lund Institute of Technology.

Ljung, L. (1975), "On Consistency and Identifiability," Proc. Intern. Symp. Stochastic Systems, Lexington, Kentucky, June 1975 (North-Holland Publishing Co.). To appear in Mathematical Programming Studies.

Ljung, L., Gustavsson, I., and Söderström, T. (1974),"Identification of Linear Multivariable Systems Operating Under Linear Feedback Control," *IEEE Trans., AC-19, No. 6* (Special Issue on System Identification and Time Series Analysis), pp. 836-841.

Lüders, G., and Narendra, K. S. (1974), "Stable Adaptive Schemes for State Estimation and Identification of Linear Systems," *IEEE Trans., AC-19, No. 6* (Special Issue on System Identification and Time Series Analysis), pp. 841-848.

Mayne, D. Q. (1972), "A Canonical Form for Identification of Multivariable Linear Systems," *IEEE Trans., AC-17, No. 5,* pp. 728-729.

Mehra, R. K., and Tyler, J. S. (1973), "Case Studies in Aircraft Parameter Identification," Preprints of 3rd IFAC Symposium on Identification and System Parameter Estimation, The Hague, pp. 117-145.

Mendel, J. M. (1973), *Discrete Techniques of Parameter Estimation: The Equation Error Formulation,* Marcel Dekker, New York.

Rissanen, J. (1974), "Basis of Invariants and Canonical Forms for Linear Dynamic Systems," *Automatica, Vol. 10, No. 2,* pp. 175-182.

Spain, D. S. (1971), "Identification and Modelling of Discrete, Stochastic Linear Systems," Tech. Report No. 6302-10, Stanford University.

Soudack, A. C., Suryanarayanan, K. L., and Rao, S. G. (1971), "A Unified Approach to Discrete-Time System Identification," *Int. J. of Control, Vol. 14, No. 6,* pp. 1009-1029.

Tse, E., and Anton, J. J. (1972), "On the Identifiability of Parameters," *IEEE Trans. AC-17, No. 5.*

Tsypkin, Ya. Z. (1973), *Foundations of the Theory of Learning Systems,* Academic Press, New York.

Wald, A. (1949), "Note on the Consistency of the Maximum Likelihood Estimate," *Ann. Math. Stat., Vol. 20,* pp. 595-601.

Woo, K. T. (1970), "Maximum Likelihood Identification of Noisy Systems," Paper 3.1, 2nd Prague IFAC Symposium on Identification and Process Parameter Estimation.

STABLE IDENTIFICATION SCHEMES

Kumpati S. Narendra
Yale University

1. INTRODUCTION

System Characterization and System Identification play a central role in Systems Theory and the past two decades have witnessed considerable research activity in these areas. System Characterization is concerned with setting up mathematical models for different classes of systems and the aim of System Identification is the determination of the characteristics of a specific model which approximates the given system in some sense. At the present time many formulations of the identification problem exist and numerous techniques have also been developed to estimate system parameters. In this paper, we describe a general identification procedure using a model reference approach developed in recent years [1-6].

For an exhaustive survey of references on the general topic of Model Reference Adaptive Systems (MRAS) the reader is referred to the paper by Landau [7]. We shall deal specifically with Model Reference Adaptive Systems using Lyapunov's direct method. In this approach, a model of the plant to be identified is assumed to be available and both model and plant have the same input u(t). The model parameters are then continuously adjusted as functions of the error between the plant and model output as well as other accessible signals in the system. The basic philosophy of the approach is to determine the adaptive laws for adjusting the model parameters so that the stability of the overall system is assured from the outset.

Since model parameters are continuously adjusted according to the adaptive laws, they also constitute the state variables of the overall adaptive system, which can therefore be represented by a nonlinear vector differential equation. The error equations describing the behavior of the errors between model states and model parameters and their desired values are, however, described by linear non-autonomous differential equations. When these equations are stable the parameter and output errors between plant and model are bounded; asymptotic stability assures that the parameters of the model converge to those of the plant while the output error tends asymptotically to zero. The identification problem can therefore be viewed in this context as the stability problem of a linear non-autonomous system.

The characteristics of the identification procedure described above can be altered by adjusting the parameters in the adaptive laws, usually referred to as adaptive parameters. The scheme described in this paper is found to be structurally stable with respect to these adaptive parameters so that changes in them do not affect the qualitative stability behavior of the overall system. Hence, optimization of the identification scheme can be carried out by adjusting these parameters without affecting stability.

The above comments indicate that the approach used in this paper differs philosophically from many other well-known approaches to the identification problem in that stability of the overall system precedes optimization. While we shall confine our attention in the following sections to adaptive identification, it is worth pointing out that the approach itself is a general one capable of being used in a variety of control situations.

2. THE PROBLEM

A multivariable plant has admissible inputs $u(t)$ {m-vector valued piecewise continuous functions} and corresponding outputs $y_p(t)$ {a p-vector valued function}. A linear system $\{C_p, A_p, B_p\}$ is said to characterize this plant if every pair $\{u(t), y_p(t)\}$ satisfies the equations

$$\begin{aligned} \dot{x}_p &= A_p\, x_p(t) + B_p\, u(t) \\ y_p(t) &= C_p\, x_p(t) \end{aligned} \tag{1}$$

where A_p, B_p and C_p are $(n \times n)$, $(n \times m)$ and $(p \times n)$ constant matrices respectively.

The identification problem may be qualitatively defined as the problem of constructing a suitable model which, when subjected to the same input $u(t)$ as the plant, produces an output $y_m(t)$ which tends towards $y_p(t)$ asymptotically. On the basis of this definition, different levels of identification may be distinguished. In the simplest case, the norm $|e(t)|$ of the output error $e(t) \stackrel{\Delta}{=} y_m(t) - y_p(t)$ may tend to zero for a given input $u(t)$; this is referred to as output identification.

Obviously, output identification for a given input $u(t)$ does not always assure output identification for all allowable inputs to the plant. This leads to the concept of transfer function (or transfer matrix) identification. When the input to the plant is sufficiently rich, (as defined in Section 3), output identification is found to imply transfer function identification.

Situations exist in which the plant transfer function (or matrix) does not uniquely specify the unknown parameters of the plant. Hence a third level, referred to as parameter identification, may be defined wherein a specific parameterization used to identify the plant leads to a unique set of parameters.

The above definitions are obviously of both theoretical and practical interest. For example, in on-line control situations, where the plant is continuously identified and controlled, an important question is to decide as to the type of identification that may be needed. In this paper we are primarily interested in transfer function identification. The other two types also arise naturally in the following sections and, in some cases, transfer function identification is found to be equivalent to parameter identification.

The problem as we have stated above assumes that (i) the plant parameters are time-invariant, (ii) the measurements are noise free, (iii) the order 'n' of the plant dynamics is known and (iv) adequate freedom exists in the model to reproduce the observed plant behavior. The results obtained in the paper can be directly extended to plants with time-varying parameters when bounds on the parameters and their time-derivatives are known. The case where observation noise is present is considerably more complex since the noise appears multiplicatively. Little theoretical work has been devoted to analyzing the stochastic stability of the overall system and in view of the practical importance of this problem it forms a potential area for future research.

To provide insight into the basic ideas and the various steps in the generation of the adaptive laws we shall first consider the case where the plant and model are described by (i) algebraic equations and (ii) scalar (first-order) differential equations. The questions of stability that arise in these two cases are seen to be closely related to those arising in more complex situations; modifications of the basic scheme to obtain improved performance as described in case (ii) also carry over directly to the vector case.

3. IDENTIFICATION OF SIMPLE SYSTEMS

A. *ALGEBRAIC EQUATIONS*

The input vector $v(t)$ and the output vector $y_p(t)$ of a plant are related by the algebraic equation

$$y_p(t) = K_p\, v(t) \tag{2}$$

where K_p is an $(m \times r)$ matrix of unknown constants. To identify the elements of the matrix K_p a model is set up which is described by

$$y_m(t) = K_m(t)\, v(t)$$

where $y_m(t)$ is the output of the model and the $(m \times r)$ time-varying matrix $K_m(t)$ can be adjusted using measurements of the vectors $y_p(t)$, $y_m(t)$ and $u(t)$. Defining the output error vector as $y_m(t) - y_p(t) \triangleq e(t)$ and the parameter error matrix as $K_m(t) - K_p \triangleq \Phi(t)$ we have

$$e(t) = \Phi(t)\, v(t) \tag{3}$$

While the elements of the parameter-error matrix $\Phi(t)$ are not known, $\dot{K}_m(t)$ and hence $\dot{\Phi}(t)$ can be adjusted. The aim of the adaptive procedure is then to adjust $\dot{\Phi}(t)$ using available input-output information so that $\Phi(t) \to 0$ and $e(t) \to 0$ as $t \to \infty$.

To determine the law for updating $\dot{\Phi}(t)$ we consider the positive definite function $V[\Phi] = 1/2 \text{ trace } [\Phi^T\Phi]$. To make the time derivative of $V[\Phi]$ negative semi-definite we choose, (Fig. 1),

$$\dot{\Phi}(t) = -\, e(t)\, v^T(t) \tag{4}$$

$$= -\, \Phi(t)\, v(t)\, v^T(t) \tag{5}$$

The updating law given by equation (4) ensures that

$$\frac{d}{dt}[V(\Phi)] = \text{trace } \dot{\Phi}^T\Phi = -v^T(t)\Phi^T\Phi v = -e^T(t)e(t) \leq 0$$

so that $V[\Phi]$ is a non-increasing function with time.

The identification problem, in this case, has been reduced to the problem of stability of the linear matrix differential equation (5) involving the symmetric semi-definite matrix $v(t)v^T(t)$. The asymptotic and uniform-asymptotic stability of such differential equations have been studied in detail by Morgan and Narendra [9].

If $v(t)$ consists of piecewise continuous bounded functions of time, then $\Phi(t)$ is bounded and $e(t) \to 0$ as $t \to \infty$. Hence the adaptive law (4) assures output identification. The fact that $e(t)$ tends to zero with time can also be used to show that the matrix $\Phi(t)$ tends to some constant matrix Φ_∞. If, however, the input vector $v(t)$ is such that real numbers $a > 0$ and b exist for which the inequality

$$\int_{t_0}^{t} \| v(\tau)\ v^T(\tau) w \|\ d\tau \geq a(t-t_0) + b \tag{6}$$

is satisfied for all unit vectors w, it is shown [9] that the linear non-autonomous system defined by equation (5) is uniformly asymptotically stable. The condition given by (6), (which is also found to be necessary), is a richness condition on the inputs and assures the asymptotic convergence of the parameter errors to zero.

The identification procedure discussed so far is seen to consist of three stages:

i) Determination of the error equations and the structure of the identifier.

ii) Choice of a suitable candidate for a Lyapunov function and the generation of an adaptive law which assures output identification.

iii) Determination of a richness condition on the input which assures the convergence of the parameter errors to zero.

In the following sections, we shall follow the same approach for the identification of more complex systems.

B. MODELS AND PLANTS DESCRIBED BY SCALAR DIFFERENTIAL EQUATIONS

We now proceed to consider the case where the plant is described by the differential equation

$$\dot{x}_p = a_p x_p + b_p u \qquad a_p < 0 \tag{7}$$

where a_p and b_p are unknown constants and $u(t)$ is the input to the system and $x_p(t)$ the corresponding output. To estimate the parameters of the plant from the input-output data we use a model described by the equation

$$\dot{x}_m = c\, x_m + (a_m(t) - c)\, x_p(t) + b_m(t)\, u \tag{8}$$

The adjustable parameters of the model are $a_m(t)$ and $b_m(t)$ and the aim of the identification procedure in this case is to adjust them in such a manner that $e(t) = x_m(t) - x_p(t) \to 0$, $a_m(t) \to a_p$ and $b_m(t) \to b_p$ as $t \to \infty$.

The principal reason for choosing the model to have the form given by equation (8) becomes evident on examining the error equation

$$\dot{e} = c\, e + \phi\, x_p + \psi\, u \tag{9}$$

where $\phi = a_m(t) - a_p$ and $\psi = b_m(t) - b_p$ are the parameter errors. Equation (9) represents the output error in terms of the parameter errors and the input and output signals of the plant. If c is a negative constant and the adaptive laws result in $\phi(t)$ and $\psi(t)$ tending to zero asymptotically, the output error $e(t)$ would also tend to zero, provided $u(t)$ and $x_p(t)$ are uniformly bounded. As in the previous case we note that while $\phi(t)$ and $\psi(t)$ are not known, $\dot{\phi}(t)\ (= \dot{a}_m(t))$ and $\dot{\psi}(t)\ (= \dot{b}_m(t))$

can be adjusted by the designer. The object of the identification scheme in this case is to adjust $\dot{a}_m(t)$ and $\dot{b}_m(t)$ using all available measurements in such a manner that $e(t)$, $\phi(t)$ and $\psi(t)$ tend to zero asymptotically. If $\dot{a}_m(t) = g_1[x_m,x_p,u]$ and $\dot{b}_m(t) = g_2[x_m,x_p,u]$, we are interested in determining the functions g_1 and g_2 such that

$$\begin{aligned}\dot{e} &= ce + \phi x_p + \psi u \\ \dot{\phi} &= g_1[x_m,x_p,u] \\ \dot{\psi} &= g_2[x_m,x_p,u]\end{aligned} \tag{10}$$

is asymptotically stable. These functions are then the adaptive laws which update the parameters $a_m(t)$ and $b_m(t)$.

To determine the stable adaptive schemes described above, we choose as a candidate for a Lyapunov function in the error space:

$$V(e,\phi,\psi) = \frac{1}{2}\,[e^2 + \phi^2 + \psi^2] \tag{11}$$

The time derivative of V is

$$\begin{aligned}\frac{dV}{dt} = \dot{V} &= ce^2 + \phi\, ex_p + \psi\, eu + \phi\dot{\phi} + \psi\dot{\psi} \\ &= ce^2 + \phi\,(\dot{\phi} + ex_p) + \psi(\dot{\psi} + eu)\end{aligned} \tag{12}$$

Since the parameters ϕ and ψ are unknown and cannot be measured, the only manner in which $\dot{V}$ can be made negative semi-definite is to choose $c < 0$ and

$$\begin{aligned}\dot{\phi} &= -\, ex_p \\ \dot{\psi} &= -\, eu\end{aligned} \tag{13}$$

Equations (13) represent the adaptive updating laws. Since V is a Lyapunov function for the system of equations (10), it is obvious that the system is stable and that ϕ and ψ are bounded while $e(t) \to 0$, provided the input $u(t)$ is uniformly bounded and the plant is asymptotically stable. For identification of the plant parameters, however, we require asymptotic stability of the overall system. Combining equations (10) and (13) we have

$$\begin{bmatrix} \dot{e} \\ \dot{\phi} \\ \dot{\psi} \end{bmatrix} = \begin{bmatrix} c & x_p & u \\ -x_p & 0 & 0 \\ -u & 0 & 0 \end{bmatrix} \begin{bmatrix} e \\ \phi \\ \psi \end{bmatrix} \tag{14}$$

which is a linear non-autonomous differential equation. The stability properties of general equations of this form have been considered by Morgan and Narendra [10] and are discussed in greater detail in the next section. In the present context, it is found that these specialize to

$$\int_{t_0}^{t} \| u^2(\tau) \| \, d\tau \geq a(t-t_0) + b \quad , \tag{15}$$

as a necessary and sufficient condition for uniform asymptotic stability. As a specific example, a periodic input containing a single frequency is found to result in uniform asymptotic stability of equation (14) and hence in the identification of the parameters a_p and b_p of the plant.

COMMENTS

From equations (13) it is seen that $\dot{\phi}(= \dot{a}_m)$ and $\dot{\psi}(= \dot{b}_m)$ are adjusted using $e(t)\, x_p(t)$ and $e(t)\, u(t)$, respectively. Several modifications of these adaptive laws are possible even while maintaining the stability of the overall identification scheme.

a. Choose $V(e,\phi,\psi)$ to have the form

$$V = \frac{1}{2}\left[e^2 + (\phi,\psi)\; R \begin{pmatrix} \phi \\ \psi \end{pmatrix} \right]$$

where R is any positive definite (2×2) matrix. In this case we obtain the following adaptive laws

$$\begin{bmatrix} \dot{\phi} \\ \dot{\psi} \end{bmatrix} = - R^{-1} \begin{bmatrix} ex_p \\ eu \end{bmatrix} \tag{16}$$

For the particular case where R^{-1} is diagonal with elements λ_1 and λ_2 we have $\dot{\phi} = -\lambda_1 ex_p$, $\dot{\psi} = -\lambda_2 eu$.

b. For the procedure outlined above to be applicable, the Lyapunov function must be chosen to be a quadratic function in ϕ and ψ. However, it need not be quadratic in the error e. If $F(e)$ is any positive function of e whose derivative $f(e)$ lies in the first and third quadrants of the $(f(e),e)$ plane then

$$V(e,\phi,\psi) = F(e) + (\text{quadratic function of } \phi \text{ and } \psi) \tag{17}$$

may be chosen as a Lyapunov function. For example, if $F(e) = e^4$, the stable adaptive laws would have the form

$$\begin{aligned} \dot{\phi} &= -e^3 x_p \\ \dot{\psi} &= -e^3 u \end{aligned} \tag{18}$$

c. All the adaptive laws used so far are seen to be first order equations; the product of the error $e(t)$ and the measured signals $x_p(t)$ (or $u(t)$) is integrated to obtain $a_m(t)$ (or $b_m(t)$). For practical reasons, particularly when observation noise is present it may be desirable to use more general adaptive laws. For example, one of the adaptive laws in (13) may be described by a second order differential equation as shown below

$$\begin{aligned} \dot{e} &= ce + \phi x_p + \psi u \\ \dot{\phi} &= -ex_p \\ \dot{\psi} + \alpha\dot{\psi} &= -\frac{d}{dt}[eu] - \beta(eu) \qquad 0 < \beta \leq \alpha \end{aligned} \tag{19}$$

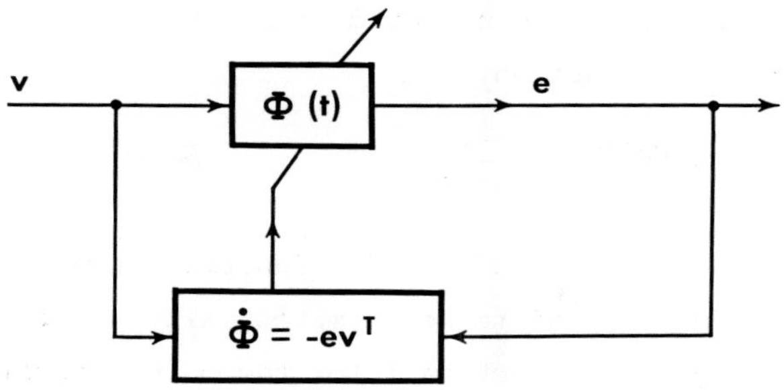

FIG. 1 ADJUSTMENT OF GAIN MATRIX

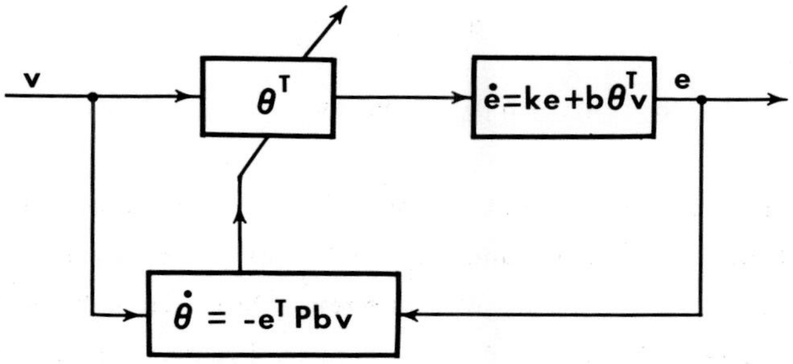

FIG. 2 GAIN VECTOR FOLLOWED BY STABLE SYSTEM
(ALL STATE VARIABLES ACCESSIBLE)

In this case $\psi(t)$ may be considered to be the output of a filter with transfer function

$$G_1(s) = \frac{s + \beta}{s(s + \alpha)} \tag{20}$$

and input $e(t)\,u(t)$. If the transfer function is positive real, the overall system is found to be asymptotically stable. The proof of this is given in Section 4 for the more general case of a system described by a vector differential equation.

d. The plant parameters a_p and b_p were assumed to be constant in the analysis carried out so far. If the plant parameters vary with time, it has been shown by Narendra and Tripathi [11] that the model parameters would track the plant parameters provided the adaptive laws are modified to have the form

$$\begin{aligned} \dot{a}_m &= -\gamma\, a_m - \lambda_1\, e x_p \\ \dot{b}_m &= -\delta\, b_m - \lambda_2\, e u \end{aligned} \tag{21}$$

where γ and δ are positive constants whose values depend on the rate at which $\dot{a}_p(t)$ and $\dot{b}_p(t)/b_p(t)$ vary with time. For non-zero values of γ and δ, however, there is a bias in the estimated values of a_p and b_p, respectively. While increasing γ and δ increases the speed of identification, this is achieved only at the expense of greater error in the final estimated values of a_p and b_p; the usual compromise between speed and accuracy is consequently called for.

4. IDENTIFICATION OF MULTIVARIABLE SYSTEMS (All States Accessible)

a. The results of the previous section can be directly extended to the case of a system with m inputs and n states when all the state variables can be measured.

Let the plant be described by the vector differential equation

$$\dot{x}_p = A_p x_p + B_p u$$

where A_p and B_p are $(n \times n)$ and $(n \times m)$ constant matrices with unknown elements. To estimate the values of these elements a model is set up which is described by

$$\dot{x}_m = C x_m + [A_m(t) - C]x_p + B_m(t) u$$

where C is a stable matrix and $A_m(t)$ and $B_m(t)$ are adjustable parameters. The objective of the identification procedure is then to determine adaptive laws to dynamically adjust these parameters so that

$$\lim_{t\to\infty} A_m(t) = A_p; \quad \lim_{t\to\infty} B_m(t) = B_p$$

and

$$\lim_{t\to\infty} [x_m(t) - x_p(t)] = \lim_{t\to\infty} e(t) = 0 .$$

The state error vector $e(t) \triangleq x_m(t) - x_p(t)$ satisfies the differential equation

$$\dot{e} = Ce + \Phi x_p + \Psi u \tag{22}$$

where the parameter error matrices are defined by

$$\Phi \triangleq [A_m - A_p] \quad \text{and} \quad \Psi \triangleq [B_m - B_p] . \tag{23}$$

As in the scalar case, we set up a Lyapunov function candidate $V[e,\Phi,\Psi]$ which has the form

$$V = e^T P e + \text{trace } [\Phi^T \Gamma_1 \Phi + \Psi^T \Gamma_2 \Psi]$$

where P is a positive definite matrix satisfying

$$C^T P + PC = -Q; \qquad Q = Q^T > 0$$

and Γ_1 and Γ_2 are symmetric positive definite matrices. The time derivative of V may be expressed as

$$\dot{V} = -e^TQe + 2e^TP\Phi x_p + 2e^TP\Phi u + 2 \text{ trace } [\Phi^T\Gamma_1\dot{\Phi} + \Psi^T\Gamma_2\dot{\Phi}] .$$

Stability is consequently assured by the adaptive laws

$$\dot{A}_m(t) = \dot{\Phi} = -\Gamma_1^{-1} P e x_p^T \tag{24}$$

$$\dot{B}_m(t) = \dot{\Psi} = -\Gamma_2^{-1} P e u^T \tag{25}$$

since they assure that $\dot{V} = -e^TQe \leq 0$ along a trajectory.

b. *Asymptotic Stability*

The existence of the quadratic Lyapunov function $V(e,\Phi,\psi)$ assures only the stability of the set of equations

$$\begin{aligned} \dot{e} &= Ce + \Phi x_p + \Psi u \\ \dot{\Phi} &= -\Gamma_1^{-1} Pex_p^T \\ \dot{\Psi} &= -\Gamma_2^{-1} Peu^T \end{aligned} \tag{26}$$

The parameter errors are bounded and since $\dot{V} = -e^TQe \leq 0$, we have output identification. From the point of view of transfer matrix identification the question of asymptotic stability of (26) is important.

As in the simpler cases considered in Section 2, it is found that the input has to be sufficiently rich to force the parameters to converge to the desired values. The class of inputs which would result in the asymptotic stability of systems equivalent to (26) has been investigated by Yuan and Wonham [12], Anderson [8], and Morgan and Narendra [10].

In [10] the asymptotic and uniform asymptotic stability of equations of the form

$$\begin{bmatrix} \dot{x} \\ \dot{y} \end{bmatrix} = \left[\begin{array}{c|c} A(t) & -B^T(t) \\ \hline B(t) & 0 \end{array}\right] \begin{bmatrix} x \\ y \end{bmatrix} \tag{27}$$

is discussed in detail. Necessary and sufficient conditions for uniform asymptotic stability are stated in terms of the time-varying matrix $B(t)$ as follows:

Equation (27) is uniformly asymptotically stable if and only if there are positive numbers T_0, ε_0 and δ_0 such that given $t_1 \geq 0$ and a unit vector $w \in R^n$, there is a t_2 $[t_1, t_1 + T_0]$ such that

$$\left| \int_{t_2}^{t_2+\delta_0} B(\tau)^T w \; d\tau \right| \geq \varepsilon_0 \tag{28}$$

This condition states that for any unit direction w, $B(t)^T w$ is periodically large. It requires that there be some fixed period T_0 such that $B(t)$ is "exciting in all directions" as t takes on values in any interval of length T_0.

Initial attempts to solve the problem of asymptotic stability centered around periodic inputs $u(t)$ into the system and the conditions were expressed in terms of the frequency content of the inputs. In particular, for an n-th order system it was shown in [22] that $u(t)$ should contain n distinct frequencies which are non-commensurate. This result can be shown to be a special case of condition (28).

c. *Comments*

i. The above identification procedure assumes that the entire state vector x_p of the plant can be measured. Hence both x_p and e used in the adaptive laws are available for practical implementation.

ii. The stability of the overall adaptive system reduces to the stability problem of a differential equation of the form

$$\dot{e} = Ke + \theta v \qquad \theta = [\Phi,\Psi], \qquad v = \begin{bmatrix} x_p \\ u \end{bmatrix}$$

where K is a stable matrix, v is a vector of specified signals (which are the inputs and outputs of the plant) and θ is a matrix whose time-derivative can be adjusted. Fig. 2 indicates this second prototype where adaptive laws can be generated using available signals.

iii. The many modifications suggested for the adaptive laws in the previous section also carry over to the vector case. In particular, general positive real transfer functions (rather than integration) can be used in the updating laws for the elements of the matrices $A_m(t)$ and $B_m(t)$.

Let the error equation have the form

$$\dot{e} = K_1 e + d_1 \eta_1 v_1 \tag{29}$$

where η_1 is the parameter to be adjusted and v_1 is a known bounded function of time. We have shown thus far that η_1 can be adjusted according to the rule $\dot{\eta}_1 = -(e^T P_1 d_1)v_1$, where P_1 is a positive definite matrix satisfying $K_1^T P_1 + P_1 K_1 = -Q$ and $Q = Q^T > 0$. We now consider the case where η_1 is the output of a dynamical system described by

$$\begin{aligned} \dot{\eta} &= K_2 \eta + d_2 r(t) \\ \eta_1 &= h_2^T \eta \end{aligned} \tag{30}$$

where $h_2^T(sI - K_2)^{-1} d_2$ is positive real. In such a case, by choosing a Lyapunov function candidate of the form $e^T P_1 e + \eta^T P_2 \eta$ it can be shown that the overall system described by equations (29) and (30) is also stable, provided

$$r(t) = -(e^T P_1 d_1) v_1(t) \tag{31}$$

The matrix P_2 in the Lyapunov function satisfies the following equations simultaneously.

$$P_2^T K_2 = K_2 P_2 = -qq^T \tag{32}$$

$$P_2 d_2 = h_2$$

The existence of such a matrix P_2 is assured by the Kalman-Yakubovich Lemma [32] since the transfer function $h_2^T(sI - K_2)^{-1}d_2$ is positive real.

d. *Identification of Discrete Systems*

Since results that hold for linear continuous systems carry over readily to the discrete case, it might be expected that such an extension would be relatively straightforward in the identification problem. This is, however, found to be not the case since the adaptive equations are nonlinear and integration of the differential equations over a discrete time interval does not directly yield the equations of the discrete system. An independent study of discrete systems however reveals that almost all the results derived for the continuous case carry over to the discrete case as well. We shall merely indicate here the adaptive laws for the two principal cases of identification when all the state variables of the plant are accessible.

A linear time-invariant discrete system is described by the vector difference equation

$$x(k+1) = Ax(k) + Bu(k) \tag{33}$$

where the elements of the $(n \times n)$ matrix A and the $(n \times m)$ matrix B are unknown but constant. The following two models have been suggested by Kudva and Narendra [13], [14], for the identification of the parameters of the system.

$$\hat{x}(k+1) = \hat{A}(k+1)\ x(k) + \hat{B}(k+1)u(k) \quad \text{(Model I)}$$

$$\hat{x}(k+1) = Cx(k) + [\hat{A}(k+1) - C]x(k) + \hat{B}(k+1)u(k) \quad \text{(Model II)} \tag{34}$$

In both cases $\hat{A}$ and $\hat{B}$ represent the estimates of A and B respectively and in equation (34) C is any constant matrix having all its eigenvalues within the unit circle. The adaptive laws for the two cases are given by

$$[\hat{A}(k+1) \mid \hat{B}(k+1)] = [\hat{A}(k) \mid \hat{B}(k)] - \frac{\alpha P e(k) q^T(k-1)}{\lambda_{max} q^T(k-1) q(k-1)} \quad (35)$$

and

$$[\hat{A}(k+1) \mid \hat{B}(k+1)] = [\hat{A}(k) \mid \hat{B}(k)] - \frac{\alpha P [e(k) - Ce(k-1) q^T(k-1)}{\lambda_{max} q^T(k-1) q(k-1)}$$

$$0 < \alpha < 2 \quad (36)$$

where P is a symmetric positive definite matrix, λ_{max} is the largest eigenvalue of P and $q^T(k) = [x_p^T(k) \; u^T(k)]$.

The Lyapunov functions $V[e(k),\Phi(k)]$ which yield the adaptive laws (33) and (34) for the two identification schemes in (32) are given by

$$V[e(k),\Phi(k)] = e^T(k)e(k) + \text{trace } \Phi^T(k)\Phi(k) \quad \text{for Model I}$$

and

$$V[e(k),\Phi(k)] = \beta z^T(k) R z(k) + \text{tr}[\Phi^T(k)\Phi(k)] \quad \text{for Model II}$$

respectively, where the parameter error matrix $\Phi(k)$ and the vector $z(k)$ are defined as

$$\Phi(k) \triangleq [\hat{A}(k) - A \mid \hat{B}(k) - B]$$

$$z(k) = e(k) - \Phi(k) q(k-1) \quad .$$

For the derivation of the adaptive laws the reader is referred to [13] and [14]. For further details regarding discrete identification schemes the reader is also referred to papers by Mendel [15, 16] and Udink Ten Cate [17].

EXAMPLES: Extensive simulations have been carried out on the digital computer to establish the effectiveness of the schemes proposed thus far. We present here very briefly two typical examples; the first example is concerned with the identification of a second order system with six unknown parameters and the second example with a fourth order system with three unknown parameters.

EXAMPLE 1: A second order system can be described by the matrices

$$A_p = \begin{bmatrix} 4 & 6 \\ -7 & -9 \end{bmatrix} \qquad B_p = \begin{bmatrix} 3 \\ 6 \end{bmatrix}$$

The matrix C in the model is chosen to be $-10I$ (where I is the unit matrix). The gain matrices Γ_1, Γ_2 and the matrix P in equations (24) and (25) are

$$\Gamma_1 = \begin{bmatrix} 30 & 3.8 \\ 3.8 & 15 \end{bmatrix} ; \quad \Gamma_2 = \begin{bmatrix} 2 & 1.5 \\ 1.5 & 3 \end{bmatrix} ; \quad P = 10I$$

A square wave input of amplitude 5 and frequency of 18 rads/sec was used for the identification procedure. The convergence of the six parameters of the model is shown in Fig. 3.

EXAMPLE 2: In this example a fourth order plant with a single input has to be identified.

$$A_p = \begin{bmatrix} -3 & 0 & 0 & 0 \\ 4 & -4 & -9 & -4 \\ -.5 & 0 & -2 & 0 \\ 5 & 8 & -11 & -8 \end{bmatrix} \qquad B_p = \begin{bmatrix} 3 \\ 1 \\ 2 \\ 4 \end{bmatrix}$$

The elements a_{21}, a_{33} and b_1 are assumed to be unknown. Using $C = -10I$ $P = 15I$, the identification procedure was carried out with the same input used in the previous example.

The parameters are seen to converge (Fig. 3b) in approximately 2 seconds.

5. THE ADAPTIVE OBSERVER

In the schemes described in Section 4, it was assumed that all the state variables of the plant are accessible for use in the adaptive laws. In many practical situations, however, all the state variables cannot be measured and, hence, may have to be generated using input-output data. The state estimation problem in systems theory is normally solved by the use of Luenberger observers or Kalman filters using models of the plant. In the identification scheme that we are considering, such a model of the plant is not available since the parameters of the plant are unknown. This leads to the concept of an adaptive observer in which the parameters as well as the states of the plant have to be estimated simultaneously.

During the period 1973-74, the adaptive observer problem for a single-input single-output dynamical system was completely resolved. Many different versions of the adaptive observer appeared in the control literature starting with the first solution given by Carroll and Lindorff [18]. Luders and Narendra [20, 21] and Kudva and Narendra [19] proposed alternate schemes and in [23] it was shown that all these results could be derived in a unified manner. During the period following the publication of [23] extensions to multivariable systems were discussed by Anderson [8] and Morse [24].

In this section, we shall follow closely the approach used in [23]. This generalized approach renders the stabilization procedure of the identification scheme transparent and thereby enables the designer to choose a convenient structure for the adaptive observer.

A. *MATHEMATICAL PRELIMINARIES*

We first state a theorem and a proposition that are crucial to the development of all the identification schemes treated

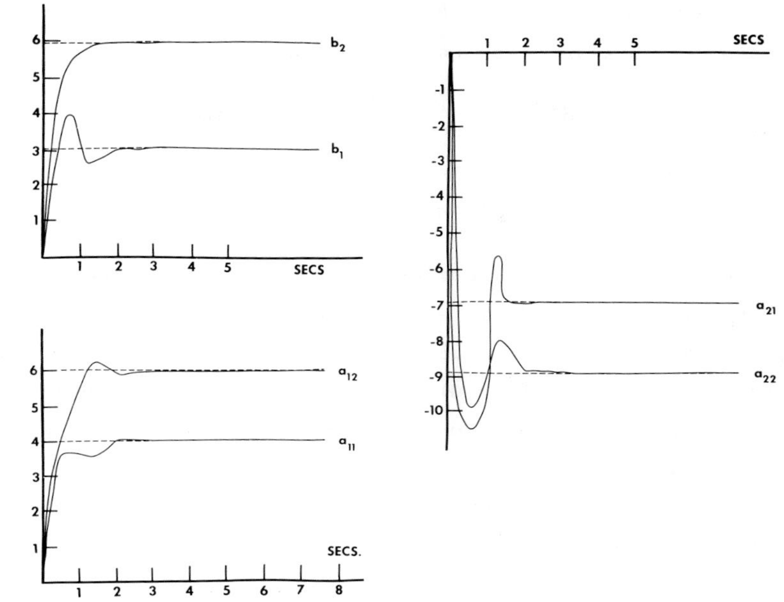

FIG. 3a IDENTIFICATION OF SECOND ORDER PLANT

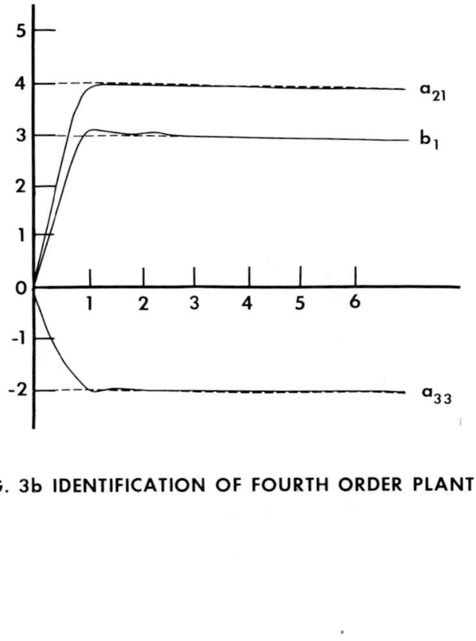

FIG. 3b IDENTIFICATION OF FOURTH ORDER PLANT

later in this section. The theorem deals with a third prototype in which adaptive laws can be determined using available signals to adjust the parameters of a given system. Together with the two other prototypes discussed in Sections 3 and 4 it constitutes the principal tool for generating adaptive identification schemes using the stability approach.

THEOREM: *Given a dynamical system (Fig. 4) represented by the completely controllable and completely observable triple* $\{h^T, K, d\}$ *where* K *is a stable matrix,* Γ *is a positive definite matrix* $(\Gamma = \Gamma^T > 0)$ *and a vector* $v(t)$ *of bounded piecewise continuous functions of time, then the system of differential equations*

$$\dot{\varepsilon} = K\varepsilon + d\phi^T v \qquad \varepsilon_1 = h^T\varepsilon \tag{37}$$

$$\dot{\phi} = -\Gamma\varepsilon_1 v \tag{38}$$

is stable provided that $H(s) \triangleq h^T(sI - K)^{-1}d$ *is strictly positive real. If, furthermore, the components of* $v(t)$ *are sufficiently rich (ref. Section 4b) the system of equations* (37)-(38) *is uniformly asymptotically stable.*

Proof. Considering the quadratic function

$$V = \frac{1}{2}[\varepsilon^T P\varepsilon + \phi^T\Gamma^{-1}\phi]$$

as a Lyapunov function, the time derivative can be expressed as

$$\dot{V} = \frac{1}{2}\varepsilon^T[K^T P + PK]\varepsilon + \varepsilon^T(Pd - h)\phi^T v .$$

$\dot{V}$ can be made negative semi-definite by choosing P such that

$$\begin{aligned} K^T P + PK &= -gg^T - \mu L \\ Pd &= h \end{aligned} \tag{39}$$

are simultaneously satisfied, where g is a vector, μ is a positive constant and L is a positive definite matrix. By the

Lefschetz version of the Kalman-Yakubovich Lemma [32], since $h^T(sI - K)^{-1}d$ is strictly positive real, such a matrix P exists. Hence $\dot{V}$ is negative semidefinite and the system is stable.

In the above theorem, the input vector $v(t)$ was assumed to be an arbitrary bounded piecewise continuous function. Hence, the theorem also holds if $v(t) = G(p)z(t)$, where $z(t)$ is a bounded function of time, p is the differential operator and $G(s)$ is a vector of stable transfer functions.

In the first prototype discussed in Section 3, the adaptive laws were derived to adjust a set of gains from input-output data. In the second prototype in Section 4, this was extended to the case where an unknown set of parameters precedes a dynamical system all of whose state variables are accessible. The above theorem generalizes this result to the case where only some of the state variables of the system can be measured; this calls for the more stringent requirement that the transfer function (or matrix) following the unknown parameters be positive real. The primary aim while developing identification schemes is to choose the identifier structure in such a manner that the error equations correspond to one of these three prototypes.

An error equation which arises often in the identification of a linear plant has the form

$$\dot{e} = Ke + \phi z \qquad e_1 = h^T e \tag{40}$$

where $z(t)$ is a bounded piecewise continuous function of time, e and ϕ are n-vectors representing state and parameter errors and e_1 is the output error between plant and model. The prototype 2 (discussed in Section 4c) gives the adaptive laws for updating $\dot{\phi}(t)$ using $z(t)$ and $e(t)$. However, these laws cannot be directly extended to the system (40) where only the output $e_1(t)$ can be measured. The following proposition which shows the output equivalence of two systems, one of which has the form of prototype 3, provides a convenient way out of the difficulty

PROPOSITION: *Given a bounded piecewise continuous function of time* z(t), *there exist vector signals* v(t) *and* w(t) *with* $v(t) = G(p)z(t)$ *and* $w = w(\dot{\phi}, v)$ *such that the following systems*

$$\dot{e} = Ke + \phi z + w \qquad e_1 = h^T e \tag{41}$$

$$\dot{\varepsilon} = K\varepsilon + d\phi^T v \qquad \varepsilon_1 = h^T \varepsilon_1 \tag{42}$$

have the same outputs (i.e. $e_1(t) \equiv \varepsilon_1(t)$*), provided the pair* (h^T, K) *is completely observable. The schematic representation of the two equivalent systems is shown in Figure* 5.

For a proof of this proposition, the reader is referred to [23]. We shall merely indicate here the manner in which the proposition finds application in the identification problem.

Equation (42) describes the error equation which is in the form of prototype 3. If $h^T(sI - K)^{-1}d$ is strictly positive real, an updating law $\dot{\phi}(t) = -\Gamma\varepsilon_1(t)v(t)$ assures the stability of the equation; furthermore; a sufficiently rich v(t) guarantees its asymptotic stability. By the proposition, since the two equations (41) and (42) are output equivalent, the same adaptive law can also be used in equation (41). The vector function v(t) is obtained using the specified z(t) and a suitable choice of the transfer function G(s); the vector w(t) is generated using the vectors ϕ and v. By adding the vector w(t) only to the input of the observer the error equations can be modified to have the form (41), as shown in the following section.

The practical realization of the updating laws involves the following steps:

i. Using z(t) as an input to generate v(t) as the state of a dynamical system.

ii. Updating $\phi(t)$ by the law $\dot{\phi}(t) = -\Gamma e_1(t)v(t)$.

iii. Generation of w(t) from $e_1(t)$ and v(t).

iv. Addition of w(t) to the input of the model.

In step (i) if $d^T = [1, d_2, d_3, \ldots, d_n]$ then v(t) may be obtained by the relation

$$v_i = \left[\frac{p^{n-i}}{p^{n-1} + d_2 p^{n-2} + \ldots + d_n}\right] z(t) \qquad i = 1,2,\ldots n \ . \tag{43}$$

$w(t)$ in step (iii) is obtained as

$$w^T = -\ e_1 [0, v^T A_2 v,\ v^T A_3 v, \ldots, v^T A_n v]\Gamma \tag{44}$$

where the matrices A_1 are suitably chosen [23]. If the overall system is asymptotically stable, $e_1(t) \to 0$ and $w(t) \to 0$. Hence, $w(t)$ represents an auxiliary stabilizing signal used in the adaptive procedure, whose effect tends to zero asymptotically.

B. *SINGLE INPUT - SINGLE OUTPUT PLANT*

Consider the linear time-invariant system described by

$$\dot{x} = Ax + bu \qquad y = h^T x \tag{45}$$

where x is an n-vector, u is a scalar input, y is a scalar output and the triple $\{h^T, A, b\}$ is completely controllable and observable. The problem then is to identify the triple $\{h^T, A, b\}$ or its equivalent and estimate the corresponding state vector from input-output measurements.

At the present time a number of models of adaptive observers exist for the above problem. We refer the reader to [23] for details regarding some of these models and merely concentrate on two approaches. In each case we will be interested in

i. a suitable representation for plant and observer

ii. derivation of the error equations and

iii. choice of the adaptive laws for adjusting model parameters to make the overall procedure asymptotically stable.

Approach 1 (Minimal Realization)

In this approach a minimal realization of the plant is adopted. For example, let the plant equations be

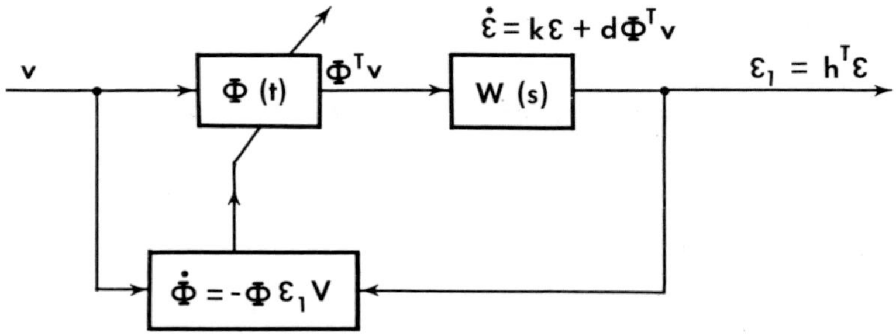

FIG. 4 GAIN VECTOR FOLLOWED BY POSITIVE REAL TRANSFER FUNCTION (SCALAR CASE)

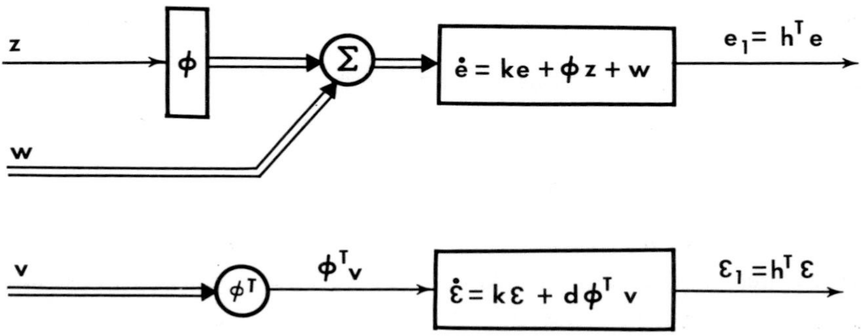

FIG. 5 EQUIVALENT SYSTEM REPRESENTATIONS

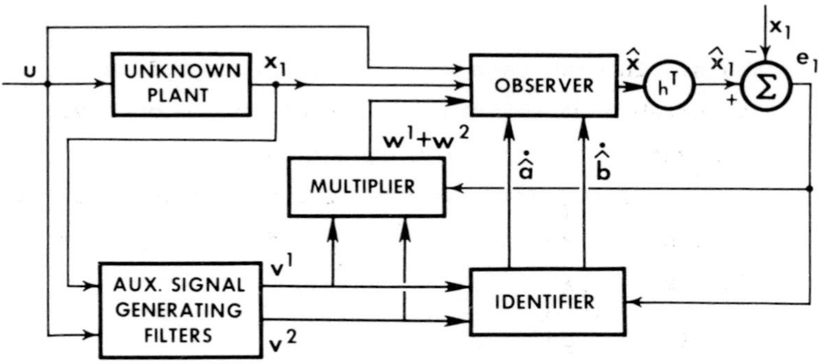

FIG. 6 THE ADAPTIVE OBSERVER

$$\dot{x} = [-a|\bar{A}]x + bu \qquad y = h^T x = x_1 \tag{46}$$

where a and b are the unknown parameter vectors to be estimated. $h^T = [1,0,0,\ldots,0]$ and $\bar{A}$ is a known $[n \times (n-1)]$ matrix. Equation (46) can be rewritten as

$$\dot{x} = Kx + (k-a)x_1 + bu \qquad y = x_1 \tag{47}$$

where

$$K = [-k|\bar{A}] \tag{48}$$

is a stable matrix.

The adaptive observer is designed to have the following structure

$$\begin{aligned} \dot{\hat{x}} &= K\hat{x} + (k-\hat{a})x_1 + \hat{b}u + w^1 + w^2 \\ \hat{y} &= h^T\hat{x} = \hat{x}_1 \end{aligned} \tag{49}$$

$\hat{a}(t)$ and $\hat{b}(t)$ are the estimates of a and b and can be continuously updated; w^1 and w^2 are the auxiliary signals discussed in the previous section which are needed to realize a stable adaptive observer.

The error vector $e \triangleq \hat{x} - x$ satisfies the differential equation

$$\dot{e} = Ke + \phi^1 x_1 + \phi^2 u + w^1 + w^2 \qquad e_1 = h^T e \tag{50}$$

where

$$\phi^1 \triangleq [a-\hat{a}], \quad \phi^2 \triangleq [\hat{b}-b] \quad . \tag{51}$$

By choosing w^1 and w^2 as shown earlier and using the adaptive laws

$$\begin{aligned} \dot{\hat{a}}(t) &= +\Gamma_1 e_1 v^1 \\ \dot{\hat{b}}(t) &= -\Gamma_2 e_2 v^2 \end{aligned} \tag{52}$$

where v^1 and v^2 are auxiliary signals generated using the output $x_1(t)$ and input $u(t)$ of the plant, the overall identification procedure may be made asymptotically stable.

The schematic diagram for this adaptive observer scheme is shown in Figure 6. The signals v^1 and v^2 play the same role in the adaptive laws as that which the state vector and input vector did in Section 4. The vector signals w^1 and w^2 added to the input of the observer are required only to assure stability. The presence of the signals w^1 and w^2, however, makes the practical realization of the adaptive observer rather difficult. The question naturally arises whether it is possible to eliminate completely the signals w^1 and w^2 by choosing a suitable representation of the plant. The second approach which utilizes a non-minimal representation of the plant provides an affirmative answer to this question.

Approach 2 (Non-Minimal Realization)

In this approach we use the input-output description of a plant using a transfer function (or transfer matrix) rather than a state variable description. While we are primarily interested in single variable systems in this section, the system representations considered here apply equally well to multivariable systems described in Section 6. The various stages in the evolution of the model used in the identification procedure are given below.

i. Let a multivariable plant be described by a matrix of unknown gains K followed by a positive real transfer matrix $W(s)$. To determine the elements of K, a model is constructed with a variable gain matrix $\hat{K}(t)$ followed by a transfer matrix $W(s)$. If $\varepsilon_0(t)$ is the error between plant and model outputs and the parameter error matrix is defined as $K - \hat{K}(t) \triangleq \Phi(t)$, Figure (7) shows the relation between $\varepsilon_0(t)$ and the input to the plant, $q(t)$.

A generalization of the results of Section 5(A) yields the adaptive law

$$\dot{\Phi}(t) = -\Gamma\varepsilon_0 q^T(t) \qquad \Gamma = \Gamma^T > 0 \tag{53}$$

which assures the convergence of $\varepsilon_0(t)$ and $\Phi(t)$ to zero asymptotically if the input $q(t)$ is sufficiently rich.

ii. Figure 8 shows a modification of the above structure when $q(t)$ is the output of a stable transfer matrix whose input is $u(t)$. The same updating scheme (53) can be used in this case also.

iii. In Figure 9 the output of the plant is fedback into both plant and model. Since this does not affect the error equations of the overall system, the updating laws (53) are still valid for this case. Hence any plant which can be represented as a feedback system with feedforward transfer matrix $W(s)KG(s)$ where $W(s)$ is a positive real transfer matrix, $G(s)$ is a stable transfer matrix and K is a matrix of unknown constants, can be identified using this procedure. Generalizations of this approach to multivariable systems were first suggested by Anderson [8] and these are considered in the next section.

iv. Any linear time-invariant dynamical system of order n with a single input and single output can be represented in the form shown in Figure 10.

Hence, the identification procedure described above can be used to estimate the unknown parameters of such a system.

In Figure 10 the scalars a_1 and b_1 and the $(n-1)$ dimensional vectors $\bar{a}$ and $\bar{b}$ are the unknown parameters of the plant. $\bar{x}_1$ and $\bar{x}_2$ are $(n-1)$ dimensional vectors which are the outputs of two systems with identical transfer matrices $(sI-\Lambda)^{-1}$ and inputs x_1 and u respectively. The model has the same structure as that shown in Figure 10 except that the output of the plant x_1 rather than $\hat{x}_1$ is used as the input in the feedback path. This results in the structure shown in Figure 11 for the identification procedure and corresponds to the observer suggested by Luders and Narendra in [21].

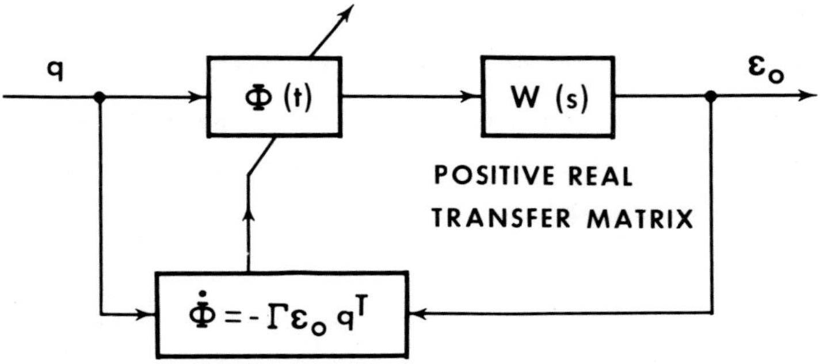

FIG. 7. IDENTIFICATION OF PARAMETERS FROM INPUT-OUTPUT DATA (MULTIVARIABLE CASE)

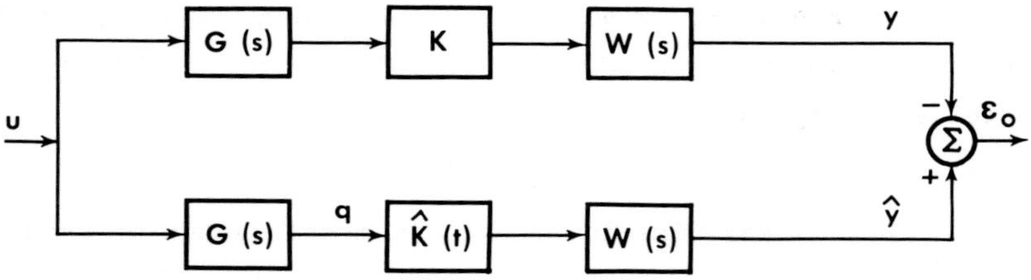

FIG.8 OPEN-LOOP MULTIVARIABLE SYSTEM IDENTIFICATION

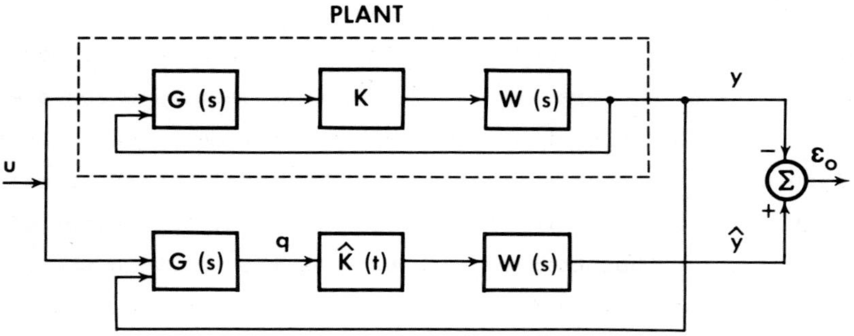

FIG. 9 CLOSED LOOP MULTIVARIABLE SYSTEM IDENTIFICATION

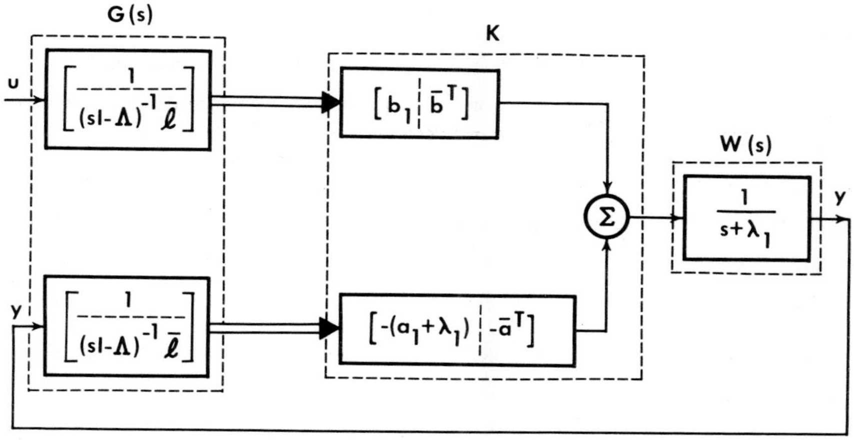

FIG. 10 NONMINIMAL REALIZATION OF PLANT

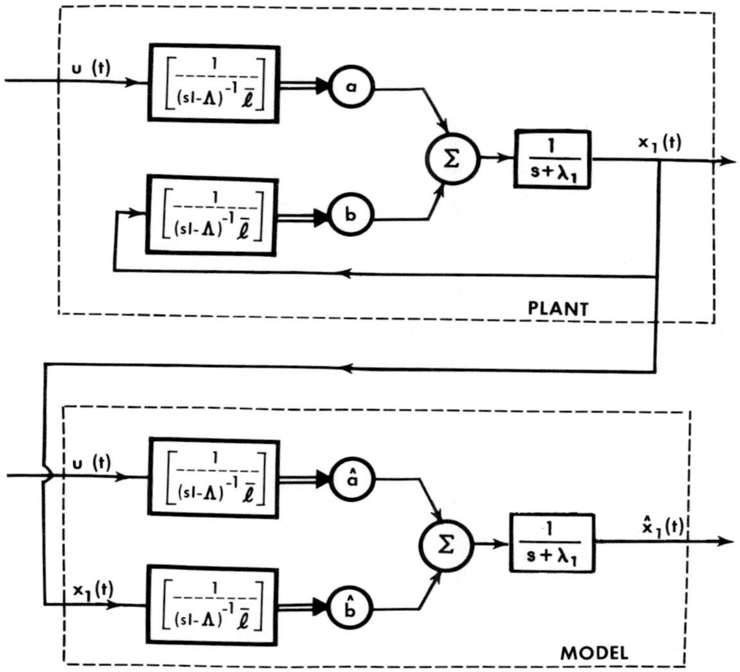

FIG.11 NON-MINIMAL REALIZATION OF SINGLE-INPUT SINGLE-OUTPUT SYSTEM

Comments

i. The identification procedure using a non-minimal realization of the plant is considerably simpler than that using a minimal realization. This is mainly due to the fact that in the former the auxiliary signals w^1 and w^2 are completely eliminated. While the error equation using the first approach has the form (42), the error equation using the second approach has the form (41).

ii. Part of the simplicity of the first approach may also be attributed to the fact that the observer also plays the role of the signal generators used in the second approach.

iii. While the state variables of the plant can be directly estimated using a minimal realization, a suitable transformation of the $(2n-1)$ observer state variables is required to estimate the state of the plant when a non-minimal realization is used.

Approach 3

Recently a modification of approach 2 was suggested [26] for the identification of the parameters of a linear system. This scheme combines the attractive features of the previous approaches and appears particularly suited for the control problem where the identified values of the parameters have to be used to generate a control signal for the plant. The plant is described by the vector equation

$$\dot{x}_p = A_p x_p + b_p u \qquad y = h^T x_p$$

The model used to identify the unknown parameters of A_p and b_p has the form

$$\dot{x}_m = A_m x_m + b_m u + d(\alpha^T v^1 + \beta^T v^2) \tag{54}$$

A_m is a stable matrix in observable canonical form. The vector d is chosen to make $h^T(sI-A_m)^{-1}d = D(s)/R(s)$ positive real and v^1 and v^2 are n-dimensional auxiliary signals generated by systems with transfer function $1/D(s)$ and the output $x_1(t)$ and input $u(t)$ as input signals.

The error equation in this case has the form

$$\begin{aligned} \dot{e}_1 &= A_m e + \phi x_1 + \psi u + d(\alpha^T v^1 + \beta^T v^2) \\ e_1 &= h^T e \end{aligned} \qquad (55)$$

where ϕ and ψ are the parameter error vectors and e_1 is the output error. The parameters α and β are adjusted as in approach 2 using the laws

$$\dot{\alpha} = - e_1 v^1$$

$$\dot{\beta} = - e_1 v^2$$

It can be readily shown that $\alpha(t) \to -\phi$ and $\beta(t) \to -\psi$ as $t \to \infty$.

For examples of approaches 1 and 2 to the identification problem, the reader is referred to works of Carroll and Lindorff [18], Kudva and Narendra [19], Narendra and Kudva [23] and Kim and Lindorff [27]. We present here a single example of a fourth order system identified using approach 3 [26].

EXAMPLE 3: A single-input single output plant is described by a fourth order differential equation with matrix A_p and vector b_p given by

$$A_p = \begin{bmatrix} -6 & 1 & 0 & 0 \\ -9 & 0 & 1 & 0 \\ -10 & 0 & 0 & 1 \\ -12 & 0 & 0 & 0 \end{bmatrix} \quad \text{and} \quad b_p = \begin{bmatrix} 1 \\ 2 \\ 2 \\ 1 \end{bmatrix}$$

where the elements $a_{i1} (i = 1,4)$ are unknown. A model described by equation (54) is used to identify the system.

$$A_m = \begin{bmatrix} -10 & 1 & 0 & 0 \\ -35 & 0 & 1 & 0 \\ -50 & 0 & 0 & 1 \\ -25 & 0 & 0 & 0 \end{bmatrix} \qquad b_m = b_p \quad .$$

A square wave of amplitude 40 and frequency 6 rads/sec was used as input to both plant and input. The evolution of the model parameters as functions of time is shown in Figure 12.

6. MULTIVARIABLE SYSTEMS

The previous section considered for the most part the parametrization of linear systems with a single output for the purpose of identification. The extension of this procedure to the multivariable case is considerably more difficult since the canonical forms for such systems have a much more complicated structure. Further, the class of all single output n-dimensional observable matrix pairs can be generated by applying to any member of the class (h^T, A) all transformations of the type $(h^T, A) \longmapsto (h^T T^{-1}, T(A + Kh^T)T^{-1})$. Almost all single-output parameter identification techniques make use of this property. However, this property fails to extend to the multi-output case.

In this section we consider three methods that have been suggested for the identification of multivariable systems. We first describe briefly some of the efforts to extend the ideas presented in the previous section directly to the multi-output case before proceeding to outline some of the basic ideas contained in the works of Morse [24] and Anderson [8].

Method I [23]

Let a multivariable system be completely observable through (say) the first output. In such a case the system can be represented by the triple $\{C, A, B\}$ where A is in observable canonical form and the first row of the matrix C is $(1,0,0,0\ldots0)$. The matrices A and B can be identified using the procedure described in Section 5. Since the estimates of the state variables are generated by this procedure, the elements of the last $(m-1)$ rows of the matrix C can also be determined simultaneously using the results of Section 3A.

In more realistic cases where the system is completely observable but not through any single output, almost any arbitrary

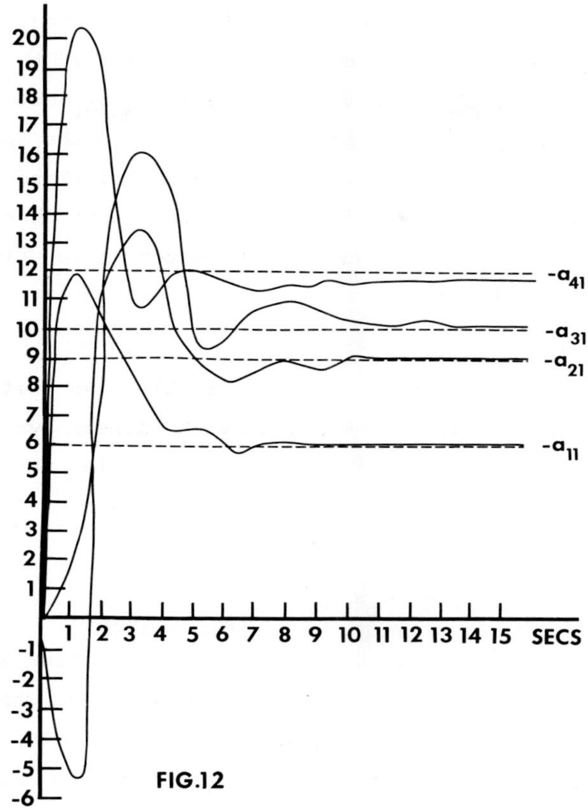

20
19
18
17
16
15
14
13
12
11
10
9
8
7
6
5
4
3
2
1
-1
-2
-3
-4
-5
-6
1 2 3 4 5 6 7 8 9 10 11 12 13 14 15 SECS
$-a_{41}$
$-a_{31}$
$-a_{21}$
$-a_{11}$

FIG.12

linear combination of the outputs yields a new output with the desired property. This follows from the fact that the pair $(c^T C, A)$ is completely observable for almost any vector c if A is cyclic. The application of this approach to the identification of the dynamics of a helicopter in the vertical plane is discussed in some detail in [30]. The single output which is used in the identification procedure is the pitch angle which can be measured relatively easily.

Method II [24]

As stated in Section 2, a multivariable plant with inputs $u(t)$ and outputs $y(t)$ is modeled by the linear system $\{C_p, A_p, B_p\}$ if every pair $\{u(t), y(t)\}$ satisfies the equations

$$\begin{aligned} \dot{x}_p &= A_p x_p + B_p u \\ y(t) &= C_p x_p \end{aligned} \tag{1}$$

The identification problem is then to determine from a knowledge of the input-output data a linear system $\{C, A, B\}$ which is equivalent to $\{C_p, A_p, B_p\}$. In [24] Morse has studied the implications of the hypothesis that a physical process can be modeled by a system of the form $(C, A + KC, B)$ where (C, A) is an observable pair and K and B are parameter matrices to be identified. If we focus attention on this special model class, the identification procedure may be described as follows:

Let the plant be modeled by the system of equations

$$\begin{aligned} \dot{x}_p &= [A_p + K_p C] x_p + B_p u(t) \\ y(t) &= C x_p(t) \end{aligned} \tag{56}$$

or, equivalently, by the system of equations

$$\dot{x}_p = [A_p + K_0 C] x_p + [K_p - K_0] C x_p + B_p u(t) \tag{57}$$

where $A_p + K_0 C$ is a stable matrix.

The model used to identify the system (55) has the form

$$\dot{\hat{x}} = [A_p + K_0C]\hat{x}(t) + [\hat{K}(t) - K_0]y(t) + \hat{B}(t)u(t) + w(t)$$

$$\hat{y}(t) + C\hat{x}(t) \tag{58}$$

where $\hat{x}(t)$ is the estimate of the state of the plant, $\hat{K}(t)$ and $\hat{B}(t)$ are estimates of the parameters and $w(t)$ is an input signal which has to be chosen to make the procedure stable.

The state and parameter errors are related by the equation

$$\begin{aligned} \dot{e} &= [A_p + K_0C]e + (\hat{K}(t) - K)y(t) + (\hat{B}(t) - B)u(t) + w(t) \\ &= [A_p + K_0C]e + \Phi(t)y(t) + \Psi(t)u(t) + w(t) \\ \hat{y}(t) - y(t) &= Ce(t) \end{aligned} \tag{59}$$

where $e(t) = \hat{x}(t) - x(t)$, $\Phi(t) = \hat{K}(t) - K$ and $\Psi(t) = \hat{B}(t) - B$.

It is seen that the error equations (59) are similar to equations (50) in Section 5 which arise in the identification of a plant using a minimal realization. The aim of the identification procedure is then to determine $\Phi(t)$, $\Psi(t)$ and $w(t)$ so that $e(t)$, $\Phi(t)$ and $\Psi(t)$ tend to zero asymptotically.

Let the $[n \mathrm{x} (mn + np)]$ matrix E be represented as

$$E = [E_1 | E_2 | \ldots . | E_{m+p}]$$

where each matrix E_i is an $(n \times n)$ matrix.

Let

$$\dot{E}_i = [A + K_0C]E_i + Iu_i \qquad (i = 1,2,\ldots,m)$$

$$\dot{E}_{m+j} = [A + K_0C]E_{m+j} + Iy_j \qquad (j = 1,2,\ldots,p)$$

where I is the unit matrix. The matrices E_i can be considered as sensitivity matrices which correspond to the auxiliary state variables in the scalar case. The input vector $w(t)$ as well as the adaptive laws for updating $\dot{\Phi}(t)$ and $\dot{\Psi}(t)$

can be expressed in terms of the sensitivity matrices as follows:

$$w(t) = -E(t)E^T(t)C^T[\hat{y}(t) - y(t)]$$

$$\dot{L} = E^T(t)C^T[\hat{y}(t) - y(t)]$$

where L is an $(nm + np)$-vector obtained by arranging the columns of $\hat{K}(t)$ and $\hat{B}(t)$ one under the other. An examination of the form of equations (58) reveals that they correspond to the equations (44) and (52) in the single variable case.

Method III [8]

The identification procedure discussed in Section 5 using a non-minimal realization of the plant can be directly extended to multivariable systems, as was first shown by Anderson. Most of the comments made in Section 5 are also relevant to the multivariable problem. The principal aim of the approach is to represent a system in such a fashion that all the unknown parameters appear as the elements of a single matrix K in the overall transfer matrix. The known elements of the transfer matrix are chosen in such a fashion that the error equation between plant and model has the form of prototype 3.

Anderson considered two representations of the plant transfer matrix $W_p(s)$. In the first

$$W_p(s) = K_p[1 - W_1(s)K_p]^{-1}W(s) \tag{60}$$

where $W(s)$ and $W_1(s)$ are known real transfer matrices with poles in the open left-half of the complex plane and K_p is an unknown matrix of constants. The transfer matrix (60) corresponds to $W(s)$ in series with a feedback loop which has K_p in the forward path and $W_1(s)$ in the feedback path. The model used for identification has an open loop structure with the output of the plant being used as the input to $W_1(s)$; the transfer matrices $W(s)$ in both plant and model have the same input

$u(t)$. The matrix $K_m(t)$ of adjustable parameters replaces K_p in the model and $\dot{K}_m(t)$ is adjusted continuously using input-output data. It is shown that with a sufficiently rich input

$$\lim_{t \to \infty} K_m(t)[1 - W_1(s)K_p]^{-1}W(s) = K_p[1 - W_1(s)K_p]^{-1}W(s)$$

for all s for which the quantities are defined.

In a second and more general structure used to represent the plant in [8],

$$W_p(s) = V(s)K_p[1 - W_1(s)V(s)K_p]^{-1}W(s) \qquad (61)$$

where a strictly positive real transfer matrix $V(s)$ is included along with the gain matrix K_p in the forward path of the feedback loop. The insertion of $V(s)$ is shown to reduce the complexity of the identifier both in the single variable and multivariable cases.

7. DISCUSSION AND CONCLUSIONS

At the present time there appears to be little doubt concerning the power and versatility of the general approach described in the previous sections and the great potential it appears to possess for the identification and control of multivariable systems. However, two major questions have to be resolved before the procedures outlined emerge as truly viable techniques for use in practical situations. The first one concerns the question of observation noise. The second is related to the problem of speed of convergence and the choice of adaptive gains.

As mentioned earlier in the paper, very little theoretical work has been done so far on the stochastic stability of the overall system when observation noise is present in the outputs. Also, relatively little that is precise is known about the speed of adaptation, though some preliminary investigations and numerous simulation studies have been reported in the literature. The latter also bring into focus the difference between off-line and on-line identification procedures as well as their dependence on the above two problems.

i. *Observation Noise*

When noise is present in the plant output measurements, the adaptive procedures described, if applied without modification, compute biased estimates of the system parameters. This is due to the fact that the noise term appears both in the plant output and the output error signals and the adaptive laws for estimating system parameters involve the multiplication of the two signals. Since no noise is present in the input to the system the adaptive procedure yields unbiased estimates of the input parameters (associated with the matrix B_p in equation 1). The effect of prefiltering the error on the adaptive procedure has been considered by Udink ten Cate [33]. As pointed out in Section 3, positive real transfer functions can be used for updating parameters without affecting the stability of the overall system and such transfer functions can be suitably designed for noise attenuation.

ii. *Speed of Convergence*

The designer has considerable freedom in choosing the parameters of the adaptive observer. In particular, he can freely locate the poles of the observer and the auxiliary signal generators and choose the adaptive gain matrices Γ_1 and Γ_2 in Section 5. The choice of these parameters together with the magnitude and frequency content of the input signals determine the convergence of the adaptive schemes. In many simulation studies, while output identification is achieved in a relatively short time, parameter identification is not completely realized. In these studies the observer time constants are chosen to be approximately those of the system and the speed of response is controlled by changing the adaptive gains.

Lindorff and Kim [27] and Kim [29] have studied the convergence properties of the adaptive observer with periodic inputs using Floquet theory and Popov's theory of hyperstability. The main reason for the unsatisfactory convergence rate of the parameter errors is attributed, in these studies, to the location of the dominant eigenvalue of an equivalent discrete system near the boundary of the unit circle. Methods are then proposed to choose the adaptive gains so that this dominant eigenvalue is within a circle of prescribed radius $(r < 1)$ in the complex plane. It is also argued in [29] that improved convergence characteristics can be achieved by making the adaptive gains time-varying. While [27] and [29] represent the first serious attempts to deal with the major problem of convergence rates, the question still remains largely open.

iii. *Off-Line Identification*

The error $e_1(t)$ between plant and model outputs is seen to play a crucial role in all the adaptive schemes discussed in Section 5. In most of the simulation studies of adaptive observers, it has been observed that this output error tends to zero very rapidly while the parameter errors are appreciably large. The fact that $e_1(t)$ and $\dot{e}_1(t)$ are small can be exploited, if off-line identification of the plant is of interest and the plant can be subjected to various inputs. The motivation for this is provided by the work of Lynch and VandeLinde [31]. The plant is forced sequentially with independent inputs and in each case the observer parameters are adjusted so that the output error becomes very small and stays in a small neighborhood of the origin. If the number K of inputs used is equal to the number of parameters to be estimated, a set of K linear algebraic equations is obtained which can then be solved by matrix inversion to obtain the parameter estimates. This modified procedure has been effectively applied to the problem of identification of a VTOL aircraft in [28]. While the above discussion is concerned primarily with off-line identification, it also indicates that second order gradient procedures will be more effective in the parameter estimation problem.

iv. *On-Line Identification and Control*

The procedure outlined in (iii) cannot be applied in an on-line situation where the results of the identification are directly used to control the plant since the designer has very little freedom in choosing the inputs. At the same time, however, the primary interest in such a case centers around making the output of the plant match a desired output rather than in matching the transfer matrix of the controlled plant to that of a desired model. Simulation studies indicate that feedback signals generated using partially identified models of plants which yield output matching are also effective in controlling the overall system. This fact will obviously have a great impact on the efforts to solve the control problem which is currently attracting considerable attention.

Acknowledgment

The research reported in this document was sponsored in part by support extended to Yale University by the U.S. Office of Naval Research under Contract N00014-67-A-0097-0020

REFERENCES

1. Grayson, L. P., "Design via Lyapunov's Second Method," in Proc. Fourth Joint Conf. on Automatic Control, 1963, pp. 589-595.

2. Parks, P. C., "Lyapunov Redesign of Model Reference Adaptive Control Systems," *IEEE Trans. Automat. Control, Vol. AC-11,* pp. 362-367, July 1966.

3. Winsor, C. A. and R. T. Roy, "Design of Model Reference Adaptive Control Systems by Lyapunov's Second Method," *IEEE Trans. Automat. Contr. (Corresp.), Vol. AC-13,* p. 204, Apr. 1968.

4. Butchart, R. L. and B. Shackloth, "Synthesis of Model Reference Adaptive Systems by Lyapunov's Second Method," in IFAC Conf. on the Theory of Self-Adaptive Control Systems, London, 1965.

5. Monopoli, R. V., "Lyapunov's Method for Adaptive Control System Design," *IEEE Trans. Automat. Contr. (Corresp.), Vol. AC-12,* pp. 334-335, June 1967.

6. Lindorff, D. P., and R. L. Carroll, "Survey of Adaptive Control Using Lyapunov Design," in Proc. 1972 Int. Conf. on Cybernetics and Society.

7. Landau, I. D., "Model Reference Adaptive Systems - A Survey (MRAS) - What is Possible and Why?" *Journal of Dynamic Systems, Measurement and Control, Vol. 94G,* pp. 119-132, June 1972.

8. Anderson, B. D. O., "Multivariable Adaptive Identification," University of Newcastle, New South Wales, Australia, Technical Report, June 1974.

9. Morgan, A. P. and K. S. Narendra, "On the Uniform Asymptotic Stability of Certain Linear Non-Autonomous Differential Equations," Yale University, New Haven, Conn., Becton Center Tech. Rep. CT-64, May 1975.

10. Morgan, A. P. and K. S. Narendra, "On the Stability of Non-Autonomous Differential Equations $x = [A + B(t)]x$, with Skew-Symmetric Matrix $B(t)$," Yale University, New Haven, Conn., Becton Center Tech. Rep. CT-66.

11. Narendra, K. S. and S. S. Tripathi, "The Choice of Adaptive Parameters in Model-Reference Control Systems," in Proc. Fifth Asilomar Conf. on Circuits and Systems, 1971.

12. Yuan, J. S. C. and W. M. Wonham, "Asymptotic Identification Using Stability Criteria," Univ. of Toronto, Toronta, Canada, Control Systems Report No. 7422, Nov. 1974.

13. Kudva, P. and K. S. Narendra, "An Identification Procedure for Discrete Multivariable Systems," *IEEE Trans. Automat. Control, Vol. AC-19,* pp. 549-552.

14. Kudva, P. and K. S. Narendra, "The Discrete Adaptive Observer," Yale Univ., New Haven, Conn., Becton Center Tech. Rep. CT-63, June 1974.

15. Mendel, J. M., "Discrete Techniques of Parameter Estimation," (M. Dekker, N.York, 1973), Chaps. 4 and 5.

16. Mendel, J. M., "Gradient Estimation Algorithms for Equation Error Formulations," *ibid.,* 1974, AC-19, pp. 820-824.

17. Udink Ten Cate, A. T., "Gradient Identification of Multivariable Discrete Systems," *Electronics Letters, Vol. 11, No. 5,* pp. 98-99, March 1975.

18. Carroll, R. L. and D. P. Lindorff, "An Adaptive Observer for Single-Input Single-Output Linear Systems," *IEEE Trans. Automat. Contr., Vol. AC-18,* pp. 428-435, Oct. 1973.

19. Kudva, P. and K. S. Narendra, "Synthesis of an Adaptive Observer Using Lyapunov's Direct Method," *Int. J. Contr., Vol. 18,* pp. 1201-1210, Dec. 1973.

20. Luders, G. and K. S. Narendra, "An Adaptive Observer and Identifier for a Linear System," *IEEE Trans. Automat. Contr., Vol. AC-18,* pp. 496-499, Oct. 1973.

21. Luders, G. and K. S. Narendra, "A New Canonical Form for an Adaptive Observer," *IEEE Trans. Automat. Contr., Vol. AC-19,* pp. 117-119, Apr. 1974.

22. Narendra, K. S. and P. Kudva, "Stable Adaptive Schemes for System Identification and Control - Part I," *IEEE Trans. on Systems, Man and Cybernetics, Vol. SMC-4,* pp. 542-551, Nov. 1974.

23. Narendra, K. S. and P. Kudva, "Stable Adaptive Schemes for System Identification and Control - Part II," *IEEE Trans. on Systems, Man and Cybernetics, Vol. SMC-4,* pp. 552-560, Nov. 1974.

24. Morse, A. S. "Representation and Parameter Identification of Multi-Output Linear Systems," Presented at the 1974 Decision and Control Conference.

25. Lefschetz,S., *Stability of Nonlinear Control Systems,* New York: Academic Press, 1965.

26. Narendra, K. S. and L. S. Valavani, "A New Procedure for the Identification of Multivariable Systems," Yale Univ., New Haven, Conn., Becton Center Technical Report.

27. Kim, C. and D. P. Lindorff, "Input Frequency Requirements for Identification Through Lyapunov Methods," *Int. J. Contr., Vol. 20, No. 1,* pp. 35-48.

28. Kudva, P., "Stable Adaptive Systems," Doctoral Dissertation, Yale Univ., New Haven, Conn., May 1975.

29. Kim, C., "Convergence Studies for an Improved Adaptive Observer, Doctoral Dissertation, Univ. of Connecticut, Storrs, Conn., 1975.

30. Luders, G. and K. S. Narendra, "Stable Adaptive Schemes for State Estimation and Identification of Linear Systems," *IEEE Trans. Automatic Control, Vol. AC-19, No. 6,* pp. 841-847, December 1974.

31. Lynch, H. M. and V. D. VandeLinde, "Multiple Equilibria of System Identifiers Using Lyapunov-Designed Model Builders," Proc. Thirteenth Annual Joint Automatic Control Conference, pp. 821-825, December 1972.

32. Narendra, K. S. and J. H. Taylor, *Frequency Domain Criteria for Absolute Stability,* Academic Press, New York, 1973.

33. Udink Ten Cate, A. J. and N. D. L. Verstoep, "Improvement of Lyapunov Model Reference Adaptive Control Systems in a Noisy Environment," to appear in *Int. Journal of Control.*

SYNTHESIS OF OPTIMAL INPUTS FOR MULTIINPUT-MULTIOUTPUT (MIMO) SYSTEMS WITH PROCESS NOISE

PART I: FREQUENCY-DOMAIN SYNTHESIS
PART II: TIME-DOMAIN SYNTHESIS

Raman K. Mehra

Division of Engineering and Applied Physics
Harvard University, Cambridge, Massachusetts

PART I: FREQUENCY-DOMAIN SYNTHESIS

1. INTRODUCTION

The problem of input design has been the subject of several recent studies [1-4].* Most of the results obtained have been confined to single-input systems without process noise. However, many of the industrial applications involve multiple inputs and process noise. In this chapter, we present a complete treatment of such systems along with practical algorithms for the computation of optimal input designs. The basic approach used is the same as in Ref.[1] but with some important differences. The procedure for obtaining the information matrix is more general and yields results not easily derived from the previous approach. The min-max conditions of Ref.[1] for single-input systems are replaced by min-max-max conditions.

---

*References for Parts I and II are given separately at the end of the chapter.

The organization of the chapter is as follows. Section 2 contains a statement of the problem. In Section 3 an asymptotic expression for the average per sample information matrix is derived. Section 4 discusses certain important properties of the set of information matrices which are used in Section 5 to show the equivalence of the D-optimal design to a min-max design. An algorithm for the computation of optimal inputs is also given. Time-domain input design is considered in Part II and results similar to frequency domain results are obtained. Extensions to other criteria and bounds are discussed in Sections 6 and 7 and the conclusions are stated in Section 8. For an overall survey of the subject, the reader is referred to Ref. [11] and for a specific application, to Ref. [12].

2. STATEMENT OF THE PROBLEM

Consider a linear discrete-time system

$$x(t+1) = \Phi x(t) + Gu(t) + \Gamma w(t) \tag{1}$$

$$y(t) = Hx(t) + v(t) \qquad t = 0,1,\ldots,N \tag{2}$$

where $x(t)$ is $n\times 1$ state vector, $u(t)$ is $q\times 1$ input vector, $w(t)$ is $r\times 1$ process noise vector, $y(t)$ is $p\times 1$ output vector, and $v(t)$ is $p\times 1$ measurement noise vector. We assume that (i) $\Phi(n\times m)$, $G(n\times q)$, $\Gamma(n\times r)$ and $H(p\times n)$ are constant matrices, (ii) Φ is stable, (iii) (Φ,G) and (Φ,Γ) are controllable pairs, and (iv) (Φ,H) is observable, (v) $w(t)$ and $v(t)$ are stationary Gaussian white noise sequences with

$$E[w(t)] = 0\ , \qquad E[w(t)w^T(\tau)] = Q\delta_{t,\tau} \tag{3}$$

$$E[v(t)] = 0\ , \qquad E\{v(t)v^T(\tau)] = R\delta_{t,\tau} \tag{4}$$

$$E[v(t)w^T(\tau)] = 0\ , \qquad \forall\, t,\tau \tag{5}$$

and (vi) $x(0) = 0$, N even. (This assumption can be easily relaxed.)

It is known [5] that only the steady-state Kalman filter representation of the system (1) and (2) is identifiable. Let us define

$$\hat{x}(t) = E[x(t)|y(1),\ldots,y(t-1)] \tag{6}$$

$$\nu(t) = (HPH^T + R)^{-1/2}(y(t) - H\hat{x}(t)) \tag{7}$$

and

$$P = E[(x(t) - \hat{x}(t))(x(t) - \hat{x}(t))^T] \tag{8}$$

The steady-state Kalman filter representation of the system (1) and (2) is

$$\hat{x}(t+1) = \Phi\hat{x}(t) + Gu(t) + K\nu(t) \tag{9}$$

$$y(t) = H\hat{x}(t) + \Sigma\nu(t) \tag{10}$$

where $\Sigma = (HPH^T + R)^{1/2}$ and where $K(n \times p)$ is a gain matrix defined by the following set of equations

$$K = \Phi PH^T(HPH^T + R)^{-1/2} \tag{11}$$

$$P = \Phi P\Phi^T - KK^T + \Gamma Q\Gamma^T \tag{12}$$

and

$$E[\nu(t)] = 0 , \qquad E[\nu(t)\nu^T(\tau)] = I\delta_{t,\tau} \tag{13}$$

Let θ denote an $m \times 1$ vector of identifiable parameters in the system representation (1)-(2) or the Kalman filter representation (9)-(12). We estimate θ from the knowledge of $\{y(t), u(t), t = 0,\ldots,N\}$ using an unbiased efficient estimator $\hat{\theta}$ with covariance M^{-1}, where M is the Fisher information matrix. It is required to select input $u(t) \in U$ such that a suitable norm of M^{-1} is minimized. We first derive an expression for M.

3. INFORMATION MATRIX

Following the approach of Ref. [1] we write all time functions in Eqs. (9)-(10) in terms of their Fourier series expansions* and

*The Fourier series expansions of random variables hold in the mean square sense, i.e., $E|x(t) - \Sigma_{n=-N/2}^{N/2-1} \tilde{x}(n)z_n^t|^2 = 0$.

obtain

$$z_n \tilde{x}(n) = \Phi \tilde{x}(n) + G\tilde{u}(n) + K\tilde{\nu}(n) \tag{14}$$

$$\tilde{y}(n) = H\tilde{x}(n) + \Sigma\tilde{\nu}(n) \tag{15}$$

where

$$n = -\frac{N}{2}, \ldots, \left(\frac{N}{2} - 1\right)$$

and

$$z_n = e^{-jn \frac{2\pi}{N}}$$

Here $\tilde{x}(n)$ denotes the Fourier series component of $x(t)$ at frequency $2\pi(n/N)$ and similarly for the other variables. From (14)-(15)

$$\tilde{y}(n) = H(z_n I-\Phi)^{-1} G\tilde{u}(n) + [H(z_n I-\Phi)^{-1} K + \Sigma]\tilde{\nu}(n)$$

$$= T_1(z_n,\theta)\tilde{u}(n) + T_2(z_n,\theta)\tilde{\nu}(n) \tag{16}$$

where

$$T_1(z_n,\theta) = H(z_n I-\Phi)^{-1} G \tag{17}$$

$$T_2(z_n,\theta) = (H(z_n I-\Phi)^{-1} K + \Sigma) \tag{18}$$

Since the Kalman filter representation (9)-(10) is invertible [6], $T_2(z_n,\theta)$ is nonsingular and one can solve Eq. (16) for $\tilde{\nu}(n)$.

$$\tilde{\nu}(n) = T_2^{-1}(z_n,\theta)[\tilde{y}(n) - T_1(z_n,\theta)\tilde{u}(n)]$$

$$= T_3(z_n,\theta)\tilde{y}(n) - T_4(z_n,\theta)\tilde{u}(n) \tag{19}$$

where

$$T_3 = T_2^{-1} \quad \text{and} \quad T_4 = T_2^{-1} T_1 \tag{20}$$

We now assume that $\nu(t)$ is periodic with period N. Later on, we let $N \to \infty$ so that $\nu(t)$ tends to a stationary process. The sequence $\tilde{\nu}(n) = \tilde{\nu}^R(n) + j\tilde{\nu}^I(n)$ consisting of real part $\tilde{\nu}^R(n)$

and imaginary part $\tilde{\nu}^I(n)$ is a complex Gaussian white sequence with properties [9]

$$E[\tilde{\nu}^R(n)] = E[\tilde{\nu}^I(n)] = 0 \tag{21}$$

$$E\left\{\begin{bmatrix}\tilde{\nu}^R(n)\\ \tilde{\nu}^I(n)\end{bmatrix}\begin{bmatrix}\tilde{\nu}^R(k)\\ \tilde{\nu}^I(k)\end{bmatrix}^T\right\} = \frac{1}{2N}\begin{bmatrix}I & O\\ O & I\end{bmatrix}\delta_{k,n} \tag{22}$$

where

$$\delta_{k,n} = \begin{cases}1, & k = n\\ 0, & k \neq n\end{cases}$$

and I is an identity matrix. Also

$$E[\tilde{\nu}(n)\tilde{\nu}^*(n)] = \frac{1}{N} I .$$

The log-likelihood function $L(\theta)$ may now be written as

$$L(\theta) = -\frac{N}{2} \operatorname{Re} \sum_{n=-N/2}^{N/2-1} \tilde{\nu}^*(n)\tilde{\nu}(n) - N \log |\Sigma| - \frac{N}{2}\log(2\pi) \tag{23}$$

where * denotes transpose and complex conjugate and Re denotes real part. Fisher information matrix M are defined as [8].

$$M_{ij} = E\left[\frac{\partial L}{\partial\theta_i}\frac{\partial L}{\partial\theta_j}\right]_{\theta=\theta_0} \qquad i = 1,\ldots,m;\; j = 1,\ldots,m \tag{24}$$

where the expectation E is taken over the sample space of observations $\{y(t), u(t), t = 0,\ldots,N\}$ and θ_0 is an a priori estimate of θ. From (23)

$$\frac{\partial L}{\partial\theta_i} = -N \operatorname{Re} \sum_{n=-N/2}^{N/2-1}\left[\tilde{\nu}^*(n)\frac{\partial\tilde{\nu}(n)}{\partial\theta_i}\right] - N \operatorname{Tr}\left(\Sigma^{-1}\frac{\partial\Sigma}{\partial\theta_i}\right) \tag{25}$$

From (19)

$$\frac{\partial \tilde{\nu}(n)}{\partial \theta_i} = \frac{\partial T_3}{\partial \theta_i} \tilde{y}(n) - \frac{\partial T_4}{\partial \theta_i} \tilde{u}(n) \tag{26}$$

Since $u(t)$ and $\nu(t)$ are uncorrelated (a property of the Kalman filter), $\tilde{u}(n)$ and $\tilde{\nu}(n)$ are also uncorrelated. Thus

$$E\left[\frac{\partial \tilde{\nu}^*(n)}{\partial \theta_i} \frac{\partial \tilde{\nu}(n)}{\partial \theta_j}\right] = \text{Tr } E\left[\frac{\partial \tilde{\nu}(n)}{\partial \theta_j} \frac{\partial \tilde{\nu}^*(n)}{\partial \theta_i}\right]$$

$$= \text{Tr } E\left[\left(\frac{\partial T_3}{\partial \theta_j} \tilde{y}(n) - \frac{\partial T_4}{\partial \theta_j} \tilde{u}(n)\right)\left(\tilde{y}^*(n) \frac{\partial T_3^*}{\partial \theta_i} - \tilde{u}^*(n) \frac{\partial T_4^*}{\partial \theta_i}\right)\right] \tag{27}$$

Let $E[\tilde{u}(n)\tilde{u}^*(n)] = \frac{1}{N} S_{uu}(n)$. From (16)

$$E[\tilde{y}(n)\tilde{y}^*(n)] = \frac{1}{N} [T_1 S_{uu} T_1^* + T_2 T_2^*] \tag{28}$$

$$E[\tilde{y}(n)\tilde{u}^*(n)] = \frac{1}{N} [T_1 S_{uu}] \tag{29}$$

$$E[\tilde{u}(n)\tilde{y}^*(n)] = \frac{1}{N} [S_{uu} T_1^*] \tag{30}$$

After some further simplifications, one obtains

$$M_{ij} = \text{Re} \sum_{n=-N/2}^{N/2-1} [\text{Tr } (B_{ij}(n) S_{uu}(n)) + A_{ij}(n)] + NC_{ij} \tag{31}$$

where

$$B_{ij} = \frac{\partial T_1^*}{\partial \theta_i} T_2^{*-1} T_2^{-1} \frac{\partial T_1}{\partial \theta_j} \tag{32}$$

and

$$A_{ij}(n) = \text{Tr}\left[T_2^{-1} \frac{\partial T_2}{\partial \theta_j} \frac{\partial T_2^*}{\partial \theta_i} T_2^{*-1}\right]_n \tag{33}$$

$$C_{ij} = 2 \text{ Tr}\left(\Sigma^{-1} \frac{\partial \Sigma}{\partial \theta_i}\right) \text{Tr}\left(\Sigma^{-1} \frac{\partial \Sigma}{\partial \theta_j}\right) \tag{34}$$

Now taking limit $N \to \infty$ as in Ref.[1], and denoting $z = e^{-j\omega}$

$$\lim_{N\to\infty} \frac{1}{N} M_{ij} = \bar{M}_{ij}$$

$$= \frac{1}{2\pi} \operatorname{Re} \int_{-\pi}^{\pi} \operatorname{Tr} \left(B_{ij}(\omega) dF_{uu}(\omega)\right) + \bar{A}_{ij} \qquad (35)$$

where

$$\bar{A}_{ij} = \frac{1}{2\pi} \operatorname{Re} \int_{-\pi}^{\pi} A_{ij}(\omega)\, d\omega + C_{ij} \qquad (36)$$

and $F_{uu}(\omega)$ is the spectral distribution function of $u(t)$.

Equation (35) is the expression for the average per sample information and it consists of two parts. The first part depends on the input $u(t)$ and the second part depends on the process noise input $w(t)$. If there is no input $u(t)$, the information matrix is $[\bar{A}_{ij}]$ and if there is no process noise $(K=0,\ T_2=1)$, the information matrix is the same as in Ref.[1].

REMARK: It is interesting to note that information matrix due to $u(t)$ alone has the same form with process noise as without process noise. The only difference is that S_{vv} of Ref.[1] is replaced by $T_2T_2^*$. Furthermore, $A_{ij} = 0$ for $\theta_i \in G$, $\theta_j \in G$. (See Ref.[2] for a similar result.)

4. PROPERTIES OF THE INFORMATION MATRIX

It was shown in Ref.[1] that in the scalar input case without process noise the information matrix is real, symmetric, and non-negative definite. Furthermore, it was shown that the set of all information matrices is a closed convex set in $\mathbb{R}^{m(m+1)/2}$ and that any information matrix can be represented by a convex combination of at most $[m(m+1)/2+1]$ point-input information matrices [1].

The next theorem states similar properties for the vector input case with process noise. As expected, the proof of the theorem is more complicated.

THEOREM 1: *(i) The information matrix* $\overline{M}$ *is a real, symmetric, nonnegative definite matrix.*

(ii) The set of information matrices $\overline{M}$ *corresponding to all normalized designs (i.e.,* $1/2\pi \int_{-\pi}^{\pi} \mathrm{Tr}\, dF_{uu}(\omega) = 1$ *) is convex and closed.*

(iii) For any normalized input design F_1 *with mixed spectrum, another design* F_2 *with a purely point spectrum of less than* $[m(m+1)/2+1]$ *points can be found such that* $\overline{M}(F_1) = \overline{M}(F_2)$.

Proof. (i) This property follows directly from the definition of the Fisher information matrix as [8]

$$\overline{M} = \frac{1}{N} E\left[\left(\frac{\partial L}{\partial \theta}\right)\left(\frac{\partial L}{\partial \theta}\right)^T\right] \tag{37}$$

(ii) Consider three normalized input designs F, F_1 and F_2 related as

$$F(\omega) = (1-\alpha)F_1(\omega) + \alpha F_2(\omega), \quad \omega \in [-\pi,\pi]; \quad 0 \leq \alpha \leq 1 \tag{38}$$

The information matrix for F is

$$\overline{M}_{ij}(F) = (1-\alpha)\overline{M}_{ij}(F_1) + \alpha\overline{M}_{ij}(F_2) \tag{39}$$

Thus the set of information matrices is convex. It is also closed since the ω-set viz $[-\pi,\pi]$ is closed and $B_{ij}(\omega)$ are continuous functions of ω for a stable system.

(iii) We only consider the case where F_1 is absolutely continuous. Then $S_1(\omega) = dF_1(\omega)/d\omega$ is the spectral density matrix and it is Hermitian nonnegative definite. Let $\{\mu_1(\omega) \cdots \mu_q(\omega)\}$ be the eigenvalues (all real) of $S_1(\omega)$ and $\{\phi_1(\omega), \cdots, \phi_q(\omega)\}$ be the corresponding orthonormal eigenvectors. Then

$$S_1(\omega) = \sum_{k=1}^{q} \mu_k(\omega)\phi_k(\omega)\phi_k^*(\omega), \qquad \mu_k(\omega) \geq 0 \tag{40}$$

The input design only effects the first term of $\overline{M}_{ij}$ (cf. Eq. (35)) viz.

$$w_{ij} = \mathrm{Re}\, \frac{1}{2\pi} \int_{-\pi}^{\pi} \mathrm{Tr}\, (B_{ij}(\omega)\, dF_{uu}(\omega)) \tag{41}$$

Substituting from (40)

$$w_{ij}(S_1) = \text{Re}\,\frac{1}{2\pi}\int_{-\pi}^{\pi}\sum_{k=1}^{q} \text{Tr}\,[B_{ij}(\omega)\phi_k(\omega)\phi_k^*(\omega)]\mu_k(\omega)\;d\omega$$

$$= \frac{1}{2\pi}\int_{-\pi}^{\pi}\sum_{k=1}^{q} w_{ijk}(\omega)\mu_k(\omega)\,d\omega \tag{42}$$

where

$$w_{ijk}(\omega) = \text{Re Tr}[B_{ij}(\omega)\phi_k(\omega)\phi_k^*(\omega)] \tag{43}$$

Furthermore, the normalization condition for S_1 implies

$$\frac{1}{2\pi}\int_{-\pi}^{\pi}\sum_{k=1}^{q} \mu_k(\omega)\;d\omega = 1 \tag{44}$$

Define

$$\bar{\mu}(\omega) = \sum_{k=1}^{q} \mu_k(\omega) \tag{45}$$

and for $\bar{\mu}(\omega) > 0$, define

$$\bar{w}_{ij}(\omega) = \frac{1}{\bar{\mu}(\omega)}\sum_{k=1}^{q} w_{ijk}(\omega)\mu_k(\omega) \tag{46}$$

If $\bar{\mu}(\omega_0) = 0$, then $\mu_k(\omega_0) = 0$ for every k and there is no contribution to the integral (42) at ω_0. Thus $\bar{w}_{ij}(\omega_0)$ may be given any arbitrary but bounded value at frequencies ω_0. We may now write*

$$[w_{ij}(S_1)] = \frac{1}{2\pi}\int_{-\pi}^{\pi} [\bar{w}_{ij}(\omega)]\bar{\mu}(\omega)\;d\omega \tag{47}$$

where $\bar{\mu}(\omega) \geq 0$ and $\frac{1}{2\pi}\int_{-\pi}^{\pi} \bar{\mu}(\omega)\;d\omega = 1$. From (47) it follows that the set of all $[w_{ij}]$ in $\mathbb{R}^{m(m+1)/2}$ is a convex hull of the set of point-input information matrices $[\bar{w}_{ij}(\omega)]$. Using the

---

* $[w_{ij}]$ denotes the matrix with elements w_{ij}.

classical theorem of Caratheodory [1], we may write

$$[w_{ij}(S_1)] = \sum_{r=1}^{\ell} \beta_r \bar{w}_{ij}(\omega_r) \tag{48}$$

$$\beta_r \geq 0, \quad \sum_{r=1}^{\ell} \beta_r = 1, \quad \text{and} \quad \ell \leq m(m+1)/2+1 \tag{49}$$

In other words, the information matrix from a continuous design F (or S_1) is the same as the information matrix from a discrete design with at most $[m(m+1)/2+1]$ points.

5. D-OPTIMAL DESIGN IN FREQUENCY DOMAIN

We now maximize $|\bar{M}|$ with respect to $\{F_{uu}(\omega), \omega \in [-\pi,\pi]\}$ subject to the constraint

$$\frac{1}{2\pi} \operatorname{Tr} \int_{-\pi}^{\pi} dF_{uu}(\omega) \leq 1 \tag{50}$$

The optimal input spectrum $\hat{F}_{uu}$ will be shown to have the following characteristics.

THEOREM 2: *For the optimal input spectrum*

$$\frac{1}{2\pi} \operatorname{Tr} \int_{-\pi}^{\pi} d\hat{F}_{uu}(\omega) = 1 \tag{51}$$

and the following are equivalent

(i) $\hat{F}_{uu}$ *maximizes* $|\bar{M}|$

(ii) $\hat{F}_{uu}$ *minimizes*

$$\max_{\omega} \lambda_{max} \left[\operatorname{Re} \sum_{i,j=1}^{m} p_{ij}(F_{uu}) B_{ij}(\omega) \right]$$

where p_{ij} *is the* (i,j) *element of* $\bar{M}^{-1}$ *and* λ_{max} *is the maximum eigenvalue of the* $q \times q$ *matrix inside the parentheses.*

(iii) $$\max_{\omega} \lambda_{max} \left[\operatorname{Re} \sum_{i,j=1}^{m} p_{ij}(\hat{F}_{uu}) B_{ij}(\omega) \right] = \operatorname{Tr}[\bar{M}^{-1}(\hat{F}_{uu}) W(\hat{F}_{uu})] \tag{52}$$

where

$$W(\hat{F}_{uu}) = \frac{1}{2\pi}\left[\mathrm{Re}\int_{-\pi}^{\pi} \mathrm{Tr}(B_{ij}(\omega)\,d\hat{F}_{uu}(\omega))\right]_{\substack{i=1,\ldots,m\\ j=1,\ldots,m}} \tag{53}$$

is the information matrix due to input u *alone.*

The information matrices of all normalized designs satisfying conditions (i)-(iii) are identical and any linear combination of these designs also satisfy (i)-(iii).

Proof. The equality

$$\frac{1}{2\pi}\,\mathrm{Tr}\int_{-\pi}^{\pi} d\hat{F}_{uu}(\omega) = 1$$

follows easily from the fact that $W(cF_{uu}) = cW(F_{uu})$ for any scalar c. By choosing $c > 1$, there is a monotonic increase in W with a monotonic increase in $\mathrm{Tr}(F_{uu})$. Thus the maximum for any F_{uu} is attained when the total power in $u(t)$ is maximum.

We prove the rest of the theorem in two parts, firstly by showing that (ii) and (iii) follow from (i) and secondly by showing that (i) and (iii) follow from (ii). We would need the following lemma in the proof.

LEMMA 1: *For all normalized designs* F_{uu},

$$\max_{\omega} \lambda_{\max}\left[\mathrm{Re}\sum_{i,j=1}^{m} p_{ij}(F_{uu})B_{ij}(\omega)\right] \geq \mathrm{Tr}[\bar{M}^{-1}W] \tag{54}$$

Proof.

$$\mathrm{Tr}(\bar{M}^{-1}W) = \sum_{i,j=1}^{m} p_{ij}w_{ij}$$

$$= \frac{1}{2\pi}\,\mathrm{Re}\int_{-\pi}^{\pi}\sum_{i,j=1}^{m} p_{ij}\,\mathrm{Tr}(B_{ij}(\omega)\,dF_{uu}(\omega))$$

$$= \frac{1}{2\pi} \text{Re} \int_{-\pi}^{\pi} \text{Tr} \left\{ \left(\sum_{i,j=1}^{m} p_{ij} B_{ij}(\omega) \right) (dF_{uu}(\omega)) \right\} \tag{55}$$

It is shown in Appendix A that for two Hermitian matrices D and $C \geq 0$,

$$\text{Tr}(DC) \leq \lambda_{max}(D)\ \text{Tr}(C) \tag{56}$$

Let

$$D(\omega) = \text{Re} \sum_{i,j=1}^{m} p_{ij} B_{ij}(\omega) \tag{57}$$

$$C = dF_{uu}(\omega)$$

Since

$$D^* = \text{Re} \sum_{i,j=1}^{m} p_{ij} B_{ji}(\omega) = \text{Re} \sum_{i,j=1}^{m} p_{ji} B_{ji}(\omega) = D$$

and C is obviously Hermitian and nonnegative definite, (56) holds. From (55)

$$\text{Tr}(\bar{M}^{-1} W) \leq \frac{1}{2\pi} \int_{-\pi}^{\pi} \lambda_{max}(D(\omega))\ \text{Tr}(dF_{uu}(\omega))$$

$$\leq \max_{\omega} \lambda_{max}(D(\omega))$$

and the lemma is proved. ■

We now show that Parts (ii) and (iii) of Theorem 2 follow from Part (i). Consider a design $\hat{F}$ perturbed from the optimal design F as follows (we omit the subscripts uu from F for ease of notation.)

$$dF = (1-\alpha)d\hat{F} + 2\pi\alpha\ \psi_{max}(\omega)\ \psi^*_{max}(\omega)\ \delta(\omega-\omega_0) \tag{58}$$

where $0 < \alpha < 1$ and $\psi_{max}(\omega)$ is a normalized eigenvector of $D(\omega)$ corresponding to $\lambda_{max}(D(\omega))$ and ω_0 maximizes $\lambda_{max}(\omega)$ over $\omega \in [-\pi,\pi]$. Then

$$\overline{M}(F) = (1-\alpha)W(\hat{F}) + \alpha W(\omega_0) + \overline{A} \tag{59}$$

where

$$W(\omega_0) = [\mathrm{Re}\ \mathrm{Tr}(B_{ij}(\omega_0)\ \psi_{max}(\omega_0)\ \psi^*_{max}(\omega_0))] \tag{60}$$

Since $\hat{F}$ is optimal,

$$\frac{\partial}{\partial\alpha} \log \left|\overline{M}(F)\right|_{\alpha=0} \leq 0 \tag{61}$$

or

$$\mathrm{Tr}\left[\overline{M}^{-1} \frac{\partial\overline{M}}{\partial\alpha}\right]_{\alpha=0} \leq 0$$

or

$$\mathrm{Tr}\{\overline{M}^{-1}(\hat{F})[-W(\hat{F}) + W(\omega_0)]\} \leq 0$$

or

$$-\mathrm{Tr}(\overline{M}^{-1} W) + \sum_{i,j=1}^{m} p_{ij} w_{ij}(\omega_0) \leq 0$$

or

$$\mathrm{Tr}\left[\mathrm{Re} \sum_{i,j=1}^{m} p_{ij} B_{ij}(\omega_0)\ \psi_{max}(\omega_0)\ \psi^*_{max}(\omega_0)\right] \leq \mathrm{Tr}(\overline{M}^{-1} W)$$

or

$$\lambda_{max}(D(\omega_0)) \leq \mathrm{Tr}(\overline{M}^{-1} W) \tag{62}$$

Equations (54) and (62) are in contradiction unless

$$\max_{\omega} \lambda_{max}(D(\omega,\hat{F})) = \mathrm{Tr}[\overline{M}^{-1}(\hat{F}) W(\hat{F})]$$

Clearly, $\hat{F}$ minimizes $\max_{\omega} \lambda_{max}(D(\omega,F))$.

We now show that Parts (i) and (iii) of the theorem follow from Part (ii). Part (iii) follows directly from Eq. (54) and the fact that $\hat{F}$ minimizes $\max_{\omega} \lambda_{max}(D(\omega,F))$. To prove that (i) follows from (ii), we assume the contrary, viz. that (ii) and (iii) hold, but (i) does not. Consider any other normalized design, $F_1 = (1-\alpha)\hat{F} + \alpha F_0$. Then for some F_0,

$$\frac{\partial}{\partial\alpha} \log \left|\bar{M}(F_1)\right|_{\alpha=0} > 0$$

or

$$\mathrm{Tr}[\bar{M}^{-1}(\hat{F})(-W(\hat{F}) + W(F_0)] > 0$$

or

$$\mathrm{Tr}[\bar{M}^{-1}(\hat{F})W(F_0)] > \mathrm{Tr}[\bar{M}^{-1}(\hat{F})W(\hat{F})] \tag{63}$$

We have shown in Theorem 1 (Eq. (48)) that any design F_0 can be replaced by a discrete design with $\ell \leq m(m+1)/2+1$ points without changing its information matrix. Let

$$W(F_0) = \sum_{r=1}^{\ell} \beta_r \bar{W}(\omega_r) \qquad \text{(From Eq. (46))}$$

$$= \sum_{r=1}^{\ell} \sum_{k=1}^{q} \beta_r W_k(\omega_r)\mu_k(\omega_r) \Big/ \sum_{k=1}^{q} \mu_k(\omega_r)$$

$$= \sum_{r=1}^{\ell} \sum_{k=1}^{q} \gamma_{rk} W_k(\omega_r) \tag{64}$$

where, from Eq. (43),

$$W_k(\omega_r) = [\mathrm{Re}\,\mathrm{Tr}(B_{ij}(\omega_r)\phi_k(\omega_r)\phi_k^*(\omega_r)] \tag{65}$$

$$\gamma_{rk} = \frac{\beta_r \mu_k(\omega_r)}{\sum_{k=1}^{q} \mu_k(\omega_r)} \tag{66}$$

and since $\Sigma_{r=1}^{\ell} \beta_r = 1$,

$$\sum_{r=1}^{\ell} \sum_{k=1}^{q} \gamma_{rk} = 1 , \qquad 0 \leq \gamma_{rk} \leq 1 \tag{67}$$

Let $\{\psi_1,\ldots,\psi_q\}$ be normalized eigenvectors of $D(\hat{F},\omega_r)$ corresponding to eigenvalues $\{\lambda_1(\omega_r),\ldots,\lambda_q(\omega_r)\}$. We can write

$$\phi_k(\omega_r) = \sum_{s=1}^{q} \zeta_{ksr}\psi_s(\omega_r) , \qquad \sum_{s=1}^{q} |\zeta_{krs}|^2 = 1 .$$

$$\begin{aligned}
\mathrm{Tr}[\bar{M}^{-1}(\hat{F})W(F_0)] &= \sum_{i,j=1}^{m} p_{ij}w_{ij}(F_0) \\
&= \sum_{i,j=1}^{m} \sum_{r=1}^{\ell} \sum_{k=1}^{q} p_{ij}\gamma_{rk}\bar{w}_{ijk}(\omega_r) \\
&= \sum_{r=1}^{\ell} \sum_{\substack{k=1\\s=1}}^{q} \gamma_{rk} \,\mathrm{Tr}\left[\mathrm{Re} \sum_{i,j=1}^{m} p_{ij}B_{ij}(\omega_r)\psi_k(\omega_r)\psi_k^*(\omega_r)\right] \cdot |\zeta_{krs}|^2 \\
&= \sum_{r=1}^{\ell} \sum_{\substack{k=1\\s=1}}^{q} \gamma_{rk}\lambda_k(\omega_r)|\zeta_{krs}|^2
\end{aligned} \tag{68}$$

Thus

$$\sum_{r=1}^{\ell} \sum_{k=1}^{q} \gamma_{rk}\lambda_k(\omega_r,\hat{F})|\zeta_{krs}|^2 > \mathrm{Tr}[\bar{M}^{-1}(\hat{F})W(\hat{F})] \tag{69}$$

But from (ii) and (iii),

$$\lambda_k(\omega,\hat{F}) \le \mathrm{Tr}[\bar{M}^{-1}(\hat{F})W(\hat{F})] \tag{70}$$

or

$$\sum_{r=1}^{\ell} \sum_{\substack{k=1\\s=1}}^{q} \gamma_{rk}\lambda_k(\omega_r,\hat{F})|\zeta_{krs}|^2 \le \mathrm{Tr}[\bar{M}^{-1}(\hat{F})W(\hat{F})] \tag{71}$$

since

$$\sum_{r=1}^{\ell} \sum_{\substack{k=1\\s=1}}^{q} \gamma_{rk}|\zeta_{krs}|^2 = 1$$

Comparing (69) and (71), there is a contradiction unless $\hat{F}$ minimizes $|\bar{M}|$.

The remaining parts of Theorem 2 follow from the concavity of the $\log |M|$ function and proof is similar to that in Ref.[1].■

We now propose an algorithm based on the above theorem for computing $\hat{F}$.

ALGORITHM 1: (a) Start with any design F_0 such that $\overline{M}(F_0)$ is nonsingular. Let $k = 0$.

(b) Compute

$$D_k = \text{Re} \sum_{i,j=1}^{m} p_{ij}(F_k) B_{ij}(\omega)$$

and find its maximum eigenvalue $\lambda_{max}^{k}(\omega)$. Find $\omega_k \in [-\pi,\pi]$ by a one dimensional search so that

$$\lambda_{max}^{k}(\omega_k) \geq \lambda_{max}(\omega) \tag{72}$$

Also compute the eigenvector ψ_{max}^{k}.

(c) If

$$\lambda_{max}^{k}(\omega_k) = \text{Tr}[\overline{M}^{-1}(F_k) W(F_k)] \tag{73}$$

stop. Otherwise proceed to (d).

(d) Update the design as follows:

$$F_{k+1} = (1-\alpha_k) F_k + \alpha_k F(\omega_k) \tag{74}$$

where $F(\omega_k)$ is a design with a single point at $\omega = \omega_k$ of size $\psi_{max}^{k}(\psi_{max}^{k})^T$. Choose $0 < \alpha_k \leq 1$ either by a one-dimensional search or any sequence such that

$$|M^{k+1}| \geq |M^k| , \qquad \sum_{k=0}^{\infty} \alpha_k = \infty , \qquad \lim_{k \to \infty} \alpha_k = 0 \tag{75}$$

(e) Go back to (b).

The convergence of the above algorithm to a global maximum can be proved in the same way as in Refs.[1,7].

6. EXTENSIONS TO OTHER CRITERIA

As shown in Ref. [1], the results obtained here can be easily generalized to handle criteria of the type $L(\bar{M}^{-1})$ or $L(\bar{M}^{-k})$ where L is a linear function, e.g., trace. In the frequency domain, one can derive the following theorem in the same fashion as Theorem 2.

THEOREM 3: *For the normalized design,* $\hat{F}_{uu}$ *the following are equivalent*

(i) $\hat{F}_{uu}$ *minimizes* $\mathrm{Tr}(\bar{M}^{-k})$, $k = 1,2,\ldots$

(ii) $\hat{F}_{uu}$ *minimizes*

$$\max \lambda_{\max} \mathrm{Re} \sum_{i,j=1}^{m} p_{ij}^{k+1} (F_{uu}) B_{ij}(\omega)$$

where p_{ij}^{k+1} *is the* (i,j) *th element of* $\bar{M}^{-(k+1)}$.

(iii)
$$\max_{\omega} \lambda_{\max} \mathrm{Re} \sum_{i,j=1}^{m} p_{ij}^{k+1} (\hat{F}_{uu}) B_{ij}(\omega) = \mathrm{Tr}[\bar{M}^{-(k+1)} W(\hat{F}_{uu})] \tag{76}$$

REMARK: It is interesting to see that $k = 0$ leads to results for D-optimal designs and $k \to \infty$ would give designs that maximize the smallest eigenvalue of M.

7. BOUNDS

In this section we derive bounds similar to those of Kiefer [10] for D-optimal designs. Consider a set of designs

$$F(\alpha) = \alpha\hat{F} + (1-\alpha)F \tag{77}$$

for $0 \le \alpha \le 1$ so that $F(0) = F$ and $F(1) = \hat{F}$.

$$M(F(\alpha)) = \alpha W(\hat{F}) + (1-\alpha)W(F) + A \tag{78}$$

$$\frac{\partial}{\partial\alpha} \log |M(F(\alpha))| = \mathrm{Tr}\{M^{-1}(F(\alpha))[W(\hat{F}) - W(F)]\} \tag{79}$$

$$\frac{\partial^2}{\partial\alpha^2} \log |M(F(\alpha))| = -\mathrm{Tr}\{M^{-1}(F(\alpha))(W(\hat{F}) - W(F))M^{-1}(F(\alpha))(W(\hat{F}) - W(F))\} \tag{80}$$

Since $M^{-1}(F(\alpha))$ is positive-definite and $(W(\hat{F}) - W(F))$ is nonnegative-definite, the eigenvalues of $M^{-1}(F(\alpha))\cdot(W(\hat{F})-W(F))$ are real and nonnegative. Thus

$$\frac{\partial^2}{\partial\alpha^2} \log |M(F(\alpha))| \leq 0 \tag{81}$$

for $\alpha \in [0,1]$. From Eq. (79),

$$\frac{\partial}{\partial\alpha} \log |M(F(\alpha))|_{\alpha=0} = \mathrm{Tr}\{M^{-1}(F)W(\hat{F}) - M^{-1}(F)W(F)\} \tag{82}$$

Now using Theorem 1 and the procedure followed in the proof of Theorem 2, $\mathrm{Tr}\{M^{-1}(F)W(\hat{F})\}$ may be written as (cf., Eq. (68))

$$\begin{aligned} \mathrm{Tr}\{M^{-1}(F)W(\hat{F})\} &= \sum_{r=1}^{\ell} \sum_{s,k=1}^{q} \gamma_{rk}|\zeta_{krs}|^2 \lambda_k(D(\omega_r,F)) \\ &\leq \max_{\omega} \lambda_{max} (D(\omega,F)) \end{aligned} \tag{83}$$

where

$$D(\omega,F) = \mathrm{Re} \sum_{i,j=1}^{m} p_{ij}(F)B_{ij}(\omega) \tag{84}$$

Thus,

$$\frac{\partial}{\partial\alpha} \log |M(F(\alpha))|_{\alpha=0} \leq \left\{\max_{\omega} \lambda_{max} (D(\omega,F) - \mathrm{Tr}(M^{-1}(F)W(F))\right\} \tag{85}$$

From Eqs. (81) and (85) it follows that the above inequality must hold for all $\alpha \in [0,1]$. Now, integrating both sides of Eq. (85) over α from 0 to 1,

$$\log \frac{|M(\hat{F})|}{|M(F)|} \leq \max_{\omega} \lambda_{max} (D(\omega,F)) - \mathrm{Tr}(M^{-1}(F)W(F))$$

or

$$\frac{|M(F)|}{|M(\hat{F})|} \geq \exp\left\{\mathrm{Tr}(M^{-1}(F)W(F)) - \max_{\omega} \lambda_{max} (D(\omega,F))\right\} \tag{86}$$

The bound (86) is easily evaluated for any design F. For single input systems without process noise, the expression on the right hand side of Eq. (86) simplifies to

$$\exp\left\{m - \max_{\omega} D(\omega,F)\right\} \quad .$$

8. CONCLUSIONS

In this chapter, earlier results on frequency domain input design of single-input systems without process noise have been extended to multiinput multioutput (MIMO) systems. The results show the generality of the approach and in all cases lead to practical algorithms for computing global optimal designs.

APPENDIX A

LEMMA: *Let* D *and* $C \geq 0$ *be* $q \times q$ *Hermitian matrices. Then*

$$\lambda_{min}(D)\,Tr(C) \leq Tr(DC) \leq \lambda_{max}(D)\,Tr(C)$$

<u>Proof</u>. Since D is Hermitian, its eigenvalues $(\lambda_1,\ldots,\lambda_q)$ are real and the column matrix P of the corresponding orthonormal eigenvectors is unitary, i.e., $P^*P = I$. Also $PDP^* = \text{Diag}[\lambda_1,\ldots,\lambda_q]$

$$\begin{aligned} Tr(DC) &= Tr(PDP^*PCP^*) \\ &= \sum_{i=1}^{q} \lambda_i (PCP^*)_{ii} \end{aligned}$$

But $(PCP^*)_{ii} \geq 0$ since C, being Hermitian and nonnegative definite can be written as $C = LL^*$. Further,

$$\sum_{i=1}^{q} (PCP^*)_{ii} = Tr(PCP^*) = Tr(C)$$

Thus

$$\begin{aligned} \lambda_{min}(D)\,Tr(C) &\leq Tr(DC) \\ &\leq \lambda_{max}(D)\,Tr(C) \end{aligned}$$

PART II: TIME DOMAIN SYNTHESIS

1. INTRODUCTION

The previous work [1-6] in time-domain input design has considered deterministic inputs for linear systems without process noise. In most cases, either simple criteria such as the trace of the Fisher information matrix are considered [1-3] or when other criteria are considered, only locally optimal inputs are obtained [4-6].

In contrast, the frequency-domain approach of the author [7,8] has produced inputs that are globally optimal in the class of both deterministic and stochastic inputs. In this paper, we present parallel and, in many cases, more general results for the design of time-domain inputs. As before, our results are based on the important work of Kiefer and Wolfowitz [9,10] in the design of statistical experiments.

We define an input design by a probability measure ξ on the space of admissible inputs. An expression for the information matrix M for linear systems with process noise is derived in Section 2. The convexity and finite support properties of M and ξ are derived in Section 3. The maximization of the determinant $|M|$ with respect to ξ is considered in Section 4 and the main theorem of the paper is proved. Computation of optimal designs is discussed in Section 5. Extensions to continuous-time systems, other criteria, nonlinear and distributed parameter systems are considered in Sections 6, 7 and 8. Conclusions are stated in Section 9.

2. TIME-DOMAIN INFORMATION MATRIX

This section gives the time domain results equivalent to those given in Part I, Section 3 for the frequency domain. Define as before, the standardized innovation process,

$$\nu(t) = \Sigma^{-1}(t)(y(t)-H\hat{x}(t)) \tag{87}$$

where $\Sigma(t)$ and $\hat{x}(t)$ are defined by Eqs. (9)-(13). The log-likelihood function $L(\theta)$ is

$$L(\theta) = -\frac{N}{2}\log(2\pi) - \frac{1}{2}\sum_{t=1}^{N}\{\nu^T(t)\nu(t) + 2\log|\Sigma(t)|\} \tag{88}$$

which is analogous to Eq. (23) in frequency domain. The Fisher information matrix M is given as [11]

$$M = E\left[\frac{\partial L}{\partial \theta}\left(\frac{\partial L}{\partial \theta}\right)^T\right] \tag{89}$$

where the expectation is taken over the sample space of observations $\{y(t),\ t=1,\ldots,N\}$. From Eq. (88),

$$\frac{\partial L}{\partial \theta_i} = -\sum_{t=1}^{N}\nu^T(t)\frac{\partial \nu(t)}{\partial \theta_i} - \sum_{t=1}^{N}\mathrm{Tr}\left(\Sigma^{-1}(t)\frac{\partial \Sigma(t)}{\partial \theta_i}\right) \tag{90}$$

$$\frac{\partial \nu(t)}{\partial \theta_i} = \frac{\partial \Sigma^{-1}(t)}{\partial \theta_i}(y - H\hat{x}(t)) - \Sigma^{-1}(t)H\frac{\partial \hat{x}(t)}{\partial \theta_i} \tag{91*}$$

From Eq. (9),

$$\frac{\partial \hat{x}(t+1)}{\partial \theta_i} = (\Phi-KH)\frac{\partial \hat{x}(t)}{\partial \theta_i} + \frac{\partial \Phi}{\partial \theta_i}\hat{x}(t) + \frac{\partial G}{\partial \theta_i}u(t) + \frac{\partial K}{\partial \theta_i}\nu(t) \tag{92}$$

---

*For simplicity, it is assumed here that H and R do not contain unknown paramters.

$$\frac{\partial \hat{x}(0)}{\partial \theta_i} = 0$$

From (91) and (92)

$$E\left[\frac{\partial \nu(t)}{\partial \theta_i} \frac{\partial \nu^T(t)}{\partial \theta_j}\right] = \Sigma^{-1}(t) \frac{\partial \Sigma(t)}{\partial \theta_i} \cdot \frac{\partial \Sigma(t)}{\partial \theta_j} \Sigma^{-1}(t)$$

$$+ \Sigma^{-1}(t) H E\left(\frac{\partial \hat{x}(t)}{\partial \theta_i} \frac{\partial \hat{x}(t)}{\partial \theta_j}^T\right) H^T \Sigma^{-1}(t) \tag{93}$$

After certain simplifications, one obtains

$$M_{ij} = \operatorname{Tr} \sum_{t=1}^{N} \left\{ \Sigma^{-1}(t) H E\left(\frac{\partial \hat{x}(t)}{\partial \theta_i} \frac{\partial \hat{x}^T(t)}{\partial \theta_j}\right) H^T \Sigma^{-1}(t) \right.$$

$$\left. + \Sigma^{-1}(t) \frac{\partial \Sigma(t)}{\partial \theta_j} \frac{\partial \Sigma(t)}{\partial \theta_i} \Sigma^{-1}(t) \right\} \tag{94}$$

$$+ 2 \sum_{t=1}^{N} \operatorname{Tr}\left(\Sigma^{-1}(t) \frac{\partial \Sigma(t)}{\partial \theta_i}\right) \cdot \operatorname{Tr}\left(\Sigma^{-1}(t) \frac{\partial \Sigma(t)}{\partial \theta_j}\right)$$

The expressions for

$$E\left[\frac{\partial \hat{x}(t)}{\partial \theta_i} \frac{\partial \hat{x}(t)}{\partial \theta_j}\right], \qquad i,j = 1,\ldots,m$$

may be obtained by defining the augmented state vector

$$\hat{x}_A(t) = \begin{bmatrix} \hat{x}(t) \\ \frac{\partial \hat{x}}{\partial \theta_1}(t) \\ \vdots \\ \frac{\partial \hat{x}}{\partial \theta_m}(t) \end{bmatrix} \tag{95}$$

and augmented system matrices.

$$\underset{n(m+1)\times n(m+1)}{\Phi_A} = \begin{bmatrix} \Phi, & 0, & \ldots, & 0 \\ \frac{\partial\Phi}{\partial\theta_1}, & \Phi - KH, & \ldots, & 0 \\ \vdots & & & \\ \frac{\partial\Phi}{\partial\theta_m}, & 0, & \ldots, & \Phi - KH \end{bmatrix}$$

$$\underset{n(m+1)\times q}{G_A} = \begin{bmatrix} G \\ \frac{\partial G}{\partial\theta_1} \\ \vdots \\ \frac{\partial G}{\partial\theta_m} \end{bmatrix}$$

$$\underset{n(m+1)\times p}{K_A(t)} = \begin{bmatrix} K(t) \\ \frac{\partial K(t)}{\partial\theta_1} \\ \vdots \\ \frac{\partial K(t)}{\partial\theta_m} \end{bmatrix}$$

Define

$$\bar{x}_A = E[\hat{x}_A] \tag{96}$$

$$\Sigma_A = E[(\hat{x}_A - \bar{x}_A)(\hat{x}_A - \bar{x}_A)^T] \tag{97}$$

The difference equation for $\hat{x}_A(t)$ may be written as

$$\hat{x}_A(t+1) = \Phi_A\hat{x}(t) + G_Au(t) + K_A\nu(t) \tag{98}$$

The mean and covariance equations for (98) are

$$\bar{x}_A(t+1) = \Phi_A\bar{x}_A(t) + G_Au(t) \tag{99}$$

$$\Sigma_A(t+1) = \Phi_A \Sigma_A(t) \Phi_A^T + K_A K_A^T \tag{100}$$

with initial conditions

$$\bar{x}_A(0) = 0$$

$$\Sigma_A(0) = \begin{bmatrix} P_0 & & \\ & 0 & \\ & & \ddots 0 \end{bmatrix}$$

The solution to Eqs. (99)-(100) may be written in terms of the multi-step transition matrix

$$\Phi_A(t,k) = \Phi_A(t-1)\Phi_A(t-2),\ldots,\Phi_A(k) \tag{101}$$

$$\bar{x}_A(t) = \sum_{k=0}^{t-1} \Phi_A(t,k) G_A u(k) \tag{102}$$

$$\Sigma_A(t) = \Phi_A(t,0)\Sigma_A(0)\Phi_A^T(t,0) + \sum_{k=0}^{t-1} \Phi_A(t,k) K_A(k) K_A^T(k) \Phi_A^T(t,k) \tag{103}$$

Now

$$E\left[\frac{\partial \hat{x}(t)}{\partial \theta_i} \quad \frac{\partial \hat{x}^T(t)}{\partial \theta_j}\right] = C_i(\bar{x}_A(t)\bar{x}_A^T(t) + \Sigma_A(t))C_j^T \tag{104}$$

where

$$C_i = [0,0,\ldots,I,0,0,\ldots,0] \quad , \quad i = 0,\ldots,m$$

C_i is of dimension $n \times n(m+1)$ and

$$\frac{\partial \hat{x}}{\partial \theta_i} = C_i x_A \quad , \qquad \hat{x} = C_0 \quad .$$

Using Eqs. (101)-(104) in Eq. (94)

$$M_{ij} = \sum_{t=1}^{N} \sum_{k=0}^{t-1} \sum_{\ell=0}^{t-1} u^T(\ell) w_{ij}(t,\ell,k) u(k) + A_{ij} \tag{105}$$

where

$$\underset{(q\times q)}{w_{ij}}(t,\ell,k) = G_A^T \Phi_A^T(t,\ell) C_j^T H^T \Sigma^{-2} H C_i \Phi_A(t,k) G_A \tag{106}$$

and A_{ij} consists of all the remaining terms in Eqs. (94) and (104) that do not depend on the input $u(t)$.

Interchanging indices in Eq. (105) using the identities,

$$\sum_{t=1}^{N} \sum_{k=0}^{t-1} \equiv \sum_{k=0}^{N-1} \sum_{t=k+1}^{N}$$

and

$$\sum_{t=k+1}^{N} \sum_{\ell=0}^{t-1} \equiv \sum_{\ell=0}^{N-1} \sum_{t=\max(k+1,\ell+1)}^{N}$$

$$M_{ij} = \sum_{k=0}^{N-1} \sum_{\ell=0}^{N-1} u^T(\ell) w_{ij}(\ell,k) u(k) + A_{ij} \tag{107}$$

where

$$\underset{(q\times q)}{w_{ij}}(\ell,k) = \sum_{t=\max(k+1,\ell+1)}^{N} w_{ij}(t,\ell,k) \tag{108}$$

Defining w_{ij} as the first term in Eq. (107)

$$M_{ij} = w_{ij} + A_{ij}$$

The above expression for the information matrix has been derived assuming that $U^T = [u^T(0),\ldots,u^T(N-1)]$ is given. For design purposes, we allow U to be chosen randomly from a compact set* $\Omega_U \subset \mathbb{R}^{Nq}$. The input design consists of a probability measure $\xi(dU)$ defined for all subsets of Ω_U including single points (i.e., probability masses are allowed). The information matrix for a randomized design of this type is

$$M(\xi) = \int_{\Omega_U} W(U)\xi(dU) + A \tag{109}$$

where

$$\int_{\Omega_U} \xi(dU) = 1 , \qquad 0 \le \xi(dU) \le 1 .$$

---

*As special cases, we would consider sets $\Omega_U = \{U: a_t \le u(t) \le b_t, t=0,\ldots,N-1\}$ and $\Omega_U = \{U: U^T U \le 1\}$.

Let

$$W(\xi) = \int_{\Omega_U} W(U)\xi(dU) \tag{110}$$

$$M(\xi) = W(\xi) + A$$

3. PROPERTIES OF THE INFORMATION MATRIX FOR RANDOMIZED DESIGNS

THEOREM 1: *1. The information matrix* $M(\xi)$ *is symmetric and positive definite.*

2. The set of all information matrices $M(\xi)$ *is convex and closed.*

3. The information matrix $M(\xi)$ *for any design* ξ *(continuous and discrete) may be achieved by another design* ξ' *that has finite support, viz.* ξ' *assigns positive probabilities to at most* $[m(m+1)/2+1]$ *points in* Ω_U.

Proof. The proof of this theorem is similar to that of an analogous theorem in frequency domain [7,8] with the only difference that the role of the spectral distribution function is replaced by the probability measure. We simply sketch the important features of the proof.

From Eq. (109), it is obvious that

$$M(\alpha\xi_1 + (1-\alpha)\xi_2) = \alpha M(\xi_1) + (1-\alpha)M(\xi_2)$$

for $0 \le \alpha \le 1$. Thus the set $M(\xi)$ is convex. It is also closed since the set Ω_U is closed and $W(U)$ is a continuous mapping from Ω_U to $\mathbb{R}^{m\times m}$.

Property 3 follows from Caratheodory's theorem on the representation of points in the convex hull of a set in $\mathbb{R}^{m(m+1)/2}$ to which $M(\xi)$ belongs. We refer the reader to Kiefer and Wolfowitz [10] for a proof of this property.

4. D-OPTIMAL DESIGN IN TIME-DOMAIN

LEMMA 1: $$\max_{U\in\Omega_U} \mathrm{Tr}[M^{-1}(\xi)W(U)] \ge \mathrm{Tr}[M^{-1}(\xi)W(\xi)] . \tag{112}$$

<u>Proof</u>. Using Eqs. (109)-(110),

$$\mathrm{Tr}[M^{-1}(\xi)W(\xi)] = \int_{\Omega_U} \mathrm{Tr}(M^{-1}(\xi)W(U))\xi(dU)$$

$$\leq \max_{U\in\Omega_U} \mathrm{Tr}(M^{-1}(\xi)W(U))$$

since $\xi(dU)$ lies between 0 and 1 and the mean value of a function is less than or equal to its maximum value. ■

THEOREM 2: *Let* ξ^* *be the optimal design. Then the following are equivalent.*

(i) ξ^* *maximizes* $|M(\xi)|$

(ii) ξ^* *minimizes* $\max_{U\in\Omega_U} \mathrm{Tr}\{M^{-1}(\xi)W(U)\}$

(iii) $$\max_{U\in\Omega_U} \mathrm{Tr}\{M^{-1}(\xi^*)W(U)\} = \mathrm{Tr}[M^{-1}(\xi^*)W(\xi^*)] \tag{113}$$

All designs satisfying (i)-(iii) and their linear combinations have the same information matrix M. *If there is no process noise and no unknown parameters in* R *implying* $A = 0$, *the saddle point value in Eq. (113) is* m, *i.e., the dimension of* θ.

<u>Proof</u>. We first show that (i) implies (ii) and (iii). Consider another design,

$$\xi = (1-\alpha)\xi^* + \alpha\xi_0 \tag{114}$$

Then

$$M(\xi) = (1-\alpha)W(\xi^*) + \alpha W(\xi_0) + A \tag{115}$$

If ξ^* maximizes $|M(\xi)|$ or $\log|M(\xi)|$, any deviation from ξ^* should result in a decrease in value, i.e.,

$$\frac{\partial}{\partial\alpha} \log |M(\xi)|_{\alpha=0} \leq 0 \tag{116}$$

or

$$\mathrm{Tr}\ [M^{-1}(\xi^*)(-W(\xi^*) + W(\xi_0))] \leq 0 \tag{117}$$

Let ξ_0 be a design that assigns all the probability mass to the value $U \in \Omega_U$ that maximizes $\mathrm{Tr}\{M^{-1}(\xi^*)W(U)\}$. Then

$$\max_{U\in\Omega_U} \mathrm{Tr}\,[M^{-1}(\xi^*)W(U)] \leq \mathrm{Tr}\,[M^{-1}(\xi^*)W(\xi^*)] \tag{118}$$

Equations (112) and (118) are in contradiction unless (ii) and (iii) hold. We now show that (i) and (iii) follow from (ii). Using Lemma 1, Part (iii) follows immediately from Part (ii). To deduc (i), assume the contrary, i.e., ξ^* satisfies (ii), but does not maximize $|M(\xi)|$. One should then be able to construct another design of the type (114) such that

$$\frac{\partial}{\partial\alpha} \log |M(\xi)|_{\alpha=0} > 0$$

or

$$\mathrm{Tr}\,[M^{-1}(\xi^*)W(\xi_0)] > \mathrm{Tr}\,[M^{-1}(\xi^*)W(\xi^*)] \tag{119}$$

From Theorem 1, any design ξ_0 may be replaced by a discrete design with $k \leq (1/2)m(m+1)+1$ points. Let

$$W(\xi_0) = \sum_{i=1}^{k} \xi_i W(U_i) \tag{120}$$

where

$$\sum_{i=1}^{k} \xi_i = 1 , \qquad 0 \leq \xi_i \leq 1 \tag{121}$$

Equations (119) and (120) imply,

$$\sum_{i=1}^{k} \xi_i \,\mathrm{Tr}\,[M^{-1}(\xi^*)W(U_i)] > \mathrm{Tr}\,[M^{-1}(\xi^*)W(\xi^*)]$$

or

$$\max_{U} \mathrm{Tr}\,[M^{-1}(\xi^*)W(U)] > \mathrm{Tr}\,[M^{-1}(\xi^*)W(\xi^*)] \tag{122}$$

which contradicts (113). Thus ξ^* must maximize $|M|$.

The remainder of the theorem follows from the fact that $|M(\xi)|$ is a concave function and if two designs ξ_1 and ξ_2 both satisfy (i)-(iii), their linear combinations would give rise to a higher value of $|M(\xi)|$ and thus violate (i). Therefore, $M(\xi_1) = M(\xi_2)$. Furthermore, when there is no process noise, $A = 0$ or $W = M$, the saddle point value is $\text{Tr}\,[M^{-1}(\xi^*)M(\xi^*)] = m$.

5. COMPUTATION OF D-OPTIMAL INPUT DESIGNS

The optimal design ξ^* may be computed as follows:

ALGORITHM 1: (1) Start with any design ξ_0 such that $M(\xi_0)$ is nonsingular. One method of choosing ξ_0 is to use a nondegenerate frequency domain input U_f [8] and assign $\xi(U_f) = 1$. In other cases, a suitable input U or combination of inputs may be chosen using physical considerations. Let $k = 0$.

(2) Compute $W(\xi_k)$, $M(\xi_k)$ and $\text{Tr}\,[M^{-1}(\xi_k)W(U)]$ using Eqs. (106)-(110).

(3) Maximize $\text{TR}\,[M^{-1}(\xi_k)W(U)]$ over $U \in \Omega_U$. Computationally, this is the most time consuming and key step in this algorithm. We consider two cases:

(a) *Bounded Energy Inputs:* In this case the set Ω_U is a hypersphere in $\mathbb{R}^{Nq}$ and without loss of generality, we may take

$$\Omega_U = \{U: U^T U \leq 1\} \tag{123}$$

Since $\text{Tr}\,[M^{-1}(\xi_k)W(U)]$ is a quadratic form in U (cf., Eq. (107)), the maximizing value U_k is an eigenvector of the related $Nq \times Nq$ Hessian matrix H corresponding to its maximimum eigenvalue. The (ℓ,k)th element of H may be written as

$$H_{\ell,k} = \sum_{i=1}^{m} \sum_{j=1}^{m} p_{ij}(\xi_k) w_{ij}(\xi,k) \tag{124}$$

where $p_{ij}(\xi_k)$ is the (i,j)th element of $M^{-1}(\xi_k)$ and $w_{ij}(\ell,k)$ is given by Eq. (108). The optimum U_k is achieved on the boundary of the set Ω_U, i.e., $U_k^T U_k = 1$.

(b) *Bounded Amplitude Input:* In this case,

$$\Omega_U = \{U: a_t \leq u(t) \leq b_t, \; i = 0,\ldots,N-1\}$$

where all a_t, b_t are specified. Such constraint sets have been considered by other authors [1,4,5], but only locally optimal non-randomized inputs have been obtained. The problem of maximizing $\text{Tr}[M^{-1}(\xi_k)W(U)]$ over $U \in \Omega_U$, in this case, is a quadratic programming problem and a number of efficient computational algorithms are available for computing the maximum. The optimizing inputs U are of the bang-bang type, i.e., $u(t)$ is either a_t or b_t.

(4) If $\text{Tr}\,\{M^{-1}(\xi_k)W(U_k)\} = \text{Tr}\,\{M^{-1}(\xi_k)W(\xi_k)\}$, stop. Otherwise, let

$$\xi_{k+1} = (1-\alpha_k)\xi_k + \alpha_k \xi(U_k) \qquad 0 < \alpha_k \leq 1 \tag{125}$$

where $\xi(U_k)$ is the design at the single point U_k. Choose α_k either by maximizing $|M(\xi_{k+1})|$ with respect to α_k or any sequence such that

$$\sum_{k=0}^{\infty} \alpha_k = \infty, \qquad \lim_{k\to\infty} \alpha_k = 0$$

and $|M(\xi_{k+1})| \geq |M(\xi_k)|$.

(5) Set $k = k+1$ and go to step (2).

THEOREM 3: *Let the conditions of Theorem 2 be satisfied. Then Algorithm 1 converges to a globally optimal design* ξ^*.

We omit the proof of this theorem since it is very similar to the proof of analogous theorems in Refs. [7-10].

REMARKS: (1) A comparison of the above algorithm with the corresponding frequency-domain algorithm [8] reveals the appealing computational simplicity of the latter. The inputs U in frequency-domain are specified by a single variable, viz., frequency ω and the set Ω_U is simply $[-\pi,\pi]$. Therefore, the step (3) of maximization is almost trivial. The price one pays for this simplification is the assumption of linearity and stationarity.

(2) The optimal design consists of a set of weights and inputs

$$\{\xi_1, U_1; \xi_2, U_2; \ldots; \xi_k, U_k\} \qquad k \le m(m+1)/2$$

which can be used in the usual manner of randomized strategies, when the experiment can be repeated. However, if the experiment cannot be repeated, one may construct a composite input by either superposition or by time concatenation of the different inputs U_i in the proportion ξ_i to approximate the randomized design.

(3) As shown in Ref. [7], the optimal design has the property that at all points of the design, $\mathrm{Tr}\,\{M^{-1}(\xi^*)W(U_i)\}$, $i = 1,\ldots,k$ has the same minmax value. In the bounded-energy case, the minmax value is the maximum eigenvalue λ_{max} of $H(\xi^*)$ (Eq. (124)). The corresponding eigenvectors U_i are orthogonal. Furthermore, any other basis in the k-dimensional subspace of eigenvectors of λ_{max} is also optimal. In addition, if

$$M\left(\sum_{i=1}^{k} \xi_i^{1/2}\, U_i\right) = M(\xi^*) ,$$

as in the frequency domain case, the nonrandomized design $\xi(U') = 1$,

$$\text{where} \qquad U' = \sum_{i=1}^{k} \xi_i^{1/2}\, U_i$$

is optimal.

(4) The quantity $\mathrm{Tr}\,[M^{-1}(\xi)W(U)]$ which is maximized with respect to $U \in \Omega_U$ and minimized with respect to ξ may be shown to be the same as

$$\mathrm{Tr}\,[\Sigma^{-2}\,\mathrm{cov}(Y_1(\theta))] = \mathrm{Tr}\left\{\Sigma^{-2} E\left[\frac{\partial Y_1}{\partial \theta}\, M^{-1}(\xi) \left(\frac{\partial Y_1}{\partial \theta}\right)^T\right]\right\}_{\theta=\theta_0}$$

where $Y_1^T = [y_1^T(1), \ldots, y_1^T(N)]$ is the output response due to input U of the Kalman filter representation of the system and the expression on the right hand is obtained by expanding $Y_1(\theta)$ to first order around the a priori value θ_0. Thus the optimal design ξ^* minimizes a scalar function of the maximum covariance of the input response function $Y_1(\theta)$ for all possible input sequences $U \in \Omega_U$. This type of optimality is known as G-optimality [9] and is a reasonable criterion in its own right. The design problem may then be thought of as a zero-sum two person game in which the designer (minimizer) chooses the probability measure ξ on subsets of Ω_U and nature (maximizer) chooses the worst input sequence $U \in \Omega_U$. The payoff function is the weighted covariance of the response $Y_1(\theta)$, viz., $\mathrm{Tr}\, [\Sigma^{-2} \mathrm{cov}(Y_1(\theta))]$.

6. CONTINUOUS-TIME SYSTEMS

Consider a continuous-time system described by the stochastic differential equation

$$dx(t) = Fx(t)dt + Gu(t)dt + \Gamma dw(t) \tag{126}$$

$$dz(t) = Hx(t)dt + dv(t) \tag{127}$$

where $\{w(t), v(t), t \in [0,T]\}$ are independent Wiener processes with incremental covariance Qdt and Rdt. Formally, $y(t) = dz(t)/dt$. The (i,j)-th element of the information matrix M for parameter set θ based on $\{z(t), t \in [0,T]\}$ can be obtained from Eq. (107) by a limit approach.

$$M_{ij} = E \int_0^T \mathrm{Tr} \left[H^T R^{-1} H \frac{\partial \hat{x}(t)}{\partial \theta_i} \frac{\partial \hat{x}^T(t)}{\partial \theta_j} \right] dt \tag{128}$$

where $\hat{x}(t)$ is the Kalman filter estimate given by

$$d\hat{x}(t) = F\hat{x}(t) + Gu(t)dt + Kd\eta(t) \tag{129}$$

$$d\eta(t) = dz(t) - H\hat{x}(t)dt \tag{130}$$

$$\hat{x}(0) = 0 \qquad \text{(an assumption)}$$

Here $\eta(t)$ is a Wiener process with incremental covariance Rdt. The innovation process is formally defined as $\nu(t) = d\eta(t)/dt$.

$K(t)$ is the Kalman filter gain given by $K(t) = P(t)H^T R^{-1}$ and
$P(t)$ is the solution to the Riccati equation.

$$\frac{d}{dt} P = FP + PF^T + \Gamma Q \Gamma^T - PH^T R^{-1} HP \quad , \quad P(0) = P_0 \tag{131}$$

$\hat{x}_0$ and P_0 are the prior mean and covariance of $x(0)$. From (129) and (130)

$$d\left(\frac{\partial \hat{x}(t)}{\partial \theta_i}\right) = (F-KH) \frac{\partial \hat{x}(t)}{\partial \theta_i} dt + \frac{\partial F}{\partial \theta_i} \hat{x}(t) dt + \frac{\partial G}{\partial \theta_i} u(t) dt + \frac{\partial K}{\partial \theta_i} d\eta(t) \tag{132}$$

Define $n(m+1)$ vector,

$$\hat{x}_A^T(t) = \left[\hat{x}^T(t), \frac{\partial \hat{x}^T(t)}{\partial \theta_1}, \ldots, \frac{\partial \hat{x}^T(t)}{\partial \theta_m}\right] \tag{133}$$

From Eqs. (83) and (85) we can write

$$d\hat{x}_A(t) = F_A \hat{x}_A(t) dt + G_A u(t) dt + K_A d\eta(t) \tag{134}$$

where F_A, G_A and K_A are defined as in the discrete-time case. Proceeding as in Section 2, one obtains

$$M_{ij} = w_{ij} + A_{ij} \tag{135}$$

where

$$A_{ij} = \int_0^T \mathrm{Tr}\, [H^T R^{-1} H C_i \Sigma_A C_j^T]\, dt \tag{136}$$

$$w_{ij} = \int_0^T \int_0^T \mathrm{Tr}\, [w_{ij}(\tau,s) u(\tau) u^T(s)]\, d\tau\, ds \tag{137}$$

and the $q \times q$ kernel $w_{ij}(\tau,s)$ is given as

$$w_{ij}(\tau,s) = \int_{\max(s,\tau)}^T G_A^T \Phi^T(t,s) C_j^T H^T R^{-1} H C_i \Phi(t,\tau) G_A\, dt \tag{138}$$

One now assigns a probability measure ξ to all Borel sets, including single points of the compact topological space Ω_U of all inputs $\{u(t),\ t \in [0,T]\}$. We would restrict attention here to inputs of bounded norm, i.e.,

$$\Omega_U = \left\{ U : \int_0^T u^T(t)u(t)\, dt \leq 1 \right\} \tag{139}$$

Under these conditions, Theorems 1 and 2 hold and the optimal design ξ^* may be computed by Algorithm 1. The only difference is in Step (3) where function space maximization has to be performed. A procedure for doing this has been given by the author [3] and involves computing the maximum eigenvalue and the corresponding eigenfunction of the following integral equation via the Riccati equation method

$$\int_0^T \sum_{i=1}^m \sum_{j=1}^m p_{ij}(\xi) w_{ij}(\tau,s) u(\tau)\, d\tau = \lambda u(s) \tag{140}$$

7. OTHER CRITERIA AND BOUNDS

We state here some results analogous to frequency-domain results, omitting proofs since they are essentially similar [8].

THEOREM 4: *The following are equivalent:*

(i) ξ^* *minimizes* $\mathrm{Tr}(M^{-k})$, $k = 1,2,\ldots$

(ii) ξ^* *minimizes* $\max_{U \in \Omega_U} \mathrm{Tr}\,\{M^{-(k+1)}(\xi)W(U)\}$

(iii) $$\max_{U \in \Omega_U} \mathrm{Tr}\,\{M^{-(k+1)}(\xi^*)W(U)\} = \mathrm{Tr}\,\{M^{-(k+1)}(\xi^*)W(U^*)\} \tag{141}$$

THEOREM 5: *A lower bound for any* D*-suboptimal design* ξ *is*

$$\frac{|M(\xi)|}{|M(\xi^*)|} \geq \exp\left\{ \mathrm{Tr}(M^{-1}(\xi)W(\xi)) - \max_{U \in \Omega_U} \mathrm{Tr}(M^{-1}(\xi)W(U)) \right\} \tag{142}$$

REMARKS: (1) Using Theorem 5, one can easily determine the degree of suboptimality of a given design ξ. The lower bound depends only on suboptimal ξ and U.

(2) In Theorem 4, $k = 1$ corresponds to an A-optimal design (average variance) and $k = \infty$ corresponds to an E-optimal design [10].

8. NONLINEAR AND DISTRIBUTED PARAMETER SYSTEMS WITHOUT PROCESS NOISE

Nonlinear Systems: Consider a nonlinear system

$$\frac{d}{dt}x(t) = f(x,u,t,\theta) \quad , \quad x(0) \text{ given} \tag{143}$$

$$y(t_i) = h(x,t_i) + v(t_i) \tag{144}$$

$0 \le t \le T$, $i = 1,\ldots,N$ and let $v(t_i)$ be zero mean Gaussian white noise sequence with known covariance $R(t_i)$. The information matrix M for θ based on observations $\{y(t_i),\ i=1,\ldots,N\}$ and a given input $u(t)$ is easily shown to be

$$M(u) = \sum_{i=1}^{N}\left(\frac{\partial x(t_i)}{\partial\theta}\right)^T H^T(x(t_i),t_i)R^{-1}(t_i)H(x(t_i),t_i)\frac{\partial x(t_i)}{\partial\theta} \tag{145}$$

where

$$H(x(t_i),t_i) = \left.\frac{\partial h(x,t_i)}{\partial x}\right|_{x=x(t_i)} \tag{146}$$

and

$$\frac{d}{dt}\frac{\partial x(t)}{\partial\theta} = \frac{\partial f}{\partial x}\cdot\frac{\partial x}{\partial\theta} + \frac{\partial f}{\partial\theta} \quad , \quad \frac{\partial x(0)}{\partial\theta} = 0 \tag{147}$$

For a randomized input $u \in \Omega_u$ with probability measure $\xi(du)$ defined for all Borel sets and points of Ω_u, the information matrix is

$$M(\xi) = \int_{\Omega_U} M(u)\cdot\xi(du) \tag{148}$$

We now consider maximization of $|M(\xi)|$ with respect to $\xi(du)$. It is clear that the only difference between this case and the linear case is in the evaluation of $M(u)$. Thus, Theorems 1 and 2 still apply and Algorithm 1 may be used to obtain ξ^*. The main difference lies in the maximization of $\mathrm{Tr}[M^{-1}(\xi_k)M(u)]$ over $u \in \Omega_u$ since $M(u)$ is no longer a quadratic functional of u. However, a calculus of variations approach or Pontryagin's maximum principle may be used [12]. In this approach, the criterion $J = \mathrm{Tr}[M^{-1}(\xi_k)M(u)]$ is maximized over $u \in \Omega_u$ subject to the constraints of Eqs. (143)-(147) using Lagrange multipliers or costate equations. In certain cases, a dynamic programming solution may also be possible, resulting in closed-form inputs [12]. It is interesting to note that by using randomization we have reduced the solution of a highly nonlinear problem, viz. maximization of $|M|$, to the solution of a sequence of simpler optimization problems. This is mainly due to the fact that randomization produces convexity.

Distributed Parameter Systems: The above approach may also be used for distributed parameter systems described by partial differential equations. For example, consider a vector PDE

$$\frac{\partial x(z,t)}{\partial t} = f\left(x, \frac{\partial x}{\partial z}, u, \theta, z, t\right)$$
$$t \in [0,T], \quad z \in \Omega_z, \quad u(z,t) \in \Omega_u \tag{149}$$

with appropriate boundary conditions for $x(z,t)$ at initial time and on the boundary $\partial\Omega_z$. Here z denotes the space variable. The measurements $y(z_j,t_i)$ are made at locations z_j and at times t_i, $j = 1,\ldots,N'$, $i = 1,\ldots,N$.

$$y(z_j,t_i) = h(x(z_j,t_i),z_j,t_i) + v(z_j,t_i) \tag{150}$$

where $v(z_j,t_i)$ is zero mean Gaussian measurement noise of covariance $R(z_j,t_i)$ and is uncorrelated in the space variable, z_j and the time variables, t_i. Then the information matrix M can be written as

$$M(u) = \sum_{j=1}^{N'} \sum_{i=1}^{N} \left(\frac{\partial x(z_j,t_i)}{\partial \theta}\right)^T H^T(z_j,t_i)R^{-1}(z_j,t_i) \cdot H(z_j,t_i)\,\frac{\partial x(z_j,t_i)}{\partial \theta} \tag{151}$$

where

$$H(z_j,t_i) = \left.\frac{\partial h(x,z,t)}{\partial x}\right|_{x=x(z_j,t_i)} \qquad (152)$$

$$\frac{\partial}{\partial t}\left(\frac{\partial x(z,t}{\partial \theta}\right) = \frac{\partial f}{\partial x}\cdot\frac{\partial x}{\partial \theta} + \frac{\partial f}{\partial(\partial x/\partial z)}\ \frac{\partial}{\partial z}\left(\frac{\partial x}{\partial \theta}\right) + \frac{\partial f}{\partial \theta} \qquad (153)$$

The boundary conditions for the linear PDE (153) may be calculated from the boundary conditions for (149).

We now assign a probability measure $\xi(du)$ to all Borel sets and points in Ω_u and proceed in the same manner as before to maximize $|M(\xi)|$ with respect to ξ. The computational problem now is that of maximizing $Tr[M^{-1}(\xi)M(u)]$ subject to the constraints of Eqs. (149)-(153). The case in which Eq. (149) is a linear PDE and the set Ω_u is a hypersphere, the maximization problem in Step (3) of Algorithm 1, is again a linear eigenvalue problem for a multidimensional Fredholm integral equation. Optimization problems of this type for linearly distributed parameter systems have been considered by Lyon [13] and Brockett [14].

9. CONCLUSIONS

A general approach for the construction of randomized D-optimal input designs for parameter estimation in dynamic systems has been presented. Previous approaches and results are shown to be special cases of the results obtained here. The optimal designs are characterized by certain minmax properties that lead to simple globally convergent algorithms for computing optimal inputs.

REFERENCES, PART I

1. Mehra, R.K., "Frequency-Domain Synthesis of Optimal Inputs for Linear System Parameter Estimates", TR 645, Division of Engineering and Applied Physics, Harvard University, July 1973. Also, *ASME Journal of Dynamic Systems, Measurement and Control*, pp. 130-138, June 1976.

2. Mehra, R.K., "Optimal Inputs for Linear System Identification", *IEEE Trans. Automatic Control, AC-19*, pp. 192-200, June 1974.

3. Gupta, N.K., R.K. Mehra, and W.E. Hall, Jr., "Application of Optimal Input Synthesis to Aircraft Parameter Identification" *ASME Journal of Dynamic Systems, Measurement and Control,* pp. 139-145, June 1976.

4. Goodwin, G.C. and R.L. Payne, "Design and Characterization of Optimal Test Signals for Linear Single Input Single Output Parameter Estimation", 3rd IFAC Symposium on Identification, Hague, June 1973.

5. Mehra, R.K., "Identification in Control and Econometrics; Similarities and Differences", *Annals of Social and Economic Measurement,* Jan. 1974.

6. Åström, K.J., *Introduction to Stochastic Control Theory,* Academic Press, New York, 1970.

7. Fedorov, V.V., *Theory of Optimal Experiments,* Academic Press, New York, 1972.

8. Rao, C.R., *Linear Statistical Inference and its Applications,* John Wiley and Sons, New York, 965.

9. Hannan, E.J., *Multiple Time Series Analysis,* John Wiley and Sons, New York, 1970.

10. Kiefer, J., "Optimal Experimental Designs V, with Application to Systematic and Rotatable Designs", Fourth Berkeley Symposium, J. Neyman, ed. (1965).

11. Mehra, R.K., "Optimal Input Signals for Parameter Estimation in Dynamic Systems--Survey and New Results", *IEEE Trans. Automatic Control, AC-19,* Dec. 1974.

12. Mehra, R.K. and N.K. Gupta, "Status of Input Design for Aircraft Parameter Identification", AGARD Conference Proceedings No. 172 on Methods for Aircraft State and Parameter Identification, Nov. 1974.

REFERENCES, PART II

1. Nahi, N.E. and D.E. Wallis, Jr., "Optimal Inputs for Parameter Estimates in Dynamic Systems with White Noise Observation Noise", Preprints, 1969, Joint Automatic Control Conference, Boulder, Colorado, August 1969.

2. Aoki, M. and R.M. Staley, "On Input Signal Synthesis in Parameter Identifications", *Automatica, 6,* 1970.

3. Mehra, R.K., "Optimal Inputs for Linear System Identification *IEEE Trans. Automatic Control, AC-19,* June 1974.

4. Goodwin, G.C., J.C. Murdock, and R.L. Payne, "Optimal Test Signal Design for Linear SISO System Identification", *Int. J. Control, 17,* 1973.

5. Reid, D.B., "Optimal Inputs for System Identification", Stanford University Guidance and Control Laboratory, SUDAAE No. 440, May 1970.

6. Gupta, N.K. and W.E. Hall, Jr., "Optimal Inputs for Identification of Stability and Control Derivatives", NASA CR-2493, Feb. 1975.

7. Mehra, R.K., "Frequency-Domain Synthesis of Optimal Inputs for Linear System Parameter Estimates", TR 645, Division of Engineering and Applied Physics, Harvard University, July 1973. Also, *ASME Journal of Dynamic Systems, Measurement and Control,* pp. 130-138, June 1976.

8. Mehra, R.K., "Synthesis of Optimal Inputs for Multiinput-Multioutput (MIMO) Systems with Process Noise: Part I - Frequency Domain Synthesis", this volume.

9. Kiefer, J., "Optimal Designs in Regression Problems, II", *Ann. Math. Stat. 32,* 1961.

10. Kiefer, J. and J. Wolfowitz, "The Equivalence of Two Extremum Problems", *Canadian J. Math. 12,* 1960.

11. Rao, C.R., *Linear Statistical Inference and its Applications,* John Wiley and Sons, New York, 1965.

12. Bryson, A.E. and Y.C. Ho, *Applied Optimal Control,* Ginn-Blaisedell, Waltham, Mass., 1969.

13. Lions, J.L., *Optimal Control of Systems Governed by Partial Differential Equations,* Springer-Verlag, 1971 (English translation by S.K. Mitter).

14. Brockett, R.W., Class Notes, Engr. 209, Harvard University, Spring 1970.

CHOICE OF SAMPLING INTERVALS

G. C. Goodwin
Department of Electrical Engineering
University of Newcastle
New South Wales, 2308, Australia

and

R. L. Payne
Department of Systems and Control
University of New South Wales
New South Wales, 2031, Australia

1. INTRODUCTION

In most experiments there are a number of variables which can be adjusted, subject to certain constraints, so that the information provided by the experiment is maximized.

Experiment design procedures have been described for a number of diverse modelling problems in many fields including agriculture, economics, social sciences, chemical engineering, industrial engineering and biological sciences [1] to [3] the main emphasis to date has been on non-dynamic systems, but recently there has been growing interest in systems where dynamic effects predominate [9] to [45]. For dynamic systems, experiment design includes choice of input and measurement ports, test signals; sampling instants and presampling filters. In general the effects of these design parameters are interrelated [46] and a complete design should be carried out. However, in practice, it is often the case that the appropriate input and measurement ports are predetermined. Furthermore, for the case of continuous observations, or discrete time systems, the problem of the choice of sampling intervals and filters does not arise. In these cases, the complete design reduces to the design of test signals.

One of the earliest publications on optimal test signal design appeared in 1960 [9], the major result being that an input having impulsive autocorrelation function is optimal for the estimation of the parameters of a moving average process in the presence of white measurement noise. Since then, more general classes of models have been considered and a number of test signal design algorithms based on optimal control principles have been proposed [13] to [15], [17], [20], [24], [25], [26], [33], [34]. Recently it has been shown that design algorithms originally developed for non-dynamic systems [4] to [8] are applicable to the design of test signals in the frequency domain, [45], [39], [40], [43], [44].

For those problems where the choice of sampling instants and pre-sampling filters does arise, it has been recognized that this choice often has a significant effect on the information return from the experiment [46] to [50]. The choice is particularly critical when the available computer storage and analysis time are limited. In these cases, the experiment should be designed to maximize the average information content of each sample.

In the sampling problem, four cases can be distinguished depending upon whether the samples are uniformly or nonuniformly spaced in time and whether the test signal is prespecified or is to be jointly designed. The purpose of this chapter is to present, what is, in the authors' opinion, the state of the art in the design of sampling intervals in each of these cases.

2. DESIGN CRITERIA

To form a basis for the comparison of different experiments a measure of the "goodness" of an experiment is required. A logical approach is to choose a measure related to the expected accuracy of the parameter estimates to be obtained from the data collected. Clearly the parameter accuracy is a function of both the experimental conditions and the form of the estimator. However, it is usual to assume that the estimator used is efficient in the sense that the Cramer Rao lower bound on the parameter covariance matrix is achieved, that is, the covariance matrix is given by the inverse of Fisher's information matrix [51]. (This property is true, at least asymptotically, for most commonly used estimators e.g., maximum likelihood [51].) A suitable measure of the "goodness" of an experiment is therefore

$$J = \phi(M) \tag{2.1}$$

where ϕ is a scalar function and M is Fisher's information matrix.

Various choices exist for the function ϕ. Possible choices are

i) $-\text{Trace}\ (M)$
ii) $\text{Det}\ (M^{-1})$ (equivalently $-\text{Det}\ (M)$ or $-\text{Log Det}\ (M)$)
iii) $\text{Trace}\ (WM^{-1})$ where W is a nonnegative definite matrix.

The choice between these alternatives depends upon the objective of the experiment. In fact it could be argued that a Bayesian decision theoretic approach is called for, that is, the measure, J, should be a Bayesian risk function which would reflect the ultimate use to which the data from the experiment will be put. This approach has been followed by a number of authors in the literature [33], [34], [38]. A useful approximation to the full Bayesian formulation is to use:

$$J = \text{Trace } [MW^{-1}] \tag{2.2}$$

where W is weighting matrix given by the second derivative of the risk function with respect to the parameters, [13], [38].

An alternative approach is to use an information measure where information is used in the mathematical sense [52]. Under the usual Gaussian assumptions the information measure reduces to

$$J = \text{Det}[M^{-1}] \tag{2.3}$$

Of the three possible criteria mentioned above, choice (i) seems to be, at first sight, particularly attractive since, for input design, use of this criteria leads to a quadratic optimization problem [17], [32], [20], [33]. However, it has been pointed out, [35], [26], [25], [13] that this criteria can lead to unidentifiability conditions and is therefore not recommended.

Other criteria, apart from the three discussed above, are possible. Most commonly met criteria (including the three above) have the following desirable property:

$$\phi(M_1) \leq \phi(M_2) \quad \text{if } M_1 - M_2 \text{ is nonnegative definite} \tag{2.4}$$

This property will be used in later sections.

3. CONSTRAINTS

In the optimization of the design criteria introduced in Section 2, it is obvious that the optimization must be carried out with due regard to the constraints on the allowable experimental conditions. Typical constraints that might be met in practice are:

(1) Input, output or state amplitude constraints,

(2) Input, output or state energy/power constraints,

(3) Total experiment time,

(4) Total number of samples,

(5) Maximum sampling rate,

(6) Total computation time for design and analysis,

(7) Availability of transducers and filters,

(8) Availability of hardware and software for analysis,

(9) Model accuracy specifications,

(10) Non-disruption of normal production.

Which of the above constraints (or others) are important in a particular experiment depends on the situation, and it is not always obvious which should be taken into account during the design. A useful approach leading to simplified designs is to work with a subset of the constraints (such as input energy and total number of samples) and to subsequently check for violation of the other constraints.

4. NONUNIFORM SAMPLING THEORY

In this section the following problem is considered:

a. The parameters in a continuous linear system are to be estimated from a fixed number of samples, N.

b. The samples are taken at times $t_1, t_2, \ldots, t_N$ where the sampling intervals

$$\Delta_k = t_{k+1} - t_k , \qquad k = 0,\ldots,N-1 \tag{4.1}$$

are not necessarily equal (t_0 corresponds to the beginning of the experiment).

c. The experimental conditions available for adjustment are the system input, the pre-sampling filter and the sampling intervals $\Delta_0, \Delta_1, \Delta_2, \ldots, \Delta_{N-1}$.

A general model for a linear time invariant system is [53]:

$$dx'(t) = A'x'(t)dt + B'u(t)dt + d\eta(t) \tag{4.2}$$

$$dy'(t) = C'x'(t)dt + d\omega(t) \tag{4.3}$$

where $y'(t) : T \to R^r$ is the output prior to filtering and sampling, $u(t) : T \to R^m$ is the input vector, $x'(t) : T \to R^{n'}$

is a state vector and $\eta(t)$, $\omega(t)$ are Wiener Processes with incremental covariance:

$$E\left\{\begin{bmatrix} d\eta(t) \\ d\omega(t) \end{bmatrix}\begin{bmatrix} d\eta^T(t)\, d\omega^T(t) \end{bmatrix}\right\} = \begin{bmatrix} R_{11} & R_{12} \\ R_{12}^T & R_{22} \end{bmatrix} dt \tag{4.4}$$

The above model is based on the usual assumptions of rational input-output transfer function and rational output noise power density spectrum.

In general, to minimize the information loss due to sampling, some form of pre-sampling filter is required. Optimization of the filter requires careful specification of the allowable filter structures as otherwise impractical solutions will result. For example, with no restrictions on the allowable structure, a filter which is in fact a complete identifier could be constructed. A common restriction is to use a linear filter of fixed dimension. The filter design then reduces to optimization of the filter parameters. A suitable linear filter is

$$\frac{d}{dt} x''(t) = A''x''(t) + B''u(t) + K''y'(t) \tag{4.5}$$

$$y(t) = C''x''(t) + D''y'(t) \tag{4.6}$$

where $y(t)$ is the filter output.

Equations (4.5), (4.6) can be combined with equations (4.2), (4.3) to produce a composite model for the system and filter of the following form:

$$dx(t) = Ax(t)dt + Bu(t)dt + Kd\varepsilon(t) \tag{4.7}$$

$$y(t) = Cx(t) \tag{4.8}$$

where $x(t)$ is the state of the composite model and $\varepsilon(t)$ is a Wiener Process with incremental covariance Σdt.

For simplicity of exposition, the class of allowable inputs is now restricted to piecewise constant functions, that is

$$u(t) = u_k, \quad t_k \leq t < t_{k+1} \tag{4.9}$$

This class of inputs has the added advantage of ease of implementation with the specified sampling scheme. However, the analysis can be readily extended to other classes of inputs, e.g., absolutely continuous functions.

As a first step towards obtaining an expression for the information matrix of the system parameter, a discrete time model for the sampled observations is now obtained.

Integration of (4.7) and (4.8) gives

$$x_{k+1} = \phi(\Delta_k, 0)x_k + \psi(\Delta_k, 0)u_k + \lambda_k \tag{4.10}$$

$$y_k = Cx_k \tag{4.11}$$

where

$$\phi(\tau_2, \tau_1) = \exp[A(\tau_2 - \tau_1)] \tag{4.12}$$

$$\psi(\tau_2, \tau_1) = \int_{\tau_1}^{\tau_2} \phi(\tau_2, \tau)B \quad d\tau \tag{4.13}$$

$$\lambda_k = \int_0^{\Delta_k} \phi(\Delta_k, \tau)K \quad d\varepsilon(\tau) \tag{4.14}$$

and

$$\Delta_k = t_{k+1} - t_k \tag{4.15}$$

The sequence $\{\lambda_k\}$ is a sequence of independent normal random variables having zero mean and covariance at time t_k given by:

$$E\{\lambda_k \lambda_k^T\} = \int_0^{\Delta_k} \phi(\Delta_k, \tau) K\Sigma K^T \phi^T(\Delta_k, \tau) \quad d\tau \tag{4.16}$$

$$= Q_k \quad \text{by definition}$$

Equations (4.10) and (4.11) can now be used to develop an expression for the information matrix of the parameters using the samples $y_1,\ldots,y_N$.

The standard expression for the information matrix [53] is:

$$M = E_{Y|\beta}\left\{\left[\frac{\partial \log p(Y|\beta)}{\partial\beta}\right]\left[\frac{\partial \log p(Y|\beta)}{\partial\beta}\right]^T\right\} \tag{4.17}$$

where Y is a random variable corresponding to the observations $y_1,\ldots,y_N$ and β is the p-vector of parameters in the system model (4.2) and (4.3). $p(Y|\beta)$ is the likelihood function for the parameters and $E_{Y|\beta}$ denotes expectation over the distribution of Y given β.

Using Bayes' rule the likelihood function can be expressed as:

$$\begin{aligned} p(Y|\beta) &= p(y_1,\ldots,y_N|\beta) \\ &= p(y_1|\beta)p(y_2|y_1,\beta)p(y_3|y_2,y_1,\beta)\ldots.p(y_N|y_{N-1}\ldots y_1,\beta) \end{aligned} \tag{4.18}$$

It follows from the assumption of Gaussian noise processes that the conditional distributions in equation (4.18) can be expressed as

$$\begin{aligned} p(y_k|y_{k-1}\ldots,y_1,\beta) = {} & ((2\pi)^r \det S_k)^{-1/2} \\ & \exp\{-1/2(y_k - \bar{y}_k)^T S_k^{-1}(y_k - \bar{y}_k)\} \end{aligned} \tag{4.19}$$

where $\bar{y}_k$ is the conditional mean and satisfies the following equations [54]:

$$\bar{y}_k = C\bar{x}_k \tag{4.20}$$

$$\bar{x}_{k+1} = \phi_k\bar{x}_k + \psi_k u_k + \Gamma_k[y_k - \bar{y}_k] \tag{4.21}$$

where $\bar{x}_k$ is the conditional mean of x_k.

In equation (4.21) the simplified notation $\phi_k = \phi(\Delta_k,0)$ and $\psi_k = \psi(\Delta_k,0)$ has been used. In equation (4.21) the matrix

Γ_k is given by:

$$\Gamma_k = \phi_k P_k C^T (C P_k C^T)^{-1} \tag{4.22}$$

and P_k is the conditional covariance of x_k and is the solution of a matrix Riccati equation:

$$P_{k+1} = \phi_k P_k \phi_k^T + Q_k - \Gamma_k (C P_k C^T)^{-1} \Gamma_k^T \tag{4.23}$$

(Note: The matrix $[CPC^T]$ in equations (4.22) and (4.23) will usually be nonsingular but if singular, then the correct results are obtained by using the pseudo inverse [59].)

The conditional covariance, S_k, of y_k is related to P_k as follows

$$S_k = C P_k C^T \tag{4.24}$$

The boundary conditions for equation (4.21) and (4.23) are assumed to be $\bar{x}_0$ and P_0 respectively where the initial state has been taken to be normally distributed with mean $\bar{x}_0$ and covariance P_0.

Substituting (4.19) to (4.18) yields

$$P(Y|\beta) = \left([2\pi]^{rN} \prod_{k=1}^{N} \det S_k \right)^{-1/2} \exp\left\{ - 1/2 \sum_{k=1}^{N} (y_k - \bar{y}_k)^T S_k^{-1} (y_k - \bar{y}_k) \right\} \tag{4.25}$$

which when substituted into (4.17) yields:

$$M_{ij} = E_{Y|\beta} \left\{ \sum_{k=1}^{N} \left(\frac{\partial \bar{y}_k}{\partial \beta_i} \right)^T S_k^{-1} \left(\frac{\partial \bar{y}_k}{\partial \beta_j} \right) \right\} + 1/2 \sum_{k=1}^{N} \text{Trace} \left\{ S_k^{-1} \frac{\partial S_k}{\partial \beta_i} \right\} \text{Trace} \left\{ S_k^{-1} \frac{\partial S_k}{\partial \beta_j} \right\} \tag{4.26}$$

The proof of equation (4.26) is straightforward, though somewhat long winded [50], and has therefore been left to the reader as an exercise.

The quantities $\partial\bar{y}_k/\partial\beta_i$ in equation (4.25) can be evaluated from the following sensitivity equations:

$$\frac{\partial\bar{y}_k}{\partial\beta_i} = -\frac{\partial C}{\partial\beta_i}\bar{x}_k - C\frac{\partial\bar{x}_k}{\partial\beta_i} \tag{4.27}$$

$$\frac{\partial\bar{x}_{k+1}}{\partial\beta_i} = (\phi_k - \Gamma_k C)\frac{\partial\bar{x}_k}{\partial\beta_i} + \left(\frac{\partial\phi_k}{\partial\beta_i} - \Gamma_k\frac{\partial C}{\partial\beta_i}\right)\bar{x}_k$$

$$+ \frac{\partial\psi_k}{\partial\beta_i}u_k + \frac{\partial\Gamma_k}{\partial\beta_i}(y_k - \bar{y}_k) \tag{4.28}$$

Equation (4.28) can be combined with equation (4.21) to give:

$$\tilde{x}_{k+1} = F_k\tilde{x}_k + G_k u_k + H_k(y_k - \bar{y}_k) \tag{4.29}$$

where

$$\tilde{x}_k = \left[\bar{x}_k^T, \frac{\partial\bar{x}_k^T}{\partial\beta_1}, \ldots, \frac{\partial\bar{x}_k^T}{\partial\beta_p}\right]^T \tag{4.30}$$

and the forms of the matrices F_k, G_k and H_k follow from (4.28) and (4.21).

Also from equation (4.27) the quantities $\partial\bar{y}_k/\partial\beta_i$ can be expressed in terms of the composite state $\tilde{x}_k$ as follows:

$$\frac{\partial\bar{y}_k}{\partial\beta_i} = \Omega_i\tilde{x}_k \tag{4.31}$$

where

$$\Omega_i = \left[\frac{\partial C}{\partial\beta_i}, 0 \ldots, 0, C, 0, \ldots, 0\right] \tag{4.32}$$

Now substituting (4.31) into (4.26) leads to the following expression for the ij-th element of the information matrix:

$$M_{ij} = E_{Y|\beta} \left\{ \sum_{k=1}^{N} \tilde{x}_k^T \Omega_i^T S_k^{-1} \Omega_j \tilde{x}_k \right\}$$

$$+ 1/2 \sum_{k=1}^{N} \text{Trace} \left\{ S_k^{-1} \frac{\partial S_k}{\partial \beta_i} \right\} \text{Trace} \left\{ S_k^{-1} \frac{\partial S_k}{\partial \beta_j} \right\} \quad (4.33)$$

Performing the expectation in equation (4.33) yields:

$$M_{ij} = \sum_{k=1}^{N} \left\{ \hat{x}_k^T \Omega_i^T S_k^{-1} \Omega_j \hat{x}_k \right\} + \sum_{k=1}^{N} \text{Trace} \left\{ \Omega_i^T S_k^{-1} \Omega_j T_k \right\}$$

$$+ 1/2 \sum_{k=1}^{N} \text{Trace} \left\{ S_k^{-1} \frac{\partial S_k}{\partial \beta_i} \right\} \text{Trace} \left\{ S_k^{-1} \frac{\partial S_k}{\partial \beta_j} \right\} \quad (4.34)$$

where $\hat{x}_k$ and T_k are the mean and covariance respectively of $\tilde{x}_k$ and are given by:

$$\hat{x}_{k+1} = F_k \hat{x}_k + G_k u_k ; \qquad \hat{x}_0 = 0 \quad (4.35)$$

$$T_{k+1} = F_k T_k F_k^T + H_k S_k H_k^T ; \qquad T_0 = 0 \quad (4.36)$$

Summing up, equation (4.34) provides a means for computing the information matrix before the experiment is performed, that is, it is possible to evaluate a scalar function ϕ of the information matrix corresponding to a particular set of experimental conditions. It follows from development of equation (4.34) that M and hence ϕ depend on the experimental conditions namely the input sequence $(u_0, \ldots, u_{N-1})$, the sampling intervals $(\Delta_0, \ldots, \Delta_{N-1})$ and the pre-sampling filter (equation (4.5) and (4.6)). Hence, it would be possible, at least in principle, to optimize ϕ with respect to these design variables.

Suitable algorithms have been developed for this type of optimization [50] but are extremely time consuming except for simple situations and unfortunately do not give a great deal of insight into the problem.

In the next section a suboptimal design scheme will be described which allows the theory developed in this section to be applied to practical systems. In subsequent sections it will be shown that, with certain constraints on the design, alternative simple design methods exist.

These alternative methods give further insight into the general sampling problem.

5. SEQUENTIAL DESIGN OF NONUNIFORM SAMPLING INTERVALS

This section describes a suboptimal design procedure for sequential determination of the input and sampling intervals.

It follows from equation (4.34) that the information matrix is given by the following recursion:

$$M_{k+1} = M_k + \mathcal{I}_{k+1} \; ; \qquad M_0 = 0 \tag{5.1}$$

where $\mathcal{I}_k$ is the information increment resulting from the k+1-th sample and has ij-th element

$$[\mathcal{I}_{k+1}]_{ij} = \hat{x}_{k+1}^T \Omega_i^T S_{k+1}^{-1} \Omega_j \hat{x}_{k+1} + \text{Trace}\left\{\Omega_i^T S_{k+1}^{-1} \Omega_j T_{k+1}\right\}$$

$$+ 1/2 \, \text{Trace}\left\{S_{k+1}^{-1} \frac{\partial S_{k+1}}{\partial \beta_i}\right\} \text{Trace}\left\{S_{k+1}^{-1} \frac{\partial S_{k+1}}{\partial \beta_j}\right\} \tag{5.2}$$

Using (4.35) equation (5.2) becomes:

$$[\mathcal{I}_{k+1}]_{ij} = (F_k \hat{x}_k + G_k u_k)^T \Omega_i^T S_{k+1}^{-1} \Omega_j (F_k \hat{x}_k + G_k u_k)$$

$$+ \text{Trace}\left\{\Omega_i^T S_{k+1}^{-1} \Omega_j T_{k+1}\right\} \tag{5.3}$$

$$+ 1/2 \, \text{Trace}\left\{S_{k+1}^{-1} \frac{\partial S_{k+1}}{\partial \beta_i}\right\} \text{Trace}\left\{S_{k+1}^{-1} \frac{\partial S_{k+1}}{\partial \beta_j}\right\}$$

Considering the situation at time k, if it is assumed that the past inputs and sampling intervals have been specified, then it follows from equations (5.1) and (5.3) that M_{k+1} is a function of only the two variables u_k and Δ_k. Hence it is a simple optimization problem to choose u_k and Δ_k to maximize the optimality criterion $\phi(M_{k+1})$. Then moving on to time $(k+1)$ the procedure can be repeated to choose u_{k+1} and Δ_{k+1} etc. This procedure is analogous to the one-step-ahead control algorithms such as minimum variance control [4.3] and will not generally be globally optimal in the sense that $\phi(M_N)$ is optimized with respect to $u_0, u_1, \ldots, u_{N-1}$ and $\Delta_0, \Delta_1, \ldots, \Delta_{N-1}$.

The procedure described above is a generalization to the sampling problem of the sequential test signal design procedures originally described by Arimoto and Kimura [16] and by Keviczky [28], [29].

The sequential design algorithm is now illustrated by a simple example. Consider the first order system [46].

$$\dot{x} = ax + bu \tag{5.4}$$

$$y = x + n \tag{5.5}$$

where the noise, $n(t)$ is assumed to have wide bandwidth (approximately "white"). Equation (5.4) can be discretized as follows:

$$x_{k+1} = \alpha_k x_k + \beta_k u_k \tag{5.6}$$

$$y_k = x_k + n_k \tag{5.7}$$

where $\alpha_k = \exp(a\Delta_k)$, $\beta_k = b/a(\alpha_k - 1)$ and n_k has zero mean and approximately constant variance. (This assumption will be valid provided the smallest sampling interval allowed is much greater than the reciprocal of the noise bandwidth; this point will be taken up again later in this section).

For this system, equation (5.1) becomes

$$M_{k+1} = M_k + \Omega\hat{x}_{k+1}\hat{x}_{k+1}^T\Omega^T \tag{5.8}$$

$$\Omega = \begin{bmatrix} 0 & 1 & 0 \\ 0 & 0 & 1 \end{bmatrix} \tag{5.9}$$

and

$$\hat{x}_{k+1} = \begin{bmatrix} \alpha_k & 0 & 0 \\ \alpha_k \Delta_k & \alpha_k & 0 \\ 0 & 0 & \alpha_k \end{bmatrix} \hat{x}_k + \begin{bmatrix} \beta_k \\ b(a\Delta_k \alpha_k - \alpha_k + 1)/a^2 \\ (\alpha_k - 1)/a \end{bmatrix} u_k \tag{5.10}$$

$$= A_k \hat{x}_k + B_k u_k \tag{5.11}$$

Here the optimality criterion is taken to be $\text{Det}(M^{-1})$ or equivalently $-\text{Det}(M)$. Hence using equation (5.8):

$$\begin{aligned} -\text{Det}(M_{k+1}) &= -\text{Det}(M_k)\text{Det}[I + M_k^{-1}\Omega\hat{x}_{k+1}\hat{x}_{k+1}^T\Omega^T] \\ &= -\text{Det}(M_k)(1 + \hat{x}_{k+1}^T\Omega^T M_k^{-1}\Omega\hat{x}_{k+1}) \\ &= -\text{Det}(M_k)(1+(A_k\hat{x}_k+B_k u_k)^T\Omega^T M_k^{-1}\Omega(A_k\hat{x}_k+B_k u_k)) \end{aligned} \tag{5.12}$$

Now assuming that u_k is constrained in amplitude and lies in the interval $-1 \leq u_k \leq 1$, then it can be seen from equation (5.12) that the optimality criterion is minimized with respect to u_k if u_k is chosen as

$$u_k^* = \text{Sign}\,(B_k^T\Omega^T M_k^{-1}\Omega A_k \hat{x}_k) \tag{5.13}$$

Substituting (5.13) back into (5.12) yields

$$\begin{aligned} -\text{Det}\, M_{k+1} = -\text{Det}\, M_k[1 &+ \hat{x}_k^T A_k \Omega^T M_k^{-1}\Omega A_k \hat{x}_k \\ &+ 2|B_k\Omega^T M_k^{-1}\Omega A_k \hat{x}_k| \\ &+ B_k^T\Omega^T M_k^{-1}\Omega B_k] \end{aligned} \tag{5.14}$$

A one dimensional optimization can now be used to find the best values for Δ_k. This value can then be substituted into (5.13) to give u_k^*.

Results of this optimization are shown in Fig. (5.1).

It can be seen from Figure (5.1) that, for this example, a significant improvement in information occurs if the input is sequentially optimized with fixed sampling intervals. However, a further improvement is achieved by a coupled design of both the input and sampling intervals. Of course the design algorithm is only optimal in the one-step-ahead sense and presumably further improvements could be obtained by extending the optimization horizon but at the cost of greatly increased computational complexity.

It should be noted that, in this example, the sampling was performed without the inclusion of a presampling filter. However, since the noise was very wide bandwidth, drastic improvements in estimation accuracy could have been achieved by the simple expedient of including a low pass filter prior to sampling. The improvement in accuracy will be of the order of the ratio of the noise bandwidth to the system bandwidth, provided that the filter is chosen appropriately. (The variance of the discrete noise on the samples is roughly proportional to the noise bandwidth at the samples.) This point will be discussed further for the case of uniform sampling intervals in the next section.

6. UNIFORM SAMPLING INTERVAL DESIGN

In this section the case of uniform sampling interval design is considered. For this case it will be shown that it is a simple matter to simultaneously find optimal test signals, pre-sampling filters and sampling rates.

For the purposes of the current discussion, consider the following model of a constant coefficient linear single input-multiple output system with coloured measurement noise:

$$\dot{x}(t) = Ax(t) + Bu(t) \tag{6.1}$$

Figure 5.1 : <u>COMPARISON OF SEQUENTIAL DESIGNS</u>

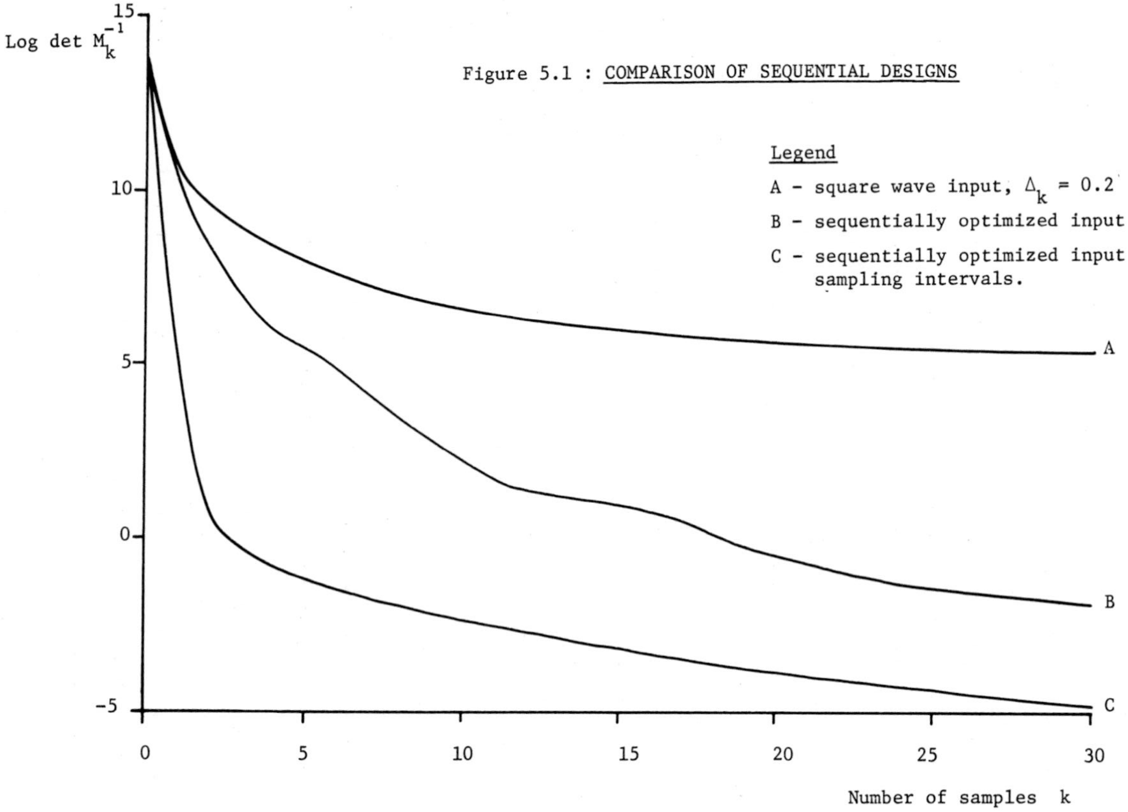

$$y(t) = Cx(t) + Du(t) + w(t) \tag{6.2}$$

where $t \in T = [0, t_f]$, $x : T \to R^n$, is the state vector, $u : T \to R^l$ is the input vector, $y : T \to R^m$ is the output vector and $w : T \to R^m$ is the measurement noise vector, assumed to have a zero mean Gaussian distribution with power density spectrum $\psi(f)$ with full rank for all $f \in (-\infty, \infty)$.

For large experiment time, t_f, it is convenient to transform the model equations (6.1), (6.2) to the frequency domain:

$$y(s) = T(s)u(s) + w(s) \; ; \quad s = j2\pi f \tag{6.3}$$

where

$$T(s) = (C(sI - A)^{-1}B + D) \tag{6.4}$$

Fisher's information for the parameters, β, in A, B, C and D (assuming for the moment continuous observations over $[0, t_f]$) is given by:

$$M_\beta = t_f \bar{M}_\beta \tag{6.5}$$

where $\bar{M}_\beta$ is the average information matrix per unit time with ik-th element given by:

$$(\bar{M}_\beta)_{ik} = \int_{-\infty}^{\infty} \frac{\partial T^T(j2\pi f)}{\partial \beta_i} \psi^{-1}(f) \frac{\partial T(j2\pi f)}{\partial \beta_k} d\xi(f) \tag{6.6}$$

where $\xi(f)$ is the input spectral distribution function [55], [56] satisfying:

$$1 = \int_{-\infty}^{\infty} d\xi(f) \tag{6.7}$$

(Note: The input power has been constrained to be unity).

Now, introducing the matrix Q having i-th column given by

$$Q_i(s) = \frac{\partial T(s)}{\partial \beta_i} \tag{6.8}$$

the average information matrix may be written as:

$$\bar{M}_\beta = \int_{-\infty}^{\infty} Q^T(-j2\pi f)\psi^{-1}(f)Q(j2\pi f) \quad d\xi(f) \tag{6.9}$$

Assuming that the test signal spectrum is band limited to $[-f_h, f_h]$, then equation (6.9) may equivalently be written as:

$$\bar{M}_\beta = \int_{-f_h}^{f_h} Q^T(-j2\pi f)\psi^{-1}(f)Q(j2\pi f) \quad d\xi(f) \tag{5.10}$$

(Note: The assumption of a band limited test signal will later be shown to be consistent with the conditions for optimality.)

The effect of sampling on the average information matrix is now investigated. Suppose that the output, $y(t)$, $t \in [0, t_f]$ is sampled at greater than the Nyquist rate, [57], for f_h (that is the sampling frequency f_s, is greater than $2f_h$). This form of sampler does not distort that part of the spectrum of y arising from the input. The part of the output spectrum arising from the noise will, however, be distorted due to aliasing. This distortion will result in an "information" loss as indicated by the following expression for the average information matrix (*per unit time*) from the *sampled* observations:

$$\bar{M}_\beta^s = \int_{-f_h}^{f_h} Q^T(-j2\pi f)\psi_s^{-1}(f)Q(j2\pi f) \quad d\xi(f) \tag{6.11}$$

where $\psi_s(f)$ is the aliased noise spectrum given by [58]:

$$\psi_s(f) = \psi(f) + \sum_{k=1}^{\infty} \{\psi(kf_s + f) + \psi(kf_s - f)\} \; ;$$

$$f \in \left(-\frac{f_s}{2}, \frac{f_s}{2}\right) \tag{6.12}$$

the fact that $\bar{M} - \bar{M}^s$ is nonnegative definite follows from equation (6.12), which indicates that $\psi_s(f) - \psi(f)$ is positive definite and hence that

$$\bar{M}_\beta - \bar{M}_\beta^s = \int_{-f_h}^{f_h} Q^T(-j2\pi f)\{\psi^{-1}(f) - \psi_s^{-1}(f)\} Q(j2\pi f)\; d\xi(f) \tag{6.13}$$

is nonnegative definite.

It follows that, for all optimality criteria ϕ satisfying (2.4), $\phi(\bar{M}_\theta^s) \leq \phi(\bar{M}_\theta)$, that is the experiment with sampled data cannot be better than the corresponding experiment with continuous observations. This result is, of course, quite expected. However, it will now be shown that, by the inclusion of an optimal presampling filter, equality can be achieved.

THEOREM 6.1: *With* ϕ *satisfying* (2.4), *and sampling rate,* f_s, *satisfying* $f_s > 2f_h$ *a presampling filter having transfer function,* F, *with the property:*

$$|F(j2\pi f)| = 0 \quad \textit{for all} \quad f \in \left(-\infty, -\frac{f_s}{2}\right] \cup \left[\frac{f_s}{2}, \infty\right) \tag{6.14}$$

and invertible for all f *to which* ξ *assigns nonzero measure, gives the same value for* ϕ *using the sampled data as with continuous observations.*

Proof. The inclusion of a presampling filter having transfer function $F(s)$ modifies the expression for the average information matrix (6.11) to:

$$\bar{M}_\beta^{FS} = \int_{-f_h}^{f_h} Q^T(-j2\pi f)F^T(-j2\pi f)\psi_{FS}^{-1}(f)F(j2\pi f)Q(j2\pi f)\; d\xi(f) \tag{6.15}$$

where

$$\psi_{FS}(f) = \psi_F(f) + \sum_{k=1}^{\infty} \{\psi_F(kf_s + f) + \psi_F(kf_s - f)\} \tag{6.16}$$

and $\psi_F(f)$ is the filtered noise spectrum given by:

$$\psi_F(f) = F(j2\pi f)\psi(f)F^T(-j2\pi f) \tag{6.17}$$

Now, for a presampling filter satisfying (6.14), $\psi_{FS}(f)$ reduces to $\psi_F(f)$ as is readily verified. Furthermore, substituting (6.17) into (6.15) yields

$$\bar{M}_\beta^{FS} = \int_{-f_h}^{f_h} Q^T(-j2\pi f)\psi^{-1}(f)Q(j2\pi f) \, d\xi(f) \qquad (6.18)$$

and comparison of (6.18) with (6.10) yields:

$$\bar{M}_\beta^{FS} = \bar{M}_\beta \qquad (6.19)$$

This completes the proof of the theorem. ■

The significance of (6.19) is that the specified presampling filter, F, eliminates the information loss due to sampling. This form of filter is often called an aliasing filter [58].

So far, only the average information matrix per unit *time* has been considered. However, if the total number of samples is constrained to be N, say, then it is more appropriate to consider the average information matrix per *sample*. If the sampling rate is f_s samples per second then the average information matrix per sample is given by:

$$\tilde{M}_\beta = \frac{1}{f_s} \bar{M}_\beta^S \qquad (6.20)$$

where $\bar{M}_\beta^S$ is the average information matrix per unit time corresponding to an input spectrum having highest frequency component f_h and samples collected at $f_s > 2f_h$. Furthermore, inclusion of an aliasing filter ensures that $\bar{M}_\beta^S$ in (6.20) can be replaced by $\bar{M}_\beta$:

$$\tilde{M}_\beta = \frac{1}{f_s} \bar{M}_\beta \qquad (6.21)$$

It would now be possible, at least in principle, to choose the test signal spectrum (or equivalently the spectral distribution function $\xi(f)$, $f \in [-f_h, f_h]$) to minimize some scalar function ϕ of, the average information matrix per unit time, $\bar{M}_\beta$. It is obvious that the tighter the constraint on the allowable input spectrum (i.e. the smaller f_h) the greater will be the "cost" ϕ. Also from (6.21) and the property (2.4) of ϕ, it can be seen that $\phi(\tilde{M}_\beta)$ is made small by increasing f_s. However, since $f_s > 2f_h$, a compromise will be reached between the improvement in ϕ due to the decrease in f_s and the degradation in ϕ due to the tightening of the constraint on the allowable input spectrum.

The optimization problem outlined above is greatly simplified by the observation that only a finite number of frequency components in the input spectrum need to be considered. This result is established in the following theorem:

THEOREM 6.2: *For any test signal spectral distribution function* $\xi_1(f)$, $f \in [-f_h, f_h]$ *with corresponding average information matrix per sample,* $\tilde{M}_\beta(\xi_1)$, *there is always a* $\xi_2(f)$, $f \in [-f_h, f_h]$, *the spectrum of which contains at most* $p(p+1)/2+1$ *lines and* $\tilde{M}_\beta(\xi_1) = \tilde{M}_\beta(\xi_2)$ *where* $\tilde{M}_\beta(\xi_2)$ *corresponds to* ξ_2.

Proof. It follows from equations (6.21), (6.7) and (6.9) that the set of all possible average information matrices per sample arising from power constrained inputs is the convex hull of the set of average information matrices per sample corresponding to inputs containing a single frequency component. Since any information matrix is symmetric, it can be represented as an element in a $p(p+1)/2$ dimensional space and hence it follows from a theorem due to Caratheodory [6] that there exist λ_k, f_k, $k = 1,\ldots,\ell$; $\ell = p(p+1)/2 + 1$ such that

$$\tilde{M}_\beta(\xi_1) = \frac{1}{f_s} \sum_{k=1}^{\ell} \lambda_k \bar{M}_\beta(f_k) \qquad (6.22)$$

where (from (6.9)):

$$\bar{M}_\beta(f_k) = 2\mathrm{Re}\,\{Q^T(-j2\pi f_k)\psi^{-1}(f_k)Q(j2\pi f_k)\} \tag{6.23}$$

and

$$\sum_{k=1}^{\ell} \lambda_k = 1 \tag{6.24}$$

Defining $\xi_2(f)$ to be the spectral distribution function which assigns measure λ_k to f_k, $k = 1,\ldots,p(p+1)/2 + 1$ and zero otherwise completes the proof of the theorem. ∎

In particular, there exists a ϕ-optimal test signal spectrum containing not more than $p(p+1)/2 + 1$ lines. The optimization problem now reduces to minimization of $\phi(\tilde{M}_\beta)$ with respect to $f_s, f_1,\ldots,f_{p(p+1)/2+1}, \lambda_1,\ldots,\lambda_{p(p+1)/2}$ subject to the constraint that $2f_k < f_s$, $k = 1,\ldots,p(p+1)2 + 1$. Obviously with the problem as stated above, the optimal value of f_s will be $2f_m$ where $f_m = \max(f_1,\ldots,f_{p(p+1)/2+1})$. In practice the aliasing filter will not be ideal and so, to avoid aliasing, f_s is taken to be $2(1+\varepsilon)f_m$ where ε is related to the cutoff characteristics of the filter and should be small. Finally, from (6.22) the experiment design may be summarized as:

$$\phi^* = \min_{\substack{\lambda_1,\ldots,\lambda_\ell \\ f_1,\ldots,f_\ell}} \phi\left\{\frac{1}{2f_m(1+\varepsilon)} \sum_{k=1}^{\ell} \lambda_k \bar{M}_\beta(f_k)\right\} \tag{6.25}$$

where

$$\ell = p(p+1)/2 + 1\ , \tag{6.26}$$

$$\lambda_k \geq 0\ ; \quad k = 1,\ldots,\ell \tag{6.27}$$

$$\sum_{k=1}^{\ell} \lambda_k = 1 \tag{6.28}$$

$$f_m = \max(f_1,\ldots,f_\ell) \tag{6.29}$$

and ϕ^* is the optimal value for the design criterion function.

The quantities $\lambda_1, \lambda_2, \ldots,$ have the physical interpretation of being the fraction of the total power at the corresponding frequencies $f_1, f_2, \ldots,$ in the test signal spectrum. The dimension of the problem is reduced even further for single input systems for which it has been proven that only $p+1$ lines are required in the test signal spectrum [44]. Furthermore, for the usual design criteria, e.g. $\det(M^{-1})$ or $\operatorname{trace}(M^{-1})$, it can be shown that the optimal information matrix is a boundary point of the convex hull, [6], and hence the maximum number of lines required is reduced to $p(p+1)/2$ in general and p for single input-single output systems.

Although the analysis in this section has been restricted to the single input case, it is a relatively simple matter to extend the results to the multiple input case. Details of the appropriate background theory may be found in references [40] and [43].

The above development has assumed that a single constant sampling rate will be used for the complete experiment. However it will be clear from equation (6.25) that it is preferable, if possible, to divide the experiment into a number of subexperiments each having a single sinewave input and each having a different aliasing filter and different Nyquist sampling rate. Equation (6.25) then becomes:

$$\phi^* = \min_{\substack{\lambda_1,\ldots,\lambda_\ell \\ f_1,\ldots,f_\ell}} \phi \left\{ \sum_{k=1}^{\ell} \frac{\lambda_k}{2f_k(1+\varepsilon)} \bar{M}_\beta(f_k) \right\} \tag{6.30}$$

The use of more than one sampling rate is often advantageous especially where there is a wide spread in time constants in the system. Further discussion on this point may be found in [60].

In the next section, the theory developed above will be applied to a number of examples leading to analytic results.

7. ANALYTIC DESIGNS FOR UNIFORM SAMPLING INTERVALS

7.1 *One Parameter Model*

Consider a first order system described by the following model:

$$y(s) = \frac{1}{\tau s+1} u(s) + e(s) \tag{7.1.1}$$

where $u(s)$, $y(s)$ are the Laplace Transforms of the input and output respectively and $e(s)$ represents coloured measurement noise having spectral density:

$$\psi(f) = \frac{1}{a^2\omega^2+1} \quad ; \qquad \omega = 2\pi f \tag{7.1.2}$$

Here the only parameter of interest is τ the system time constant, and hence a suitable design criterion is simply $\det(M^{-1}) = 1/M$ where M is the (1×1) information matrix.

Following the discussion of Theorem 6.2, an optimal one line input spectrum can be found (since $p = 1$ for this example).

The information matrix per unit time corresponding to a single frequency input at $f = \omega/2\pi$ is given by:

$$\bar{M} = 2 \frac{(a^2\omega^2+1)\omega^2}{(\tau^2\omega^2+1)^2} \tag{7.1.3}$$

Hence it follows that, for continuous observations, the input which minimizes $\det(\bar{M}^{-1})$ has frequency $\bar{\omega}$ given by

$$\bar{\omega} = (\tau^2 - 2a)^{-1/2} \qquad \text{if} \quad 2a^2 < \tau^2 \tag{7.1.4}$$

$$= \infty \qquad \text{if} \quad 2a^2 \geq \tau^2 \tag{7.1.5}$$

That is, if the noise is wide band $(a \to 0)$ then the following appealing result is obtained:

$$\bar{\omega} = \frac{1}{\tau} \tag{7.1.6}$$

The above results apply only to the case where there are no restrictions on the allowable number of samples (i.e., continuous observations are possible). The case where the total number of samples is fixed will now be considered.

Assuming the optimal aliasing filter has been incorporated, the average information matrix per sample, (6.25), is:

$$\tilde{M} = \frac{2\pi(a^2\omega^2+1)\omega}{(1+\varepsilon)(\tau^2\omega^2+1)^2} \tag{7.1.7}$$

where ε is the aliasing filter constant. From (7.1.7) the input frequency $\tilde{\omega}$ which minimizes $\det(\tilde{M}^{-1})$ satisfies the equation:

$$- a^2\tau^2\tilde{\omega}^4 + 3(a^2 - \tau^2)\tilde{\omega}^2 + 1 = 0 \tag{7.1.8}$$

which for wide band noise $(a \to 0)$ yields:

$$\tilde{\omega} = \frac{1}{\tau\sqrt{3}} \tag{7.1.9}$$

Note that $\tilde{\omega}$ (sampled observations) is less than $\bar{\omega}$ (continuous observations) as expected. The sampling rate corresponding to $\tilde{\omega}$ is

$$f_s = \frac{1+\varepsilon}{\pi}\tilde{\omega}$$

7.2 *First Order Two Parameter Model*

The system of (7.1) is generalized to include a gain parameter:

$$y(s) = \frac{K}{\tau s+1} u(s) + e(s) \tag{7.2.1}$$

where $u(s)$, $y(s)$ and $e(s)$ are as for subsection 7.1.

From equation (6.23) the average information matrix per unit time corresponding to a single line is:

$$\bar{M}_\beta\left(\frac{\omega}{2\pi}\right) = 2(a^2\omega^2+1)\begin{bmatrix} \frac{1}{(\tau^2\omega^2+1)} & -\frac{K\tau\omega^2}{(\tau^2\omega^2+1)^2} \\ -\frac{K\tau\omega^2}{(\tau^2\omega^2+1)^2} & \frac{K^2\omega^2}{(\tau^2\omega^2+1)} \end{bmatrix} \tag{7.2.2}$$

where $\beta^T = (K,\tau)$.

For the determinant optimality criterion, it would appear from the discussion in Section 6, that the optimal input spectrum will contain two lines since $p = 2$. However, it can be shown that, in fact, one line is sufficient for this problem. The verification of this result is straightforward depending on the detailed structure of the convex hull and is left as an exercise for the reader.

Thus, for continuous observations, the optimal input frequency, $\bar{\omega}$, minimizing $\det(\bar{M}^{-1})$ satisfies:

$$- a^2\tau^2\bar{\omega}^4 + 3(a^2 - \tau^2)\bar{\omega}^2 + 1 = 0 \tag{7.2.3}$$

which for wide band noise $(a \to 0)$ yields:

$$\bar{\omega} = \frac{1}{\tau\sqrt{3}} \tag{7.2.4}$$

The variation of $\det(\bar{M}_\beta)$ with ω is shown in Fig. (7.1).

Now considering the case where the total number of samples is fixed, the average information matrix per sample is given by:

$$\tilde{M}_\beta = \frac{2\pi(a^2\omega^2+1)}{\omega(1+\varepsilon)}\begin{bmatrix} \frac{1}{(\tau^2\omega^2+1)} & -\frac{-K\tau\omega^2}{(\tau^2\omega^2+1)^2} \\ -\frac{-K\tau\omega^2}{(\tau^2\omega^2+1)^2} & \frac{K^2\omega^2}{(\tau^2\omega^2+1)^2} \end{bmatrix} \tag{7.2.5}$$

Hence the optimal input frequency with a fixed number of samples is

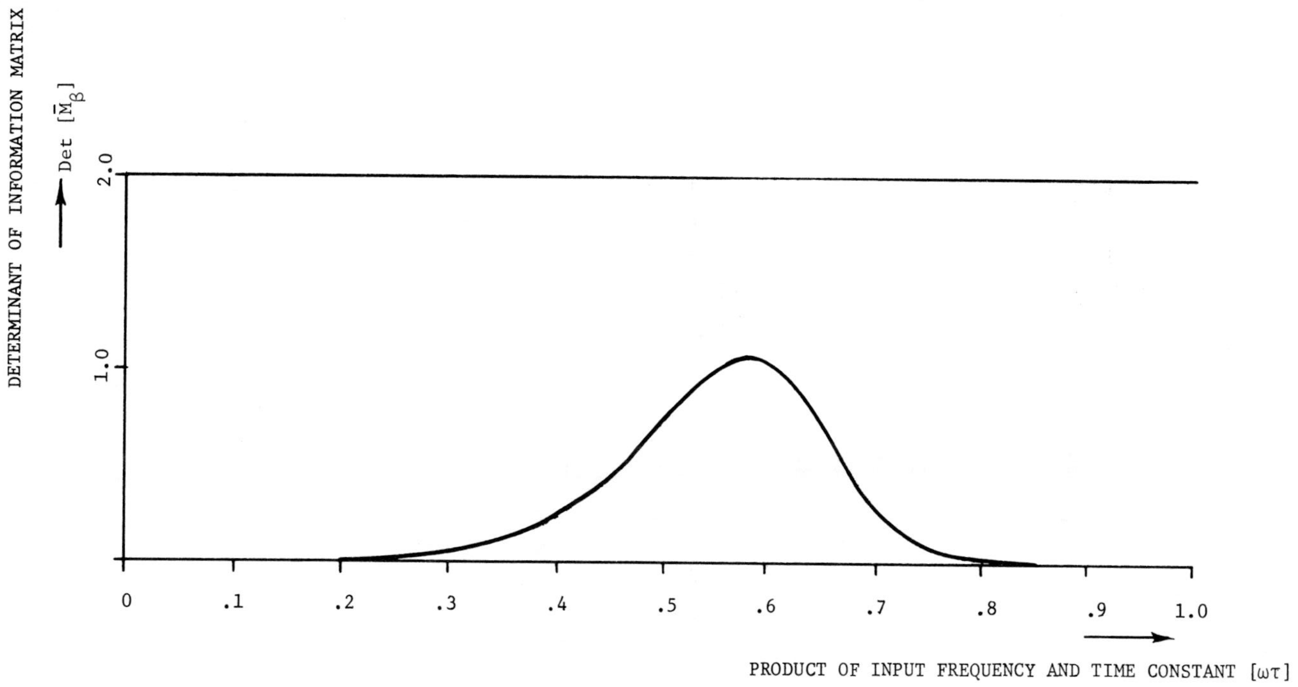

Figure 7.1

Variation of Information Matrix with Input Frequency for First Order System with White Output Noise and Fixed Experiment Time

$$\tilde{\omega} = \left(\frac{1}{\tau^2} - \frac{2}{a^2}\right)^{1/2} \quad \text{if} \quad a > \tau\sqrt{2} \tag{7.2.6}$$

$$= 0 \quad \text{if} \quad a \leq \tau\sqrt{2} \tag{7.2.7}$$

The corresponding optimal sampling frequency is

$$f_s = \frac{1+\varepsilon}{2\pi} \tilde{\omega} \tag{7.2.8}$$

The result in (7.2.7) for wide band noise is, of course, impractical but can correctly be interpreted as meaning "apply a low frequency input and sample slowly." Just how slowly will depend on the total allowable experiment time.

7.3 *Second Order Two Parameter Model*

Consider a second order system having undamped natural frequency ω_0 and damping ratio ξ:

$$y(s) = \frac{\omega_0^2}{s^2+2\xi\,\omega_0 s+\omega_0^2} u(s) + n(s) \tag{7.3.1}$$

where $u(s)$, $y(s)$ are the Laplace Transform of the input and output respectively and $n(s)$ denotes wide band noise.

Again using equation (6.23), the average information matrix per unit time corresponding to an input having a single frequency component is

$$\bar{M}_\beta\left(\frac{\omega}{2\pi}\right) = \frac{2}{\left\{\left[1-\left(\frac{\omega}{\omega_0}\right)^2\right]^2+4\xi^2\left[\frac{\omega}{\omega_0}\right]^2\right\}^2} \begin{bmatrix} 4\left(\frac{\omega^4}{\omega_0^6}\right) + 4\xi^2\left(\frac{\omega^2}{\omega_0^4}\right) & -4\xi\left[\frac{\omega^2}{\omega_0^3}\right] \\ -4\xi\left[\frac{\omega^2}{\omega_0^3}\right] & 4\left[\frac{\omega^2}{\omega_0^2}\right] \end{bmatrix} \tag{7.3.3}$$

where $\beta^T = [\omega_0, \xi]$.

For this problem it can again be shown that the optimal input spectra need only contain one line for the determinant criterion. The optimal value of the input frequency with continuous observations is:

$$\bar{\omega} = \frac{\omega_0}{\sqrt{5}} \{(1-2\xi^2) + [(1-2\xi^2)^2 + 15]^{1/2}\}^{1/2} \qquad (7.3.4)$$

The variation of $\bar{\omega}$ with the damping ratio ξ is shown in Figure 7.3.1. It can be seen that the optimal input frequency drops as the damping increases. This result might have been expected from intuitive reasoning.

Proceeding as in the previous examples, the optimal input frequency, $\tilde{\omega}$, with sampled observations is

$$\tilde{\omega} = \frac{\omega_0}{\sqrt{3}} \{(1-2\xi^2) + [(1-2\xi^2)^2 + 3]^{1/2}\}^{1/2} \qquad (7.3.5)$$

The corresponding optimal sampling rate is

$$f_s = \left(\frac{1+\varepsilon}{2\pi}\right) \tilde{\omega} \qquad (7.3.6)$$

The variation of $\tilde{\omega}$ with the damping ratio is also shown in Figure 7.3.1. It can be seen from the figure that the optimal input frequency with sampled observations is always less than or equal to the optimal input frequency with continuous observations (equality occurs only when the system is marginally stable). This is in line with the general observations made in Section 6.

8. CONCLUSIONS

This chapter has considered the problem of design of experiments for dynamic system identification. The main emphasis has been on the design of the sampling system, namely the choice of presampling filters and sampling intervals. It has been demonstrated that, in order to achieve maximal return from an

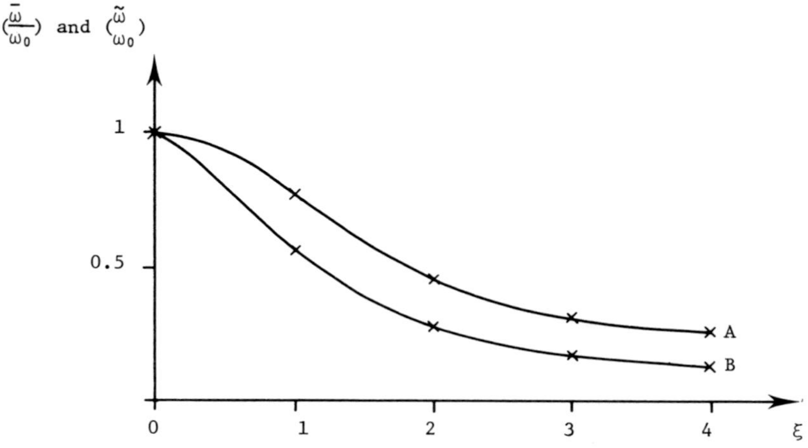

Figure 7.3.1

Variation of Optimal Input Frequency versus Damping Ratio For Second Order System with "White" Output Noise

Legend: A - Continuous Observations $(\frac{\bar{\omega}}{\omega_0})$

B - Sampled Observations $(\frac{\tilde{\omega}}{\omega_0})$

experiment, a coupled design of the sampling system and test signal is required.

The general problem, in which nonuniform sampling intervals are allowed, has been formulated and expressions developed for the information return from an experiment. Special identification algorithms are, of course, required to analyze data collected at nonuniform sampling intervals but an appropriate maximum likelihood algorithm follows from the expressions developed for the likelihood function in the text. The general experiment design problem with nonuniform sampling requires a very sophisticated optimization to be carried out. To simplify the computations a suboptimal one-step-ahead design algorithm has been described. The algorithm would be suitable for on-line experiment design. Examples have been presented which show that significant improvements in identification accuracy can be achieved with this simply optimization procedure. Presumably, further improvements could be achieved by extending the optimization horizon but the complexity of the design would be very greatly increased. For the case of uniform sampling intervals and power constrained test signals it has been shown that the experiment design problem has a simple interpretation in the frequency domain. The usual aliasing filter is optimal in a well defined sense and the optimal input spectra need only contain a finite number of lines. These facts lead to an optimization in a space of low dimension. Analytic designs are possible for simple cases.

The final choice between the different design methods outlined in this chapter (and other methods also) depends upon the nature of the constraints on the allowable experimental conditions. Care must be exercised in stating the constraints as otherwise impractical or trivial designs can result. However, with judicious choice of the constraints, then significant improvements in estimation accuracy can be achieved from relatively simple experiment design procedures similar to those outlined in this chapter.

It has been the authors' intention that this chapter should not only provide methods for experiment design but, perhaps more importantly, also provide insight into the ramifications of the information structure in identification problems.

Acknowledgments

The work reported in this chapter grew largely from joint work of the authors whilst they were in the Department of Computing and Control at Imperial College. The authors wish to acknowledge many helpful suggestions made by the staff and students of Imperial College. Special credit is due to Professor David Mayne who was a constant source of inspiration and to Martin Zarrop who originally developed some of the techniques described in Sections 4 and 5 as part of his doctoral research.

REFERENCES

The References have been collected under the following headings for the reader's convenience:

1. *Experiment Design for Non-Dynamic Systems*
2. *Test Signal Design for Dynamic System Identification*
3. *Design of Sampling Intervals for Dynamic System Identification*
4. *Mathematical Background*

---

1. *EXPERIMENT DESIGN FOR NON-DYNAMIC SYSTEMS*

1. Cox, D. R., *Planning of Experiments*, Wiley, 1958.

2. Davies, O. L., *Design and Analysis of Industrial Experiments*, Oliver and Boyd, 1954.

3. Kempthorne, O., *Design and Analysis of Experiments*, Wiley, New York, 1952.

4. Kiefer, J. and J. Wolfowitz, "The Equivalence of Two Extremum Problems," *Canadian Journal of Maths., Vol. 12,* 1966, pp. 363-366.

5. Karlin, S. and W. J. Studden, "Optimal Experimental Designs," *Annals of Math. Stat., Vol. 37,* 1966, pp. 783-815.

6. Federov, V. V., *Theory of Optimal Experiments,* Academic Press, New York and London, 1971.

7. Whittle, P., "Some General Points in the Theory of Optimal Experimental Design," *J. R. Statist. Soc., B, 35, No. 1,* 1973, pp. 123-130.

8. Wynn, H. P., "Results in the Theory and Construction of D-Optimum Experimental Designs, *J. R. Statist. Soc., B, 34, No. 2,* 1972, pp. 133-147.

2. *TEST SIGNAL DESIGN FOR DYNAMIC SYSTEM IDENTIFICATION*

9. Levin, M. J., "Optimal Estimation of Impulse Response in the Presence of Noise," *IRE Trans. on Circuit Theory, Vol. CT-7,* March 1960, pp. 50-56.

10. Litman, S. and W. H. Huggins, "Growing Exponentials as a Probing Signal for System Identification," *Proc. IEEE, Vol. 51,* June 1963, pp. 917-923.

11. Levadi, V. S., "Design of Input Signals for Parameter Estimation," *IEEE Trans. Automatic Control, Vol. AC-11, No. 2,* April 1966, pp. 205-211.

12. Gagliardi, R. M., "Input Selection for Parameter Identification in Discrete Systems," *IEEE Trans. Automatic Control, Vol. AC-12, No. 5,* October 1967.

13. Goodwin, G. C. and R. L. Payne, "Design and Characterization of Optimal Test Signals for Linear Single Input-Single Output Parameter Estimation," Paper TT-1, 3rd IFAC Symposium, The Hague/Delft, June 1973.

14. Goodwin, G. C., R. L. Payne and J. C. Murdoch, "Optimal Test Signal Design for Linear Single Input-Single Output Closed Loop Identification," CACSD Conference, Cambridge, 2-4th April 1973.

15. Goodwin, G. C., J. C. Murdoch and R. L. Payne, "Optimal Test Signal Design for Linear Single Input-Single Output System Identification," *Int. J. Control, Vol. 17, No. 1,* 1973, pp. 45-55.

16. Arimoto, S. and H. Kimura, "Optimal Input Test Signals for System Identification - An Information Theoretic Approach," *Int. J. Systems Science, Vol. 1, No. 3,* 1971, pp. 279-290.

17. Mehra, R. K., "Optimal Inputs for System Identification," *IEEE Trans. Automatic Control, Vol. AC-19,* June 1974, pp. 192-200.

18. Smith, P. L., "Test Input Evaluation for Optimal Adaptive Filtering," Preprints 3rd IFAC Symposium, The Hague/Delft, Paper TT-5, 1973.

19. Box, G. E. P. and G. M. Jenkins, *Time Series Analysis Forecasting and Control,* Holden Day, San Francisco, 1970, pp. 416-420.

20. Nahi, N. E. and D. E. Wallis, Jr., "Optimal Inputs for Parameter Estimation in Dynamic Systems with White Observation Noise," Paper IV-A5, Proc. JACC, Boulder, Colorado, 1969, pp. 506-512.

21. Inoue, K., K. Ogino and Y. Sawaragi, "Sensitivity Synthesis of Optimal Inputs for Parameter Identification," Paper 9-7, IFAC Symposium, Prague, June 1970.

22. Rault, A., R. Pouliquen and J. Richalet, "Sensitivizing Inputs and Identification," Preprints 4th IFAC Congress, Warsaw, Paper 26.2, 1969.

23. Van den Bos, A., "Selection of Periodic Test Signals for Estimation of Linear System Dynamics," Paper TT-3, Preprints of 3rd IFAC Symposium, The Hague/Delft, 1973.

24. Goodwin, G. C., "Optimal Input Signals for Nonlinear System Identification," *Proc. IEEE, Vol. 118, No. 7,* 1971, pp. 922-926.

25. Goodwin, G. C., "Input Synthesis for Minimum Covariance State and Parameter Estimation," *Electronics Letters, Vol. 5, No. 21,* 1969.

26. Reid, D. B., "Optimal Inputs for System Identification," Stanford University Rpt. No. SUDAAR 440, May 1972.

27. Sawaragi, Y. and K. Ogino, "Game Theoretic Design of Input Signals for Parameter Identification," Paper TT-6, Preprints 3rd IFAC Symposium, The Hague/Delft, June 1973.

28. Keviczky, L. and C. S. Banyasz, "On Input Signal Synthesis for Linear Discrete-Time Systems," Paper TT-2, Preprints 3rd IFAC Symposium, The Hague/Delft, June 1973.

29. Keviczky, L., "On Some Questions of Input Signal Synthesis," Report 7226(B), Lund Institute of Technology, Division of Automatic Control, November 1972.

30. Aoki, M., R. M. Staley, "On Approximate Input Signal Synthesis in Plant Parameter Identification," presented at 1st Hawaii International Conference on System Sciences, University of Hawaii, January 1968.

31. Aoki, M. and R. M. Staley, "Some Computational Considerations in Input Signal Synthesis Problems," 2nd International Conference on Computing Methods in Optimization Problems, sponsored by SIAM, San Remo, Italy, September 1968.

32. Aoki, M. and R. M. Staley, "On Input Signal Synthesis in Parameter Identification," *Automatica, Vol. 6,* 1970, pp. 431-440. Originally presented at 4th IFAC Congress, Warsaw, June 1969.

33. Nahi, N. E. and G. A. Napjus, "Design of Optimal Probing Signals for Vector Parameter Estimation," Paper W9-5, Preprints IEEE Conference on Decision and Control, Miami, December 1971.

34. Napjus, G. A., "Design of Optimal Inputs for Parameter Estimation," Ph.D. Dissertation, University of Southern California, August, 1971.

35. Tse, E., "Information Matrix and Local Identifiability of Parameters," Paper 20-3, *JACC,* Columbus, Ohio, June 1973.

36. Minnich, G. M., "Some Considerations in the Choice of Experimental Designs for Dynamic Models," Ph.D. Dissertation, University of Wisconsin, 1972.

37. Kalaba, R. E. and I. Spingarn, "Optimal Inputs and Sensitivities for Parameter Estimation," *JOTA, 11, 1,* 1973, pp. 56-67.

38. Payne, R. L. and G. C. Goodwin, "A Bayesian Approach to Experiment Design with Applications to Linear Multivariable Dynamic Systems," IMA Conference on Computational Problems in Statistics, Univ. of Essex, July 1973.

39. Mehra, R. K. "Frequency Domain Synthesis of Optimal Inputs for Parameter Estimation," Techn. Report No. 645, Div. of Eng. and Appl. Physics, Harvard University, July 1973.

40. Mehra, R. K., "Synthesis of Optimal Inputs for Multiinput-Multioutput (MIMO) Systems with Process Noise," Techn. Report No. 649, Div. of Eng. and Appl. Physics, Harvard University, February 1974.

41. Van den Bos, A., "Construction of Binary Multifrequency Test Signals," Paper 4-6, IFAC Symposium on Identification, Prague, June 1967.

42. Mehra, R. K., "Optimal Measurement Policies and Sensor Designs for State and Parameter Estimation," Milwaukee Symposium on Automatic Controls, March 1974.

43. Goodwin, G. C., R. L. Payne and M. B. Zarrop, "Frequency Domain Experiment Design for Multiple Input Systems," Report No. 75/17, Department of Computing and Control, Imperial College, London.

44. Payne, R. L. and G. C. Goodwin, "Simplification of Frequency Domain Experiment Design for Single Input - Single Output Systems," Report No. 74/3, Department of Computing and Control, Imperial College, London.

45. Viort, B., "D-Optimal Designs for Dynamic Models - Part I-Theory," Techn. Report 314, Department of Statistics, University of Wisconsin, October 1972.

3. DESIGN OF SAMPLING INTERVALS FOR DYNAMIC SYSTEM IDENTIFICATION

46. Goodwin, G. C., M. B. Zarrop and R. L. Payne, "Coupled Design of Test Signal, Sampling Intervals and Filters for System Identification," *IEEE Trans. AC Special Issue on Identification,* December 1974.

47. Payne, R. L., G. C. Goodwin and M. B. Zarrop, "Frequency Domain Approach for Designing Sampling Rates for System Identification," *Automatica,* March 1975.

48. Astrom, K. J., "On the Choice of Sampling Rates in Parameter Identification of Time Series," *Inf. Sci., Vol. 1,* 1969, pp. 273-287.

49. Gustavsson, I., "Choice of Sampling Intervals for Parametric Identification," Report 7103, Lund Institute of Technology, 1971.

50. Zarrop, M.B., "Experimental Design for System Identification," M.Sc. Dissertation, Imperial College of Science and Technology, 1973.

4. MATHEMATICAL BACKGROUND

51. Silvey, S. D., *Statistical Inference,* Penguin, 1970, pp. 35-44.

52. Kullback, S., *Information Theory and Statistics,* Dover, New York, 1968.

53. Astrom, K. J., *Introduction to Stochastic Control Theory,* Academic Press, 1970.

54. Kailath, T., "An Innovation Approach to Least Squares Estimation - Part 1: Linear Filtering in Additive White Noise," *IEEE Trans. A-13, No. 6,* December 1968.

55. Cox, D. R. and H. D. Miller, *The Theory of Stochastic Processes,* Chapman Hall, London, 1967.

56. Wong, E., *Stochastic Processes in Information and Dynamical Systems,* McGraw Hill, 1970.

57. Yen, J. L., "On Nonuniform Sampling of Bandwidth Limited Signals," *IRE Trans. Circuit Theory,* December 1956, pp. 251-257.

58. Bendat, J. S. and A. G. Piersol, *Measurement and Analysis of Random Data,* Wiley, New York, 1966.

59. Anderson, B. D. O., K. L. Hitz and N. D. Diem, "Recursive Algorithm for Spectral Factorization," *IEEE Trans. Circuits and Systems, Vol. CAS-21, No. 6,* 1974.

60. Goodwin, G. C. and R. L. Payne, "Dynamic System Identification: Experiment Design and Data Analysis," Book to appear 1977.

MODELLING AND RECURSIVE ESTIMATION FOR DOUBLE INDEXED SEQUENCES

Samer Attasi
Iria-Laboria
78150 Rocquencourt, France

1. INTRODUCTION

It is commonly agreed that linear recursive processing is a practical solution to the main drawback of digital technology, which is its slowness for many real time signal processing applications.

A universally known theory has been developed in the last decades with important contributions of namely Faurre [8], Kailath [14], [15], Kalman [17], [18], [19], [20], Mehra [22],

[23], Youla [32]..., and provides various tools for recursive processing of signals considered as gaussian stationary time series. The common mathematical back-ground developed and used by this theory consists of the results on recursive linear transformations. Various characterizations (spectral, state-space, autoregressive...) of these transformations have been used to define various processing algorithms which are therefore theoretically equivalent. Later, the practical advantages and drawbacks of each of the implementations were analyzed and led to finer descriptions (canonical forms...).

In the case of image processing the data is naturally represented by a double indexed sequence. So, one has to start again and create first the equivalent of the mathematical background that we just discussed, that is, introduce a general model of invariant double indexed recursive transformations. This model would then be used as a tool to follow the above theory with the ultimate goal of deriving tractable and mathematically coherent recursive solutions to estimation problems involving double indexed data.

But, how should one start? It seems reasonable to try natural generalizations of the various equivalent characterizations of linear recursive time invariant transformations, which are mainly:

— The algebraic characterization as finite rank linear operators.

— The "behavioral" characterization as a transformation described by a finite order autoregressive form relating the inputs and the outputs.

It turns out however that the generalizations of these two characterizations do not define any more the same class of linear space invariant transformations. Our approach was then to generalize the algebraic characterization which as the following sections will show, turns to be very fertile and allows to build a self-contained theory for statistical recursive processing for a large class of double indexed sequences.

Section 2 introduces a model for double indexed space invariant linear transformations. A realization theory is developed leading to an algebraic characterization of those transformations, an approximation property and a minimal realization algorithm.

Section 3 introduces a class of double indexed gaussian sequences built as "output" of such linear transformations "driven" by white noise. After an approximation theorem proving the generality of this class of sequences, the study of their correlation function leads to stochastic identification algorithms and throws some light on spectral factorization properties of double indexed sequences.

Finally Section 4 yields a recursive solution to filtering and smoothing problems involving gaussian sequences defined in Section 3.

2. A MODEL FOR RECURSIVE LINEAR SPACE INVARIANT TRANSFORMATIONS

The most general space invariant linear transformations of an input sequence $\{u_{i,j};\ (i,j) \in Z^2,\ u_{i,j} \in \mathbb{R}^m\}$ into an output sequence $\{y_{i,j};\ (i,j) \in Z^2,\ y_{i,j} \in \mathbb{R}^p\}$ can be written as:

$$y_{i,j} = \sum_{(k,l) \in Z^2} A_{k,l}\, u_{i-k,j-l} \tag{2.1}$$

where the sequence of $(p \times m)$ matrices $\{A_{k,l};\ (k,l) \in Z^2\}$ called the impulse response, is supposed to be square summable to avoid any ambiguity in the definition of the above summation.*

The problem is then to define a class of linear double indexed recursive transformations which:

---

*All the double indexed summations involved in this work, deal with absolutely (or normally) convergent series so that the ordering of the terms in the summation is indifferent.

i) is large enough, in the sense that the impulse response $\{A_{k,l};\ (k,l) \in Z^2\}$ of any transformation of the type (2.1) can be approximated (in the space of square summable sequences) by the impulse response of an element of this class.

ii) contains a simple algebraic structure, that is to say we wish to be able to characterize easily the elements of the class in terms of their impulse response in order to derive simple and efficient identification algorithms.

The class of transformations described below satisfies both conditions i) and ii).

2.1 DEFINITION

A linear space invariant transformation belongs to the considered class σ if its input-output relation (2.1) can be rewritten as:

$$\begin{cases} y_{i,j} = H\,X_{i,j} \\ X_{i,j} = F_1X_{i-1,j} + F_2X_{i,j-1} - F_1F_2X_{i-1,j-1} + Gu_{i+N,j+M} \end{cases} \tag{2.2}$$

where

$X_{k,l} \in \mathbb{R}^n$ is the "state vector"

N, M are two integers

H, F_1, F_2, G are matrices of appropriate dimensions which satisfy the commutativity condition:

$$F_1F_2 = F_2F_1 . \tag{2.3}$$

The impulse response $\{A_{i,j};\ (i,j) \in Z^2\}$ of such a transformation is given by:

$$A_{i,j} = HF_1^{i+N}\,F_2^{j+M}\,G \qquad i \geq -N,\ j \geq -M \tag{2.4}$$

$$A_{i,j} = 0 \qquad \text{otherwise} \tag{2.4'}$$

Note that integers N and M will play no role in the

realization theory that we are developing next; their introduction has the only aim of emphasizing the fact that "causality" is not intrinsically related to a recursive equation of type (2.2). Moreover, without entering into a stability theory briefly discussed in Attasi [1], we shall assume throughout this work that matrices F_1 and F_2 are both asymptotically stable, that is all their eigenvalues are of modulus less than one.

2.2 *THE EQUIVALENCE CLASS OF MINIMAL REALIZATIONS OF A GIVEN SEQUENCE*

In this paragraph we study the set (supposing it is not empty) of quadruples $\{H,F_1,F_2,G\}$ associated to a given sequence $\{A_{i,j};\ i \geq -N,\ j \geq -N\}$ by equations (2.3) and (2.4), and develop a theory very similar to that of standard dynamical systems, leading to the important concept of minimal realizations.

DEFINITION 2.1: *The quadruple* $\{H,F_1,F_2,G\}_n$ *is a realization of the given sequence* $\{A_{i,j};\ i \geq -N,\ j \geq -M\}$ *if it satisfies equations (2.3) and (2.4). The subscript* n *denotes the dimension of matrices* F_1 *and* F_2.

Although it is possible [Attasi] to give an interpretation in terms of the recurrent equation (2.2), to the concepts of observability and controllability of realizations, formal definitions are sufficient for our present purpose.

So let us define the sequences of matrices C_k and O_k:

$$\forall k > 0 \quad \begin{cases} C_k = [G,\ldots,F_1^iF_2^jG,\ldots,F_1^{k-1}F_2^{k-1}G]\ ; & 0 \leq i \leq k-1, \\ & 0 \leq j \leq k-1 \\ O'_k = [H',\ldots,F_1'^{i}F_2'^{j}H',\ldots,F_1'^{k-1}F_2'^{k-1}H']; & 0 \leq i \leq k-1,^{*} \\ & 0 \leq j \leq k-1, \end{cases}$$

*For any matrix M, M' stands for the transpose of M. In the case of finite l^2 norm operators (Section 4), M' stands for the dual of M.

such that the ordering of blocks $F_1^i F_2^j G$ and $H F_1^i F_2^j$ in C_k and O_k respectively, corresponds to the same bijection $b : \mathbb{N}^2 \to \mathbb{N}$.

If n is the dimension of matrices F_1 and F_2, it follows from Hamilton Cayley theorem that:

$$\begin{aligned} &\forall k \geq n \quad \text{rank } (C_k) = \text{rank } (C_n) = n_1 \leq n \\ &\forall k \geq n \quad \text{rank } (O_k) = \text{rank } (O_n) = n_2 \leq n \end{aligned} \tag{2.5}$$

DEFINITION 2.2: *The realization* $\{H,F_1,F_2,G\}_n$ *(or the triple* $\{F_1,F_2,G\}_n$*) is said to be controllable if the corresponding controllability matrix* C_n *is of full rank* n.

DEFINITION 2.3: *The realization* $\{H,F_1,F_2,G\}_n$ *(or the triple* $\{H,F_1,F_2\}_n$*) is said to be observable if the corresponding observability matrix* O_n *is of full rank* n.

One more definition is needed before proving the existence of controllable and observable realizations.

DEFINITION 2.4: *Two realizations* $\{H,F_1,F_2,G\}_n$ *and* $\{\bar{H},\bar{F}_1,\bar{F}_2,\bar{G}\}_n$ *of the sequence are said to be equivalent if moreover there exists an invertible matrix* T *(therefore defining a change of coordinates in the state space) such that:*

$$\begin{cases} \bar{H} = H\,T \\ \bar{F}_i = T^{-1} F_i\, T \qquad i = 1,2 \\ \bar{G} = T^{-1} G \end{cases} \tag{2.6}$$

Notice that for two such realizations we also have:

$$\begin{cases} \bar{C}_k = T^{-1} C_k \\ \bar{O}_k = O_k\, T \end{cases} \tag{2.7}$$

THEOREM 2.1: *Any realization* $\{H,F_1,F_2,G\}_n$ *of the sequence* $\{A_{i,j};\ i \geq -N,\ j \geq -M\}$, *where we pose rank* $(O_n\ C_n) = n_1 \leq n$, *is equivalent to a realization* $\{\bar{H},\bar{F}_1,\bar{F}_2,\bar{G}\}_n$ *in which:*

$$\bar{F}_i = \begin{bmatrix} \bar{F}_{11,i} & 0 & \bar{F}_{13,i} \\ \bar{F}_{21,i} & \bar{F}_{22,i} & \bar{F}_{23,i} \\ 0 & 0 & \bar{F}_{33,i} \end{bmatrix} \quad i = 1,2; \quad \bar{G} = \begin{bmatrix} \bar{G}_1 \\ \bar{G}_2 \\ 0 \end{bmatrix}; \quad \bar{H} = [\bar{H}_1,0,\bar{H}_2] \ . \tag{2.8}$$

Moreover the quadruple $\{\bar{H}_1,\bar{F}_{11,1},\bar{F}_{11,2}\bar{G}_1\}_{n_1}$ *is a controllable and observable realization of the same sequence* $\{A_{i,j};\ i \geq -N,\ j \geq -M\}$.

Proof. We follow without any difficulty the steps of the standard proof for realizations of dynamical systems. This is the outline of the proof.

i) Build a realization $\{H^c,F_1^c,F_2^c,G^c\}_n$ equivalent to $\{H,F_1,F_2,G\}_n$, of the form:

$$F_i^c = \begin{bmatrix} F_{11,i}^c & F_{12,i}^c \\ 0 & F_{22,i}^c \end{bmatrix} \quad i = 1,2; \quad H^c = (H_1^c,H_2^c); \quad G^c = \begin{pmatrix} G_1^c \\ 0 \end{pmatrix} \tag{2.9}$$

such that $\{H_1^c,F_{11,1}^c,F_{11,2}^c,G_1^c\}_{n_2}$ is a controllable realization of the sequence $\{A_{i,j};\ i \geq -N,\ j \geq -M\}$ where $n_2 = \text{rank}\ C_n$.

This is achieved by applying equation (2.6) on the initial realization $\{H,F_1,F_2,G\}_n$, with an invertible matrix T_2 whose first n_2 columns are "chosen" among the independent column vectors of C_n.

ii) Let $O_{n_2}^c$ be the observability matrix associated to the controllable realization $\{H_1^c,F_{11,1}^c,F_{11,2}^c,G_1^c\}_{n_2}$; and let $n_3 = \text{rank}\ O_{n_2}^c$. Then, the final realization $(\bar{H},\bar{F}_1,\bar{F}_2,\bar{G})$ with the form (2.8) is obtained by applying equations (2.6) on the intermediate realization $\{H^c,F_1^c,F_2^c,G^c\}_n$ with an invertible matrix T_3 built as follows:

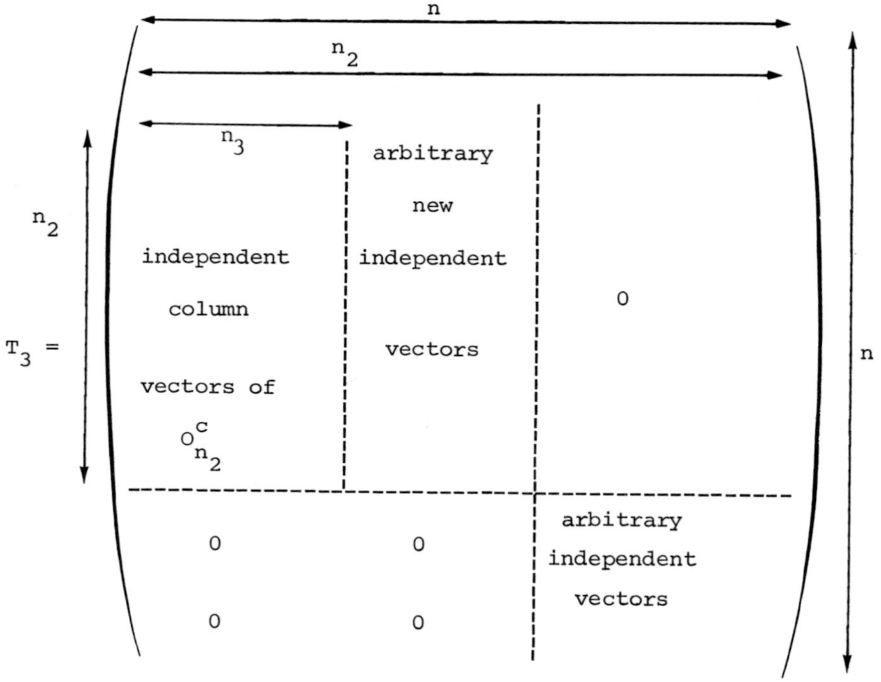

(2.10)

iii) Prove trivially that $\{\bar{H}_1, \bar{F}_{11,1}, \bar{F}_{11,2}, G_1\}_{n_3}$ is a controllable and observable realization of the sequence $\{A_{i,j};\ i \geq -N,\ j \geq -M\}$ for example by using a very simple lemma saying: two quadruples $\{H, F_1, F_2, G\}_n$ and $\{\bar{H}, \bar{F}_1, \bar{F}_2, \bar{G}\}_{\bar{n}}$ are realizations of the same sequence if and only if $O_k\ C_l = \bar{O}_k\ \bar{C}_l$ for all k and l. Finally one proves by contradiction that $n_3 = n$. ■

Note that the commutativity condition remains always verified because a change of coordinates does not alter commutativity of matrices.

A natural consequence (proved by contradiction) of Theorem (3.1) is:

COROLLARY 2.1: *A controllable and observable realization of a given sequence* $\{A_{i,j};\ i \geq -N,\ j \geq -N\}$ *is a minimal realization (the dimension of matrices* F_1 *and* F_2 *is minimal).*

Having proved the existence of observable and controllable realizations which are moreover minimal realizations, we now determine the structure of the set of minimal realizations, associated to a given sequence $\{A_{i,j};\ i \geq -N,\ j \geq -M\}$:

THEOREM 2.2: *All minimal realizations of a sequence* $\{A_{i,j};\ i \geq -N,\ j \geq -M\}$ *are equivalent in the sense of definition* (2.4) *thus constituting a single equivalence class.*

Proof. Let us consider two minimal realizations $\{H,F_1,F_2,G\}_n$ and $(\bar{H},\bar{F}_1,\bar{F}_2,\bar{G})_n$ with their respective observability and controllability matrices O_n, C_n and $\bar{O}_n$, $\bar{C}_n$.

Since these are realizations of the same sequence, we have:

$$\begin{cases} \bar{O}_n\, \bar{C}_n = O_n\, C_n \\ \bar{O}_n\, \bar{C}_{n+1} = O_n\, C_{n+1} \end{cases} . \tag{2.11}$$

But all these matrices are of full rank, so, using the pseudo inverses $\bar{O}_n^+$ and C_n^+ of matrices $\bar{O}_n$ and C_n, we get:

$$\bar{C}_n = T\, C_n \tag{2.12}$$

$$\bar{C}_{n+1} = T\, C_{n+1} \tag{2.13}$$

$$\bar{O}_n = O_n\, T^{-1} \tag{2.14}$$

where matrix T is given by:

$$T = \bar{C}_n\, C_n^+ = \bar{O}_n^+\, O_n \quad . \tag{2.15}$$

Considering equation (2.13) and noticing that, *thanks to commutativity,* $F_i\, C_n$ $(i = 1,2)$ and $\bar{F}_i\, \bar{C}_n$ $(i = 1,2)$ are submatrices of C_{n+1} and $\bar{C}_{n+1}$ respectively, we get:

$$\bar{F}_i \bar{C}_n = T F_i C_n \qquad i = 1,2 \quad . \tag{2.16}$$

Taking into account equation (2.15), we get:

$$\bar{F}_i = T F_i T^{-1} \qquad i = (1,2) \quad . \tag{2.17}$$

Finally equations (2.12) and (2.14) imply trivially that:

$$\begin{cases} \bar{G} = T G \\ \bar{H} = H T^{-1} \end{cases} \tag{2.18}$$

This completes the proof. ■

The set of realizations of a double indexed sequence $\{A_{i,j};\ i \geq N,\ j \geq -M\}$ has therefore all the nice properties of the set of realizations for discrete dynamical systems.

2.3 REALIZATION THEORY - MINIMAL REALIZATION ALGORITHM

We now proceed to an algebraic characterization of σ - transformations, in terms of linear space morphisms, that will lead to a minimal realization algorithm based on a generalized version of Hankel matrices. For the sake of simplicity of notations we shall consider the scalar input scalar output case ($p = m = 1$).

2.3.1 Algebraic Representation of σ-Transformations and Its Consequences

Let:

$$\Omega = \{\omega = \{\omega_{i,j};\ i \leq 0,\ j \leq 0,\ \omega_{i,j} \in \mathbb{R}\}$$

$$_N\Gamma_M = \{\gamma = \{\gamma_{i,j};\ i \geq -N,\ j \geq -M,\ \gamma_{i,j} \in \mathbb{R}\}$$

To each sequence $\{A_{i,j};\ i \geq -N,\ j \geq -M\}$ we associate a linear mapping $f_{N,M} : \Omega \rightarrow \Gamma_{N,M}$ characterized by:

$$\forall i \geq -N,\ j \geq -M,\ f_{N,M}(1_{-N-i,-M-j}) = \begin{bmatrix} A_{i,j} & A_{i+1,j} & \cdots \\ A_{i,j+1} & \cdots & \\ \vdots & & \end{bmatrix} \quad (2.19)$$

where $1_{\alpha,\beta}$ represents the sequence in Ω, given by:

$$1_{\alpha,\beta} = \{\omega_{k,l} = \delta_{k-\alpha,l-\beta}\}$$

and

$$\delta_{k-\alpha,l-\beta} = 1 \quad \text{if } k = \alpha,\ l = \beta$$
$$= 0 \quad \text{otherwise} \quad .$$

If we now define the canonical shift operators in the following manner:

$$\forall \{\omega_{k,l}\} \in \Omega \qquad \sigma\Omega_1\{\omega_{k,l}\} = \{\omega_{k+1,l}\}$$
$$\sigma\Omega_2\{\omega_{k,l}\} = \{\omega_{k,l+1}\}$$

$$\forall \{\gamma_{k,l}\} \in \Gamma_{N,M} \qquad \sigma\Gamma_1\{\gamma_{k,l}\} = \{\gamma_{k+1,l}\}$$
$$\sigma\Gamma_2\{\gamma_{k,l}\} = \{\gamma_{k,l+1}\}$$

equation (2.19) then implies:

$$f \circ \sigma\Omega_i = \sigma\Gamma_i \circ f \qquad i = 1,2 \quad . \qquad (2.20)$$

DEFINITION 2.5: *The linear mapping* $f_{N,M}$ *defined by equations* (2.19) *is said to be realizable if there exists a* σ-*transformation with an impulse response equal to* $\{A_{i,j};\ i \geq -N,\ j \geq -M\}$.

Then it follows from linearity and "space invariancy" of σ-transformations that these are entirely described by the corresponding $f_{N,M}$ linear mappings.

THEOREM 2.3: *The linear mapping* $f_{N,M}$ *defined by equations* (2.19) *is realizable if and only if it is of finite rank. Moreover if* $f_{N,M}$ *is realizable we have:*

$$\text{rank } f_{N,M} = \text{dimension of the minimal realization.}$$

Proof. *The condition is necessary:*

If $f_{N,M}$ is realizable, there exists a quadruple $\{H,F_1,F_2,G\}_n$ with $F_1F_2 = F_2F_1$, such that:

$$\forall i \geq -N, \quad j \geq -M; \quad A_{i,j} = H\, F_1^{i+N}\, F_2^{j+M}\, G \quad .$$

Let $b : \mathbb{N}^2 \to \mathbb{N}$ be a bijection of $\mathbb{N}^2$ in $\mathbb{N}$; $f_{N,M}$ can then be represented by a Hankel matrix $\mathcal{A}$ defined by:

$$\forall i \geq 0, \quad j \geq 0, \quad k \geq 0, \quad l \geq 0$$

$$\mathcal{A}(b(i,j),\, b(k,l)) = A_{i+k-N,j+l-M} = H\, F_1^{i+k}\, F_2^{j+l}\, G \quad . \tag{2.21}$$

Using the *commutativity between* F_1 *and* F_2, (2.21) becomes:

$$\mathcal{A}(b(i,j), \quad b(k,l)) \quad = \quad (H\, F_1^i\, F_2^j)(F_1^k\, F_2^l\, G) \quad .$$

This implies that matrix $\mathcal{A}$ can be written as a product:

$$\mathcal{A} = O_\infty C_\infty \tag{2.22}$$

where O_∞ and C_∞ are the infinite observability and controllability matrices of the realization $\{H,F_1,F_2,G\}_n$.

Moreover, use of Cayley Hamilton theorem implies:

$$\text{rank } \mathcal{A} = \text{rank}(O_\infty C_\infty) = \text{rank }(O_n C_\infty) = \text{rank}(O_n C_n) \quad .$$

Therefore rank $\mathcal{A}$ is finite, and as a consequence of Theorem (2.1), is equal to the dimension of the minimal realizations.

The condition is sufficient:

We first establish a new model for σ-transformations.

LEMMA: *A σ-transformation characterized by a realization $\{H,F_1,F_2,G\}_n$ can equivalently be represented by the following recursive equations:*

$$\begin{cases} y_{i,j} = H\,X_{i,j} \\ X_{i-1,j} = F_2\,X_{i-1,j-1} + \sigma_1(i-1,j) \\ X_{i,j-1} = F_1\,X_{i-1,j-1} + \sigma_2(i,j-1) \\ X_{i,j} = F_1\,F_2\,X_{i-1,j-1} + F_1\,\sigma_1(i-1,j) + F_2\,\sigma_2(i,j-1) \\ \qquad + Gu_{i+N,j+M} \end{cases} \tag{2.23}$$

where sequences $\sigma_1(i,j)$ *and* $\sigma_2(i,j)$ *are defined by:*

$$\begin{cases} \sigma_1(i,j) = F_1\,\sigma_1(i-1,j) + Gu_{i+N,j+M} \\ \sigma_2(i,j) = F_2\,\sigma_2(i,j-1) + Gu_{i+N,j+M} \end{cases} \tag{2.24}$$

This result is trivially proved by showing that both recursive models (2.2) and (2.23) define a unique sequence of "states" and outputs given by:

$$\begin{cases} X_{i,j} = \sum\limits_{\substack{k\geq 0 \\ l\geq 0}} F_1^k\,F_2^l\,Gu_{i-k+N,j-l+M} \\ y_{i,j} = H\,X_{i,j} \end{cases} \tag{2.25}$$

Back to the sufficient condition, let us consider a linear mapping $f_{N,M} : \Omega \to \Gamma_{N,M}$, of finite rank n, we shall show that such a mapping can be represented by a model of type (2.23), where matrices F_1 and F_2 are also of dimension n.

It is known [17], [27] that such a mapping can be factorized through a linear space X isomorphic to the range of $f_{N,M}$ in $\Gamma_{N,M}$ according to the algebraic scheme:

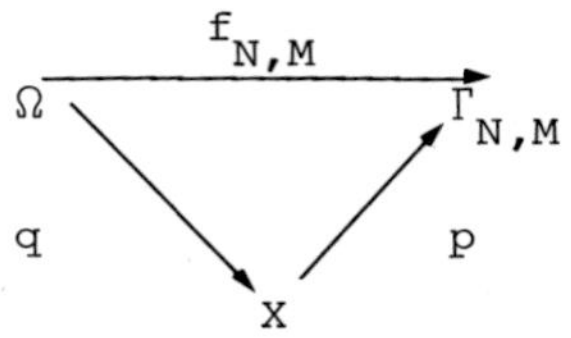

where

p is a linear injection,

q is a linear surjection.

This factorization is called canonical factorization of $f_{N,M}$. Consider then this factorization of $f_{N,M}$ and let:

$\mathcal{U}_1$ = the space of sequences of the form $[\ldots u_{-n}, \ldots, u_0]$; $u_k \in \mathbb{R}$,

$\mathcal{U}_2$ = the space of sequences of the form $\begin{bmatrix} u_0 \\ \vdots \\ u_{-n} \\ \vdots \end{bmatrix}$

j_1 = the canonical injection of u_1 in Ω, that is to say:

$$j_1 \, [\ldots u_{-n}, \ldots, u_0] = \left\{ \begin{matrix} \ldots u_{-n}, & \ldots & \ldots & \ldots & u_0 \\ 0 & 0 & 0 & 0 & 0 \\ \vdots & \vdots & \vdots & \vdots & \vdots \end{matrix} \right\}$$

j_2 = the canonical injection of u_2 in Ω, that is to say:

$$j_2 \begin{bmatrix} u_0 \\ \vdots \\ u_{-n} \\ \vdots \end{bmatrix} = \left\{ \begin{matrix} \cdots\cdots\cdots & 0 & u_0 \\ & & \vdots \\ \cdots\cdots\cdots & 0 & u_{-n} \\ & & \vdots \\ \cdots\cdots\cdots & 0 & \vdots \end{matrix} \right\}$$

j = the canonical injection of R in Ω, that is to say:

$$j(u) = \left\{ \begin{matrix} 0 & \cdots\cdots\cdots & 0 & u \\ & \cdots\cdots\cdots & 0 & 0 \\ & \cdots\cdots\cdots & 0 & 0 \\ & & \vdots & \vdots \\ & \cdots\cdots\cdots & \vdots & \vdots \end{matrix} \right\}$$

we then introduce the following algebraic transition diagrams.

i) First diagram.*

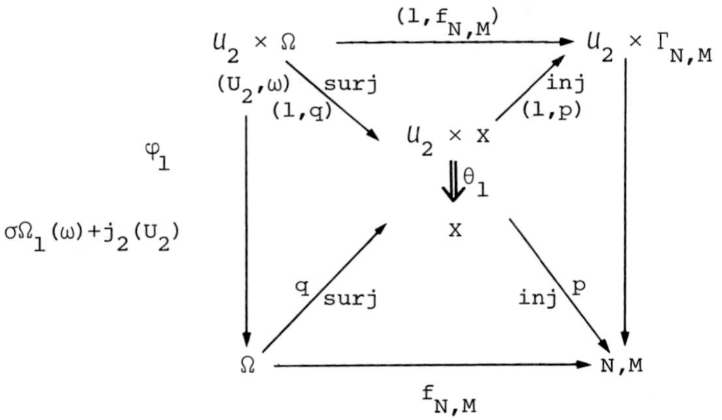

This diagram induces a unique mapping $\theta_1 : U_2 \times X \to X$ given by:

---

*In these algebraic diagrams 1 represents the identity indifferently in $\mathbb{R}$, U_1, U_2.

$$\forall (U_2,X) \in (\mathcal{U}_2 \times \mathcal{X}); \quad \theta_1(U_2,X) = p^{-1} \circ \sigma\Gamma_1 \circ p(X) + q \circ j_2(U_2), \tag{2.26}$$

where p^{-1} is defined on the range of $f_{N,M}$ in $\Gamma_{N,M}$ and satisfies:

$$p^{-1} \circ f_{N,M} = q \qquad (\text{or } p^{-1} \circ p = \text{identity in } \mathcal{X}). \tag{2.27}$$

This result is obtained by analyzing the above diagram in the following manner.

Since q is surjective, for any couple (U_2,X) in $(\mathcal{U}_2,\mathcal{X})$ there exists a set of $\omega \in \Omega$ such that:

$$(U_2,X) = (U_2,q(\omega)) . \tag{2.28}$$

One must then prove that for all (U_2,ω) verifying (2.28) (with fixed X), the mapping:

$$q \circ \varphi_1 = p^{-1} \circ f_{N,M} \circ \varphi_1 \tag{2.29}$$

determines a unique $X \in \mathcal{X}$, given by (2.26).

This results from the following:

$$p^{-1} \circ f_{N,M} \circ \varphi_1(U_2,\omega) \triangleq p^{-1} f_{N,M}(\sigma\Omega_1(\omega) + j_2(U_2)) . \tag{2.30}$$

Using equation (2.20) and linearity of $f_{N,M}$ we get:

$$p^{-1} \circ f_{N,M} \circ \varphi_1(U_2,\omega) = p^{-1} \circ \sigma\Gamma_1 \circ f_{N,M}(\omega) + p^{-1} \circ f_{N,M} \circ j_2(U_2) \tag{2.31}$$

but

$$\begin{cases} f_{N,M}(\omega) = p \circ q(\omega) = p(X) \\ p^{-1} \circ f_{N,M} \circ j_2(U_2) = q \circ j_2(U_2) \end{cases} . \tag{2.32}$$

Substituting (2.32) in (2.31) yields 2.26.

ii) Second diagram.

By a permutation of index 1 and 2, in diagram i), we define a unique mapping $\theta_2 : U_1 \times \mathcal{X} \to \mathcal{X}$ given by:

$$\forall (U_1, X) \in (\mathcal{U}_1 \times \mathcal{X});\quad \theta_2(U_1, X) = p^{-1} \circ \sigma\Gamma_2 \circ p(X) + q \circ j_1(U_1). \tag{2.34}$$

iii) Third diagram.

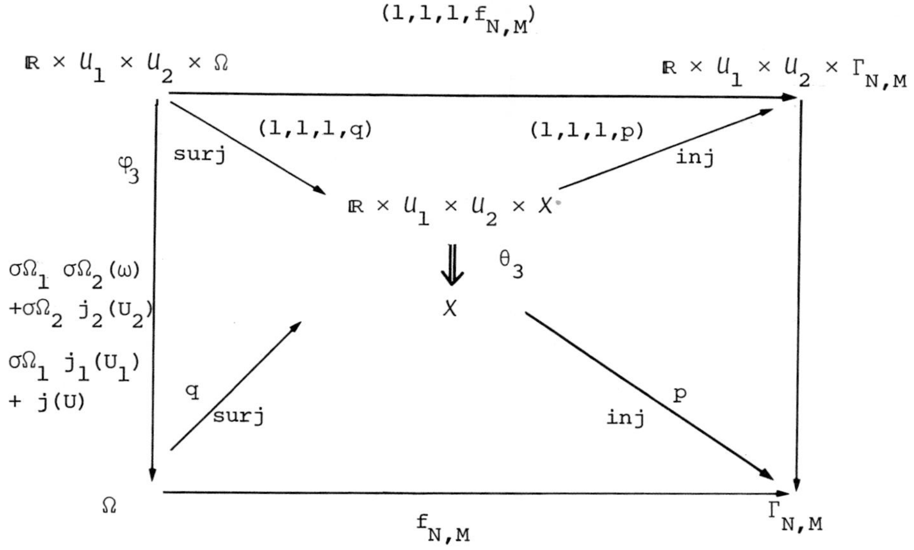

with the same proof as i), this diagram induces a unique mapping $\theta_3 : \mathbb{R} \times \mathcal{U}_1 \times \mathcal{U}_2 \times \mathcal{X} \to \mathcal{X}$ given by:

$$\forall (U, U_1, U_2, X) \in \mathbb{R} \times \mathcal{U}_1 \times \mathcal{U}_2 \times \mathcal{X}:$$

$$\begin{aligned}\theta_3(U, U_1, U_2, X) = {} & p^{-1} \circ \sigma\Gamma_1 \circ \sigma\Gamma_2 \circ p(X) + (p^{-1} \circ \sigma\Gamma_1 \circ p) \circ (q \circ j_1)(U_1) \\ & + (p^{-1} \circ \sigma\Gamma_2 \circ p)(q \circ j_2)(U_2) \\ & + q \circ j(U) \quad .\end{aligned} \tag{2.35}$$

Now define:

$$\begin{cases} F_1 = p^{-1} \circ \sigma\Gamma_1 \circ p \\ F_2 = p^{-1} \circ \sigma\Gamma_2 \circ p \\ G = q \circ j \\ H = \Pi \circ p \quad \text{where } \Pi \text{ is the canonical projection} \\ \qquad \text{operator of } \Gamma_{N,M} \text{ onto } \mathbb{R}\text{, that is to say:} \end{cases} \tag{2.36}$$

$$\Pi\{\gamma_{i,j};\ i \geq -N,\ j \geq -M\} = \gamma_{-N,-M} .$$

To prove that F_1 and F_2 commute and that:

$$F_1F_2 = F_2F_1 = p^{-1}\circ\sigma\Gamma_1\circ\sigma\Gamma_2\circ p = p^{-1}\circ\sigma\Gamma_2\circ\sigma\Gamma_1 \circ p ,$$

we just notice that $pp^{-1}(y) = y$ for all y belonging to the range of p in $\Gamma_{N,M}$ and that $\sigma\Gamma_1 \circ p(X)$ and $\sigma\Gamma_2 \circ p(X)$ precisely are in the range of p in $\Gamma_{N,M}$, for all $X \in \mathcal{X}$. This last property is proved by using equation (2.20) together with the fact that the mapping q is surjective.

Defining finally:

$$\begin{aligned} \sigma_2(U_2) &= q \circ j_2(U_2) = \theta_1(0,U_2) \\ \sigma_1(U_1) &= q \circ j_1(U_1) = \theta_2(0,U_1) \end{aligned} \tag{2.37}$$

and substituting equations (2.36) and (2.37) in equations (2.26), (2.34) and (2.35) we get a unique sequence of states (since for any $\omega \in \Omega$, the corresponding state X verifies $X = q(\omega)$ in all three diagrams) and outputs verifying the recursive equations (2.23).

The quadruple $\{H,F_1,F_2,G\}_n$ thus defined is therefore a realization of a σ-transformation represented by the mapping $f_{N,M}$.

Finally the necessary condition of this theorem implies that the realization $\{H,F_1,F_2,G\}$ is of minimal dimensions since its dimension is equal to the rank of $f_{N,M}$. ■

As a direct consequence of this theorem we get the practically very important approximation result.

THEOREM 3.4: *σ-transformations can approximate any linear transformation with an impulse response* $\{A_{i,j};\ (i,j) \in Z^2\}$ *verifying* $\sum_{(i,j)\in Z^2} \|A_{i,j}\|^2 < \infty$.

Proof. We use Theorem (3.3) to prove this result in the scalar case $(p = m = 1)$. The suspicious reader will find a direct proof of the result in the multivariable case in [1].

Theorem (3.3) has proved that any mapping $f_{N,M}$ of finite rank is realizable. It is moreover clear that any mapping $f_{N,M}$ associated to a "finite" sequence $\{A_{i,j};\ -N \le 0 \le N,\ -M \le 0 \le M\}$ is of maximum rank 4NM since the number of nonzero vectors in the associated Hankel matrix $\mathcal{A}$ is 4NM.

Therefore, all transformations with a "finite" impulse response $\{A_{i,j};\ -N \le i \le N,\ -M \le j \le M\}$ are σ-transformations.

Finally since any sequence verifying $\sum_{(i,j)\in Z^2} A_{i,j}^2 < \infty$ is trivially approximated by the finite sequences $\{A_{i,j};\ -N \le i \le N,\ -M \le j \le M\}$, the proof is completed. ■

Theorem (2.3) also leads to a minimal realization algorithm described below.

2.3.2 Minimal Realization Algorithm

It follows from Theorem (2.3) that a minimal realization of a given sequence $\{A_{i,j};\ i \ge -N,\ j \ge -M\}$ is obtained from equations (2.36) once a canonical factorization of the corresponding mapping $f_{N,M}$ has been performed.

The mapping $f_{N,M}$ is represented by a "Hankel" matrix $\mathcal{A}$ defined by:

$$\forall i \geq 0,\ k \geq 0,\ l \geq 0,\ j \geq 0$$

$$A(b(i,j),\ b(k,l)) = A_{i+k-N,j+l-M} \tag{2.38}$$

where b is any bijection $b : \mathbb{N}^2 \to \mathbb{N}$.

A row vector $A_{b(i,j)}$ (or more simply $A_{i,j}$) of matrix A, is defined by:

$$A_{i,j} = \{A(b(i,j),\ b(k,l));\ k \geq 0,\ l \geq 0\} . \tag{2.39}$$

We must first have a procedure for determining all the independent row vectors with a minimum number of steps.

LEMMA: *If for any couple* $(i,j) \in \mathbb{N}^2$, *we have:*

$$A_{i,j} = \sum_{(p,q)\in D} c_{p,q} A_{p,q} \qquad D \textit{ any subset of } \mathbb{N}^2, \tag{2.40}$$

then

$$\forall \alpha \geq 0,\ A_{i+\alpha,j+\beta} = \sum_{(p,q)\ D} c_{p,q} A_{p+\alpha,q+\beta} \tag{2.41}$$

Proof. Equation (2.40) translates, in terms of the sequence $\{A_{i,j};\ i \geq -N,\ j \geq -M\}$, into the set of equations:

$$\forall k \geq 0,\ l \geq 0\ A_{i+k-N,j+l-M} = \sum_{(p,q)\in D} c_{p,q} A_{p+k-N,q+l-N}$$

A subset of these equations is:

$$\forall k \geq \alpha,\ l \geq \beta\ A_{i+k-N,j+l-M} = \sum_{(p,q)\in D} c_{p,q} A_{p+k-N,q+l-M}$$

which is nothing but a translation of equation (2.41). ■

This lemma leads to the following procedure (illustrated in Fig. 2.1) for determining all the independent row vectors:

— Step $i = 0$.

Let a_0 be the first index for which A_{0,a_0} is linearly dependent of the rows $\{A_{o,p};\ p < a_0\}$. The above lemma implies then that $\forall k \geq a_0$, $\forall i \geq 0$, the row $A_{i,k}$ is linearly dependent of rows $\{A_{i,p};\ p < a_0\}$.

— Step i.

Let a_i be the first index such that A_{i,a_i} is linearly dependent of the rows $\{A_{p,q};\ p \leq i,\ q \leq a_p\}$. The above lemma also implies that $\forall n \geq 0,\ m \geq 0$ the rows $\{A_{n,m};\ m \geq a_i\}$ are linearly dependent of the rows $\{A_{i,j};\ i \leq n,\ j < a_i\}$.

Therefore the set of independent rows is $\{A_{i,j};\ i \leq \alpha,\ j \leq a_i\}$ where a_α is the last nonzero value of the decreasing sequence $a_1, a_2, \ldots$.

To illustrate that procedure we give the following hypothetical example:

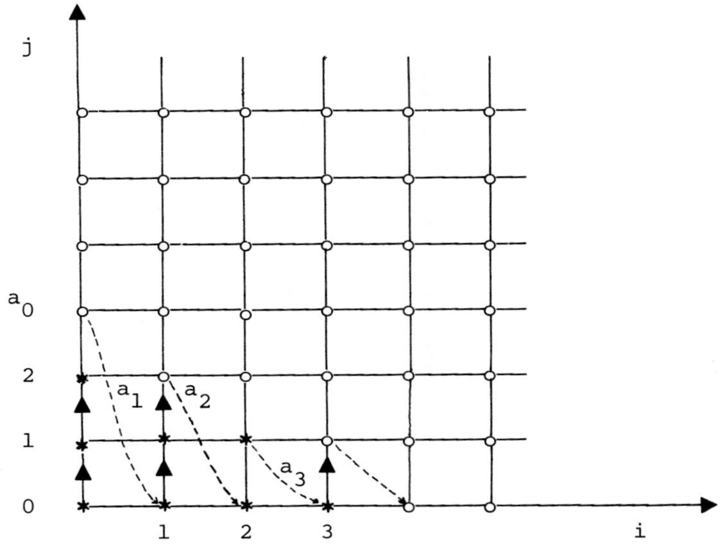

Figure 2.1

The points (i,j) with a cross represent the independent rows $\mathcal{A}_{i,j}$. The points with a circle represent the rows that are linearly dependent of the first ones. The full and dotted arrows show the order in which the rows are investigated.

Now we can propose the general factorization procedure:

We have to define two full rank matrices $\mathcal{P}$ (independent columns) and $\mathcal{Q}$ (independent rows) satisfying:

$$\mathcal{A}(b(i,j),\, b(k,l)) = \sum_{r} \mathcal{P}(b(i,j),r)\; \mathcal{Q}(r,b(k,l)). \qquad (2.42)$$

i) Construct the diagram of Fig. (2.1) by the method indicated above. Let I (for independent) be the set of couples (i,j) with a cross, C_I the complementary of I in $\mathbb{N}^2$.

ii) For each couple $(i,j) \in I$, introduce a new row $\mathcal{Q}_r$ of $\mathcal{Q}$ given by:

$$\forall k \geq 0,\ l \geq 0 \qquad \mathcal{Q}(r,b(k,l)) = \mathcal{A}(b(i,j),\, b(k,l)).$$

The corresponding row of $\mathcal{P}$ is then uniquely determined by:

$$\begin{aligned} \mathcal{P}(b(i,j),s) &= 1 \qquad \text{if } s = r, \\ &= 0 \qquad \text{otherwise.} \end{aligned}$$

iii) The other rows of matrix $\mathcal{P}$ are composed of the coefficients of the linear dependence characterizing each row $\mathcal{A}_{i,j}$ corresponding to a couple $(i,j) \in C_I$ (in fact, only the coefficients of rows $\mathcal{A}_{i,a_i}$ have to be computed).

The general form of matrices $\mathcal{P}$ and $\mathcal{Q}$ will be then:

$$\mathcal{P} = \underset{\longleftarrow\qquad n \qquad\longrightarrow}{\begin{bmatrix} 1 & 0 & \overset{\downarrow r}{\cdots} & \cdots & \cdots & \cdots \\ 0 & 1 & 0 & \cdots & \cdots & \cdots \\ \alpha_1 & \alpha_2 & 0 & \cdots & \cdots & \cdots \\ \gamma_1 & \gamma_2 & 0 & \cdots & \cdots & \cdots \\ 0 & 0 & 1 & \cdots & \cdots & \cdots \\ 0 & 0 & 0 & 1 & \cdots & \cdots \\ \beta_1 & \beta_2 & \beta_3 & \beta_4 & 0 & \cdots \end{bmatrix}} \begin{matrix} \\ \\ \\ \longleftarrow b(i,j) \\ \\ \\ \\ \end{matrix}$$

$$\mathcal{Q} = n \updownarrow \left(\begin{array}{c} \downarrow b(k,\ell) \\ \text{independent rows of} \\ \text{matrix } \mathcal{A} \end{array} \right) \xleftarrow{\ r\ }$$

The minimal realization $\{H,F_1,F_2,G\}_n$ associated to this factorization is then obtained very easily:

*Matrix H is the first row of $\mathcal{P}$, thus:

$$H = (1 \quad 0 \ldots\ldots 0)$$

*Matrix G is the first column of $\mathcal{Q}$.

*Mappings $\sigma\Gamma_1 p$ and $\sigma\Gamma_2 p$ are represented by the matrices:

$$\mathcal{P}_{*,1}(\sigma(i,j),r) = \mathcal{P}(\sigma(i+1,j),r)$$

$$\mathcal{P}_{*,2}(\sigma(i,j),r) = \mathcal{P}(\sigma(i,j+1),r) \ .$$

The form of matrix $\mathcal{P}$ them implies that F_1 and F_2 are respectively submatrices of $\mathcal{P}_{*,1}$ and $\mathcal{P}_{*,2}$ composed of the rows $\sigma(i,j)$ of these matrices for the couples $(i,j) \in I$.

The use of this algorithm for stochastic identification is discussed in the next section.

3. A MODEL FOR GAUSSIAN HOMOGENEOUS DOUBLE INDEXED SEQUENCES

Proceeding in the general spirit outlined in the introduction we shall now consider Gaussian homogeneous,† zero mean sequences $\{y_{i,j}; (i,j) \in Z^2\}$ taking values in $\mathbb{R}^p$, characterized therefore by their correlation function:

$$\Lambda(k,l) = E[y_{i+k,j+l} y'_{i,j}] \tag{3.1}$$

and study the features of the class Σ of gaussian sequences which can be represented by the model:

---

†A sequence of random variables is said to be homogeneous if its correlation function verifies $f(i,j,k,l)=E[y_{i,j}y'_{k,l}]=\Lambda(i-k,j-l)$.

$$\begin{cases} y_{i,j} = H\,X_{i,j} \\ X_{i,j} = F_1 X_{i-1,j} + F_2 X_{i,j-1} - F_1 F_2 X_{i-1,j-1} + v_{i-1,j-1} \end{cases} \tag{3.2}$$

where

i) $\{v_{i,j}; (i,j) \in Z^2\}$ is white noise taking values in $\mathbb{R}^n$, with $E[v_{i+k,j+l}\ v'_{i,j}] = Q\ \delta_{k,l} = LL'\ \delta_{k,l}$ (in the factorization of Q, L is taken of full rank).

ii) H, F_1, F_2 are matrices of appropriate dimensions. Moreover F_1 and F_2 commute and have all their eigenvalues of modulus less than one.

Such a sequence can equivalently be represented by:

$$y_{i,j} = \sum_{\substack{k \geq 0 \\ l \geq 0}} H\, F_1^k\, F_2^l\, v_{i-k-1,j-l-1}$$

First, it is easy to prove that there is no loss in generality in considering models where $\{H, F_1, F_2, L\}$ is an *observable and controllable quadruple*. We shall then consider only such quadruples and try to answer positively the two following important questions:

1 - How general is the class Σ.

2 - Can we define stochastic identification algorithms - which starting from the correlation values $\Lambda(k,l)$ determine the corresponding set of models $\{H, F_1, F_2, Q\}_n$ or $\{H, F_1, F_2, L\}_n$, which will be called stochastic realizations.

3.1 *GENERALITY OF THE CLASS* Σ

A fundamental property of purely nondeterministic gaussian [6], [9] time series is that to any sequence $\{y_k; k \in Z\}$ taking values in $\mathbb{R}^p$, one can associate an innovation sequence $\{v_k;\ k \in Z\}$ taking values in $\mathbb{R}^p$, defined by:

$$v_k = y_k - E[y_k / y_{k-1}, y_{k-2} \dots] \tag{3.3}$$

and verifying:

i) The sequence $\{v_k;\ k \in Z\}$ is a stationary white noise.

ii) There exist two square summable sequences of $(p \times p)$ matrices, $\{\alpha_i;\ i \geq 0\}$ and $\{\beta_i;\ i \geq 0\}$ such that:

$$v_k = \sum_{i \geq 0} \alpha_i\, y_{k-i} \tag{3.4}$$

$$y_k = \sum_{i \geq 0} \beta_i\, v_{k-i} \quad . \tag{3.5}$$

These properties imply that any nondeterministic stationary Gaussian sequence $\{y_k; k \in Z\}$ is the output of a time invariant causal filter (and causally invertible) whose input is the innovation. Unfortunately, such properties are related to the *total order* on the time axis and do not generalize to double indexed sequences.

However, the most general class of double indexed Gaussian sequences that one can "imagine" as a generalization of equation (3.5) is:

$$y_{k,l} = \sum_{(i,j) \in Z^2} \beta_{i,j}\, v_{k-i,l-j} \tag{3.6}$$

where

- $\{v_{i,j}; (i,j) \in Z^2\}$ is homogeneous white noise of *arbitrary dimension*.

- The sequence $\{\beta_{i,j}; (i,j) \in Z^2\}$ is square summable but does *not necessarily define a causal transformation*.

It is difficult to evaluate the restriction, if any, introduced by a model of type (3.6) with respect to the whole class of double indexed homogeneous Gaussian sequences. Let us just notice that the definition of a Gaussian sequence* already

---

*A Gaussian sequence is more than a set of Gaussian variables; any collection of these variables must also define a Gaussian vector.

contains (using the innovation approach) the assumption that it can be represented by *linear transformations* on white noise *taking values in* $\mathbb{R}^p$. The additional assumptions in (3.6) which are the space invariancy of the linear transformation and the homogeneity of the white noise, are balanced by the fact that the white noise $\{v_{i,j}; (i,j) \in Z^2\}$ is of arbitrary dimension. At this point we believe it is very important to emphasize the fact that unlike the case of Gaussian time series, one gets a much larger class of sequences of type (3.6) by allowing the noise $\{v_{i,j}; (i,j) \in Z^2\}$ to be of arbitrary dimension than restraining it to the dimension of $\{y_{k,l}; (k,l) \in Z^2\}$. We shall illustrate this with an example in 3.2.2.

Finally we would say that the concept of Gaussian time sequence was very convenient for time series because it led to equation (3.5); equation (3.6) defines probably the most general tractable generalization of that concept, to double indexed sequences.

Back to Σ-sequences we now show that they can approximate any sequence of type (3.6).

THEOREM 3.1: *The class* Σ *is dense in the set of sequences of type* (3.6) *(convergence in* $l^\infty(Z;L^2(\Omega, \mathbb{R}^p))$*)*.

Proof. Given a sequence:

$$y_{i,j} = \sum_{(k,l)\in Z^2} \beta_{k,l}\, v_{i-k,j-l} \tag{3.7}$$

let

$$Q = E[v_{i,j}\, v'_{i,j}] \tag{3.8}$$

and M a positive integer such that:

$$0 \leq Q \leq M\, I \quad (I = \text{identity matrix}). \tag{3.9}$$

Define the sequences $\{y^{(N)}_{i,j};\ (i,j) \in Z^2\}$ as:

$$y_{i,j}^{(N)} = \sum_{\substack{-N\leq k\leq N \\ -N\leq l\leq N}} \beta_{k,l}\, v_{i-k,j-l} \,. \tag{3.10}$$

We first prove that $\{y_{i,j}^{(N)}; (i,j) \in Z^2\}$ are Σ-sequences. Indeed, since the sequence $\{\beta_{k,l}; -N\leq k\leq N, \ -N\leq l\leq N\}$ is finite, it follows from Theorem (2.4) that there exists a minimal quadruple $\{H_N, F_{1,N}, F_{2,N}, G_N\}$ such that:

$$\begin{cases} H_N\, F_{1,N}^{k+N}\, F_{2,N}^{l+N}\, G_N = \beta_{k,l} & \text{if} \quad -N\leq k\leq N;\ -N\leq l\leq N \\ H_N\, F_{1,N}^{k+N}\, F_{2,N}^{l+N}\, G_N = 0 & \text{if} \quad k > N \ \text{ or } \ l > N \,. \end{cases} \tag{3.11}$$

Equation (3.10) then becomes:

$$\begin{cases} y_{i,j}^{(N)} = \sum_{\substack{k\geq 0 \\ l\geq 0}} H_N\, F_{1,N}^{k}\, F_{2,N}^{l}\, u_{i-k-1,j-l-1}^{(N)} \\ u_{i,j}^{(N)} = G_N\, v_{i-N+1,j-N+1} \,. \end{cases} \tag{3.12}$$

The quadruple $\{H_N, F_{1,N}, F_{2,N}, G_N Q G_N'\}$ is therefore a stochastic realization of $\{y_{i,j}^{(N)}; (i,j) \in Z^2\}$.

Now we prove that the sequences $\{y_{i,j}^{(N)}; (i,j) \in Z^2\}$ converge uniformly in the quadratic mean to $\{y_{i,j}; (i,j) \in Z^2\}$.

Define:

$$\varepsilon_{i,j}^{(N)} = y_{i,j} - y_{i,j}^{(N)} = \sum_{(k,l)\in C_N} \beta_{k,l}\, v_{i-k,j-l}$$

where C_N is the complementary in Z^2 of the set $(-N,N) \times (-N,N)$. We have:

$$E\left[\varepsilon_{i,j}^{(N)}\ \varepsilon_{i,j}^{\prime(N)}\right] = \sum_{(k,l)\in C_N} \beta_{k,l}\ Q\ \beta'_{k,l} \leq M \sum_{(k,l)\in C_N} \|\beta_{k,l}\ \beta'_{k,l}\| \ . \tag{3.12}$$

Since the sequence $\{\beta_{k,l};(k,l)\in Z^2\}$ is square summable, the right hand side of equation (3.12) goes to zero as N tends to infinity. Moreover this convergence is uniform with respect to (i,j).

$$\sup_{(i,j)\in Z^2} E\ \varepsilon_{i,j}^{(N)}\ \varepsilon_{i,j}^{\prime(N)} \to 0 \ . \ \blacksquare$$

The following corollary shows that the above convergence is well adapted to practical purposes where an estimated correlation function has to be approximated with a correlation function of an element of Σ.

COROLLARY 3.1: *The correlation function* $\Lambda(k,l)$ *of a sequence* $\{y_{i,j};(i,j)\in Z^2\}$ *of type* (3.6) *is the uniform limit of correlation functions* $\Lambda^{(N)}(i,j)$ *of the sequences* $\{y_{i,j}^{(N)};$ $(i,j)\in Z^2\}$ *defined in Theorem* (3.1).

Proof. This is a direct consequence of the application of Schwartz inequality in the Hilbert space of random variables taking values in $\mathbb{R}^p$, to equations (3.12).

3.2 STATISTICAL PROPERTIES OF Σ-SEQUENCES

Now that the generality of the class Σ is established, we wish to characterize correlation functions of Σ-sequences and define stochastic realization algorithms.

3.2.1 The Stochastic Realization Problem

THEOREM 3.2: *The set, if not empty, of stochastic realizations* $\{H,F_1,F_2,L\}_n$ *associated to a given correlation function* $\Lambda(.,.)$ *(verifying* $\Lambda(-i,-j) = \Lambda'(i,j)$ *because of homogeneity) is given by the solutions of the system:*

$$\Lambda(i,j) = H F_1^i F_2^j P H' \qquad (3.13)$$

$$\forall (i,j) \in \mathbb{N}^2$$

$$\Lambda(-i,j) = H F_2^j P F_1'^i H' \qquad (3.13)'$$

$$P > 0 \qquad (3.14)$$

$$P - F_1 P F_1' - F_2 P F_2' + F_1 F_2 P F_1' F_2' = Q = LL' \geq 0 \qquad (3.14)'$$

Proof. To prove the theorem it is sufficient to show that equations (3.13) and (3.14) define the correlation function of a Σ-sequence in terms of any given stochastic realization $\{H,F_1,F_2,L\}$.

So, given a stochastic realization $\{H,F_1,F_2,L\}$ or $\{H,F_1,F_2,Q\}$ of a Σ-sequence $\{y_{k,l}; (k,l) \in Z^2\}$, we have:

$$y_{k,l} = H \sum_{\substack{p>0 \\ q>0}} F_1^{p-1} F_2^{q-1} v_{k-p,l-q} \qquad (3.15)$$

with

$$E[v_{k-p,l-q} v_{k,l}'] = Q \delta_{p,q} \quad . \qquad (3.16)$$

Using expressions (3.15) and (3.16) into the definition:

$$\Lambda(i,j) \triangleq E[y_{k+i,l+j} y_{k,l}'] \qquad (3.17)$$

we get after some manipulations:

$$\begin{aligned} \Lambda(i,j) &= H F_1^i F_2^j P H' \\ \Lambda(-i,j) &= H F_2^j P F_1'^i H' \end{aligned} \qquad (3.18)$$

where matrix P is defined as:

$$P = \sum_{\substack{n \geq 0 \\ m \geq 0}} F_1^n F_2^m Q F_1'^n F_2'^m \quad . \qquad (3.19)$$

The equivalence between (3.19) and (3.14)' can be shown for example by rewriting (3.14)' in the form of a "compound" Lyapunov equation:

$$(P - F_1 P F_1') - F_2 (P - F_1 P F_1') F_2' = Q \qquad (3.20)$$

and expanding it twice:

$$P - F_1 P F_1' = \sum_{m \geq 0} F_2^m Q F_2'^m = Q_1 \qquad (3.21)$$

$$P = \sum_{n \geq 0} F_1^n Q_1 F_1'^n = \sum_{\substack{n \geq 0 \\ m \geq 0}} F_1^n F_2^m Q F_1'^n F_2'^m \quad . \qquad (3.22)$$

Finally, the strict positivity of P given by (3.19) is a consequence of the controllability of $\{F_1, F_2, L\}$ since if C_∞ is the infinite controllability matrix of $\{F_1, F_2, L\}$ we have:

$$P = C_\infty C_\infty' \quad .$$

Notice that P can be interpreted as:

$$P = E[X_{i,j} X_{i,j}']$$

where $X_{i,j}$ is the "state" defined in the recurrent model (3.2). ■

The stochastic realization problem is then to determine (given a function $\Lambda(.,.)$ verifying $\Lambda(i,j) = \Lambda'(-i,-j)$) the set of solutions $\{H, F_1, F_2, L\}$ to equations (3.13) and (3.14).

We shall do this under very weak and realistic assumptions on these realizations.

A1 Matrices F_1 and F_2 have no multiple eigenvalues. In this case [10], the commutativity condition implies that F_2 expresses as a polynomial in terms of F_1 and vice versa. This in turn implies by the use of Cayley Hamilton theorem, that observability of the triple $\{H,F_1,F_2\}$ is equivalent to the standard observability of the couples $\{H,F_1\}$ and $\{H,F_2\}$.

A2 If we associate to a stochastic realization $\{H,F_1,F_2,L\}$, the quadruple $\{H,F_1,F_2,PH'\}$ where P is given by (3.14)' or (3.19), then this latter quadruple is also observable and controllable. It is easy to see that this condition amounts practically to saying that when all the modes of a linear system are excited ($\{F_1,F_2,L\}$ controllable $\rightarrow P > 0$) and observed ($\{H,F_1,F_2\}$ observable), then they appear in the correlation function of the output of the linear system.

Mathematically speaking both assumptions i) and ii) are verified for all minimal quadruples $\{H,F_1,F_2,L\}$ outside a zero measure set.

THEOREM 3.3: *Under assumptions* A1 *and* A2, *the set of stochastic realizations* $\{H,F_1,F_2,L\}$ *associated to a given function* $\Lambda(.,.)$ *(verifying* $\Lambda(-i,-j) = \Lambda'(i,j)$*) is a single equivalence class defined by the coordinate transformation* $\{H,F_1,F_2,L\} \longrightarrow \{HT^{-1},TF_1T^{-1},TF_2T^{-1},TL\}$. *Moreover, any element* $\{H,F_1,F_2,L\}$ *of this equivalence class can be obtained as follows.*

a) Search for the minimal dimension quadruples $\{H,F_1,F_2,G\}$ *verifying:*

$$\forall i \geq 0,\ j \geq 0 \qquad \Lambda(i,j) = H\,F_1^i\,F_2^j\,G. \tag{3.23}$$

b) For each such quadruple the set of equations:

$$\begin{cases} P\,H' = G \\ P - F_1PF_1' - F_2PF_2' + F_1F_2PF_1'F_2' = Q = LL' \geq 0 \\ HF_2^jPF_1'^iH' = \Lambda(-i,j), \quad \forall i \geq 0,\ j \geq 0 \end{cases} \qquad (3.24)$$

has a unique solution P *leading to a unique solution* Q ...

Proof. Under assumption A2 any two quadruples $\{H,F_1,F_2,L\}$ and $\{\bar{H},\bar{F}_1,\bar{F}_2,\bar{P}\bar{H}'\}$ associated to two stochastic realizations $\{H,F_1,F_2,L\}$ and $\{\bar{H},\bar{F}_1,\bar{F}_2,\bar{L}\}$, are observable and controllable and are therefore minimal realizations (in the sense of Section 2) of the sequence $\{\Lambda(i,j);\ i \geq 0,\ j \geq 0\}$. They are therefore related by a coordinate transformation T:

$$\begin{cases} \bar{H} = H\,T^{-1} \\ \bar{F}_i = T\,F_i\,T^{-1},\ i = 1,2\ . \\ \bar{P}\,\bar{H}' = T\,P\,H' \end{cases} \qquad (3.25)$$

Equations (3.13)' then imply:

$$\Lambda(-i,j) = \bar{H}\bar{F}_2^j\bar{P}\bar{F}_1'^i\bar{H}' = HF_2^jPF_1'^iH' \quad . \qquad (3.26)$$

Using equations (3.25), we get:

$$\bar{H}\bar{F}_2^j\bar{P}\bar{F}_1'^i\bar{H}' = \bar{H}\bar{F}_2^jTPT'\bar{F}_1'^i\bar{H}' \quad . \qquad (3.27)$$

Under assumption A1 $((H,F_1)$ and (H,F_2) observable), equation (3.27) leads to:

$$\bar{P} = T\,P\,T' \qquad (3.28)$$

Feeding equation (3.28) into equation (3.14) written for both $\bar{P}$ and P yields:

$$\bar{T} = T\,L \quad . \qquad (3.29)$$

This proves the unicity modulo a coordinate transformation of the set of stochastic realizations associated to a given correlation function.

The procedure for determining a stochastic realization follows immediately. Note that the condition $PH' = G$ is redundant in equations (3.24) since it is included in the set of equations:

$$\forall i \geq 0,\ j \geq 0 \qquad HF_2^j PF_1'^i H' = \Lambda(-i,j) \ \blacksquare \tag{3.30}$$

Theorem (3.3) *suggests the following practical stochastic identification procedure.*

Given a sequence $\{y_{i,j}; (i,j) \in Z^2\}$ assumed to be produced by some double indexed ergodic Gaussian sequence.

i) Estimate the correlation function $\Lambda(.,.)$ for couples (i,j):

$$-N \leq i \leq N$$

$$0 \leq j \leq N$$

ii) Use the minimal realization algorithm of Section 2 to determine a quadruple verifying:

$$\forall\, 0 \leq i \leq N;\ \Lambda(i,j) = H\, F_1^i\, F_2^j\, G$$

iii) Although equations (3.24) define generally a unique stochastic realization, the algebraic determination of matrices P and Q may be sensitive to errors in the estimation of $\Lambda(.,.)$. A determination of matrix P by a least squares method is therefore more reliable. Stochastic realization results published in [4] determine after step ii) the symmetric matrix P which minimizes the expression:

$$\sum_{\substack{i\geq 0\\ j\geq 0}} (\Lambda(i,j) - HF_1^i F_2^j PH')^2 + (\Lambda(-i,j) - HF_2^j PF_1'^i H')^2$$

$$= \|\tilde{\Lambda}_{i,j}\|_{1^2}$$

This stochastic identification procedure is probably not the best practical solution. In private conversations Professor Mehra drew our attention to the existence of canonical forms associated to Σ models and to the possibility of using a spectral approach that would probably yield a maximum likelihood (or least squares) estimate of the parameters of the canonical form. Investigation of this problem is probably worthwhile.

3.2.2 An Example of a Scalar Σ-Sequence Which Cannot Be Generated by a Scalar White Noise

We are now able to illustrate the generality of Σ-sequences with an example.

The unicity (or the very strong constraint of equation (3.24)) of the covariance Q of the driving noise is an outstanding feature of Σ-sequences compared to the known case of state space representation of Gaussian time series. Beyond its theoretical aspects, this fact contributes to the richness of class Σ: a scalar Σ-sequence, with a state space representation of dimension $n > 1$ cannot generally be generated by a scalar white noise. We illustrate this fact with an example given in [30] of a scalar correlation function $\Lambda(.,.)$ which the author analyzed and concluded it could not be associated to a recursive model driven by white noise. This conclusion is only partially true because as we shall see, it is possible to associate to this correlation function a Σ-sequence generated by *two white noise components* (rank $Q = 2$).

The considered correlation function is defined by:

$$\Lambda(0,0) = 1$$

$$\Lambda(1,0) = \Lambda(-1,0) = \Lambda(1,1) = \Lambda(-1,-1) = 1/4$$

$$\Lambda(i,j) = 0 \quad \text{otherwise} \quad .$$

Following the procedure indicated in Theorem (3.3), we use the minimal realization algorithm of Section 2, to get a quadruple:

$$H = \begin{pmatrix} 1 & 0 & 0 & 0 \end{pmatrix}$$

$$G' = \begin{pmatrix} 1 & 0 & 1/4 & 1/4 \end{pmatrix}$$

$$F_1 = \begin{pmatrix} 0 & 0 & 1 & 0 \\ 0 & 0 & 0 & 1 \\ 0 & 0 & 0 & 0 \\ 0 & 0 & 0 & 0 \end{pmatrix}$$

$$F_2 = \begin{pmatrix} 0 & 1 & 0 & 0 \\ 0 & 0 & 0 & 0 \\ 0 & 0 & 0 & 1 \\ 0 & 0 & 0 & 0 \end{pmatrix}$$

verifying

$$\forall i \geq 0,\ j \geq 0 \qquad \Lambda(i,j) = H\, F_1^i\, F_2^j\, G \quad .$$

We must now produce matrices P, Q, L satisfying equations (3.24). These matrices are not necessarily unique since assumption A1 of Theorem (3.3) (eigenvalues of F_1 and F_2 are not simple in this academic example) is not satisfied.

To verify the algebraic conditions:

$$\begin{aligned} &P\, H' = G \\ &H\, F_2^j\, P\, F_1'^{\,i}\, H' = \Lambda(-i,j) \qquad \forall i \geq 0,\ j \geq 0 \end{aligned} \tag{3.31}$$

matrix P must be of the form:

$$P = \begin{pmatrix} 1 & 0 & 1/4 & 1/4 \\ 0 & a & 0 & \beta \\ 1/4 & 0 & b & \gamma \\ 1/4 & \beta & \gamma & c \end{pmatrix}$$

Matrix Q related to P by:

$$P - F_1PF_1' - F_2PF_2' + F_1F_2PF_1'F_2' = Q \tag{3.33}$$

must be of the form:

$$Q = \begin{pmatrix} 1-a-b+c & -\gamma & 1/4-\beta & 1/4 \\ -\gamma & a-c & 0 & \beta \\ 1/4-\beta & 0 & b-c & \gamma \\ 1/4 & \beta & \gamma & c \end{pmatrix} = LL' \geq 0 \tag{3.34}$$

Investigation of the rank of Q leads to the following conclusions:

— rank $Q > 1$ for all parameters a,b,c,γ,β. Therefore we cannot find a Σ-sequence generated by a scalar white noise, and associated to the given function $\Lambda(.,.)$.

— for $\beta = \gamma = 0$, $a = c = 1/4$, $b = 1/2$, we get:

$$Q = \begin{pmatrix} 1/2 & 0 & 1/4 & 1/4 \\ 0 & 0 & 0 & 0 \\ 1/4 & 0 & 1/4 & 0 \\ 1/4 & 0 & 0 & 1/4 \end{pmatrix} = LL' \geq 0 \quad (\text{rank } Q = 2) \tag{3.35}$$

with

$$L = \frac{1}{2\sqrt{2}} \begin{pmatrix} 2 & 0 \\ 0 & 0 \\ 1 & -1 \\ 1 & 1 \end{pmatrix} \tag{3.36}$$

The quadruple $\{H,F_1,F_2,G\}$ thus defined is therefore a stochastic realization of a -sequence associated to the considered correlation function.

3.2.3 Spectral Analysis of Σ-Sequences

The statistical property previously discussed can also be expressed in terms of spectral factorization properties.

DEFINITION 3.1: *The spectrum of a Σ-sequence* $\{y_{i,j}; (i,j) \in Z^2\}$ *to which we associate a stochastic realization* $\{H,F_1,F_2,L\}$ *is the function of two complex variables* (z_1,z_2):

$$S(z_1,z_2) = \sum_{(i,j)\in Z^2} \Lambda(i,j)\; z_1^{-i}\; z_2^{-j}$$

defined for

$$\lambda_{max,1} < |z_1| < \frac{1}{\lambda_{max,1}}$$
$$\lambda_{max,2} < |z_2| < \frac{1}{\lambda_{max,2}} \qquad (3.37)$$

where $\lambda_{max,1}$ *and* $\lambda_{max,2}$ *stand respectively for the highest module of the eigenvalues of matrices* F_1 *and* F_2, *and where* $\Lambda(i,j)$ *is the correlation function of the* Σ*-sequence* $\{y_{i,j}; (i,j) \in Z^2\}$.

Using the four partial sums

$$\sum_{\substack{i \geq 0\\ j\geq 0}}\,, \quad \sum_{\substack{i \geq 0\\ j< 0}}\,, \quad \sum_{\substack{i < 0\\ j\geq 0}}\,, \quad \sum_{\substack{i < 0\\ j< 0}}$$

respectively defined and analytical (product of two analytical functions of one variable) in the domains $\{|z_1| > \lambda_{max,1}, |z_2| > \lambda_{max,2}\}$, $\{|z_1| > \lambda_{max,1}\ |z_2| < 1/\lambda_{max,2}\}$, $\{|z_1| < 1/\lambda_{max,1}, |z_2| > \lambda_{max,2}\}$, $\{|z_1| < 1/\lambda_{max,1}, |z_2| < 1/\lambda_{max,2}\}$, we get an expression of the spectrum $S(z_1,z_2)$ in terms of the stochastic realization $\{H,F_1,F_2,L\}$:

$$S(z_1,z_2) = H(I-z_1^{-1}F_1)^{-1}(I-z_2^{-1}F_2)^{-1}PH' + H(I-z_1^{-1}F_1)^{-1} PF_2'z_2(I-z_2F_2')^{-1}H'$$

$$+ H(I-z_2^{-1}F_2)^{-1}PF_1'z_1(I-z_1F_1')^{-1}H'$$

$$+ HPF_1'z_1(I-z_1F_1')^{-1}F_2'z_2(I-z_2F_2')^{-1}H' \qquad (3.38)$$

where the matrix P is the solution of:

$$P - F_1PF_1' - F_2PF_2' + F_1F_2PF_1'F_2' = Q = LL' \qquad (3.39)$$

THEOREM 3.4: *To each stochastic realization* $\{H,F_1,F_2,L\}$ *of a* Σ*-sequence* $\{y_{i,j}; (i,j) \in Z^2\}$, *corresponds a "spectral factorization" of the form:*

$$\begin{cases} S(z_1,z_2) = \mathcal{H}(z_1,z_2)\ \mathcal{H}'(1/z_1,\ 1/z_2) \\ \mathcal{H}(z_1,z_2) = H(I-z_1^{-1}F_1)^{-1}(I-z_2^{-1}F_2)^{-1}L \end{cases} \qquad (3.40)$$

and vice versa.

<u>Proof</u>. If $\{H,F_1,F_2,L\}$ is a stochastic realization, the spectrum $S(z_1,z_2)$ verifies equations (3.38) and (3.39). Equation (3.39) is equivalent to the identity:

$$Q = P(I-F_1'z_1)(I-F_2'z_2) + (I-z_1^{-1}F_1)PF_1'z_1(I-z_2F_2')$$

$$+ (I-z_2^{-1}F_2)PF_2'z_2(I-z_1F_1')$$

$$+ (I-z_1^{-1}F_1)(I-z_2^{-1}F_2)PF_1'F_2'z_1z_2 \quad . \qquad (3.41)$$

Use of this identity in equation (3.38) leads to equation (3.40). Conversely if:

$$S(z_1,z_2) = \sum_{(i,j)\in Z^2} \Lambda(i,j)z_1^{-i}z_2^{-j}$$

verifies (3.40); defining matrix P by equation (3.39) leads through the use of identity (3.41) to equation (3.38).

Considering the analycity domains of each term of the right hand side of equation (3.38), we can identify each of these terms to a partial sum in equation (3.42):

$$\begin{cases} H(I-z_1^{-1}F_1)(I-z_2^{-1}F_2)PH' = \sum\limits_{\substack{i\geq 0\\ j\geq 0}} \Lambda(i,j)z_1^{-i}z_2^{-j} \\ H(I-z_1^{-1}F_1)^{-1}PF_2'z_2(I-z_2F_2')^{-1}H' = \sum\limits_{\substack{i\geq 0\\ j<0}} \Lambda(i,j)z_1^{-i}z_2^{-j} \end{cases} \quad (3.43)$$

(the other two terms correspond to the identity $\Lambda(-i,-j) = \Lambda'(i,j)$). Equations (3.43) lead to:

$$\Lambda(i,j) = HF_1^i F_2^j P H'$$

$$\Lambda(-i,j) = HF_2^j PF_1'^i H'$$

The quadruple $\{H,F_1,F_2,L\}$ is therefore a stochastic realization associated to $\Lambda(.,.)$. ∎

Using this theorem together with Theorem (3.3) leads immediately to:

COROLLARY: *Under assumptions* A1 *and* A2 *of Theorem* (3.3) *the spectrum of a* Σ*-sequence admits one and only one factorization of the form* (3.40) *(modulo a coordinate transformation).*

Note that, in the case of a scalar Σ-sequence for example, the dimensions of the factor $H(z_1,z_2)$ in this factorization are $(1 \times r)$, where r is the rank of matrix Q. This is another illustration to the fact that a scalar white noise may not be enough to generate a Σ-sequence.

4. LEAST SQUARES RECURSIVE ESTIMATION ALGORITHMS

We shall consider situations where we have observation models of the type:

$$y_{i,j} = z_{i,j} + w_{i,j} \tag{4.1}$$

where

— $\{z_{i,j}\}$ is a Σ-sequence and represents the "image" to be estimated.

— $\{w_{i,j}\}$ is an additive "perturbation" belonging to a class of "admissible" perturbations.

By admissible perturbations we mean double indexed Gaussian sequences whose statistics do not alter the algebraic structure of the overall correlation function of $\{y_{i,j}\}$, such that one could determine without any ambiguity the joint statistics of $\{z_{i,j}\}$ and $\{w_{i,j}\}$.

An example of a class of admissible perturbations, (which we shall use for the estimation problem) is when $\{w_{i,j}\}$ is white noise (with covariance $R > 0$) independent from the Σ-sequence $\{z_{i,j}\}$. Indeed, in such a case, if $\{H,F_1,F_2,L\}$ is a stochastic realization associated to $\{z_{i,j}\}$, the correlation function of $\{y_{i,j}\}$ is given by:

$$\begin{aligned} \Lambda(i,j) &= HF_1^iF_2^jPH' \quad \forall i \geq 0,\ j \geq 0 \quad (i,j) \neq (0,0) \\ \Lambda(-i,j) &= HF_2^jPF_1'^iH' \quad \forall i \geq 0,\ j \geq 0 \quad (i,j) \neq (0,0) \end{aligned} \tag{4.2}$$

$$\Lambda(0,0) = HPH' + R \tag{4.3}$$

where P is the solution of:

$$P - F_1PF_1' - F_2PF_2' + F_1F_2PF_1'F_2' = LL' \tag{4.4}$$

It is easy to see then that, under assumptions A1 and A2 of Theorem (3.3), (plus assumption that F_1 and F_2 have no zero eigenvalues) equations (4.2) define the unique (modulo a coordinate transformation) stochastic realization $\{H,F_1,F_2,P\}$ associated to $\{z_{i,j}\}$ and that equation (4.3) defines then a unique matrix R covariance of $\{w_{i,j}\}$.

But this example is not exclusive, another class of admissible perturbations [3] consists of sequences $\{w_{i,j}\}$ that are correlated only along the line, that is:

$$E[w_{i+k,j+l}\ w'_{i,j}] \quad = \quad f(l)\ \delta_k \tag{4.5}$$

The purpose of the previous discussion is just to emphasize the fact that the additive white noise perturbation model is not the only one that can be dealt with. The interested reader is invited to figure, as an exercise, what happens in the filtering and smoothing algorithms if we replace the additive white noise model by an additive model with a correlation function given by (4.5).

4.1 *FORMULATION OF THE ESTIMATION PROBLEMS*

Consider the observation model:

$$y_{i,j} \quad = \quad z_{i,j} + w_{i,j} \tag{4.6}$$

where

$$\begin{aligned} z_{i,j} &= H\ x_{i,j} \\ x_{i,j} &= F_1 x_{i-1,j} + F_2 x_{i,j-1} - F_1 F_2 x_{i-1,j-1} \\ &\quad + v_{i-1,j-1} \end{aligned} \tag{4.7}$$

$$\begin{cases} E[w_{i+k,j+l}\ w'_{i,j}] = R\ \delta_{k,l} \\ E[v_{i+k,j+l}\ v'_{i,j}] = Q\ \delta_{k,l} \\ E[v_{i+k,j+l}\ w'_{i,j}] = 0 \end{cases} \tag{4.8}$$

and define the estimates:

$$\hat{x}_{i,j} = E[x_{i,j}/y_{k,l};\ k < i,\ l \in Z];\ \hat{x}_{0,j} = 0 \qquad (4.9)$$

$$x^*_{i,j} = E[x_{i,j}/y_{k,l};\ k \leq i,\ l \in Z] \qquad (4.10)$$

$$\hat{\hat{x}}_{i,j} = E[x_{i,j}/y_{k,l};\ 0 \leq k \leq N,\ l \in Z] \ . \qquad (4.11)$$

Thus the "physical" image is supposed to consist of $N + 1$ "lines" $(i = 0, \ldots N)$ each consisting of a large number of "points" $(l \in Z)$.

A natural and mathematically coherent interpretation of the above estimates in terms of prediction, filtering and smoothing of an "infinite dimension time series" can be obtained as follows. Let:

$$u_{i,j} = x_{i,j} - F_1\, x_{i-1,j} \qquad (4.12)$$

Using (4.6) and (4.12) we get:

$$u_{i,j} = F_2\, u_{i,j-1} + v_{i-1,j-1} \ . \qquad (4.13)$$

Defining, for fixed i, the new "infinite" variables $\{y_{i,j}; j \in Z\}$, $\{x_{i,j}; j \in Z\}$, $\{u_{i,j}; j \in Z\}$, $\{v_{i,j}; j \in Z\}$, $\{w_{i,j}; j \in Z\}$; these are related by:

$$\{y_{i,j};\ j \in Z\} = \text{diag}\,(H)\{x_{i,j};\ j \in Z\} + \{w_{i,j};\ j \in Z\}^* \qquad (4.14)$$

$$\{x_{i,j};\ j \in Z\} = \text{diag}(F_1)\{x_{i-1,j};\ j \in Z\} + \{u_{i,j};\ j \in Z\}. \qquad (4.14)'$$

---

*For any matrix M of dimensions $(m \times n)$, diag(M) stands for the operator $\{u_k;\ u \in Z,\ u_k \in \mathbb{R}^n\} \xrightarrow{\text{diag}(M)} \{Mu_k,\ k \in Z\}$.

Considering that for fixed i, the variables $\{u_{i,j};\ j \in Z\}$ and $\{w_{i,j}; j \in Z\}$ are purely nondeterministic ((F_2, L) controllable in (4.13)) Gauss Markov "time" series in j, their "covariances" are therefore finite l^2 norm Toeplitz Operators given by:

$$\text{cov}\ \{w_{i,j};\ j \in Z\} = \text{diag}\ R$$

$$\text{cov}\ \{u_{i,j};\ j \in Z\} = Q = \text{Toep}(I, F_2, Q_2)^*$$

with

$$Q_2 - F_2\, Q_2\, F_2' = Q = LL' \quad .$$

Considering moreover that due to its "internal correlation":

$$E[x_{i,j+1}\ x_{i,j}'] = F_2^1\ P$$

the variable $\{x_{i,j};\ j \in Z\}$ can also be represented by a Gauss Markov purely nondeterministic "time" series in j, the "covariance" of $\{x_{i,j};\ j \in Z\}$ is also a finite l^2 norm Toeplitz Operator given by:

$$P_0 = \text{cov}\ \{x_{i,j};\ j \in Z\} = \text{Toep}(I, F_2, P) \quad .$$

Finally, noticing that finite l^2 norm Toeplitz Operators form an Algebra (closed by linear operations and by multiplication) containing all operators of the type diag(M), we conclude that equations (4.14) define coherent Markov representation of the sequence $\{y_{i,j}\}$. Thus, the estimates defined by equations (4.9), (4.10), (4.11) can be obtained *formally* with the standard

---

*Toep (A,B,C) stands for the Toeplitz Operator where $\alpha_k = AB^kC$. We say then that the Toeplitz Operator Toep (A,B,C) is generated by the sequence $\{\alpha_l; l \in Z\}$.

$$\begin{pmatrix} \cdots\alpha_0\ \alpha_1' & & & & \\ \cdots\alpha_1\ \alpha_0\ \alpha_1'\ \alpha_2' & & & & \\ \cdots\alpha_2\ \alpha_1\ \alpha_0\ \alpha_1'\ \alpha_2' & & & & \\ & & & & \alpha_0 \end{pmatrix}$$

techniques for the recursive estimation of time series [14], [15], [20], by replacing all matrices by the corresponding finite l^2 norm Toeplitz Operators.

The point in the next paragraphs is to show how a tractable solution to those formal equations, can be produced.

4.2 *THE PREDICTIVE AND THE FILTERED ESTIMATES*

In order to allow a better understanding of the procedure we shall use the conclusions of the above discussion only when they are technically necessary.

THEOREM 4.1: *The estimates* $\hat{x}_{i,j}$ *and* $x^*_{i,j}$ *defined by equations* (4.9) *and* (4.10) *verify equations:*

$$\hat{x}_{i,j} = F_1(\hat{x}_{i-1,j} + \hat{\tilde{x}}_{i-1,j}) \tag{4.15}$$

$$x^*_{i,j} = \hat{x}_{i,j} + \hat{\tilde{x}}_{i,j} \tag{4.16}$$

where $\{\hat{\tilde{x}}_{i,j};\ j \in Z\}$ *is defined as:*

$$\hat{\tilde{x}}_{i,j} = E[x_{i,j} - \hat{x}_{i,j}/y_{i,l} - H\,\hat{x}_{i,l};\ l \in Z]\ . \tag{4.17}$$

<u>Proof</u>. Substituting equation (4.12) in equation (4.9) and noticing that $u_{i,j}$ is independent of the observations $\{y_{k,l};\ 0 \leq k < i,\ l \in Z\}$, we get:

$$\hat{x}_{i,j} = F_1\ E[x_{i-1,j}/y_{k,l};\ 0 \leq k < i,\ l \in Z]\ . \tag{4.18}$$

Introducing the following "line innovation":

$$I_{i-1,j} = y_{i-1,j} - E[y_{i-1,j}/y_{k,l};\ 0 \leq k < i-1,\ l \in Z]$$

$$= y_{i-1,j} - H\,\hat{x}_{i-1,j} \tag{4.19}$$

equation (4.18) becomes:

$$\hat{x}_{i,j} = F_1(\hat{x}_{i-1,j} + E[x_{i-1,j}/I_{i-1,l};\ l \in Z]. \quad (4.20)$$

Introducing the error $\tilde{x}_{i,j} = x_{i,j} - \hat{x}_{i,j}$, and applying the orthogonality conditions, equation (4.20) becomes:

$$\hat{x}_{i,j} = F_1(\hat{x}_{i-1,j} + E[\tilde{x}_{i-1,j}/I_{i-1,l};\ l \in Z]). \quad (4.21)$$

This establishes equations (4.15) and (4.17). As for equation (4.16), it is a direct consequence of introducing the innovation. ∎

This theorem suggests that the prediction and filtering solutions are easily determined whenever the solution of equation (4.17) *is determined.*

Noticing that:

$$\forall j \in Z \qquad I_{i,j} = y_{i,j} - H\,\hat{x}_{i,j} = H\,\tilde{x}_{i,j} + w_{i,j} \qquad (4.22)$$

shows that equation (4.17) defines simply a "one dimensional" smoothing problem the solution of which is easily derived once the correlation function:

$$\Lambda^i(l) = E[\tilde{x}_{i,j+l}\,\tilde{x}'_{i,j}] \qquad (4.23)$$

or the spectrum:

$$S_i(z) = \sum_{l \in Z} \Lambda^i(l)\, z^{-l} \qquad |z| = 1 \qquad (4.24)$$

have been determined.

THEOREM 4.2: *The spectras* $S_i(z)$ *verify the "spectral" Riccati equation:*

$$\begin{cases} S_{i+1}(z) = F_1(S_i(z) - S_i(z)H'(HS_i(z)H'+R)^{-1}HS_i(z))F_1'+Q(z) \\ S_0(z) = \sum_{-\infty}^{+\infty} \Lambda_j^0 z^{-i} = \sum_{j\geq 0} F_2^j Pz^{-i} + \sum_{j>0} PF_2'^j z^i \\ Q(z) = S_0(z) - F_1 S_0(z) F_1' \quad . \end{cases} \tag{4.25}$$

And the correlation function $\Lambda^i(l)$ *is the sequence of Fourier coefficients associated to the periodic function of* θ*:*

$$\mathcal{S}_i(\theta) = S_i(e^{\theta\sqrt{-1}}) \quad . \tag{4.26}$$

Proof. It was shown in paragraph 4.1 that equations (4.14) define a coherent Gauss Markov representation for the sequence in i, of infinite variables $\{y_{i,j};\ j \in Z\}$, provided we replace standard matrices by the corresponding finite l^2 norm Toeplitz Operators. Thus, identifying the predictive estimate:

$$\hat{x}_{i,j} = E[x_{i,j}/y_{k,l};\ 0 \leq k < i,\ l \in Z]$$

as the j-th component of the predictive estimate:

$$\{\hat{x}_{i,j};\ j \in Z\} = E[x_{i,j};\ j \in Z/\{y_{k,l};\ l \in Z\};\ 0 \leq k < i] \tag{4.27}$$

it becomes clear that the predictive error "covariance" $\mathcal{P}_i$ in model (4.14) is nothing but the finite l^2 norm Toeplitz Operator generated by the sequence $\{\Lambda^i(l);\ l \in Z\}$.

Equations (4.14) and (4.27) define a regular prediction problem in the sense that the "covariance" of the observation noise, diag(R), is a coercive (generalization of positive definite) finite l^2 norm Toeplitz Operator. Hence the $\mathcal{P}_i$ are given by the successive iterations of the Riccati equation:

$$
\begin{cases}
\forall i > 0 \\
\mathcal{P}_i = \operatorname{diag}(F_1)(\mathcal{P}_{i-1} - \mathcal{P}_{i-1}\operatorname{diag}(H')(\operatorname{diag}(H)\mathcal{P}_{i-1}\operatorname{diag}(H') \\
\quad + \operatorname{diag}(R))^{-1}\operatorname{diag}(H)\mathcal{P}_{i-1})\operatorname{diag}(F_1') + \mathcal{Q} \\
\mathcal{P}_0 = \operatorname{Toep}(I, F_2, P) \\
\mathcal{Q} = \mathcal{P}_0 - \operatorname{diag}(F_1)\,\mathcal{P}_0\operatorname{diag}(F_1')
\end{cases}
\tag{4.28}
$$

Consider finally the mapping:

$$\mathcal{P}_i \longrightarrow S_i(z) = \sum_{l \in Z} \Lambda^i(l) z^{-l}, \; |z| = 1$$

from the Algebra of finite l^2 norm Toeplitz Operators into the Algebra of bounded functions of one complex variable defined on the unit circle ($|z| = 1$). It is known (Doetsch, Horst, Schoute, Grenander, Szego) that this mapping is a continuous Algebra isomorphism which will transform equations (4.29) into equation (4.25) preserving also the steady state solutions. ■

The general *theoretical procedure* for determining the predictive and the filtered estimates is then

THEOREM 4.3: *Using the previous notations, the predictive and the filtered estimates are obtained as follows.*

a. *The correlation functions* $\Lambda^i(l)$ *are the Fourier coefficients of the spectras* $S_i(z)$ *given by equations* (4.25). *At line* i, *the spectrum* $S_i(z)$ *is a rational matrix of degree bounded by* ni.

b. *The "one dimensional" smoothed estimate* $\{\hat{\hat{x}}_{i,j};\ j \in Z\}$ *defined by equations* (4.17), *is obtained by the recurrent equations:*

$$\begin{cases} \hat{\hat{x}}_{i,j} = H_i \eta_{i,j} + (G_i' - H_i P_{*,i}) \gamma_{i,j} \\ \eta_{i,j} = F_i \eta_{i,j-1} + S_{*,i} R_{*,i}^{-1} (I_{i,j-1} - HH_i \eta_{i,j-1}) \\ \gamma_{i,j} = F_{*,i}' \gamma_{i,j+1} + H_i' H' R_{*,i}^{-1} (I_{i,j} - HH_i \eta_{i,j}) \end{cases} \tag{4.29}$$

where

- *The triple* $\{H_i, F_i, G_i\}$ *is a minimal triple verifying:*

$$\Lambda^i(l) = H_i F_i^l G_i \tag{4.30}$$

The dimension of this triple is bounded by ni *(degree of* $S_i(z)$*).*

- $P_{*,i}$ *is the steady state solution of the Riccati equation:*

$$\begin{cases} P_{0,1} = 0 \\ P_{k+1,i} = F_i P_{k,i} F_i' + (F_i G_i H' - F_i P_{k,i} H_i' H')(HH_i G_i H' + R - \\ \qquad HH_i P_{k,i} H_i' H')^{-1} (F_i G_i H' - F_i P_{k,i} H_i' H')' \end{cases} \tag{4.31}$$

- $S_{*,i}$, $R_{*,i}$, $F_{*,i}$ *are given by*

$$\begin{cases} S_{*,i} = F_i G_i H' - F_i P_{*,i} H_i' H' \\ R_{*,i} = HH_i (G_i - P_{*,i} H_i') H' + R \\ F_{*,i} = F_i - S_{*,i} R_{*,i}^{-1} HH_i \end{cases} \tag{4.32}$$

c. *The predictive and filtered estimates are then obtained recursively by equations* (4.15) *and* (4.16).

Proof. This theorem simply gathers the results of Theorems (4.1) and (4.2). Point b is merely the result of applying the "one dimensional" smoothing techniques to the sequence in j:

$$\forall j \in Z \qquad I_{i,j} = H\,\tilde{x}_{i,j} + w_{i,j}$$

$$E[\tilde{x}_{i,j+1}\,\tilde{x}'_{i,j}] = \Lambda^{i}(1) \quad .$$

The *practical procedure* for determining the predictive and the filtered estimates, follows immediately from Theorem (4.3), using the following remarks:

1. In step -a- of Theorem (4.3) it is practically prohibitive to determine the spectras $S_i(z)$ in the form of rational functions of increasing degree. Practically $S_0(z)$ and $\Lambda^0(1)$ (which are related by fast Fourier transformation) are necessarily defined for a finite number of values of their arguments. Equations (4.25) then yield the numerical values of $S_i(z)$ for each value of z defining $S_0(z)$. The correlation function $\Lambda^i(1)$ is then related to $S_i(z)$ by fast Fourier transformation. Finally a standard realization algorithm will allow to approximate uniformly and independently all the correlation functions $\Lambda^i(1)$ with triples $\{H_i, F_i, G_i\}$ of reasonable dimensions.

2. Moreover, equations (4.25) converge for each value of z (since all convergence conditions are satisfied: $Q(z)$ and R positive, H, F_1 observable...) to a steady state solution $S_\infty(z)$. The corresponding correlation function $\Lambda^\infty(1)$ is obtained by fast Fourier transformation. A minimal realization algorithm is then applied to define a triple $\{H_\infty, F_\infty, G_\infty\}$ of reasonable dimensions. This will yield the steady state solutions of the filtering and prediction problems.

4.3 THE SMOOTHED ESTIMATE

Using model (4.14) and applying smoothing techniques where standard matrices are replaced by finite l^2 norm Toeplitz Operators, we get the following formal expression for the smoothed estimate (with the previous notations).

THEOREM 4.4: *The smoothed estimate*

$$\hat{\hat{x}}_{i,j} = E[x_{i,j}/y_{k,l};\ 0 \leq k \leq N,\ l \quad Z] \tag{4.33}$$

is given formally by

$$\hat{\hat{x}}_{i,j} = \hat{x}_{i,j} + \xi_{i,j} \tag{4.34}$$

$$\begin{cases} \{\xi_{i,j};\ j \in Z\} = \mathcal{P}_i\{\tau_{i,j};\ j \in Z\} \\ \{\tau_{i,j};\ j \in Z\} = (\mathrm{diag}(F_1) - \mathcal{L}_i\ \mathrm{diag}(H))'\{\tau_{i+1,j};j\in Z\} \\ \qquad\qquad + \mathrm{diag}(H')\ \mathcal{M}_i\ \{I_{i,j};\ j \in Z\} \\ \{\tau_{N,j};\ j \in Z\} = \mathrm{diag}(H')\ \mathcal{M}_N\ \{I_{N,j};\ j \in Z\} \end{cases} \tag{4.35}$$

where

$$\mathcal{M}_i = [\mathrm{diag}(H)\ \mathcal{P}_i\ \mathrm{diag}(H') + \mathrm{diag}(R)]^{-1}$$

$\mathcal{L}_i = \mathrm{diag}(F_1)\ \mathcal{P}_i\ \mathrm{diag}(H')\mathcal{M}_i$ *is the "one dimensional" smoothing operator transforming the sequence* $\{I_{i,j};\ l \in Z\}$ *into the sequence* $\{F_1\hat{\hat{x}}_{i,l};\ l \in Z\}$ *according to equations* (4.29) *in Theorem* (3.3).

<u>Proof</u>. Using the "line innovations" $\{I_{i,j};\ j \in Z\}$, equation (4.33) gives immediately

$$\hat{\hat{x}}_{i,j} = \hat{x}_{i,j} + \xi_{i,j}$$

with

$$\xi_{i,j} = E[x_{i,j}/\{I_{k,l};\ l \in Z\}, \quad i \leq k \leq N]$$

which in turn becomes, using orthogonality properties:

$$\xi_{i,j} = \sum_{i \leq k \leq N} E[\tilde{x}_{i,j}/I_{k,l};\ l \in Z]$$

$$= \sum_{i \leq k \leq N} E[\tilde{x}_{i,j}/H\ \tilde{x}_{k,l} + w_{k,l};\ l \in Z\ ] \tag{4.36}$$

The problem is now to show that equation (4.36) can be transformed into (4.35).

By definition:

$$\{\hat{\tilde{x}}_{i,j};\ j \in Z\} = E[\tilde{x}_{i,j};\ j \in Z/I_{i,l};\ l \in Z] \tag{4.37}$$

$$\{\hat{\tilde{x}}_{i,j};\ j \in Z\} = P_i\ \mathrm{diag}(H')\ M_i\{I_{i,l};\ l \in Z\} \tag{4.38}$$

since this last expression verifies the necessary and sufficient measurability and orthogonality conditions of the conditional expectation in (4.37).

Therefore:

$$\mathrm{diag}(F_1)\{\hat{\tilde{x}}_{i,j};\ j \in Z\} = L_i\{I_{i,j};\ j \in Z\}$$

$$= L_i\ \mathrm{diag}(H)\{\tilde{x}_{i,j};\ j \in Z\}$$

$$+ L_i\{w_{i,j};\ j \in Z\}\ . \tag{4.39}$$

The left hand side of equation (4.39) can be transformed using (4.13) and (4.15) into:

$$\mathrm{diag}(F_1)\{\hat{\tilde{x}}_{i,j};j\in Z\} = -\{\tilde{x}_{i+1,j};j\in Z\} + \mathrm{diag}(F_1)\{\tilde{x}_{i,j};j\in Z\} + \{u_{i+1,j};j\in Z\} \quad . \tag{4.40}$$

Substituting equation (4.40) into equation (4.39) yields a model for $\{\tilde{x}_{i,j};j\in Z\}$:

$$\{\tilde{x}_{i+1,j};j\in Z\} = [\mathrm{diag}(F_1) - L_i\ \mathrm{diag}(H)]\ \{\tilde{x}_{i,j};\ j\in Z\} + \{u_{i+1,j};j\in Z\} - L_i\ \{w_{i,j};j\in Z\} \quad . \tag{4.41}$$

Consider now the finite l^2 norm Toeplitz Operators $J_{i,k}$, $k \geq 0$ generated by the cross correlation functions between $\{\tilde{x}_{i+k,j};j\in Z\}$ and $\{\tilde{x}_{i,j};j\in Z\}$.

One easily verifies that equation (4.36) transform into:

$$\{\xi_{i,j};j\in Z\} = \sum_{k=i}^{n} J'_{i,k-i}\ \mathrm{diag}(H')\ M_k\{I_{k,j};j\in Z\} \tag{4.42}$$

Since this expression verifies the necessary and sufficient measurability and orthogonality conditions of the conditional expectation defining $\xi_{i,j}$.

It follows then from model (4.41) where the driving terms, $\{u_{i+k,j};\ 0 < k \leq N-i;j\in Z\}$ and $L_i\{w_{i+k,j};0 \leq k \leq N-i;j\in Z\}$, are independent from $\{\tilde{x}_{i,j};j\ \ Z\}$, that operators $J_{i,k}$ are given by

$$\forall 0 < k \leq N-i \qquad J_{i,k} = [\mathrm{diag}(F_1) - L_{i+k-1}\ \mathrm{diag}(H)]J_{i,k-1}$$

$$J_{i,0} = P_i \tag{4.43}$$

The announced equations (4.35) are just the recursive form of equation (4.42) taking into account (4.43). ■

Equations (4.34) and (4.35) yield the smoothed estimate in terms of finite l^2 norm Toeplitz Operators.

The final point is now to specify to computations involved in equations (4.35) and show that they all amount to standard recursive linear transformations.

Notice first that equations (4.35) can be rewritten in the form:

$$\begin{cases} \{\xi_{i,j};j \in Z\} = P_i\{\tau_{i,j};j \in Z\} \\ \{\tau_{i,j};j \in Z\} = \mathrm{diag}(F_1')\tau_{i+1,j} + \mathrm{diag}(H')L_i'\{\tau_{i+1,j};j \in Z\} \\ \qquad\qquad + \mathrm{diag}(H')M_i\{I_{i,j};j \in Z\} \\ \{\tau_{N,j};j \in Z\} = \mathrm{diag}(H')M_N\{I_{N,j};j \in Z\} \quad . \end{cases} \tag{4.44}$$

Operators $\mathrm{diag}(F_1)$ and $\mathrm{diag}(H')$ are of course trivial. Only the recursive form of operators P_i, M_i, L_i' need therefore to be specified.

THEOREM 4.5: *Operators* P_i, M_i, L_i' *involved in equations (4.44) are respectively characterized by the following recursive equations:*

i) $$P_i\{\tau_{i,j};j \in Z\} = \{\mu_{1,i,j};j \in Z\} + \{\mu_{2,i,j};j \in Z\} \tag{4.45}$$

$$\begin{cases} \mu_{1,i,j} = H_i\ \nu_{1,i,j} \\ \nu_{1,i,j+1} = F_i\ \nu_{1,i,j} + G_i\ \tau_{i,j+1} \end{cases} \tag{4.46}$$

$$\begin{cases} \mu_{2,i,j} = G_i'\ \nu_{2,i,j} \\ \nu_{2,i,j} = F_i'\ \nu_{2,i,j+1} + H_i'\ \tau_{i,j+1} \end{cases} \tag{4.47}$$

P_i *is thus the sum of two recursive dual systems.*

ii) *In terms of matrices* Π_i *and* M_i *and* T_i *given by*

$$\begin{cases} T_i = S_* R_*^{-1} \\ \Pi_i - F'_{*,i} \Pi_i F_{*,i} = H'_i H' R_{*,i}^{-1} H H_i \\ M_i = - F_{*,i} \Pi_i T_i + H'_i H' R_{*,i}^{-1} \end{cases} \tag{4.48}$$

We have

$$M_i\{I_{i,j}; j \in Z\} = \{(T'_i \Pi_i T_i + R_{*,i}^{-1}) I_{i,j}; j \in Z\} + \{\alpha_{1,i,j}; j \in Z\} + \{\alpha_{2,i,j}; j \in Z\} \tag{4.49}$$

$$\begin{cases} \alpha_{1,i,j} = M'_i \beta_{1,i,j} \\ \beta_{1,i,j+1} = F_{*,i} \beta_{1,i,j} - T_i I_{i,j} \end{cases} \tag{4.50}$$

$$\begin{cases} \alpha_{2,i,j} = - T'_i \beta_{2,i,j} \\ \beta_{2,i,j} = F'_{*,i} \beta_{2,i,j+1} + M_i I_{i,j+1} \end{cases} \tag{4.51}$$

iii) $L'_i\{\tau_{i,j}; j \in Z\} = \{\theta_{1,i,j}; j \in Z\} + \{\theta_{2,i,j}; j \in Z\}$

$$\begin{cases} \theta_{1,i,j} = R_{*,i}^{-1} H H_i \eta_{1,i,j} \\ \eta_{1,i,j+1} = F_{*,i} \eta_{1,i,j} + (G_i - P_{*,i} H'_i) F'_1 \tau_{i,j+1} \end{cases} \tag{4.52}$$

$$\begin{cases} \theta_{2,i,j} = T'_i \eta_{2,i,j} \\ \eta_{2,i,j} = F'_{*,i} \eta_{2,i,j+1} + H'_i (F'_1 \tau_{i,j+1} - H' \theta_{1,i,j+1}) \, . \end{cases} \tag{4.53}$$

Proof. i) Equations (4.30) imply

$$P_i\{\tau_{i,j}; j \in Z\} = \left\{ \sum_{l>0} G_i' F_i'^{\,l} H_i' \tau_{i,j+1} + \sum_{l\geq 0} H_i F_i^l G_i \tau_{i,j-1}; j \in Z \right\} \tag{4.54}$$

Equations (4.45) to (4.47) are merely the recursive form of (4.54).

ii) First notice that keeping the notations of Theorem (4.3) the spectrum $I_i(z)$ associated to the Toeplitz Operator

$$\mathrm{cov}\{I_{i,j}; j \in Z\} = \mathrm{diag}(H) P_i \ \mathrm{diag}(H') + \mathrm{diag}(R) = M_i^{-1} \tag{4.55}$$

is factorizable [11], [14], [15], [16] into the form

$$I_i(z) = [I + HH_i(zI - F_i)^{-1} T_i] R_{*,i} [I + T_i'(z^{-1}I - F_i')^{-1} H_i' H'] \tag{4.56}$$

where T_i is given by (4.48).

Moreover the spectrum associated to M_i is nothing but $I_i^{-1}(z)$; one must therefore determine $I_i^{-1}(z)$ in the form of a standard rational transfer function.

The algebraic techniques related to the inversion of rational spectras [11] applied to $I_i(z)$, then, lead to the following expressions of $I_i^{-1}(z)$

$$I_i^{-1}(z^{-1}) = [I - T_i'(zI - F_{*,i}')^{-1} H_i' \ H'] \\ R_{*,i}^{-1} \ [I - H \ H_i(z^{-1}I - F_{*,i})^{-1} T_i] \ , \tag{4.57}$$

$$I_i^{-1}(z^{-1}) = - T_i'(zI - F_{*,i}')^{-1} M_i - M_i'(z^{-1}I - F_{*,i})^{-1} \ T_i \\ + T_i' \ \Pi_i \ T_i + R_{*,i}^{-1} \ , \tag{4.58}$$

where Π_i and M_i are given by (4.48). Equation (4.58) yields straightforwardly $I_i^{-1}(2)$ which is exactly the "transfer function" associated to the recursive equations (4.49) to (4.51).

iii) First notice that any Toeplitz Operator of the form Toep(H,F,G) is uniquely decomposed into:

$$\mathrm{Toep}(H,F,G) \;=\; \mathrm{Toep}_c(h,F,G) + \mathrm{Toep}_{nc}(H,F,G) + \mathrm{diag}(HG)\ , \tag{4.59}$$

where $\mathrm{Toep}_c(H,F,G)$ and $\mathrm{Toep}_{nc}(H,F,G)$ are *dual operators* corresponding respectively to the causal and noncausal part of Toep(H,F,G), whose impulse responses are respectively

$$\begin{cases} A_l \;=\; HF^{l-1}G & l > 0 \\ A_l \;=\; 0 & l \leq 0 \end{cases} \qquad \begin{cases} A_l = G'F'^{-l-1}H' & l < 0 \\ A_l = 0 & l \geq 0 \end{cases} \tag{4.60}$$

Moreover in the case of a Toeplitz Operator associated to a correlation function, we have

$$H\,G \;=\; G'\,H'$$

and diag(HG) can be included in either $\mathrm{Toep}_c(H,F,G)$ or $\mathrm{Toep}_{nc}(H,F,G)$, so it will be neglected (but not forgotten!) in the following equations.

Using equations (4.29), operator L_i can be rewritten in terms of Toeplitz Operators

$$L_i \;=\; \mathrm{diag}(F_1)\left\{\mathrm{Toep}_c(H_i,F_{*,i},T_i) + \mathrm{Toep}_{nc}(R_{*,i}^{-1}\,HH_i,F_{*,i},\right. \\ \left. G_i - P_{*,i}\,H_i')\,[I - \mathrm{diag}(H)\,\mathrm{Toep}_c(H_i,F_{*,i},T_i)]\right\} \tag{4.61}$$

The dual operator L_i' is then:

$$L_i' = \left\{ \text{Toep}_{nc}(H_i, F_{*,i}, T_i) + [I - \text{Toep}_{nc}(H_i, F_{*,i}, T_i)\ \text{diag}(H')] \right.$$

$$\left. \text{Toep}_c(R_{*,i}^{-1}\ HH_i, F_{*,i}, G_i - P_{*,i}\ H_i') \quad \text{diag}(F_1') \right\}. \qquad (4.62)$$

Rewriting (4.62) in terms of recursive equations yields (4.52) and (4.53). ■

Of course, the remarks on the implementation of the predictive and filtered estimates are still valid. Thus the steady state smoothing solution is obtained by replacing all subscripts i in the matrices defining the recursive equations, by the subscript ∞.

5. COMMENTS AND CONCLUSIONS

A glance at the literature on recursive statistical image processing shows a great dispersion, mainly due to the lack of appropriate tools in dealing with data defined over a two dimensional frame. To our knowledge, only Wong and Zakai [30] have been trying with very sophisticated mathematics to provide such tools through the introduction of a new stochastic calculus which would be applied to modelize and process continuous images. Although mathematically very attractive, this approach has an inherent drawback which is to introduce an artifical causality in the processing algorithms. The other authors apply various techniques ranging from the direct use of Kalman filtering on separate parts of the image, considered as vector valued time sequences [24], [28], to the use of heuristic two dimensional filtering structures [7], [13].

The present developments try to define the framework of a theory for the recursive processing of double indexed sequences. New adaptations can be made in the perspective of practical implementations. As in the case of time series [23] more reliable stochastic identification algorithms can probably be defined using a maximum likelihood ratio approach (see paragraph 3). "Fast algorithms" [16], [21] can be used instead of computing explicitly the solutions of the various Riccati equations.

The material presented here is minimal. It meant only to prove the feasibility of recursive least squares estimation for a large class of double indexed data. Actually large economies in computing time should be possible by using the -σ- and -Σ- modelizations, in classical algorithms: the adapted filter in detection can be represented by a -σ- recursive transformation; the spectral inversion in two dimensional Wiener filtering [25] is considerably simplified, as is shown in [3] if a Σ-model is used to represent the image to be processed etc... .

Finally, we should point out that in many practical image processing applications where the final "sensor" is the human visual system, complementary algorithms taking into account some features of this visual system should be included in the processing chain.

REFERENCES

1. Attasi, S., "Systèmes linéaires homogènes à deux indices," Rapport Laboria No. 31, Sept. 1973.

2. Attasi, S., "Modèles stochastiques de suites gaussiennes à deux indices," Rapport Laboria No. 56, Feb. 1974.

3. Attasi, S., "Modélisation et traitement des suites à deux indices," Rapport Laboria, Oct. 1975.

4. Attasi, S. and J. P. Chieze, "Mise en oeuvre expérimentale d'algorithmes d'identification stochastique," Rapport Laboria No. 99, Jan. 1975.

5. Doetsch, G., *Handbuch der Laplace Transformationen,* Vol. 1, 2, 3, Birkhauser Verlag Basel.

6. Doob, J. L., "The Elementary Gaussian Processes," *Ann. Math. Stat. 15*, 1944, pp. 229-282.

7. Durrani, T. S., C. G. Hart, and E. M. Stafford, "Digital Signal Processing for Image Deconvolution and Enhancement," Proceeding of NATO Institute on New Directions in Signal Processing in Communication and Control, 5-17 Aug. 1974, U.K.

8. Faurre, P., "Réalisations Markoviennes de Processus Stationnaires," Rapport Laboria No. 13, March 1973.

9. Faurre, P., "Representation of Stochastic Processes," Ph.D. Dissertation, Stanford University, 1967.

10. Gantmacher, F. R., "Théorie des matrices," Tome 1, Dunod, 1966.

11. Germain, F., "Algorithmes de Calcul des Réalisations Makoviennes, Cas Singuliers," Rapport Laboria No. 66, April 1974.

12. Grenander, U. and G. Szego, "Toeplitz Forms and Their Applications," California Monographs in Mathematical Sciences University of California Press, Berkeley, Los Angeles, 58.

13. Habibi, A., "Two Dimensional Bayesian Estimate of Images," Proc. of IEEE, Vol. 00, No. 7, July 72, pp. 878-883.

14. Kailath, T., "An Innovation Approach to Least Squares Estimation," *IEEE Trans. AC 13, No. 6,* Dec. 68, pp. 646-660. *IEEE Trans. AC 18, No. 6,* Dec. 73, pp. 588-600.

15. Kailath, T., "A View of Three Decades of Linear Filtering Theory," *IEEE Trans. IT/20, No. 2,* March 1971, pp. 146-181.

16. Kailath, T., "Some New Algorithms for Recursive Estimation in Constant and all Linear, Discrete-Time Systems," *IEEE Trans. AC 19, No. 4,* Aug. 1974.

17. Kalman, R. E., P. L. Falb, and M. A. Arbib, *Topics in Mathematical Systems Theory,* McGraw Hill, 1969.

18. Kalman, R. E., "Mathematical Description of Linear Dynamical Systems," *SIAM J. Control, Vol. 1,* 1963, pp. 152-192.

19. Kalman, R. E. and J. E. Bertram, "Control System Design Via the Second Method of Lyapunov," Part 2, *ASME Journal of Basic Engineering,* June 1960, pp. 394-400.

20. Kalman, R. E. and R. S. Bucy, "New Results in Linear Filtering and Prediction Theory," *Trans. ASME (J. Basic Eng.) Ser. D., Vol. 83,* Dec. 61, pp. 95-107.

21. Lindquist, A., "A New Algorithm for Optimal Filtering of Discrete-Time Stationary Processes," *SIAM Journal of Control,* Nov. 74.

22. Mehra, R. K., "On Line Identification of Linear Dynamic Systems with Applications to Kalman Filtering," *IEEE Trans. AC 16, No. 1,* 1971, pp. 12-21.

23. Mehra, R. K. and N. K. Gupta, "Computational Aspects of Maximum Likelihood Estimation and Reduction in Sensitivity Function Calculations," *IEEE Trans. AC,* Dec. 1974.

24. Nahi, N. E., "Role of Recursive Estimation in Statistical Image Enhancement," Proc. IEEE, Vol. 60, July 1972, pp. 872-878.

25. Pratt, W. K., "Generalized Wiener Filtering Computation Techniques," *IEEE Trans. on Computers, Vol. C-21, No. 7,* July 1972 (Special Issue), pp. 636-641.

26. Rissanen, J., "Recursive Identification of Linear Dynamical Systems," *SIAM J. Control, Vol. 9, No. 3,* Aug. 1971, pp. 420-430.

27. Silverman, L. M., "Realization of Linear Dynamical Systems," *IEEE Trans. AC 14, No. 6,* Dec. 1971, pp. 554-567.

28. Silverman, L. M. and S. R. Powell, "Modeling of Two Dimensional Covariance Functions with Applications to Image Restoration," *IEEE Trans. AC 19, No. 1,* Feb. 1974, pp. 8-13.

29. Woods, J. W., "Two Dimensional Discrete Markovian Fields," *IEEE Trans. IT/18, No. 2,* March 1972, pp. 232-240.

30. Wong, E., "Recursive Filtering and Detection for Two Parameter Random Fields," Proceedings International Symposium on Control Theory, Numerical Methods and Computer System Modeling, 17-21 June 1974, Iria, Rocquencourt, France.

31. Youla, D. C., "On the Factorization of Rational Matrices," *IRE Trans. on Information Theory,* July 1961, pp. 172-189.

ESTIMATION, IDENTIFICATION AND FEEDBACK

P. E. Caines
Systems Control Group
Department of Electrical Engineering
University of Toronto
Toronto, Canada

and

C. W. Chan
Systems Engineering Section
Unilever Research Laboratories
Port Sunlight, England

1. INTRODUCTION

The analysis of a complex system frequently entails the construction of a signal flow graph. The nodes of such a graph denote processes and the directed edges denote transformations of the processes which flow through those edges. In engineering, economic and biological modelling the existence and sense of direction of the edges of the signal flow graph are often chosen on *a priori* grounds. In this chapter we present a concept which distinguishes those cases where a unique direction of influence exists between two given processes. This notion is formalized in the definition of *feedback free* processes given in Section 2.

The equivalent formulations of the definition of feedback free processes yields a set of techniques for the identification of statistically significant directed relationships between the processes observed in a complex system. Among the many applications of our approach are the following: (1) the direction of influence between gross domestic product and unemployment in post war economic time series, (2) the identification of the dynamics of subsections of regional power networks and (3) the detection of driving centers amongst a group of alpha rhythm measurement points in the human brain.

The definition of feedback between two observed processes which we give in Section 2 involves the structure of the innovations of the joint observed process. This notion was implicit in the work of Granger [1-4] who extended an idea due to Wiener in order to define causality between time series. In [5] Sims showed that Granger's definition could be formulated in terms of the representation of the joint process, as in Definition 1 below, and in terms of the representation of one process with respect to the other. Sims also used statistical tests equivalent to the filtering criterion presented here to test for the direction of causality between two series. This circle of ideas is completely described in our main theorem in Section 2.

In [6] and [7] the authors used a multivariate statistical technique for the detection of feedback for both a simulated example and for the G.D.P.-unemployment relation of the U.K. economy. In Section 3 we describe a set of three multivariate techniques for the detection of feedback that follow from the theorem in Section 2. Wall has given another useful multivariate statistical criterion for the detection of feedback in [8] and we also outline his method.

Parameter estimation for dynamic systems under feedback regulation is a common problem in systems engineering. Consequently a significant amount of work has appeared in the engineering literature concerned with this question (see for instance [9-14]).

Principally, Ljung (see [12] and chapter by Ljung in this volume) has shown that prediction error methods identify the feedforward loop of a feedback system when the feedforward loop contains a delay and the feedback loop contains a noise disturbance. Further, if the feedback loop is noiseless only an equivalence class of feedback systems may be identified. Earlier Bohlin [9] had discussed the analogs of these two results for the maximum likelihood estimates of the parameters of a feedback system. In Section 3 we present a simple proof that (asymptotically) a class of pseudo-maximum likelihood estimators are minimum prediction error estimators and hence that the two results above also hold for this important case. If a delay does not exist in the feedforward or feedback loop we argue that one should resort to identifying the innovations representation of the joint input-output process. A similar idea has also been suggested by Phadke and Wu [13] who applied it to the identification of a blast furnace under feedback control.

The penultimate section of this chapter contains a description of applications of the methods presented in Sections 2 and 3 to economics, power systems and physiology.

2. MAIN RESULTS

Let $\underline{\underline{H}}$ denote the Hilbert space which is the completion with respect to the norm $(Ex^Tx)^{1/2}$ of the linear space of random variables with zero mean and finite second order moments (see [15, 16]). Consider the r component process ζ lying in $\underline{\underline{H}}$, i.e., $\zeta_t \in \underline{\underline{H}}$ for $t \in Z$, where Z denotes the set of integers. Let $\underline{\underline{H}}_m^n$ denote the closed subspace of $\underline{\underline{H}}$ generated by $\{\zeta_t; m \leq t \leq n\}$ and let $\eta|\underline{\underline{H}}_m^n$ denote the projection of the random variable η on $\underline{\underline{H}}_m^n$. When $\lim_{n\to-\infty} \zeta_t|\underline{\underline{H}}_{-\infty}^n = 0$ the process ζ has zero projection on its infinite past and is termed regular. Further, ζ is said to have full rank when $E\tilde{\zeta}_n\tilde{\zeta}_n^T = \Sigma_n > 0$ for all $n \in Z$, when $\tilde{\zeta}_n \triangleq \zeta_n - \zeta_n|\underline{\underline{H}}_{-\infty}^{n-1}$. Wold showed (see [15] and [17]) that regular full rank stationary processes possess canonical one sided moving-average representations of the form

$$\zeta_t = \Phi_0\varepsilon_t + \Phi_1\varepsilon_{t-1} + \Phi_2\varepsilon_{t-2} + \dots, \qquad t \in Z, \tag{2.1}$$

where ε is an r component stationary orthonormal process (i.e. $E\varepsilon_t\varepsilon_s^T = I\delta_{t,s}$, where $\delta_{t,s} = 1$ if $t = s$ and 0 otherwise), Φ_0 is invertible and the subspaces spanned by $\{\zeta_t; t \leq n\}$ and $\{\varepsilon_t; t \leq n\}$ are identical. In order to make the representation (2.1) unique we take Φ_0 to be upper triangular with positive elements on the diagonal i.e., Φ_0 is the Cholesky factor of $\Phi_0\Phi_0^T$. If this condition is not imposed the sequence $\{\Phi_0 U, {}_1U, \dots\}$ together with the orthonormal process ε yields an alternative canonical representation of the process ζ in (2.1) when U is an $r \times r$ constant matrix such that $UU^T = I$.

The sequence of matrices $\{\Phi_i; i \geq 0\}$ in (2.1) are square summable. In this paper we make the useful technical assumption that the linear representation of ε with respect to ζ also has this summability property (see [15]). This assumption merely excludes processes whose spectra have zeroes on the unit circle.

In this paper we shall deal only with finitely generated stationary processes; these constitute the subclass of regular full rank stationary processes for which there exist two sets of real matrices $\Delta = \{\Delta_1, \dots, \Delta_n\}$ and $\Gamma = \{\Gamma_0, \Gamma_1, \dots, \Gamma_n\}$ such that

$$\zeta_t + \Delta_1\zeta_{t-1} + \dots + \Delta_n\zeta_{t-n} = \Gamma_0\varepsilon_t + \dots + \Gamma_n\varepsilon_{t-n}, \qquad t \in Z \quad . \tag{2.2}$$

Although the results of the main theorem are true without this restriction, it has the effect of simplifying the discussion at several places and further makes our results directly interpretable for applications.

It is emphasized at this point that the properties of the innovations representation (2.1), the assumption on the zeroes of the corresponding spectra and the finite generation assumption together imply that whenever the term 'stability' is used in this chapter it invariably denotes 'asymptotic stability'.

Finally, we remark that from here on the phrase 'stationary stochastic process' shall denote only a regular full rank finitely generated stationary process subject to the assumption above on the zeroes of its spectrum.

Let

$$\Delta(z) \underline{\Delta} I + \Delta_1 z + \ldots + \Delta_n z^n ,$$

and

$$\Gamma(z) \underline{\Delta} \Gamma_0 + \Gamma_1 z + \ldots + \Gamma_n z^n ,$$

where $z \in \mathbb{C}$ the field of complex numbers. Clearly $\Delta^{-1}(z)$ exists at all except a finite number of values of the argument and

$$\Delta^{-1}(z)\Gamma(z) = \Phi(z) \underline{\Delta} \Phi_0 + \Phi_1 z + \Phi_2 z^2 + \ldots .$$

In order to present Definition 1 we use the following block decomposition of the matrix of rational functions $\Delta^{-1}(z)\Gamma(z)$. We write

$$\Delta^{-1}(z)\Gamma(z) = \begin{bmatrix} A(z) & B(z) \\ C(z) & D(z) \end{bmatrix} \qquad (2.3)$$

where $A(z)$, $B(z)$, $C(z)$, $D(z)$ are matrices of rational functions of dimension $p \times p$, $p \times q$, $q \times p$ and $q \times q$ respectively, where $p + q = r$.

DEFINITION 1: *Let* ζ *be an* r *component stationary stochastic process whose first* p *components consists of the stochastic process* y *and whose remaining* q *components consist of the process* u. *Then the ordered pair of processes* $\langle y,u \rangle$ *is feedback free if and only if the matrix* $\Phi(z)$ *has the form*

$$\begin{bmatrix} A(z) & B(z) \\ 0 & D(z) \end{bmatrix} . \tag{2.4}$$

If the matrix C(z) *in* (2.3) *is not zero we say there is feed-back from* y *to* u.

THEOREM 2.1: *The following statements about the stationary stochastic process* $\zeta^T = [y^T, u^T]$ *are equivalent:*

1. *The ordered pair of processes* $\langle y,u \rangle$ *is feedback free*

2. *Let* $\Psi(z) = \Phi(z)\Phi^T(z^{-1})$ *denote the spectral density matrix of the stationary stochastic process* ζ. *Then there exist matrices of rational functions* A(z), B(z), D(z) *of dimension* $p \times p$, $p \times q$, $q \times q$ *respectively, with* A(z), D(z) *of full normal rank (i.e., full rank except at a finite number of points in* $\mathbb{C}$*), with the poles of* A(z), B(z), D(z), $A^{-1}(z)$ *and* $D^{-1}(z)$ *lying outside the closed unit disc, and with* $A_0 \triangleq A(0)$ *and* $D_0 \triangleq D(0)$ *upper triangular with positive elements on the diagonal, such that*

$$\Psi(z) = \begin{bmatrix} A(z)A^*(z) + B(z)B^*(z) & B(z)D^*(z) \\ D(z)B^*(z) & D(z)D^*(z) \end{bmatrix},$$

where $A^*(z)$ *denotes* $A^T(z^{-1})$ *etc.*

3. *The least squares estimate* $\hat{y}_{t|t}$ *of* y_t *given the observations* $\{\ldots u_{t-1}, u_t\}$ *is identical to the estimate* $\hat{y}_{t|\infty}$ *of* y_t *given the observations* $\{\ldots u_{t-1}, u_t . u_{t+1}, \ldots\}$, *for all* $t \in Z$; *in other words, the anticipative and non-anticipative filters for the estimation of* y *from* u *are identical.*

4. *There exists a unique representation of* y *with respect to* u *of the form*

$$y_t = \sum_{i=0}^{\infty} K_i u_{t-i} + \sum_{i=0}^{\infty} L_i v_{t-i} ,$$

where

$$K(z) \triangleq \sum_{i=0}^{\infty} K_i z^i \quad \text{and} \quad L(z) \triangleq \sum_{i=0}^{\infty} L_i z^i$$

are matrices of rational functions such that $L(z)$ *has full normal rank,* $K(z)$, $L(z)$ *and* $L^{-1}(z)$ *have all their poles outside the closed disc,* L_0 *is upper triangular with positive elements on the diagonal, the processes* u *and* v *are orthogonal (i.e.,* $Eu_t v_s^T = 0;\ t, s \in Z$) *and* v *is an orthonormal process.*

5. *The least squares estimate* $\hat{u}_{t+k}|\underline{\underline{U}}^t$ *of* u_{t+k} *given the observations* $\{\ldots u_{t-1}, u_t\}$ *is identical to the estimate* $u_{t+k}|\underline{\underline{U}}^t + \underline{\underline{Y}}^t$ *of* u_{t+k} *given the observations* $\{\ldots u_{t-1}, u_t\} \cup \{\ldots y_{t-1}, y_t\}$ *for all* $t \in Z$, $k \geq 1$. *(This is Granger's definition* [1-4].)

Proof. We show (1) $\Leftrightarrow$ (2) $\Rightarrow$ (3) $\Rightarrow$ (4) $\Rightarrow$ (5) $\Rightarrow$ (1) and we also show that (1) $\Rightarrow$ (3) directly.

(1) $\Leftrightarrow$ (2):

(1) $\Rightarrow$ (2) is obtained by multiplying $\Phi(z)$ given by (2.4) by its adjoint $\Phi^T(z^{-1})$. $A(z)$, $B(z)$ and $D(z)$ have the stability properties claimed for them since the collection of their poles constitutes the set of poles of $\Phi(z)$. It remains to show $A^{-1}(z)$ and $D^{-1}(z)$ exist and are stable. However, this is an immediate consequence of our technical assumption on the zeroes of the spectrum of the process ζ.

To show (2) $\Rightarrow$ (1) assume the stationary stochastic process ζ has a spectral density matrix $\Psi(z)$ which satisfies the conditions described in (2) above. Clearly in this case $\Psi(z)$ has a spectral factor $\Phi(z)$ given by (2.4). By the given properties of $A(z)$, $B(z)$ and $D(z)$ it follows that $\Phi(z)$ is the unique stable and inverse stable spectral factor of $\Psi(z)$ (see [18]). Hence there exists an orthonormal process ε which, together with $\Phi(z)$, constitutes the unique innovations representation of ζ. Clearly this representation possesses the structure specified as feedback free by Definition 1.

This proves the desired result.

(2) $\Rightarrow$ (3):

Let $\Psi_{yu}(z)$ and $\Psi_u(z)$ denote the cross spectrum of y and u and the spectrum of u respectively. If the process ζ has a spectrum with the properties described in (2) we have the representations

$$\Psi_{yu}(z) = B(z)D^*(z) \quad ,$$

$$\Psi_u(z) = D(z)D^*(z) \quad .$$

Let $H_+(z)$ and $H(z)$ respectively denote the z-transforms of the impulse response sequences of the non-anticipative and anticipative filters for the least squares estimation of y from u. Further let $[\cdot]_+$ denote the operation of taking the causal summand of a Laurent expansion of the argument in the square brackets i.e., the extraction of that part of the expansion that converges inside the closed unit disc in $\mathbb{C}$ (notice no series in this chapter diverge on the unit circle in $\mathbb{C}$). Then employing standard formulae (see, e.g. [19]) one obtains

$$\begin{aligned} H_+(z) &= [B(z)D^*(z)(D^*(z))^{-1}]_+D^{-1}(z) \\ &= B(z)D^{-1}(z) \\ &= \Psi_{yu}(z)(\Psi_u(z))^{-1} \\ &= H(z) \end{aligned}$$

which establishes (2) $\Rightarrow$ (3).

We shall now give a second proof that statement (2) implies statement (3). This will be given in terms of an elementary Hilbert space argument of the type we shall also use to demonstrate all of the other implications. Since we have shown earlier that (2) $\Rightarrow$ (1) it is sufficient to prove (1) $\Rightarrow$ (3).

Let U^t and E_u^t denote the spaces spanned by $\{\ldots u_{t-1}, u_t\}$ and by $\{\ldots e_{u,t-1}, e_{u,t}\}$ respectively where e_u denotes the innovations process of u. By construction $U^t = E_u^t$ for all t.

It is well known that the linear least squares estimate of y_t given $\{\ldots u_{k-1}, u_k\}$ is given by the orthogonal projection $y_t|U^k$. Now let $<y,u>$ be feedback free and let e_1, e_2, denote the first p and the remaining q components of the innovations representation of the joint process $[y^T, u^T]$. It is clear from (2.4) that $e_u = e_2$ and so $U^t = E_2^t$ for all t where E_2^t denotes the span of $\{\ldots e_{2,t-1}, e_{2,t}\}$. Since the processes e_1 and e_2 are orthogonal, and since

$$\left(\sum_{i=0}^{\infty} B_i e_{2,t-i} | E_2^{\infty}\right) = \sum_{i=0}^{\infty} B_i e_{2,t-i} = \left(\sum_{i=0}^{\infty} B_i e_{2,t-i} | E_2^t\right)$$

it follows that

$$\begin{aligned}
y_t|U^t &= y_t|E_u^t \\
&= \left(\left\{\sum_{i=0}^{\infty} A_i e_{1,t-i} + \sum_{i=0}^{\infty} B_i e_{2,t-i}\right\} | E_2^t\right) \\
&= \left(\left\{\sum_{i=0}^{\infty} A_i e_{1,t-i} + \sum_{i=0}^{\infty} B_i e_{2,t-i}\right\} | E_2^{\infty}\right) = y_t|E_u^{\infty} .
\end{aligned}$$

Hence the causal and non-causal estimates of y_t are identical. Further,

$$\begin{aligned}
y_t|E_u^{\infty} &= \sum_{i=0}^{\infty} B_i e_{2,t-i} \\
&= \sum_{i=0}^{\infty} \sum_{j=0}^{\infty} B_i \hat{D}_j u_{t-i-j} ,
\end{aligned}$$

where

$$\hat{D}(z) = \sum_{i=0}^{\infty} \hat{D}_i z^i$$

is the inverse of the matrix $D(z)$. Consequently both filters are given by $B(z)D^{-1}(z)$ which is, of course, in agreement with the formula for $H_+(z)$ given earlier.

(3) ⇒ (4):

Let $\underline{\underline{U}}^t$ denote the subspace of $\underline{\underline{H}}$ spanned by $\{u_s; s \leq t\}$. Defining the stochastic process w by

$$w_t = y_t - y_t|\underline{\underline{U}}^t, \qquad t \in Z ,$$

we obtain

$$y_t = \sum_{i=0}^{\infty} K_i u_{t-i} + w_t , \qquad t \in Z .$$

Observe that

$$K(z) \underline{\Delta} \sum_{i=0}^{\infty} K_i z^i = H_+(z) ,$$

and so, by the calculation in the demonstration of (2) ⇒ (3), it follows that K(z) is the matrix of rational functions $B(z)(D(z))^{-1}$. It remains to show w is orthogonal to u.

According to (3) the non-anticipative and anticipative estimates of y_t are identical. Then $y_t|\underline{\underline{U}}^t = y_t|\underline{\underline{U}}^\infty$, and since $\underline{\underline{U}}^t \subset \underline{\underline{U}}^{t+k} \subset \underline{\underline{U}}^\infty$ for $k \geq 0$, we have

$$\left(\left\{\sum_{i=0}^{\infty} K_i u_{t-i} + w_t\right\}|\underline{\underline{U}}^t\right) = y_t|\underline{\underline{U}}^t = y_t|\underline{\underline{U}}^{t+k}$$

$$= \left(\left\{\sum_{i=0}^{\infty} K_i u_{t-i} + w_t\right\}|\underline{\underline{U}}^{t+k}\right), \quad k \geq 0.$$

Hence $w_t|U^{t+k} = w_t|U^t = 0$, $k \geq 0$. But by construction w_t is orthogonal to $\underline{\underline{U}}^t$ and hence to $\underline{\underline{U}}^{t-k}$ for all $k \geq 0$. We conclude that $\{w_t|\underline{\underline{U}}^{t+k} = 0;\ k, t \in Z\}$ and so w and u are orthogonal processes. Let

$$w_t = \sum_{i=0}^{\infty} L_i v_{t-i}$$

denote the innovations representation of w_t. Then y has the representation

$$y_t = \sum_{i=0}^{\infty} K_i u_{t-i} + \sum_{i=0}^{\infty} L_i v_{t-i} \quad . \tag{2.5}$$

By the finite generation assumption on ζ, w is seen to be the difference between two processes finitely generated with respect to their innovations processes. Consequently w is also finitely generated. The uniqueness of (2.5) is shown as follows. Suppose

$$\sum_{i=0}^{\infty} K_i u_{t-i} + \sum_{i=0}^{\infty} L_i v_{t-i} = y_t = \sum_{i=0}^{\infty} K'_i u_{t-i}$$

$$+ \sum_{i=0}^{\infty} L'_i v'_{t-i} \ , \qquad t \in Z \ .$$

Now $K_i = K'_i$, $i \in Z_+$, since the impulse response of the non-anticipative filter $H(z)$ is uniquely specified by the spectrum of ζ. (Here Z_+ denotes the non-negative integers.) Since

$$\sum_{i=0}^{\infty} L_i v_{t-i} \qquad \text{and} \qquad \sum_{i=0}^{\infty} L'_i v'_{t-i}$$

are just innovations representations of w they are identical in the sense that $L_i = L'_i$, $i \in Z_+$, and $v = v'$.

It follows that the representation (2.5) has all the properties described in the statement (4) of the theorem and so the required demonstration is complete.

(4) $\Rightarrow$ (5):

According to Granger's definition there is no feedback from y to u if $u_{t+k}|\underline{\underline{U}}^t + \underline{\underline{Y}}^t = u_{t+k}|\underline{\underline{U}}^t$, for all $k \geq 1$, where $\underline{\underline{U}}^t$ denotes the space spanned by $\{u_s; s \leq t\}$ and $\underline{\underline{U}}^t + \underline{\underline{Y}}^t$ denotes the space spanned by $\{u_r, y_s; r, s \leq t\}$. (Notice $u_{t-k}|\underline{\underline{U}}^t = u_t$ for all $k \geq 0$.) Now let $\underline{\underline{W}}^t$ denote the space spanned by $\{w_s; s \leq t\}$ where w was defined in the proof of the previous implication. Then $\underline{\underline{U}}^t + \underline{\underline{Y}}^t = \underline{\underline{U}}^t + \underline{\underline{W}}^t$ and $u_{t+k}|\underline{\underline{W}}^t = 0$, $t, k \in Z$. Hence $u_{t+k}|\underline{\underline{U}}^t + \underline{\underline{Y}}^t = u_{t+k}|\underline{\underline{U}}^t$ for all $k \geq 1$, which is Granger's criterion.

(5) $\Rightarrow$ (1):

As before construct the process w by defining $w_t = y_t - y_t|\underline{\underline{U}}^t$. Further let

$$u_t = \sum_{i=0}^{\infty} D_i \delta_{t-i} \quad , \qquad t \in Z \quad ,$$

where δ is an orthonormal process, be the innovations representation of u. Write

$$D(z) = \sum_{i=0}^{\infty} D_i z^i \quad .$$

$D(z)$ is necessarily a matrix of rational functions by virtue of the assumption on ζ. Assume statement (5) holds. Then

$$u_{t+k}|\underline{\underline{U}}^t = u_{t+k}|\underline{\underline{U}}^t + \underline{\underline{Y}}^t = u_{t+k}|\underline{\underline{U}}^t \oplus \underline{\underline{W}}^t \; , \quad t \in Z, \;\; k \geq 1 \; ,$$

where $\oplus$, as usual, denotes the direct sum of subspaces. Consequently $u_{t+k}|\underline{\underline{W}}^t = 0$ for $t \in Z$, $k \geq 1$. But by the definition of w, $u_{t+k}|\underline{\underline{W}}^t = 0$ for $t \in Z$, $k \leq 0$ and it follows that u and w are orthogonal.

Let

$$w_t = \sum_{i=0}^{\infty} A_i \gamma_{t-i} \quad , \qquad t \in Z \quad ,$$

be the innovations representation of w. Then

$$y_t = \sum_{i=0}^{\infty} A_1 \gamma_{t-i} + \sum_{i=0}^{\infty} B_i \delta_{t-i}$$

$$u_t = \sum_{i=0}^{\infty} D_i \delta_{t-i}$$

where

$$B(z) = \sum_{i=0}^{\infty} B_i z^i$$

is given by $H_+(z)D(z)$. Writing

$$\Phi(z) = \begin{bmatrix} A(z) & B(z) \\ 0 & D(z) \end{bmatrix} \quad \text{and} \quad \varepsilon = \begin{bmatrix} \gamma \\ \delta \end{bmatrix}$$

we see that $<y, u>$ has an innovations representation satisfying all the requirements of Definition 1. This completes the proof of (1) $\Leftrightarrow$ (2) $\Rightarrow$ (3) $\Rightarrow$ (4) $\Rightarrow$ (5) $\Rightarrow$ (1).

There is a short demonstration that (1) $\Rightarrow$ (3) directly. It goes as follows: let ε^y denote the first p components of ε in (2.1) and ε^u denote the remaining q components. Let $\underline{\underline{E}}_y^t$ and $\underline{\underline{E}}_u^t$ denote the subspaces of $\underline{\underline{H}}$ generated by the components of ε_s^y, $s \leq t$, and ε_s^u, $s \leq t$, respectively. Then $\underline{\underline{E}}_u^t = \underline{\underline{U}}^t$, $t \in Z$. Since e^y and e^u are independent and since

$$\sum_{i=0}^{\infty} B_i e_{t-i}^u | \underline{\underline{E}}_u^t = \sum_{i=0}^{\infty} B_i e_{t-i}^u = \sum_{i=0}^{\infty} B_i e_{t-i}^u | \underline{\underline{E}}_u^{\infty}$$

we have

$$\begin{aligned} y_t | \underline{\underline{U}}^t = y_t | \underline{\underline{E}}_u^t &= \left(\left\{ \sum_{i=0}^{\infty} A_i e_{t-i}^y + \sum_{i=0}^{\infty} B_i e_{t-i}^u \right\} | \underline{\underline{E}}_u^t \right) \\ &= \sum_{i=0}^{\infty} B_i e_{t-i}^u \\ &= \left(\left\{ \sum_{i=0}^{\infty} A_i e_{t-i}^y + \sum_{i=0}^{\infty} B_i e_{t-i}^u \right\} | \underline{\underline{E}}_u^{\infty} \right) \\ &= y_t | \underline{\underline{U}}^{\infty} \end{aligned},$$

i.e., the anticipative and non-anticipative estimates of y are identical. It is clear from this calculation that

$$y_t | \underline{\underline{U}}^t = \sum_{i=0}^{\infty} H_i u_{t-i} ,$$

where

$$H(z) = \sum_{i=0}^{\infty} H_i z^i = B(z)(D(z))^{-1} .$$

3. DETECTION OF FEEDBACK AND THE IDENTIFICATION OF CLOSED LOOP SYSTEMS

The set of equivalent formulations for the notion of feedback in Section 2 yields a battery of tests for its presence between any two given multivariate stochastic processes. We shall describe these techniques in subsections A through D in an order which corresponds to the list of properties given in the main theorem. Then in subsection E we treat the identification of closed loop systems.

A. *MAXIMUM LIKELIHOOD IDENTIFICATION OF ARMA MODELS*

1. Theoretical Results

Before presenting some basic theoretical results pertaining to this method we describe a convenient alternative canonical form to (2.1). During an identification experiment it is not feasible to constrain the sequence of residuals to have unit covariance. Consequently it is convenient to employ the alternative unique innovations representation

$$\zeta_t = \varepsilon_t + H_1\varepsilon_{t-1} + H_2\varepsilon_{t-2} + \cdots, \qquad t \in Z, \quad (3.1)$$

where the covariance of the innovations sequence ε is given by $E\varepsilon_{t+k}\varepsilon_t^T = \Sigma\delta_{t,k} \triangleq \Phi_0\Phi_0^T\delta_{t,k}$; $t, k \in Z$. Observe that $H(0) = I$ when

$$H(z) \triangleq \sum_{i=0}^{\infty} H_i z^i$$

with $H_0 \triangleq I$. Since Φ_0 in (2.1) is upper triangular and has positive elements on the diagonal it is clear that ζ is feedback free if and only if $H(z)$ is upper block triangular with blocks whose dimensions are compatible with the dimensions of the constituent processes y and u. We now introduce two parameterizations of the process ζ:

1. $\mathbb{P}_1 = \{\theta = (\Delta,\Gamma) \in \mathbb{R}^{(2n+1)r^2}$. where $(\Delta,\Gamma) \triangleq (\Delta_1,\ldots,\Delta_n, \Gamma_0,\ldots,\Gamma_n)$, $\det \Delta(z)$ and $\det \Gamma(z)$ have all zeroes outside the closed unit disc and Γ_0 is upper triangular with positive elements on the diagonal$\}$,

$\mathbb{P}_2 = \{\theta = (\Sigma,\nabla,\Lambda) \in \mathbb{R}^{(2n+1)r^2}$, where $(\nabla,\Lambda) \triangleq (\nabla_1,\ldots,\nabla_n,$ $\Lambda_1,\ldots,\Lambda_n)$, $\Lambda_0 \triangleq I$, Σ is symmetric positive definite and $\det \nabla(z)$ and $\det \Lambda(z)$ have all zeroes outside the closed unit disc$\}$.

These parameterizations correspond to the representations

$$\zeta_t + \Delta_1\zeta_{t-1} + \cdots + \Delta_n\zeta_{t-n} = \Gamma_0\varepsilon_t + \Gamma_1\varepsilon_{t-1} + \cdots + \Gamma_n\varepsilon_{t-n},$$

$$E\varepsilon_t\varepsilon_s^T = I\delta_{t,s}, \qquad t,s \in Z, \tag{3.2}$$

and

$$\zeta_t + \nabla_1\zeta_{t-1} + \cdots + \nabla_n\zeta_{t-n} = \varepsilon_t + \Lambda_1\varepsilon_{t-1} + \cdots + \Lambda_n\varepsilon_{t-n},$$

$$E\varepsilon_t\varepsilon_s^T = \Sigma\delta_{t,s}, \qquad t,s \in Z, \tag{3.3}$$

respectively.

It should be emphasized that without further technical restrictions on $\mathbb{P}_1$ and $\mathbb{P}_2$ there does not necessarily exist a unique parameter $\mathring{\theta}$ in either $\mathbb{P}_1$ or $\mathbb{P}_2$ generating an observed process ζ. This contrasts with the situation for the sequences $\{\Phi_0,\Phi_1,\ldots\}$ and $\{\Sigma, H_1,H_2,\ldots\}$ appear in the unique innovations representations (2.1) and (3.1) respectively.

In order to use the maximum likelihood method we impose upon ζ the assumption that it is a Gaussian process. As a result, the innovations process of ζ, denoted by e, is also a Gaussian process. Then for the probability density of the observed process we obtain

$$p_\theta(\zeta^N) = \prod_{t=1}^{N} p_\theta(\zeta_t|\zeta^{t-1})$$

$$= \prod_{t=1}^{N} \frac{1}{(2\pi)^{p/2}} \frac{1}{(\det \Sigma_t)^{1/2}} \exp - \frac{1}{2}(e_t^T\Sigma_t^{-1}e_t),$$

where $(\zeta^N)^T \triangleq (\zeta_1^T,\ldots,\zeta_N^T)$, and the filtering error e_t is distributed $N(0,\Sigma_t)$ for $t \geq 1$. It follows that, up to a constant, the log-likelihood function is given by

$$L^N(\zeta^N,\theta) = -\frac{1}{2}\sum_{t=1}^{N}\log\det\Sigma_t - \frac{1}{2}\sum_{t=1}^{N} e_t^T\Sigma_t^{-1}e_t\ . \qquad (3.4)$$

Assume the observed process ζ is generated with $\theta = \overset{\circ}{\theta}$ in either of the spaces $\mathbb{P}_1$ or $\mathbb{P}_2$. Then the maximum likelihood estimate $\theta^N(\zeta^N)$ of $\overset{\circ}{\theta}$ is generated by maximizing (3.4) with respect to θ over $\mathbb{P}_1$ or $\mathbb{P}_2$. It is well known [21,22] that for the case of independent observations the maximum likelihood estimates (MLEs) have the desirable properties of strong consistency, asymptotic efficiency and asymptotic normality. The extension of these classical results to scalar autoregressive moving average models was first carried out by Åström and Bohlin [23,24]. Åström and Bohlin used the law of large numbers in conjunction with a technique of Kendall and Stuart [25] to prove the strong consistency of the sequence of maximum likelihood estimates $\{\theta^N, N = 1,2,...\}$. Their proof was incomplete but Rissanen and Caines [26, 27] used their suggestion of employing the Ergodic Theorem to produce another demonstration of the strong consistency of the estimates. It turns out that without knowledge of the Kronecker indices [28], which describe the structure of the rational matrix $\Phi(z)$, neither the parameters of the ARMA representations (3.3), (3.4), nor the parameters of a Markovian state space representation, may be consistently estimated (see [26-32]). However, without this knowledge the following result [26,27] still holds. (For simplicity of presentation we use $\mathbb{P}_1$ in the theorem statement below. However an obvious transcription yields the corresponding result for $\mathbb{P}_2$.)

THEOREM 3.1: *Let* S *be a closed and bounded subset of* $\mathbb{P}_1$ *containing a true parameter* $\overset{\circ}{\theta}$ *generating the observed process* ζ. *Let* $\theta^N = (\Delta^N,\Gamma^N)$, *maximize* $L_N(\zeta^N,\theta)$ *over* S,

$$\Phi^N(z) \,\underline{\Delta}\, \sum_{i=0}^{\infty} \Phi_i^N z^i = (\Delta^N(z))^{-1}(\Gamma^N(z))$$

and

$$\overset{\circ}{\Phi}(z) \underline{\Delta} \sum_{i=0}^{\infty} \overset{\circ}{\Phi}_i z^i = (\overset{\circ}{\Delta}(z))^{-1} \overset{\circ}{\Gamma}(z) \quad .$$

Then

$$\sum_{i=0}^{\infty} \| \Phi_i^N - \overset{\circ}{\Phi}_i \| \rightarrow 0 \qquad a.s. \quad as \quad N \rightarrow \infty \quad ,$$

where $\|\cdot\|$ *denotes any matrix norm.*

In other words the maximum likelihood method produces strongly consistent estimates of the true impulse response.

The asymptotic distribution of θ^N is described in Theorem 3.2 below. This result may be obtained by elaborating Åström and Bohlin's [23] extension of the classical result [22] in this area. Assume that the Kronecker indices of $\Phi(z)$ are known. It is then possible to specify a subset $S \subset \mathbb{P}_1$ (resp. $\mathbb{P}_2$) which constitutes the candidate set of parameters constrained in such a way that (Δ, Γ) $\big($resp. $(\Sigma, \nabla, \Lambda)\big)$ for $\theta \in S$ is a family of representatives of $\Phi(z)$ in canonical form, i.e., there is a one to one correspondence between $\{\Phi_i; \, i \in Z_+\}$ and $\theta \in S$. Further this correspondence will have the crucial property that it is a homeomorphism between the parameter sets in question when the space of impulse responses is given the ℓ_1 norm. In this case, for θ^N ranging over S, we have

THEOREM 3.2: (a) *The random variable* $L_{\theta\theta} \underline{\Delta} \lim_{N\to\infty} 1/N \, L^N_{\theta\theta}(\zeta^N, \theta)$ *exists with probability one and* $\lim_{N\to\infty} 1/N \, L^N_{\theta\theta}(\zeta^N, \theta) = \lim_{N\to\infty} 1/N \, EL^N_{\theta\theta}(\zeta^N, \theta)$ *with probability one.*

(b) $L_{\overset{\circ}{\theta}\theta}(\overset{\circ}{\theta})\sqrt{N}(\theta^N - \overset{\circ}{\theta})$ *is asymptotically distributed* $N(0, - L_{\overset{\circ}{\theta}\theta}(\overset{\circ}{\theta}))$.

(c) If $L_{\overset{\circ}{\theta}\theta}(\overset{\circ}{\theta})$ *is nonsingular then* θ^N *is asymptotically distributed* $N(\overset{\circ}{\theta}, - 1/N \, L^{-1}_{\theta\theta}(\overset{\circ}{\theta}))$.

2. Computer Implementation and Structure Determination

There are two main computational problems in designing algorithms to compute MLEs of parametric ARMA models for an observed multivariate process. These are respectively the choice of a structure for the multivariate ARMA model of the observed process, this includes the problem of model order determination, and the choice of optimization procedures for maximizing the likelihood function. We shall not deal with the second question here since it has been treated at length elsewhere, see for instance [7, 33, 34].

Consider first the identification the matrix parameters in the model (3.2) with $\nabla(z)$ and $\Lambda(z)$ free excepting that n is fixed and the zeroes of $\det \nabla(z)$ and $\det \Lambda(z)$ are constrained to lie outside the closed unit disc. Despite the vast amount of model identification work that has been carried out in engineering and econometrics there appears to be, as yet, no reliable body of experiments with "free" models of the form (3.2) from which to draw conclusions. On the other hand, analogous problems of the behavior of "free" models arise when Markovian state models are used in parametric identification, and for this case there exists some computational experience [35]. It shows that state space models provided with the correct structural information,but lacking a specified state space basis,produce estimates that wander over sets equivalent (in the sense of the state space isomorphism theorem [36]) to the parameter of the system that generated the observations. These estimates however were observed to give rise to acceptable estimates of the true transfer function matrix. When unique state space models are used [29, 37], and the system structure is known,the computed estimates are seen to converge to the true parameter. These results are clearly consistent with Theorem 3.1. However, as far as we are aware, the interesting case where the state space dimension is specified, but the remaining structural information is unspecified, has not been adequately investigated.

The situation described above leads one to seek convenient unique parameterized models. Such models will have the additional property that they allow one to invoke statistical theorems concerning the estimates such as Theorem 3.2. At present there seem to be only two ways to attempt this: First, for any given set of Kronecker indices one may specify $\Delta(z)$ and $\Gamma(z)$ (resp. $\nabla(z)$ and $\Lambda(z)$) to take a canonical form such as the echelon form introduced by Popov [28] (see also [38]). In terms of computer implementation this is rather an inconvenient method, but a related technique [39] has been successfully employed in econometric identification experiments (see [33]). Second, one may specify individually the orders $\{n_{ij};\ 1 \le i \le r,\ 1 \le j \le r\}$ and $\{d_{ij};\ 1 \le i \le r,\ 1 \le j \le r\}$ of the numerators and the denominators of the entries of the rational matrix $\Phi(z) = \Delta^{-1}(z)\Gamma(z)$. Because $\Phi(z)$ is regular, we have $n_{ij} \le d_{ij}$, $1 \le i \le r$, $1 \le j \le r$. (Notice also that in this chapter we always take the number of rows of $\Phi(z)$ equal to the number of columns because this fits the problem of identifying the innovations representation of a full rank stochastic process). Now it is clearly possible to constrain $\Phi(z)$ so that $\Phi(0) = I$ and, as in $\mathbb{P}_2$, we let Σ denote the covariance matrix of the innovations process. Then z is a factor of any numerator polynomial for which $i \ne j$. When cancellations are forbidden between the numerators and denominators of the entries of $\Phi(z)$ we have a unique parameterization of the innovations representation of the observed ARMA process. We shall denote this representation by (Φ,Σ) and call it the rational transfer function (RTF) model. This model is often used in practice and is the one we employ in the examples in this chapter. In particular it allows for direct manipulation of the orders of the elements of $\Phi(z)$ and hence permits a combinatoric search for the most acceptable model with respect to some criterion.

Once a canonical form has been chosen for the identification algorithm one is faced with the statistical problem of structure determination. This includes as a special case the problem of model order determination for scalar ARMA models which at present is an active research area, see for instance [40-44]. We also point out that the chapter by H. Akaike in this volume contains some recent results in this area for multivariate ARMA models and their Markovian representations. In so far as we use RTF models for both the detection of feedback and model structure determination, the problems reduce to the same task, namely, the use of statistical methods to determine the orders of the terms in the RTF model $\Phi(z)$. Before we describe the hypothesis testing procedures we employ for structure determination we should mention some important topics which are omitted from this section. First, before conducting any identification experiment the given data sequence must be examined for stationarity and possibly prefiltered to produce an acceptably stationary sample sequence. Second, for any given model structure, the resulting residual sequence should undergo diagonistic tests for its acceptability as a white noise sequence. A standard test for whiteness involves computing the first few cross correlations of the residual sequence. We shall not describe any of these diagnostic tests here since descriptions of their theory and use may be found in [45-48] and [33].

Our general feedback detection procedure involves two steps: first, the selection of the best model in the class of feedback and feedback free systems respectively and, second, the choice between the two resulting competing models. We briefly describe below two techniques for judging the acceptability of any given model. The first consists of an F-test which is used to perform pairwise comparisons of the estimated models. The second uses the estimated covariance matrix to compute confidence regions for the estimated parameters.

(a) F-*Test*

Let θ_u denote the parameter vector of an RTF model and let θ_0 denote the parameter vector for the same model with a certain set of parameters constrained to take zero values. Let N denote the total sample length, n_0 denote the number of parameters in the constrained model, n_u the corresponding number in the unconstrained model and let H_0 denote the hypothesis that the true values of the constrained terms in the θ_0 model are all zero. Finally, let $V(\theta)$ denote the determinant of the covariance of the residual sequence generated using the parameter θ. Define

$$t = \frac{(N-n_u)}{(n_u-n_0)} \frac{V(\theta_0) - V(\theta_u)}{V(\theta_u)} .$$

For univariate autoregressive processes it is known [22] that t is an F-distributed statistic with $(N-n_u, n_u-n_0)$ degrees of freedom. However, in [49] Åström proposed its use for the ARMA case and reported satisfactory results. We proceed analogously in the multivariate case. Define

$$s = \frac{N-n_u-1}{n_u-n_0} \cdot \frac{V^{1/2}(\theta_0) - V^{1/2}(\theta_u)}{V^{1/2}(\theta_u)}$$

When $V(\theta)$ denotes the determinant of the covariance matrix of the residual sequence for an *autoregressive* system whose output process has dimension 2 this statistic is F distributed with $(2(N-n_u-1), 2(n_u-n_0))$ degrees of freedom [50]. Using the s statistic for a two-dimensional ARMA process yields the decision rule: Accept H_0 if $s \leq s_\alpha$. Reject H_0 if $s > s_\alpha$, where s is the 100α% point of the distribution $F(2(N-n_u+1), 2(n_u-n_0))$ given by Prob $(s > s_\alpha) = \alpha$ when H_0 is true. For ARMA processes with higher dimension the use of the F test becomes increasingly more complex.

A second test is available for the comparison of two models if it is assumed that the number of samples N is large enough to justify an asymptotic analysis of the results (notice we have already used this assumption in stating Theorems 3.1 and 3.2 earlier in this section). Let λ denote the likelihood ratio given by $\lambda = (V(\theta_u)/V(\theta_0))^{N/2}$ when ζ has any finite dimension. Then asymptotically $-2 \log \lambda$ is χ^2 distributed with $n_c - n_0$ degrees of freedom [22]. The test is then: Accept H_0 if

$-2 \log \lambda \leq \gamma_\alpha$; Reject H_0 if $-2 \log \lambda > \gamma_\alpha$, where γ_α is the 100α% point of χ^2 distribution with $n_c - n_0$ degrees of freedom.

(b) *Confidence Regions*

Given an r component random variable ζ distributed $N(\overset{\circ}{\zeta},\Sigma)$ we may define the random variable ζ by $\xi = (\zeta-\overset{\circ}{\zeta})^T \Sigma^{-1} (\zeta-\overset{\circ}{\zeta})$; ξ will then have the χ^2 distribution with r degrees of freedom. Since θ^N is asymptotically normally distributed $N(\overset{\circ}{\theta},-L^{-1}_{\theta\theta}(\overset{\circ}{\theta}))$ we may formulate the following test. Let ϕ denote the m vector of entries of θ which are under test for significant difference from $0 \in R^m$ and let $\overset{\circ}{\phi}$ be the true value of ϕ. By Theorem 3.2 $\xi_\phi \triangleq (\phi^N-\overset{\circ}{\phi})^T(\Sigma^N_\phi)^{-1}(\phi^N-\overset{\circ}{\phi})$ has asymptotically the χ^2 distribution with m degrees of freedom, where ϕ^N is the estimate of $\overset{\circ}{\phi}$ and Σ^N_ϕ is the $(m \times m)$ submatrix of the estimate of $-L^{-1}_{\theta\theta}(\overset{\circ}{\theta})$ corresponding to ϕ. Then the decision procedure is as follows: Reject H_1 if $\xi_\phi < \gamma_\alpha$; Accept H_1 if $\xi_\phi \geq \gamma_\alpha$, where ξ_ϕ is computed with $\overset{\circ}{\phi} = 0$ and γ_α is the 100α% limit for the χ^2 distribution with m degrees of freedom. In other words we construct the $(1-\alpha)$% confidence region around ϕ^N and reject H_1 if 0 lies within this region.

3. Simulation Example

We illustrate the techniques described above with a simple simulation example. 200 pairs of univariate input and output observation data were generated with the model

$$[y] = A[\varepsilon_1] + B[u] \quad ,$$
$$[u] = [\varepsilon_2] \quad ,$$

where $[p]$ denotes the z-transform of a given process p,

$$A(z) = \frac{1 + 0.4z}{1 + 0.6z} \ , \qquad B(z) = \frac{0.7z}{1 + 0.9z} \quad ,$$

and where ε_1, ε_2 were serially and mutually independent Gaussian random variables (with distributions $N(0, 1.0^2)$ and $N(0, 0.5^2)$ respectively) which were generated by a standard computer subroutine.

First we assume feedback is present between the ordered pair <y,u>. A search through a family of candidate structures and the application of the diagnostic test described above yielded the model (3.5-6) shown below as the most acceptable model in the feedback class. The estimated standard deviations of the estimated parameters are shown in brackets.

$$\begin{bmatrix} y \\ u \end{bmatrix} = \begin{bmatrix} \dfrac{\underset{(\pm 0.157)}{1 + 0.414z}}{\underset{(\pm 0.138)}{1 + 0.591z}} & \dfrac{\underset{(\pm 0.232)}{0.723z}}{\underset{(\pm 0.292)}{1 + 0.913z}} \\ \underset{(\pm 0.032)}{-0.014z} & 1 \end{bmatrix} \begin{bmatrix} \varepsilon_1 \\ \varepsilon_2 \end{bmatrix} \qquad (3.5)$$

and

$$\text{Cov} \begin{bmatrix} \varepsilon_1 \\ \varepsilon_2 \end{bmatrix} = \begin{bmatrix} 1.044^2 & -0.015 \\ -0.015 & 0.496^2 \end{bmatrix} \qquad (3.6)$$

Next we assumed <y, u> was feedback free and the most acceptable identified model was as follows:

$$\begin{bmatrix} y \\ u \end{bmatrix} = \begin{bmatrix} \dfrac{\underset{(\pm 0.309)}{1 + 0.429z}}{\underset{(\pm 0.269)}{1 + 0.610z}} & \dfrac{\underset{(\pm 0.101)}{0.722z}}{\underset{(\pm 0.018)}{1 + 0.913z}} \\ 0 & 1 \end{bmatrix} \begin{bmatrix} \varepsilon_1 \\ \varepsilon_2 \end{bmatrix} \qquad (3.7)$$

$$\text{Cov} \begin{bmatrix} \varepsilon_1 \\ \varepsilon_2 \end{bmatrix} = \begin{bmatrix} 1.047^2 & -0.015 \\ -0.015 & 0.496^2 \end{bmatrix} . \qquad (3.8)$$

It seems reasonable from inspection of (3.5-8) that the observed joint process was feedback free. The hypothesis testing procedures proposed in Subsection 2 above confirmed this by giving the following results.

Computing the s statistic for this experiment yields a value of 0.578. Since the 5% level for $F(388, 2)$ is 3.00 we accept the feedback free hypothesis by the F test. Next, applying the likelihood ratio test, we obtain

$$-2 \log \lambda = -200 \log \frac{0.541}{0.544} = 0.480;$$

but the 5% level for the χ^2 distribution with one degree of freedom is 3.84 and so we accept the feedback free hypothesis by this test also. Finally, in the example above the inversion of the $(n_c - n_0) \times (n_c - n_0)$ submatrix $\hat{\Sigma}_c$ is simple since $\hat{\Sigma}_c$ is merely the scalar $(0.032)^2$. This gives $\xi_c = (\hat{\emptyset}_c - 0)\Sigma_c^{-1}(\hat{\emptyset}_c - 0) = (0.014)^2(0.032)^{-2} = 0.191$. Since the 95% confidence region is given by $\xi_c \leq 3.84$ we reject the feedback hypothesis at the 5% risk level.

To summarize, we see that in this simulated example all four of the proposed tests come to the correct conclusion that $\langle y,u \rangle$ is feedback free.

B. SPECTRUM ESTIMATION AND FACTORIZATION

1. Theoretical Development

The second part of the theorem in Section 2 gives a characterization of feedback free processes in terms of the spectrum of the joint process. Granger has also discussed [3] the properties of the cross spectrum $\Psi_{yu}(z)$ for feedback free processes, but did not devise any statistical tests based on its structure. Indeed, out the context of the stable inverse-stable factorization of $\Psi_\zeta(z)$, it is difficult to see how to use $\Psi_{yu}(z)$ as a test for feedback, since this involves a decision between $\Psi_{yu}(z) = A(z)C^*(z) + B(z)D^*(z)$ and $\Psi_{yu}(z) = B(z)D^*(z)$, where the matrices

$A(z)$, $B(z)$, $C(z)$, $D(z)$ are given in terms of the joint spectrum $\Psi_\zeta(z)$ (see Section 2). On the other hand, the fact that multivariate spectrum factorization techniques yield alternative methods for the identification of the innovations representation of an observed process shows that these techniques may be used for the detection of feedback.

It is well known that estimating the spectrum of a stationary process with the periodogram does not give consistent estimates in mean square or almost surely, and there is a substantial statistical literature describing smoothing techniques to circumvent this problem (see e.g. [45-47]). One result of this situation is that MLE estimation methods for ARMA models are attractive precisely because they yield statistically consistent estimates for the factors of the spectrum of the observed process (see the accompanying chapter by Akaike).

The spectral factorization techniques we describe use the truncated sequences of covariance matrix estimates $\{\hat{R}_\tau;\ 0 \leq \tau \leq M\}$, where $M << N$ when N denotes the data record length. It is known that

$$\hat{R}_\tau \triangleq \frac{1}{N} \sum_{t=1}^{N-\tau} \zeta_{t+\tau}\zeta_t^T \quad , \qquad 0 \leq \tau \leq M \quad ,$$

are asymptotically unbiased and strongly consistent when ζ is a stationary ergodic process. However, any stationary Gaussian process with a summable covariance sequence is ergodic and the estimates are then also mean square consistent. Clearly this summability property holds for the processes generated by (2.2). It should be mentioned that a disadvantage in using the estimates $\{\hat{R}_\tau;\ 0 \leq \tau \leq M\}$ above is that the variance of the estimates increases with τ, for fixed N, and $\hat{R}_\tau$, $\hat{R}_{\tau+\nu}$ are correlated for all τ and ν (see [45-47] for the appropriate formulae). For these and other reasons smoothing techniques are often used in estimating $\{R_\tau; \tau \in Z\}$.

In this subsection we proceed on the assumption that acceptable estimates of the truncated covariance sequence

$\{R_\tau;\ -M \le \tau \le M\}$ are available. In this case the technique we propose to test for the presence of feedback consists of factoring $\Phi^M(z)$, constructed from $\{\hat{R}_0, \ldots, \hat{R}_M\}$ as a substitute for $\Phi(z)$, and examining the resulting factor for the stipulated feedback free structure.

The factorization of matrix spectral functions has been the subject of intense research by workers in the areas of time series analysis and system theory for many years. We shall not attempt to survey this field. The development we present follows Anderson [51] and Faurre [52, 53] (see also [54, 55]).

Consider a spectral density matrix $\Psi(z)$ of a process ζ. By the standing assumption that ζ is finitely generated $\Psi(z)$ is a matrix of rational functions. Furthermore $\Psi(z)$ may be decomposed as

$$\Psi(z) = W(z) + W^T(z^{-1}) = \left(H(z^{-1}I-F)^{-1}G + \frac{R_0}{2}\right) + \left(H(zI-F)^{-1}G + \frac{R_0}{2}\right)^T ,$$

where $W(z)$ is a (discrete) positive real matrix and

$$\left(H,\ F,\ G,\ \frac{R_0}{2}\right)$$

is a minimal realization of $W(z)$. Now the Discrete Positive Real Lemma [53] states that there exists matrices P, L, J, with P symmetric positive semi-definite such that

$$\begin{aligned} P - FPF^T &= LL^T , \\ G - FPH^T &= LJ^T , \\ R_0 - HPH^T &= JJ^T . \end{aligned} \tag{3.9}$$

Any solution set (P, L, J) together with (H, F) yields a Markovian realization of ζ via

$$x(k+1) = Fx(k) + w(k)$$

$$\zeta(k) = Hx(k) + v(k)$$

where $\begin{bmatrix} w \\ v \end{bmatrix}$ is a Gaussian i.i.d. process with

$$E \begin{bmatrix} w(k) \\ v(k) \end{bmatrix} [w^T(k) \quad v^T(k)] = \begin{bmatrix} LL^T & LJ^T \\ JL^T & JJ^T \end{bmatrix} \geq 0, \quad k \in Z,$$

and

$$Ex(k)x^T(k) = P \qquad , \quad k \in Z.$$

Denote LJ^T by S and JJ^T by R. Then clearly $\Phi(z) = (H(z^{-1}I - F)^{-1}SR^{-1} + I)R^{1/2}$ constitutes a stable spectral factor of $\Psi(z)$. Let $\mathbb{P}$ denote the family of solution matrices to (3.9). Then it is easy to see that P satisfies

$$P = FPF^T + (G-FPH^T)(R_0-HPH^T)^{-1}(G^T-HPF^T) \quad . \tag{3.10}$$

In [53] it is shown that $\mathbb{P}$ constitutes a closed convex bounded set possessing a maximum P^* and minimum P_*, i.e., $P \in \mathbb{P}$ implies $P_* \leq P \leq P^*$. Further P_* is shown to yield a spectral factor such that $\det \Phi(z)$ has no zeroes in the region $|z| < 1$. Given the assumptions on the process ζ in (2.2) it follows that we may find the (asymptotically) stable and (asymptotically) inverse stable spectral factor $(H(Iz^{-1}-F)^{-1}SR^{-1}+I)R^{1/2}$ of $\Psi(z)$ via the computation of P_* for (3.10). A reasonable numerical procedure to find P_* is to iterate the equation

$$P_{n+1} = FP_nF^T + (G-FP_nH^T)(R_0-HP_nH^T)^{-1}(G^T-HP_nF^T)$$
$$P_0 = 0 \tag{3.11}$$

This method is used in the simulation example below.

Let us assume that the quadruple $\{H^M, F^M, G^M, \hat{R}_0\}$ has been derived from a given sequence of estimated covariances $\{\hat{R}_0, \hat{R}_1, \ldots, \hat{R}_M\}$ and the corresponding P_* computed. Referring to the representation (3.1), we see that to detect feedback from the first p to the remaining q components of ζ it is sufficient to check $\Phi^M(z) \underline{\Delta} H^M(Iz^{-1}-F^M)^{-1}S^M(R^M)^{-1} + I$ for the corresponding upper block triangular structure where S^M and R^M are computed from (3.9-11) using $\{H^M, F^M, G^M, \hat{R}_0\}$.

Now it appears difficult to connect the statistical properties of the estimated covariance matrices with the statistical properties of the coefficients of the resulting spectral factor $\Phi^M(z)$. This, of course, is one of the main advantages of the computationally more expensive maximum likelihood estimation method. However, reasonable *ad hoc* criteria may be proposed such as the following: $\langle y, u \rangle$ is feedback free at the (N,ε) level if

$$\max\left\{c^n_{ijk}, c^d_{ijk}\right\}/\max\left\{a^n_{ijk}, a^d_{ijk}, b^n_{ijk}, b^d_{ijk}, d^n_{ijk}, d^d_{ijk}\right\} < \varepsilon$$

where the entries in the expression above correspond in an obvious way to the coefficients of the numerators and denominators of the entries of $C(z)$, $A(z)$, $B(z)$ and $D(z)$ respectively and N denotes the length of the sample sequence. Other methods would be simply to use the criterion

$$\max\left\{c^n_{ijk}, c^d_{ijk}\right\} < \varepsilon \quad ,$$

or the direct inspection of the Markov matrices of the computed spectral factor.

2. Simulation Experiments

Computational experiments implementing the procedure described above have been conducted by S. Sinha at the University of Toronto. The system described in Subsection A3 above was used to generate sample sequences of 500, 1,000, 2,500 and 10,000 points. In a second set of experiments $A(z)$ and $B(z)$ (see A3) had poles at (-0.6) and (-0.7), respectively, and, in a third set, had poles at (-0.2) and (-0.3), respectively.

The partial realization algorithm of Rissanen [54,55] was used for the first step of the factorization procedure. Since the estimated covariance sequence forms a generic set of data the algorithm may be verified to constitute a continuous map from $\{\hat{R}_1, \ldots, \hat{R}_M\}$ to $\{H^M, F^M, G^M\}$ and the state dimension of the resulting realization will take the maximum possible value for each M. However, our experiments show that small variations in the covariance data produce large variations in the spectrum of the state transition matrix F^M which often results in F^M being unstable. This is a serious difficulty since (3.11) can be seen to converge only for stable state transition matrices. Contrary to the hopes expressed in [55], increasing the value of M or N does not increase the likelihood of obtaining a stable matrix. However, smoothing the covariance data by various methods was found to increase the possibility of obtaining a stable matrix F^M, for each value of M. It would be most valuable to conduct experiments to see whether the matrix fraction partial realization algorithm of Dickinson, Morf and Kailath [74] does not suffer from the "unstable partial realization" problem described above, when used on real data, or whether the problem can be avoided by using a Chandrasekhar type algorithm instead of (3.11).

The iteration of (3.11) was found to be numerically satisfactory for matrices F^M with eigenvalues of modulus less than approximately 0.85. For state covariance matrices of dimension less than 10 convergence was usually found to occur in less than 50 iterations with a halting rule specifying a final error of 10^{-6} per entry. This reinforces our conclusion that the partial realization step forms the weak link in the procedure presented in Subsection B1.

We present below the second through fifth Markov matrices of the true and estimated spectral factors for two of our experiments. By construction the first Markov matrix is always the identity matrix. Since these results are representative of the best results obtained to date they illustrate the unsatisfactory nature of this procedure at the present moment. (We should, perhaps, point out that, despite the difference between $\Phi^M(z)$ and $\Phi(z)$ in these examples, the product $\Phi^M(z)R^M(\Phi^M(z))^*$ will exactly fit the estimated covariance sequence up to $\hat{R}_{\pm M}$. This follows directly from the definition of $\Phi^M(z)$).

EXAMPLE 1: Pole $A(z)$ at -0.6, pole $B(z)$ at -0.7, 10,000 sample points, smoothed covariance data, partial realization state dimension 2.

Exact Factor, Markov Matrices 2 through 5.

$$\begin{bmatrix} -0.2 & 0.7 \\ 0.0 & 0.0 \end{bmatrix} \quad \begin{bmatrix} 0.12 & -0.49 \\ 0.0 & 0.0 \end{bmatrix} \quad \begin{bmatrix} -0.072 & 0.343 \\ 0.0 & 0.0 \end{bmatrix} \quad \begin{bmatrix} 0.043 & -0.24 \\ 0.0 & 0.0 \end{bmatrix}$$

Estimated Factor, Markov Matrices 2 through 5

$$\begin{bmatrix} -0.270 & -0.093 \\ 0.105 & 0.137 \end{bmatrix} \begin{bmatrix} 0.182 & 0.119 \\ -0.077 & -0.04 \end{bmatrix} \begin{bmatrix} -0.126 & -0.076 \\ 0.523 & 0.033 \end{bmatrix} \begin{bmatrix} 0.087 & 0.053 \\ -0.036 & -0.022 \end{bmatrix}$$

EXAMPLE 2: Pole $A(z)$ at -0.6, pole $B(z)$ at -0.9, 10,000 sample points, partial realization state dimension 4 (this is the same system as used in A3).

Exact Factor, Markov Matrices 2 through 5.

$$\begin{bmatrix} -0.2 & 0.7 \\ 0.0 & 0.0 \end{bmatrix} \quad \begin{bmatrix} 0.12 & -0.63 \\ 0.0 & 0.0 \end{bmatrix} \quad \begin{bmatrix} -0.072 & 0.567 \\ 0.0 & 0.0 \end{bmatrix} \quad \begin{bmatrix} 0.043 & -0.51 \\ 0.0 & 0.0 \end{bmatrix}$$

Estimated Factor, Markov Matrices 2 through 5.

$$\begin{bmatrix} -0.069 & -0.229 \\ 0.051 & 0.338 \end{bmatrix} \begin{bmatrix} 0.151 & 0.184 \\ -0.026 & -0.046 \end{bmatrix} \begin{bmatrix} -0.133 & -0.169 \\ 0.023 & 0.046 \end{bmatrix} \begin{bmatrix} 0.113 & 0.135 \\ -0.034 & -0.038 \end{bmatrix}$$

(All entries above are given to three decimal places.) Satisfactory results using this technique have recently been presented in [75].

C. FILTERING AND REGRESSION METHODS

Part 3 of our main theorem states that the pair $\langle y, u \rangle$ is feedback free if and only if the anticipative and non-anticipative filters for the estimation of y from u are identical. This implies that a regression of y on past and future u observations,

$$y_t = \sum_{i=-\infty}^{\infty} K_i u_{t-i} + w_t , \quad t \in Z , \tag{3.12}$$

yields a sequence of matrix coefficients with $K_i = 0$, $i < 0$, and a residual sequence w independent of u. This immediately suggests that a test for feedback is to regress y against u and apply an appropriate statistical test for the significance of the "future" coefficients $\{K_i, i < 0\}$. This is precisely what Sims [5] does in his analysis of the U.S. money supply--GNP relation and we very briefly describe his results below.

Sims took quarterly data of the money supply, measured as currency plus reserves and as currency plus demand deposits, and quarterly data of the GNP from 1947-69. (Data was prepared by the Federal Reserve Bank of St. Louis.) Logarithms were taken of all data values and all the series were prefiltered. Then least squares regressions were computed of GNP against 8 past, and 4 future and 8 past values of each measure of the money supply. Application of the F-test showed GNP "driven" by money supply, that is, computations which included future lagged variables yielded insignificant coefficients for these variables. On the other hand, the same regressions of money on GNP showed that money was not simply driven by past GNP. Stated in our terminology, Sims draws the conclusion that for U.S. time series the pair <GNP, money> is feedback free. However other workers using this technique [56] did not draw the same conclusion from analogous time series for the U.K.

D. *A TEST FOR FEEDBACK USING RESIDUAL SEQUENCES*

Wall [8] has proposed a test for feedback between stochastic processes in terms of the whiteness properties of the residuals generated in the identification of a specified factorization of $\Phi(z)$. In the brief description of Wall's technique presented here we make some small changes from his original presentation.

Let $\Phi(z)$ be uniquely factored as

$$\Phi(z) = \begin{bmatrix} A(z) & 0 \\ 0 & B(z) \end{bmatrix}^{-1} \begin{bmatrix} I & E(z) \\ F(z) & I \end{bmatrix} ,$$

where $A(z)$, $B(z)$, $E(z)$ and $F(z)$ are polynomial matrices. Then, when v is a serially uncorrelated zero mean stationary process of covariance Σ, we have a representation of $[y^T, u^T]$ as

$$\begin{bmatrix} y \\ u \end{bmatrix} = \begin{bmatrix} A(z) & 0 \\ 0 & B(z) \end{bmatrix} [w] \qquad (3.13)$$

when

$$[w] \underline{\Delta} \begin{bmatrix} I & E(z) \\ F(z) & I \end{bmatrix} [v] ,$$

where $[p]$ denotes the z-transform of a given process p.

Now without loss of generality we may take $F(0) = F_0 = 0$ and $E(0) = E_0 = 0$. (We do this by absorbing

$$\begin{bmatrix} I & E(0) \\ F(0) & I \end{bmatrix}$$

into a noise covariance matrix Σ). Then it is clear that $\langle y, u \rangle$ is feedback free if and only if $F(z) \equiv 0$. Now observe that for $\ell \geq 1$,

$$\begin{aligned} Ew_u(k+\ell)w_y^T(k) &= E(v_u(k+\ell) + F_1 v_y(k+\ell-1) + F_2 v_y(k+\ell-2) + \ldots) \times \\ &\qquad (v_y(k) + E_1 v_u(k-1) + E_2 v_u(k-1) + \ldots)^T \\ &= F_\ell \Sigma_{yy} , \end{aligned}$$

where w_y and w_u (resp. v_y, v_u) denote the first p and remaining q components respectively of the process w (resp. v)

and $\Sigma_{yy} \triangleq \text{Cov}(v_y)$. Consequently a test for $\langle y, u\rangle$ to be feedback free may be posed in terms of the statistical significance from zero of the estimates of $Ew_u(k+\ell)w_y^T(k)$, $\ell \geq 1$, generated by a model of the form (3.13) displayed above. Carrying out tests of this type Wall concluded that the pair of process $\langle$income, money$\rangle$ contained feedback at statistically significant levels.

E. *IDENTIFICATION OF CLOSED LOOP SYSTEMS*

Consider a feedback system described by the equations

$$y = Ku + Lv \quad , \tag{3.14}$$

$$u = My + Nw \quad , \tag{3.15}$$

where v and w are orthogonal stationary processes which are assumed mutually orthogonal, $L_0 = N_0 = I$ and the notation Ap denotes the process

$$\left\{ \sum_{i=0}^{\infty} A_i p_{t-i} \; ; \quad t \in Z \right\}$$

when p denotes the process $\{p_t; \;\; t \in Z\}$. We remark in passing that solving (3.14-15) for the joint (y,u) process shows there is feedback from y to u (resp. u to y), according to our formal Definition 1, if and only if M (resp. K) is nonzero (see [7]).

Clearly the presence of the feedback loop (3.15) causes the processes u and Lv in the feedforward loop (3.14) to be correlated. As a consequence the generalized least squares and instrumental variable identification methods will not yield consistent or asymptotically unbiased estimates. Another feature of systems such as (3.14-15) is that there is an inherent non-uniqueness in the representation of the y and u processes. For example, for any non-anticipative operator $\tilde{K}$, such that $(I-\tilde{K}M)$ is a stable non-anticipative operator, y has a representation

$$y = (I - \tilde{K}M)^{-1}(K - \tilde{K})u + \tilde{w} \quad ,$$

where the spectral density of $\tilde{w}$ is $(I-\tilde{K}M)^{-1}[\tilde{K}NN^*\tilde{K}^*+LL^*](I-\tilde{K}M)^{*-1}$.

Other aspects of the representations of feedback systems are discussed in Akaike [57] and Chan [58].

Prediction Error Identification Method

Under various assumptions on the structure of the system (3.14-15) and the information $\{I^t, t \in Z\}$ supplied to a prediction algorithm it is possible to prove important results [12] about the identification of (3.14) by the prediction error method. This method involves searching over constant stable linear predictors generating the prediction process $\{\hat{y}_t(\theta|I^t); t \in Z\}$ in order that

$$\frac{1}{N}\sum_{t=1}^{N}(y_t - \hat{y}_t(\theta))^T(y_t - \hat{y}_t(\theta))$$

is minimized for any given sample (y_1^N, u_1^N), where $\hat{y}_t(\theta|I^t)$ is denoted $\hat{y}_t(\theta)$ for brevity.

In this discussion we shall take linear least squares predictors to be stationary and will further assume that the prediction process $\hat{y}$ is stationary. This implies that we assume observations of the y and u processes are available from the infinite past and this allows us to give the simplest statement of the results below. We remark that the analogous time varying results i.e., those which take into account initial conditions for the prediction process and the time varying nature of the optimal predictor, are also true. The appropriate elaboration of the analysis below goes along the lines of the study of the MLE in [26] and [27], where formulae for the time varying predictors will also be found.

Assume that the processes v and w in (3.14) and (3.15) are Gaussian, the operators K and M are stable and rational, i.e. have rational z-transforms, the operators L and N are stable, inverse stable and rational and the closed loop system is stable. It follows that y, u and $\hat{y}$ are ergodic and

$$\lim_N \frac{1}{N}\sum_{t=1}^{N}(y_t-\hat{y}_t(\theta))(y_t-\hat{y}_t(\theta)) = E(y_0-\hat{y}_0(\theta))^T(y_0-\hat{y}_0(\theta)) \text{ a.s.}$$

where θ parameterizes some linear predictor. Further, denote by S a compact subset of the parameterized space of linear predictors with the same Kronecker indices as the optimal linear predictor, and further let S contain a parameter corresponding to the true optimal predictor. Next, following [12], we define

$$D = \{\theta \,|\, \theta \in S,\ E\|F_\theta y - G_\theta u\|^2 = 0\} ,$$

where $F_\theta(z) \triangleq L_\theta^{-1}(z) - L^{-1}(z)$ and $G_\theta(z) \triangleq L_\theta^{-1}(z)K_\theta(z) - L^{-1}(z)K(z)$. Finally, we define the following alternative conditions on the system and information structures of our set up:

A. The feedforward loop (3.14) contains a delay of at least one time unit and I^t, the information supplied to the prediction algorithm, is given by $I^t = (y^{t-1}, u^{t-1})$, where p^t denotes $p^t_{-\infty}$, $t \in Z$, for a given process p. (This is essentially the condition used by Ljung [12].)

B. The feedback loop (3.15) contains a delay of at least one time unit and $I^t = (y^{t-1}, u^t)$.

In the following discussion the symbol I^t is assumed to be prescribed by whichever of the conditions A or B is in force.

The reason for adopting the conditions A or B is that each imply the predictor is computed only in terms of the feedforward loop dynamics and hence has the stationary form $\hat{y} = (I-L^{-1})y + L^{-1}Ku$. In both theorems in this subsection we also assume that $\{y^t;\ t \in Z\}$ is full rank with respect to $\{y^{t-1}, u^t;\ t \in Z\}$ i.e., the prediction error matrix is full rank.

Now by specializing and slightly modifying the more general result of Ljung (see [12] and the chapter by Ljung in this volume) we obtain the following

THEOREM 3.3: *Assume either condition* A *or* B *holds for the system* (3.14-15). *Let* θ^N *minimize*

$$\frac{1}{N} \sum_{t=1}^{N} (y_t - \hat{y}_t(\theta))^T (y_t - \hat{y}_t(\theta))$$

over S. *Then* θ^N *converges into the set* D *almost surely as* N *tends to infinity.*

By a straightforward calculation Theorem 3.3 implies that a system under noiseless feedback control (N = 0) can only be identified by prediction error methods up to the equivalence class of feedforward transfer functions $(I-\tilde{K}M)^{-1}(K-\tilde{K})$, $(I-\tilde{K}M)^{-1}L$; where $(I-\tilde{K}M)$ is rational and inverse stable . However, it may also be easily verified [12] that if the feedback loop contains disturbances of the type described in (3.15) the feedforward loop transfer functions are uniquely identifiable.

Maximum Likelihood Identification Method

We present below a result which grows out of the investigations of Bohlin*, Chan and Ljung. To be more specific, in [58] Chan studied the form of the likelihood function and the role of delays in closed loop system identification, and in [12] Ljung discussed the connection between MLEs and prediction error estimators. We also remark in this connection that in [27] Rissanen and Caines indicated that the MLE techniques of [26,27] could be used to treat the prediction error identification method.

Let $\ell_N(y^N,\theta)$ denote the pseudo-likelihood function $\Pi_{t=1}^N p_\theta(y_t|I^t)$ for the system (3.14) under the conditions A or B, where θ parameterizes the predictors (equivalent systems) within the set S. (Notice $\ell_N(y^N,\theta)$ differs from the usual likelihood functions $p_\theta(y^N,u^N)$, $p_\theta(y^N|u^N)$ and $p_\theta(y^N|u^{N-1})$.) Let the pseudo-maximum likelihood estimator θ^N maximize $\ell(y^N,\theta)$ over S. Then we have the following result due to Caines:

THEOREM 3.4: *Assume either condition* A *or* B *holds for the system* (3.14-15). *Then the pseudo-maximum likelihood estimator* θ^N *converges into the set* D *almost surely as* N *tends to infinity.*

---

*Private Correspondence

Proof. Let $V(\theta)$ denote $Ee_0(\theta)\, e_0^T(\theta)$, where $\{e_t(\theta) \triangleq y_t - \hat{y}_t(\theta);\ t \in Z,\ \theta \in S\}$ and let $\hat{\theta}$ parameterize a linear least squares predictor i.e., let $\mathrm{Tr}\, V(\theta) \geq \mathrm{Tr}\, V(\hat{\theta})$ for all $\theta \in S$. Now for any stochastic process

$$\lambda^T(E(y_t-\tilde{y})(y_t-\tilde{y})^T | I^T))\lambda \;=\; E(\lambda^T(y_t-\tilde{y}))^2 | I^t),\quad \lambda \in \mathbb{R}^p\ ,$$

is minimized, for all λ, at $\tilde{y} = Ey_t | I^t)$ [60]. Hence, by virtue of the Gaussian assumption on y, the linear least squares predictor $\tilde{y}_t(\hat{\theta})$ is identical to $Ey_t|I^t)$. Consequently

$$\mathrm{Tr}\, V(\theta) \geq \mathrm{Tr}\, V(\hat{\theta}),\quad \forall\theta \in S,\quad \Rightarrow V(\theta) \geq V(\hat{\theta}) \qquad \forall\theta \in S\ . \quad (3.16)$$

By definition, the prediction error estimate $\hat{\theta}^N$ minimizes

$$Q_N(y^N,\theta) \triangleq \frac{1}{N} \sum_{t=1}^{N} (y_t - \hat{y}_t(\theta))^T(y_t - \hat{y}_t(\theta))$$

over S. We shall now demonstrate that $\hat{\theta}^N$ converges a.s. as N tends to infinity into the set of parameters which minimizes the determinant of $V(\theta)$.

By the assumption that the predictor and prediction process are stationary and by the Gaussian assumption in all the processes in (3.14-15) we may employ the Ergodic Theorem to obtain

$$Q_N(y^N,\theta) \to \mathrm{Tr}\, E\varepsilon_0(\theta)\varepsilon_0^T(\theta),\qquad \text{a.s.} \qquad \text{as} \quad N \to \infty\ .$$

Further it may be shown, in a manner exactly analogous to that of [27, Appendix B], that this convergence is uniform in θ over S. Now let (y,u) be any sample for which the indicated convergence above occurs and denote the limit by $Q(\theta)$. Further let $\{\hat{\theta}^M\}$ denote a convergent subsequence of $\{\hat{\theta}^N\}$ in the compact set S with accumulation point $\hat{\theta}^*$. Then

$$Q(\theta) \;=\; \lim_{M\to\infty} Q_M(y^M,\theta) \geq \lim_{M\to\infty} Q_M(y^M,\theta^M) \;=\; Q(\hat{\theta}^*),\quad \forall\theta \in S,$$

and so we obtain $\mathrm{Tr}\, V(\theta) \geq \mathrm{Tr}\, V(\hat{\theta}^*)$ for all $\theta \in S$. However, we proved earlier that under the assumptions of our theorem any

θ^* with this property also has the property that $V(\theta) \geq V(\hat{\theta}^*)$ for all $\theta \in S$.

Now it is straightforward to show that if for some $\bar{\theta}$, $V(\theta) \geq V(\bar{\theta})$, for all $\theta \in S$, then it is also true that $|V(\theta)| \geq |V(\bar{\theta})|$, for all $\theta \in S$. The demonstration is as follows. For a symmetric positive definite matrix V let $V^{1/2}$ denote the unique symmetric positive definite matrix such that, $V^{1/2}V^{1/2} = V$. Further let diag(V) denote any diagonal matrix of the eigenvalues of V. Now observe that $V(\bar{\theta}) > 0$ by the full rank assumption on $\{y^t;\ t \in Z\}$ with respect to $\{y^{t-1}, u^t;\ t \in Z\}$. Then $V(\theta) \geq V(\bar{\theta})$ yields $I \geq M(\theta,\bar{\theta})$, where $M(\theta, \bar{\theta}) \underline{\Delta}\ V^{-1/2}(\theta)V(\bar{\theta})V^{-1/2}(\theta)$, and hence $I \geq \text{diag}\ M(\theta,\bar{\theta})$. It follows that $1 \geq |\text{diag}\ M(\theta,\bar{\theta})| = |M(\theta,\bar{\theta})| = |V^{-1}(\theta)V(\bar{\theta})|$. Consequently

$$V(\theta) \geq V(\bar{\theta}),\quad \forall\theta \in S,\quad \Rightarrow\ |V(\theta)| \geq |V(\bar{\theta})|,\quad \forall\theta \in S\ . \tag{3.17}$$

To conclude this first part of the proof we use $V(\theta) \geq V(\hat{\theta}^*)$ for all $\theta \in S$ and each $\hat{\theta}^* \in S$ and (3.17), to obtain $|V(\theta)| \geq |V(\hat{\theta}^*)|$ for all $\theta \in S$ and each $\hat{\theta}^* \in S$ i.e., the determinant of $V(\theta)$ is minimized at each $\hat{\theta}^* \in S$.

For the second part of the proof we need to establish the familiar notion that asymptotically the MLE (pseudo-MLE in the present case) also minimizes the determinant of $V(\theta)$.

Observe that since condition A or B holds

$$L_N(y^N,\theta)\ \underline{\Delta}\ -\frac{2}{N}\log \ell_N(y^M,\theta) - p\log 2\pi$$

$$= \frac{1}{N}\sum_{t=1}^{N}\left(\log|\Sigma_0(\theta)| + e_t^T(\theta)\Sigma_0^{-1}(\theta)e_t(\theta)\right) \quad ,$$

where $e_t(\theta)$ is precisely the prediction error $y_t - \hat{y}_t(\theta)$ of the linear least squares predictor, computed using only (3.14), and $\Sigma_0(\theta)$ is the steady state error covariance computed from the Riccati equation or from the corresponding ARMA algorithm [59]. Using the Ergodic Theorem again we have

$$L_N(y^N,\theta) \to \log|\Sigma(\theta)| + \mathrm{Tr}[(Ee_0(\theta)e_0^T(\theta))\Sigma^{-1}(\theta)], \text{ a.s.}$$
$$\text{as} \quad N \to \infty. \qquad (3.18)$$

Denote the right hand side of (3.18) by $L(\theta)$ i.e., let $L(\theta) \triangleq \log|\Sigma(\theta)| + \mathrm{Tr}[V(\theta)\Sigma^{-1}(\theta)]$. Now consider any sample path (y,u) such that (3.18) holds and let $\{\theta^M\}$ be a convergent subsequence of $\{\theta^N\}$. Call the limit point $\tilde{\theta}$. It may be shown ([27], Appendix B) that the indicated convergence in (3.18) is uniform with respect to θ over S. Consequently by the defining property of the pseudo-MLE θ^N:

$$L(\theta) = \lim_{M\to\infty} L_M(y^M,\theta) \geq \lim_{M\to\infty} L_M(y^M,\theta^M) = L(\tilde{\theta}), \quad \forall\theta \in S. \quad (3.19)$$

Now let $\overset{\circ}{\theta}$ parameterize the optimum steady state predictor for the process y i.e., $\hat{y}_t(\overset{\circ}{\theta}) = Ey_t|I^t)$. Notice that $\mathrm{Tr}\, V(\theta) \geq \mathrm{Tr}\, V(\overset{\circ}{\theta})$ for all $\theta \in S$ and indeed by the arguments above $V(\theta) \geq V(\overset{\circ}{\theta})$ and $|V(\theta)| \geq |V(\overset{\circ}{\theta})|$ for all $\theta \in S$. Further $\overset{\circ}{\theta}$ clearly has the property that $V(\overset{\circ}{\theta}) = \Sigma(\overset{\circ}{\theta})$. We then have the following string of inequalities, where the third holds by virtue of (3.19) and the fourth is a standard inequality:

$$\begin{aligned}\log|V(\theta)| + p \geq \log|V(\overset{\circ}{\theta})| + p &= \log|\Sigma(\overset{\circ}{\theta})| + \mathrm{Tr}[V(\overset{\circ}{\theta})\Sigma^{-1}(\overset{\circ}{\theta})] \\ &\geq \log|\Sigma(\tilde{\theta})| + \mathrm{Tr}[V(\tilde{\theta})\Sigma^{-1}(\tilde{\theta})] \\ &\geq \log|V(\tilde{\theta})| + p, \qquad \forall\theta \in S \quad .\end{aligned}$$

From this inequality and $|V(\theta)| \geq |V(\hat{\theta}^*)|$, for all $\theta \in S$, which we established earlier, we deduce $|V(\tilde{\theta})| = |V(\hat{\theta}^*)|$ for each $\tilde{\theta}$, $\hat{\theta}^*$ in S.

Now from $V(\tilde{\theta}) \geq V(\hat{\theta}^*)$ it follows that $M(\hat{\theta}^*,\tilde{\theta}) \geq I$ and so $\mathrm{diag}\, M(\hat{\theta}^*,\tilde{\theta}) \geq I$. But $|V(\hat{\theta}^*)| = |V(\tilde{\theta})|$ implies $|M(\hat{\theta}^*,\tilde{\theta})| = 1$ and hence each eigenvalue of $\mathrm{diag}\, M(\hat{\theta}^*,\tilde{\theta})$ is 1. Consequently $V(\hat{\theta}^*) = V(\tilde{\theta})$ and so $\mathrm{Tr}\, V(\hat{\theta}^*) = \mathrm{Tr}\, V(\tilde{\theta})$. We conclude that, for almost all sample paths, all subsequential limits of the sequence of pseudo-maximum likelihood estimators lie in the subset of S which minimizes $\mathrm{Tr}\, V(\theta)$. But this subset is readily seen to be the set D which proves the desired result.

General Feedback Systems

At present it appears necessary to retain the assumptions A or B for Theorems 3.3 and 3.4 to hold. Otherwise it is possible to construct simple examples where the prediction error for (3.14) may be reduced by "trading-off" u for its corresponding representation in (3.15). It should be remarked that although the assumptions A and B are reasonable for technological systems this is not the case when the dynamics of the system are appreciably faster than the observation sampling rate. This situation occurs, for instance, in the quarterly sampling of socio-economic quantities for econometric modelling. In this case the variables involved are averages of the true variables over three months and one is reluctant to include artificial delays or artificial ignorance in the econometric model.

In [61] Caines and Wall proposed the identification of the innovations representation of the joint input-output process for econometric systems containing feedback. (See also Chan [58] for a discussion of this topic). Phadke and Wu [13] have used an analogous technique for the identification of a blast furnace under feedback control.

We close this section by remarking that it is possible to compute unique estimates for the feedforward and feedback loops from the joint innovations represent without any extra conditions. This is simply shown as follows: the unique innovations representation

$$\begin{bmatrix} y \\ u \end{bmatrix} = \begin{bmatrix} A & B \\ C & D \end{bmatrix} \begin{bmatrix} e_1 \\ e_2 \end{bmatrix}, \quad \begin{bmatrix} e_1 \\ e_2 \end{bmatrix} \sim N(0,\ I) \quad ,$$

yields, in an obvious notation,

$$\begin{bmatrix} A & B \\ C & D \end{bmatrix}^{-1} \begin{bmatrix} y \\ u \end{bmatrix} = \begin{bmatrix} P & Q \\ R & S \end{bmatrix} \begin{bmatrix} y \\ u \end{bmatrix} = \begin{bmatrix} e_1 \\ e_2 \end{bmatrix} .$$

Consequently

$$\begin{aligned} y &= - P^{-1}Qu + P^{-1}e_1 , \\ u &= - S^{-1}Ry + S^{-1}e_2 , \end{aligned} \qquad (3.21)$$

forms a unique representation of the feedforward and feedback loops of the form (3.14-15) and may be computed from an estimate of the innovations representation. The authors have not yet conducted any computational experiments using this method.

4. APPLICATIONS TO ECONOMIC, POWER AND PHYSIOLOGICAL SYSTEMS

In this section we present three areas for the application of the feedback detection techniques described earlier in this chapter. The first consists of an exercise to detect feedback between the post-war unemployment and gross domestic product time series for the U.K., the second describes the formulation of a feedback detection problem in power system identification and the third describes a problem in the analysis of electrophysiological signals recorded in the cat and human brain. The application of the techniques of Section 3 to the latter two problems are still at present under development.

A. *ECONOMIC APPLICATION*

Sixty-five values of the gross domestic product (GDP) and unemployment (UN) time series for the United Kingdom from the first quarter of 1955 (1955 I) to 1971 I were used in this experiment. The GDP data was at 1963 factor cost and seasonally adjusted; data for 1955-1967 is that quoted by Bray [62] from Treasury sources and that for 1968-1971 is taken from *Economic Trends*, July 1971, Table 4, pxii, col. 6. Unemployment denotes wholly unemployed, excluding school leavers and is seasonally adjusted: 1955-1970 II is that quoted by Bray, while 1970 III-1971 I is from *Trade and Industry*, No. 28, October, p. 200. Sixty-four pairs of normalized difference data were generated by computing $D_k = (d_k - d_{k-1})/d_k$, $k = 2,\ldots,65$, for the GDP and UN time series. Then two zero mean processes were obtained by subtracting from each series its average value over the sampling period.

Let $[\mathrm{UN}, \mathrm{GDP}]^T$ and $[\varepsilon_1, \varepsilon_2]^T$ denote the column vectors whose entries are the z-transforms of the UN and GDP time series and the joint innovations time series respectively.

We first assumed that feedback was present in the ordered pair of processes $\langle \mathrm{UN}, \mathrm{GDP} \rangle$. Using the method described in Section IIIA we found that the most acceptable innovations representation model for the joint observed time series was

$$\begin{bmatrix} \mathrm{UN} \\ \mathrm{GDP} \end{bmatrix} = \begin{bmatrix} \dfrac{1}{1-\underset{(\pm 0.073)}{0.337}z^1} & \dfrac{-\overset{(\pm 0.315)}{1.973}z^1-\overset{(\pm 0.079)}{3.812}z^2}{1-\underset{(\pm 0.331)}{0.098}z^1-\underset{(\pm 0.087)}{0.355}z^2+\underset{(\pm 0.182)}{0.113}z^3} \\ -\underset{(\pm 0.069)}{0.018}z^1 & \dfrac{1}{1+\underset{(\pm 0.119)}{0.178}z^1} \end{bmatrix} \begin{bmatrix} \varepsilon_1 \\ \varepsilon_2 \end{bmatrix} \tag{4.1}$$

$$\mathrm{Cov} \begin{bmatrix} \varepsilon_1 \\ \varepsilon_2 \end{bmatrix} = \begin{bmatrix} 5.848^2 & -2.901 \\ -2.901 & 1.054^2 \end{bmatrix} . \tag{4.2}$$

Second we assumed that the pair UN, GDP was feedback free. This yielded the model:

$$\begin{bmatrix} UN \\ GDP \end{bmatrix} = \begin{bmatrix} \dfrac{1}{1-0.312z^1} & \dfrac{\underset{(\pm 0.059)}{-2.013z^1} \quad \underset{(\pm 0.313)}{-3.197z^2}}{1-0.104z^1-0.332z^2+0.117z^3} \\ (\pm 0.284) & (\pm 0.079) \quad (\pm 0.134) \quad (\pm 0.065) \\ 0 & \dfrac{1}{1+0.116z^1} \\ & (\pm 0.094) \end{bmatrix} \begin{bmatrix} \varepsilon_1 \\ \varepsilon_2 \end{bmatrix} \tag{4.3}$$

$$\text{Cov} \begin{bmatrix} \varepsilon_1 \\ \varepsilon_2 \end{bmatrix} = \begin{bmatrix} 5.912^2 & -2.948 \\ -2.948 & 1.052^2 \end{bmatrix} . \tag{4.4}$$

We now carry out the tests described in Section 3 for the presence of feedback. Computing the s statistic for this example yields $s = 0.391$. But the 5% level for $F\ (112,\ 2)$ is 3.07 and consequently we accept the feedback free hypothesis (H_0) at this level. On applying the asymptotic likelihood ratio test we obtain $-2 \log \lambda = 0.382$ and since $\gamma_{.05} = 3.84$ for the χ^2 distribution with one degree of freedom we also accept H_0 by this test. Finally computing ξ in the manner described earlier we obtain a value of 0.068. Since $\xi < 3.84$ we see that the feedback hypothesis is rejected by the past χ^2 test.

In the light of these results we claim that the ordered pair of processes UN, GDP is feedback free. Consequently part 4 of the theorem in Section 2 permits the direct identification of the gross domestic product--unemployment relation. In [63] this is given as

$$[UN] = \frac{\underset{(\pm 0.5457)}{-2.3275} \quad \underset{(\pm 0.6124)}{+1.6396z^1}}{\underset{(\pm 0.0773)}{1-1.7094z^1} + \underset{(\pm 0.0658)}{0.8074z^2}} [GDP] + \frac{5.560}{\underset{(\pm 0.1268)}{1-0.224z^1}} [E]$$

where [E] denotes the z-transform of an N(0,1) Gaussian noise process.*

B. POWER SYSTEMS LOAD MODEL IDENTIFICATION

The identification of power systems presents several problems where we believe the techniques described earlier in this chapter may be usefully applied. The formulation of the power system identification problem presented in this subsection follows the analysis of Semlyen [64] and Sinha [65].

Suppose that we wish to construct a dynamical model for the electromechanical transients of a typical power system as seen at a load node when it is making small oscillations about an operating point. Let the symbols listed in the left hand column below represent the z-transforms of the discrete time series of the quantities which are described in the right hand column:

$[\Delta v_p]$	in phase component of incremental voltage
$[\Delta v_q]$	quadrature component of incremental voltage
$[\Delta \dot{v}_q]$	rate of change of quadrature component of incremental voltage
$[\Delta i_p]$	in phase component of incremental current
$[\Delta i_q]$	quadrature component of incremental voltage.

It is shown in [65] that a composite load consisting of passive elements, induction and synchronous motors possesses a linearized dynamical model of the form

$$\begin{bmatrix} \Delta i_p \\ \Delta i_q \end{bmatrix} = \begin{bmatrix} Y_{11} & Y_{12} & Y_{13} \\ Y_{21} & Y_{22} & Y_{23} \end{bmatrix} \begin{bmatrix} \Delta v_p \\ \Delta v_q \\ \Delta \dot{v}_q \end{bmatrix}, \tag{4.5}$$

where the dynamic admittance transfer function matrix $Y = [Y_{ij}]$

---

*The standard deviation of the numerator of the noise transfer function is not recorded in [64].

is a matrix of regular $(Y(\infty) < \infty)$ transfer functions whose denominator polynomial has degree not greater than 3.

Relatively little work has been done on the problem of dynamic load model identification from normal operating records. However two pieces of previous research should be mentioned. First, Stanton [66, 67], in 1963, collected normal operating data for a 50 megawatt (MW) turboalternator (synchronous generator) operating in parallel with an interconnected network having a capacity of 5000 MW. He computed a third order transfer function between the inputs Δi_p, Δi_q to the outputs Δv_p, Δv_q using a spectral factorization technique. The reader is referred to Jenkins and Watts [45, pp. 498-508] for an account of Stanton's results. Second, Deville and Schweppe [68] have described the modelling of a network as an equivalent real power admittance matrix using observations from power lines during periods when generation load changes were observed. (See also [69-70]).

It appears that dynamic load modelling is necessarily a closed loop system identification problem. Due to the dynamics of the load voltage variations in the network exterior to the load produce current variations at the load node. To these are added random current variations generated in the external network. This situation may be described by

$$\Delta i = Y \Delta v + \eta \quad , \tag{4.6}$$

where Δi and Δv are the vector random processes $\Delta i^T = [\Delta i_p, \Delta i_q]$ and $\Delta v^T = [\Delta v_p, \Delta v_q, \Delta \dot{v}_q]$ respectively, η is a two component random process independent of Δv and Y is the 2×3 dynamic admittance operator corresponding to the transfer function described earlier. Now it is reasonable to assume that any current variation at the load node will produce a voltage regulation effect via the nearest source system i.e., the nearest generating station. In addition various voltage variations produced randomly in the external network will be detected at the measurement bus at the load node. This may be expressed by the equation

$$\Delta v = Z\, \Delta i + \zeta \tag{4.7}$$

where ζ is a three component random process independent of Δi and Z is a (3×2) dynamic impedance operator.

Clearly the equations (4.6) and (4.7) describe a feedback system. Of course for small loads one would expect that the current to voltage feedback effect described in (4.7) would be insignificant.

In contrast to economics, long records of operating data are available for power systems. Consideration of the highest frequencies of voltage and current oscillations in a power system due to its electromechanical components leads to a Nyquist sampling frequency of 10-20 $\sec^{-1}$. It is quite feasible to collect measurements on Δi and Δv at this rate. Further, it is known [70] that measurements taken at a load node will be stationary over periods of about 5 minutes, although presumably this will not be the case for data collected over significantly longer periods.

As a result of the considerations above, the application of the techniques of Section 3 are at present under investigation for the detection of feedback in power systems and the identification of subsections of power systems exhibiting closed loop behavior.

C. *EEG DATA ANALYSIS*

1. Gersch *et al.* [71-73] have considered two interesting problems arising in physiological time series analysis: (1) the identification of the site of an epileptic focus in the cat brain during a generalized seizure and (2) the examination of human alpha rhythm data for the identification of driving α rhythm generators.

Gersch [72] has proposed that a particular time series z is causal to a given set of time series $(x_1,\ldots,x_n)$ if (i) all the observed time series have pairwise significant spectral coherence functions over the frequency domain of interest and (ii) the partial spectral coherence between all signal pairs conditioned on z is not significantly different from zero over this same interval. (See e.g., [45] for treatments of spectral coherence functions). Gersch has given examples where z would be called causal relative to (x_1, x_2) by the criteria of Section 2 ($\langle y, (x_1, x_2)\rangle$ feedback free) but not by the spectral coherence criterion just introduced. We give a characterization of this notion of causality in Subsection C2 below.

The following is a brief outline of the results obtained by Gersch *et al.* [71-73] in two sets of experiments. We present these results because they forcefully illustrate the interest in establishing driving relationships between certain physiological time series and because the data from these experiments is presently being reanalysed by the methods described in Section 3.

In the first set of experiments [71, 72] epileptic seizures were induced in a cat by daily 5-second electrical stimulation in the piriform cortex. Recordings of six simultaneous channels of activity were taken from bipolar electrodes implanted into sites deep in the brain of normal, ictal (seizure) and postictal activity. After several weeks of such stimulation, each stimulation produced epileptic spike and wave activity. During the ictal phase, data from the septal area, piriform cortex, mesencephalic reticular formation, putamen, nucleus lateralis posterior of the thalamus, and the motor cortex were simultaneously recorded. 800 data points were collected by sampling each channel digitally at 10 millisecond intervals for 8 seconds. By autoregressive model fitting spectral coherence functions and partial spectral coherence functions were computed for the 20 distinct triples of data sets taken from the six simultaneous channels. An examination of the resulting spectral functions showed that the piriform cortex exclusively was driving the putamen, the reticular formation and the nucleus lateralis posterior of the thalamus.

Consequently it was concluded that the piriform cortex was the site of the epileptic focus, i.e., that region of the brain driving the epileptic seizure, out of the six possible sites in this analysis.

In the second set of experiments [73] ten channels of scalp EEG (electroencephalogram) data were recorded from a relaxed human subject whose eyes were closed. The recording was obtained by bipolar chaining of the electrodes which were placed according to the standard clinical "10-20" system. Portions of the recorded data (of duration 9.6 seconds) which contained alpha rhythm activity were selected and digitized at a rate of 100 samples/second. The spectral functions for all triples of data channels were computed as described previously. Analysis of the resulting spectra around the alpha frequency of 9 hertz showed that two channels in the parietal-occipital region in the right hemisphere drove all other channels in the right hemisphere. Similarly two driving channels were detected in the same region on the right hand side. Furthermore it was shown that right hemisphere alpha activity drove the left hemisphere alpha activity. For more details and further analysis of this experiment the reader is referred to [73].

2.* The original criterion proposed by Gersch [72] for a scalar process u to drive two scalar processes y_1 and y_2 at a frequency ω was that $W^2_{y_1y_2;u}(\omega) = 0$, where

$$\Psi_{y_iy_i\cdot u}(\omega) \triangleq \Psi_{y_iy_i}\left(1 - \frac{|\Psi_{y_1y_2}(\omega)|^2}{\Psi_{y_1y_1}(\omega)\Psi_{y_2y_2}(\omega)}\right), \quad i = 1,2,$$

---

*The first author of this chapter would like to acknowledge many constructive conversations on this subsection with Will Gersch.

$$\Psi_{y_1y_2\cdot u}(\omega) \triangleq \Psi_{y_1y_2}\left(1 - \frac{\Psi_{y_1u}(\omega)\Psi_{uy_2}(\omega)}{\Psi_{y_1y_2}(\omega)\Psi_{uu}(\omega)}\right) ,$$

$$W^2_{y_1y_2\cdot u}(\omega) \triangleq \frac{|\Psi_{y_1y_2\cdot u}(\omega)|^2}{\Psi_{y_1y_1\cdot u}(\omega)\Psi_{y_2y_2\cdot u}(\omega)} .$$

We make the following obvious extension to multivariate processes:

DEFINITION 2: *Let* y *and* u *be* p *and* q *component jointly full rank multivariate stationary zero mean stochastic processes which jointly satisfy the conditions on* ζ *in Section* 2. *Then* u *drives* y *at the frequency* ω *if and only if*

$$\Psi_{y\cdot u}(\omega) \triangleq \Psi_{yy}(\omega) - \Psi_{yu}(\omega)\Psi^{-1}_{uu}(\omega)\Psi^*_{yu}(\omega) = \Psi_{diag}(\omega) ,$$

where $\Psi_{diag}(u)$ *denotes a* $p \times p$ *spectral density matrix which has zero off-diagonal entries at the frequency* ω.

We now have the following

THEOREM 4.1 (Caines): *(i) The full rank process* u *drives the process* y *at* ω *if and only if the process* Mu *drives the process* y *at* ω *for all stationary operators* M. *(NOTE: the operator* M *need not be non-anticipative).*

(ii) Given any pair of stationary processes (y,u) *and any* ω *there exists a constant matrix* K_ω *such that* u *drives* $K_\omega y$ *at* ω.

(iii) Given any pair of stationary processes (y,u), *where is full rank, there exists a non-anticipative operator* K *such that* u *drives* K_y *at all frequencies* .

Proof. The first part is established by direct substitution in the definition of driving. The only technicality is the invertibility of $\Psi_{MuMu}(\omega)$ and this is guaranteed by the conditions on (y, u).

Let $\underline{\underline{H}}_u$ denote the Hilbert space generated by all values of the process u. Let $[Ly]_t$ denote the orthogonal projection $y_t|\underline{\underline{H}}_u$ of y_t on $\underline{\underline{H}}_u$. Let $v_t \underline{\Delta} y_t - [Ly]_t$. Clearly v is orthogonal to $\underline{\underline{H}}_u$ and hence the processes u and v are orthogonal. This yields the representation

$$y_t = \sum_{i=-\infty}^{\infty} L_i u_{t-i} + v_t , \qquad t \in Z , \tag{4.8}$$

with the processes u and v orthogonal.

Let $\Psi_{vv}(\omega)$ denote the spectral density matrix of the process v at ω. Then using (4.8) we obtain

$$\Psi_{y \cdot u} = \Psi_{yy}(\omega) - \Psi_{yu}(\omega)\Psi_{uu}^{-1}(\omega)\Psi_{yu}^{*}(\omega) = \Psi_{vv}(\omega) .$$

Let $K_\omega^{-1}(K_\omega^{-1})^* = \Psi_{vv}(\omega)$. Then clearly $\Psi_{K_\omega y \cdot u} = I$ and we conclude u drives $K_\omega y$. This proves (ii). By taking $K^{-1}(z)$ as the stable inverse stable factor of $\Psi_{vv}(z)$ we obtain (iii).

It would seem from (ii) and (iii) of the proposition above that Definition 2 is not as strong a characterization of driving between stochastic processes as it might at first appear. However, there still seems to be scope for more analysis and investigation of these various formulations of driving between time series.

5. CONCLUSION

In this chapter we have presented a conceptualization of the notion of feedback between stationary stochastic processes and developed a sequence of equivalent formulations of this idea. These equivalent notions give rise to a set of techniques for the detection of feedback. Further, the theory presented here gives a useful framework for the discussion and results we present on the identification of closed loop systems. We have also given applications of these ideas in the areas of economics, power systems and physiology respectively. It is believed there still exist many exciting potential applications in these and other areas where judgements must be made concerning complex systems observed under normal operating conditions.

REFERENCES

1. Granger, C. W. J., "Economic Processes Involving Feedback," *Information and Control, Vol. 6,* pp. 28-48, 1963.

2. Granger, C. W. J. and M. Hatanaka, *Spectral Analysis of Economic Time Series,* Princeton University Press, Princeton, New Jersey, 1964.

3. Granger, C. W. J., "Investigating Causal Relations by Econometric Models and Cross-Spectral Methods," *Econometrica, Vol. 37, No. 3,* July 1969.

4. Granger, C. W. J., Proc. IFAC/IFORS International Conference on Dynamic Modelling and Control of National Economies, held at the University of Warwick, Conventry, England, July 1973.

5. Sims, C. A., "Money, Income and Causality," *American Economic Review, Vol. 62,* pp. 540-552, 1972.

6. Chan, C. W., "The Identification of Closed Loop Systems with Application to Econometric Problems," M.Sc. Dissertation, University of Manchester Institute of Science and Technology, 1972.

7. Caines, P. E. and C. W. Chan, "Feedback between Stationary Stochastic Processes," University of Toronto Control Systems Report No. 7421. To appear *IEEE Trans. Automatic Control, Vol. AC-20, No. 4,* August 1975. (A version of this paper was presented at the Conference on Information Sciences and Systems, The Johns Hopkins University, April 1975).

8. Wall, K. D., "An Application of Simultaneous Estimation to the Determination of Causality between Money and Income," Discussion Paper No. 8, Program for Research into Econometric Methods, University of London, April 1974.

9. Bohlin, T., "On the Problem of Ambiguities in Maximum Likelihood Identification," *Automatica, Vol. 7,* pp. 199-210, 1971.

10. Gustavsson, I., L. Lung, T. Söderström, "Identification of Linear Multivariable Process Dynamics Using Closed Loop Experiments," Report No. 7401, January 1974, Lund Institute of Technology, Sweden.

11. Gustavsson, I., L. Ljung, T. Söderström, "Identification of Linear Multivariable Systems Operating under Linear Feedback Control," *IEEE Trans. Automatic Control, Vol. AC-19, No. 6,* December 1974.

12. Ljung, L., "On Consistency for Prediction Error Identification Methods," Report 7405, Division of Automatic Control, Lund Institute of Technology, March 1974.

13. Phadke, M. S. and S. M. Wu, "Identification of Multi-Input, Multi-Output Transfer Function and Noise Model of a Blast Furnace from Closed-Loop Data," *IEEE Trans. Automatic Control, Vol. AC-19, No. 6,* pp. 944-951, December 1975.

14. Wellstead, P. E., "Least Squares Identification of Systems Involving Feedback," University of Manchester Institute of Science and Technology, Control Systems Centre Report No. 261, August 1974.

15. Wiener, N., P. Masani, "The Prediction Theory of Multivariate Stochastic Processes," V, *Acta Mathematica, Part I, 98,* pp. 111-150, 1957; *Part II, 99,* pp. 93-137, 1958.

16. Cramér, H. and M. R. Leadbetter, *Stationary and Related Stochastic Process,* John Wiley, New York, 1967.

17. Wold, H., *A Study in the Analysis of Stationary Time Series,* (2 ed.) Almquist and Wiksell, Stockholm, 1954.

18. Youla, D. C., "On the Factorization of Rational Matrices," *IRE Trans. Information Theory, IT-7(3),* pp. 172-189, July 1961.

19. Davenport, W. B. and W. L. Root, *An Introduction to the Theory of Random Signals and Noise,* McGraw-Hill, 1958.

20. Whittle, P., *Prediction and Regulation,* Van Nostrand, 1963.

21. Wald, A., "Note on the Consistency of the Maximum Likelihood Estimate," *Ann. Math. Stat., Vol. 20,* pp. 595-601, 1949.

22. Wilks, S. S., *Mathematical Statistics,* John Wiley, 1962.

23. Åström, K. J., T. Bohlin, S. Wensmark, "Automatic Construction of Linear Stochastic Dynamic Models for Stationary Industrial

Processes with Random Disturbances Using Operating Records," Rep. TP 18.150, IBM Nordic Laboratories, Lindingo, Sweden, 1965.

24. Åström, K. J. and T. Bohlin, "Numerical Identification of Linear Dynamic Systems from Normal Operating Records," Proc. IFAC Symp. On the Theory of Self-Adaptive Control Systems 1965, Plenum Press, 1966.

25. Kendall, M. G., A. Stuart, *The Advanced Theory of Statistics,* Vol. 2, Hafner Publishing Co., New York, 1967.

26. Caines, P. E. and J. Rissanen, "Maximum Likelihood Estimation of Parameters in Multivariate Gaussian Stochastic Processes," *IEEE Trans. Information Theory, Vol. IT-20, No. 1,* Jan. 1974.

27. Rissanen, J. and P. E. Caines, "Consistency of Maximum Likelihood Estimators for ARMA Processes," Submitted to *Annals of Statistics.*

28. Popov, V. M., "Some Properties of the Control Systems with Irreducible Matrix-Transfer Functions," *Lecture Notes in Math.,* No. 144, Springer-Verlag, Berlin, 1969.

29. Caines, P. E., "The Paramter Estimation of State Variable Models of Multivariable Linear Systems," Proc. Fourth UKAC Control Convention, Manchester, *IEEE Conference Publication No. 78,* September 1971.

30. Mayne, D. Q., "A Canonical Model for Identification of Multivariable Linear Systems," *IEEE Trans. Automat. Contr.* (Corresp.), Vol. AC-17, No. 5, pp. 728-729, 1972.

31. Glover, K. and J. C. Willems, "Parameterization of Linear Dynamical Systems: Canonical Forms and Identifiability," *IEEE Trans. Automat. Contr., Vol. AC-19, No. 6,* pp. 640-645, December 1974.

32. Denham, M. J., "Canonical Forms for the Identification of Multivariable Linear Systems," *IEEE Trans. Automatic Control, Vol. AC-19, No. 6,* pp. 646-656, December 1974.

33. Wall, K. D. and J. H. Westcott, "Macroeconomic Modelling for Control," *IEEE Trans. Automat. Contr., Vol. AC-19, No. 6,* pp. 862-873, December 1974.

34. Gupta, N. K. and R. K. Mehra, "Computational Aspects of Maximum Likelihood Estimation and Reduction in Sensitivity Function Calculations," *IEEE Trans. Automatic Control, Vol. AC-19, No. 6,* pp. 774-783, December 1974.

35. Cummings, R. A., "The Computer Identification of Discrete Time Linear Multivariable Stochastic Systems," M.Sc. Dissertation, University of Manchester Institute of Science and Technology, December 1971.

36. Kalman, R. E., P. L. Falb, M. A. Arbib, *Topics in Mathematical System Theory,* McGraw-Hill, New York, 1969.

37. Caines, P. E., "The Minimal Realization of Transfer Function Matrices," *Int. J. Control, Vol. 13, No. 3,* pp.529-547, 1971.

38. Dickinson, B. W., T. Kailath, M. Morf, "Canonical Matrix Fraction and State-Space Descriptions for Deterministic and Stochastic Linear Systems," *IEEE Trans. Automatic Control, Vol. AC-19, No. 6,* pp. 656-667, December 1974.

39. Preston, A. J. and K. D. Wall, "An Extended Identification Problem for State Space Representation of Econometric Models," Discussion Paper No. 6, June 1973. Presented at IFAC/IFORS International Conference on Dynamic Modelling and Control of National Economies, held at the University of Warwick, Conventry, England, July 1973.

40. Akaike, H., "Autoregressive Model Fitting for Control," *Ann. Inst. Statistic. Math., Vol. 23,* pp. 163-180, 1971.

41. Akaike, H., "Information Theory and an Extension of the Maximum Likelihood Principle," presented at 2nd Int. Symp. Information Theory, Tsahkadsor, Armenian SSR, Sept. 2-8, 1971, also in *Problems of Control and Information Theory,* Hungary, Akademiai Kiado.

42. Akaike, H., "Use of an Information Theoretic Quantity for Statistical Model Identification," in Proc. 5th Hawaii Int. Conf. System Sciences, pp. 249-250, 1972.

43. Parzen, E., "Some Recent Advances in Time Series Modelling," *IEEE Trans. Automatic Control, Vol. AC-19, No. 6,* pp. 716-723, December 1974.

44. Chan, C. W., C. J. Harris, P. E. Wellstead, "An Order Testing Criterion for Mixed Autoregressive Moving Average Processes," *Int. J. Control, Vol. 20, No. 5,* pp. 817-834, 1974.

45. Jenkins, G. M. and D. G. Watts, *Spectral Analysis and its Applications,* Holden-Day, San Francisco, 1968.

46. Box, G. E. P. and G. M. Jenkins, *Time Series Analysis, Forecasting and Control,* Holden-Day, San Francisco, 1970.

47. Hannan, E. J., *Time Series Analysis*, Methuen & Co. Ltd., London, 1960.

48. Hannan, E. J., *Multiple Time Series*, John Wiley, New York, 1970.

49. Åström, "On the Achievable Accuracy in Identification Problems," IFAC Symposium on Identification in Automatic Control Systems, Prague, Czechoslovakia, 1967.

50. Anderson, T. W., *Introduction to Multivariate Statistical Analysis*, John Wiley, New York, 1958.

51. Anderson, B. D. O., "An Algebraic Solution to the Spectral Factorization Problem," *IEEE Trans. Automatic Control, Vol. AC-12*, pp. 410-414, August 1967.

52. Faurre, P., J. P. Marmorat, "Un Algorithme de Réalisation Stochastic," *C. R. Acad. Sc. Paris, 268, Ser. A.*, pp. 978-981, April, 1969.

53. Faurre, P. "Realisations Markoviennes de Processus Stationaires," IRIA Rapport de Recherche No. 13, March 1975.

54. Rissanen, J., "Recursive Identification of Linear Systems," *SIAM J. Control, Vol. 10, No. 2*, pp. 252-264, 1972.

55. Rissanen, J., and T. Kailath, "Partial Realization of Stochastic Systems," *Automatica, Vol. 8*, pp. 389-396, 1972.

56. Goodhart, D., D. Williams, and D. Gowland, "Money, Income and Causality: The U.K. Experience," submitted to American Economic Review.

57. Akaike, H., "On the Use of a Linear Model for the Identification of Feedback Systems," *Ann. Inst. Statist. Math., Vol. 20, No. 3*, 1968.

58. Chan, C. W., "The Identification of Linear Stochastic Feedback Systems," Control Systems Centre Report No. 247, University of Manchester Institute of Science and Technology, June 1974.

59. Rissanen, J. and L. Barbosa, "Properties of Infinite Covariance Matrices and Stability of Optimum Predictors," *Information Sciences, Vol. 1*, pp. 221-236, 1969.

60. Bucy, R. S. and P. D. Joseph, *Filtering for Stochastic Processes with Applications to Guidance*, Interscience, 1968.

61. Caines, P. E. and K. D. Wall, "Theoretical Foundations for Methods of Estimation and Control in Economic Policy Optimization," Chapter 13 in *Modelling the Economy*, ed. G. A. Renton, Crane Russak and Co. Inc., New York, 1975.

62. Bray, J., "Dynamic Equations for Econometric Forecasting with the G.D.P.-Unemployment Relation and the Growth of G.D.P. in the U.K. as an Example," *J. Royal Stat. Soc., Series A, Vol. 134, Part 2,* 1971.

63. Wall, K. D., A. J. Preston, J. W. Bray and M. H. Pestan, "Estimates for a Simple Control Model of the U.K. Economy," Proc. SSRC Conf., London, July 1972. Chapter 14 in *Modelling the Economy,* publ. by Crane Russak & Co. Inc., New York, Dec. 1974.

64. Semlyen, A., "Identification of Power System Components: Methods of Measurements," *IEECE Conf. Digest,* Toronto, pp. 14-15, 1973.

65. Sinha, S., "Load Modelling and Identification in Power Systems," M.A.Sc. Thesis, University of Toronto, Jan. 1975.

66. Stanton, K. N., "Measurement of Turbo-Alternator Transfer Functions Using Normal Operating Data," *Proc. Inst. Electr. Engrs., Vol. 110, No. 11,* 1963.

67. Stanton, K. N., "Estimation of Turbo-Alternator Transfer Functions Using Normal Operating Data," *Proc. Inst. Electr. Engrs., Vol. 112, No. 9,* 1965.

68. Deville, T. G. and F. C. Shweppe, "On-Line Identification of Interconnected Network Equivalents," presented at IEEE PES Summer Power Meeting 1972.

69. Price, W. W., F. C. Schweppe, E. M. Gulachenski, R. F. Silva, "Maximum Likelihood Identification of Power System Dynamic Equivalents," Paper No. THP4.2, pp. 579-586, *Proc. 1974 Decisiom and Control Conference,* Phoenix, Arizona.

70. Price, W. W., D. N. Ewart, E. M. Gulachenski, R. F. Silva, "Dynamic Equivalents from On-Line Measurement," presented at IEEE PES Winter Meeting, New York, January 26-31, 1975.

71. Gersch, W., G. V. Goddard, "Epileptic Focus Location: Spectral Analysis Method," *Science, Vol.169,* pp. 701-702, 14 Aug. 1970.

72. Gersch, W., "Causality or Driving in Electrophysiological Signal Analysis," *Math. Biosciences, Vol. 14,* 177-196, 1972.

73. Gersch, W., A. Midkiff, B. Tharp, "Alpha Rhythm Generators," Proc. Fifth International Conference in System Sciences, Computers in Biomedicine Supplement, Honolulu, Hawaii, Jan. 1972.

74. Dickinson, B. W., M. Morf, T. Kailath, "A Minimal Realization Algorithm for Matrix Sequences," *IEEE Trans. Automatic Control, Vol. AC-19, No. 1,* pp. 31-38, Feburary 1974.

75. Caines, P. E., S. Sinha, "An Application of the Statistical Theory of Feedback to Power System Identification," Proceedings of *IEEE Conference on Decision and Control,* Houston, Texas, December 1975.

SOME PROBLEMS IN THE IDENTIFICATION AND ESTIMATION OF CONTINUOUS TIME SYSTEMS FROM DISCRETE TIME SERIES

P. M. Robinson
Harvard University
Cambridge, Massachusetts

1. INTRODUCTION

We shall be concerned with the identification and estimation of open-loop multi-input, multi-output stochastic systems of the form

$$\underline{y}(t) = \int_{-\infty}^{\infty} \underline{h}(s)\underline{u}(t-s)\ ds + \underline{w}(t),\ -\infty < t < \infty . \qquad (1)$$

In this system, $y(t)$ is a $q \times 1$ vector of outputs; $\underline{u}(t)$ is an $r \times 1$ vector of inputs; $\underline{w}(t)$ is a $q \times 1$ vector of system noises; $\underline{h}(s)$ is a $q \times r$ matrix of impulse response functions, such that

$$\int_{-\infty}^{\infty} \|\underline{h}(s)\| \, ds < \infty , \tag{2}$$

where $\|\underline{h}(s)\|^2$ is the greatest eigenvalue of $\underline{h}(s)\underline{h}(s)^\tau$, $\underline{h}(s)^\tau$ being the transpose of $\underline{h}(s)$. Sometimes one would regard (1) as a solution of a stochastic functional equation system of the form

$$\int_{-\infty}^{\infty} \underline{h}_1(s)\underline{y}(t-s) \, ds = \int_{-\infty}^{\infty} \underline{h}_2(s)\underline{u}(t-s) \, ds + \underline{x}(t) , \tag{3}$$

where $\underline{x}(t)$ is a new noise vector related to $\underline{w}(t)$, $\underline{h}_1(s)$ is a $q \times q$ matrix and $\underline{h}_2(s)$ is a $q \times r$ matrix. We shall be principally concerned with (1), but under appropriate assumptions one can sometimes estimate (3) directly.

The system (1) is a stochastic linear approximation to the mechanism that generates $\underline{y}(t)$ from $\underline{u}(t)$ and $\underline{w}(t)$. Knowledge of $\underline{h}(s)$ is likely in practice to be severely limited, but it is identifiable on the basis of records such as $\{\underline{y}(t);\ 0 < t < T\}$, $\{\underline{u}(t);\ 0 < t < T\}$, and knowledge of the relationship between $\underline{u}(t)$ and $\underline{w}(t)$ (see Akaike, 1). When $\underline{h}(s)$ is expressed as a uniquely defined function of a finite number of parameters, one can estimate the latter by, in effect, a mapping from the estimates of $\underline{h}(s)$. Parzen [2] has considered the regression analysis of continuous time series in a wide context. The properties of estimators of the regression matrix $\underline{B}$, arising when

$$\underline{h}(s) = \underline{B}\delta(s), \ -\infty < s < \infty$$

(where $\delta(s)$ is the Dirac delta function

$$\delta(s) = \infty, \quad s = 0; \quad = 0, \quad s \neq 0)$$

have been studied by Hannan [3], Kholevo [4], Heble [5]. In this connection, we mention also work by Bartlett [6], Ibragimov [7], Dzhaparidze [8,9] on systems that are explicitly closed-loop, so that $\underline{h}_2(s) \equiv \underline{0}$ in (3), and the parameters of $\underline{h}_1(s)$ are to be determined. A case of interest is that where (3) is a linear differential equation with constant coefficients and driven by white noise, so that

$$\underline{h}_1(s) = \sum_j \underline{a}_j \frac{d^j\delta(s)}{ds^j} \quad , \; E\{\underline{x}(t)\underline{x}(t+s)^\tau\} \propto \delta(s) \quad ,$$

where "$E\{\cdot\}$" is an expectation over the space of all possible realizations of $\underline{x}(t)$. We shall be exclusively concerned, however, with systems that include an observable input.

The use of continuous time records raises problems, however. In the first place, such data may well be expensive to collect and to handle. In the second place, no man-made measuring or displaying device would be sensitive to oscillations of arbitrarily high frequency, and may serve as a low-pass filter; moreover, the signal-to-noise ratio may be very low at high frequencies. Third, some of the estimates for finite parameter systems that have been suggested seem difficult to compute. A sensible alternative approach would involve sampling the continuous record at positive equally-spaced intervals Δ, choosing Δ to be as large as possible, consistent with the aim of losing no more than a negligible quantity of information, so that the spectral density of the discrete record appears to closely approximate that of the continuous record. Under the assumption that $\underline{u}(t)$ and $\underline{w}(t)$ are incoherent wide-sense stationary stochastic processes, one can then generally make use of the so-called *cross-spectral method*, based principally on the easily computed estimates of the spectral density matrix of $\{\underline{u}(t);\ t = 0, \pm\Delta, \ldots\}$, and the cross-spectral density matrix of $\{\underline{u}(t);\ t = 0, \pm\Delta, \ldots,\}$ and $\{\underline{y}(t);\ t = 0, \pm\Delta, \ldots,\}$. The estimation of the Fourier transform of $\underline{h}(s)$ by this method has been considered by Akaike and Yamanouchi [10], Jenkins [11]. The estimation problem in the context of a finite-parameter $\underline{h}(s)$ has been considered by Hannan and Robinson [12], Robinson [13].

In many practical situations, however, the discretely sampled $\underline{y}(t)$ and $\underline{u}(t)$ are all that one has to start off with, because continuous sampling is exhorbitantly expensive or else entirely infeasible. The latter situation often obtains outside the confines of a laboratory experiment. Indeed, in many circumstances one's choice of Δ is severely limited by institutional factors so that one has no prior grounds for believing that it is small relative to the dynamics of the continuous processes. To some extent one can still justify use of the cross-spectral method, for one can often argue that only a small proportion of the spectral mass is likely to lie beyond Nyquist frequency. On the other hand, there are undoubtedly many practical situations in which there are factors causing fluctuations that cannot be detected by the sampling schedule available. Moreover, there is a need to reconcile the cross-spectral method with the importance in the theory of stochastic processes of spectral densities that never vanish, such as spectra of "regular" processes. Closed-loop systems are frequently modelled in terms of such processes, and it would be of interest to investigate how the cross-spectral method performs when $\underline{u}(t)$ is of this type, and whether it can be easily modified to produce better results in such cases.

In the sequel we shall review and discuss the use of the cross-spectral method in estimating frequency and time domain characteristics of both unparameterized and parameterized systems of the form (1); we shall investigate the biases that may occur in the estimates, or the identification error, when $\underline{u}(t)$ has spectral mass beyond Nyquist frequency; we shall suggest modifications that should produce less biased estimates under such circumstances. There are of course other circumstances in which the cross-spectral method is invalid or liable to produce poor results, and requires further study and modification. We briefly describe below some problems which are of practical concern.

A. *THE SPECTRAL ESTIMATION PROBLEM*

Many different spectral estimators have been suggested, having different bias and variance properties, and possibly producing rather different results in practice. The methods most commonly used nowadays for reasonably long series are based on the fast Fourier transform. The ones quoted in the following section use the Fourier-transformed data in a direct fashion, but it may be preferable to use estimators which are based directly on the sample autocovariances, and in large samples these are most cheaply computed by forming and fast Fourier transforming the periodograms $\underline{I}_{uu}$, $\underline{I}_{yy}$, $\underline{I}_{yu}$ introduced below. Each method in any case requires the choice of a "bandwidth" parameter, which crucially affects the bias and variance of the estimator. As the record length increases, the bandwidth must decrease at a suitable rate, in order that the estimators should have neat asymptotic properties. Little is known of the finite-sample properties of spectral estimators.

B. *LACK OF TIME-INVARIANCE*

The property of time-invariance of the system (1), is one that may not be unreasonable in the context of many natural and man-made phenomena, at least over relatively long periods of time that extend beyond the span of the available record. Moreover, certain types of departure from the underlying assumption that the processes are wide-sense stationary seem acceptable in the context of the cross-spectral method. (See Granger and Hatanaka [14], Hannan [15]. However, some types of non-stationarity that one comes across in practice may seriously invalidate the method, and in applications where a relatively sparse record is spread over a relatively long period it would often be much more realistic to allow the kernel in (1) to have the more general form $\underline{h}(s,t)$, or perhaps allow the parameters of the system to be themselves time-varying stochastic processes.

C. PROCESS NOISE

In (1) measurement noise is incorporated as a component of the system noise, $\underline{w}(t)$, but it may well be that instead of observing $\underline{u}(t)$ we observe

$$\underline{v}(t) = \underline{u}(t) + \underline{z}(t) \quad ,$$

or, more generally,

$$\underline{v}(t) = \int_{-\infty}^{\infty} \underline{g}(s)\underline{u}(t-s) \, ds + \underline{z}(t) \quad ,$$

where $\underline{z}(t)$ is an $r \times 1$ process noise vector and $\underline{g}(s)$ is a square matrix. Use of $\underline{v}(t)$ instead of $\underline{u}(t)$ in the cross-spectral method will then produce biased estimates. This is a version of the "errors-in-variables" problem.

D. FEEDBACK

The system (1) is not necessarily "realizable", in the sense that $\underline{h}(s)$ need not vanish over the entire negative real line. (Indeed, the cross-spectral method generally produces an "unrealizable" estimate even when the system itself is realizable.) In the unrealizable case, the implication is that $\underline{y}(t)$ is related to future values of $\underline{u}(t)$, as well as past ones. This could be interpreted to mean that causality is bidirectional, so that there is some feedback from $\underline{y}(t)$ to $\underline{u}(t)$, in which case a more suitable model would involve supplementing (1) by

$$\underline{u}(t) = \int_{-\infty}^{\infty} \underline{k}(s)\underline{y}(t-s) \, ds + \underline{z}(t),$$

where $\underline{k}(s)$ is $r \times q$ and $\underline{z}(t)$ is a $r \times 1$ noise vector, and possibly $\underline{h}(s) \equiv \underline{0}$, $\underline{k}(s) \equiv \underline{0}$, $s < 0$. As is well known, it is then appropriate to represent $\underline{u}(t)$ and $\underline{y}(t)$ as the joint output of a closed-loop system (see Akaike, 1) with unobservable input $\underline{w}(t)$ and $\underline{z}(t)$. Then the cross-spectral method would produce biased estimates, and indeed $\underline{h}(s)$ would become unidentifiable unless a good deal of further knowledge is available.

While bearing in mind these four important problems, we shall concentrate here on the discrete-continuous time problem, because it is also important and seems to have received relatively little attention.

2. IDENTIFICATION AND ESTIMATION OF FREQUENCY DOMAIN CHARACTERISTICS

If $\underline{y}(t)$, $\underline{u}(t)$, $\underline{w}(t)$ are jointly wide-sense stationary stochastic processes with means

$$\underline{\mu}_y = E\{\underline{y}(t)\} , \quad \underline{\mu}_u = E\{\underline{u}(t)\} , \quad \underline{\mu}_w = E\{\underline{w}(t)\} = 0 .$$

The matrix

$$\underline{C}(s) = \begin{bmatrix} \underline{C}_{yy}(s) & \underline{C}_{yu}(s) & \underline{C}_{yw}(s) \\ \underline{C}_{uy}(s) & \underline{C}_{uu}(s) & \underline{C}_{uw}(s) \\ \underline{C}_{wy}(s) & \underline{C}_{wu}(s) & \underline{C}_{ww}(s) \end{bmatrix} = E\left\{ \begin{bmatrix} \underline{y}(t)-\underline{\mu}_y \\ \underline{u}(t)-\underline{\mu}_u \\ \underline{w}(t) \end{bmatrix} \begin{bmatrix} \underline{y}(t)-\underline{\mu}_y \\ \underline{u}(t)-\underline{\mu}_u \\ \underline{w}(t) \end{bmatrix}^{\tau} \right\},$$

is defined for $-\infty < s < \infty$. We have $\underline{C}(-s) = \underline{C}(s)^{\tau}$. We assume $\underline{C}_{wu}(s) \equiv \underline{0}$. (Cf. our discussion of process noise in Section 1.) Appropriate operations on (1) then yield the integral equations

$$\underline{C}_{yu}(s) = \int_{-\infty}^{\infty} \underline{h}(t)\underline{C}_{uu}(s-t)\, dt , \tag{4}$$

$$\underline{C}_{yy}(s) = \int_{-\infty}^{\infty}\int_{-\infty}^{\infty} \underline{h}(t)\underline{C}_{uu}(s+t-r)\underline{h}(r)^{\tau}\, dt\, dr + \underline{C}_{ww}(s) , \tag{5}$$

$-\infty < s < \infty$. We assume the existence of the spectral density matrix

$$\underline{P}(f) = \int_{-\infty}^{\infty} \underline{C}(s)e^{-2\pi i f s}\, ds, \quad -\infty < s < \infty ,$$

and partition it corresponding to our partition of $\underline{C}(s)$:

$$\underline{P}(f) = \begin{bmatrix} \underline{P}_{yy}(f) & \underline{P}_{yu}(f) & \underline{P}_{yw}(f) \\ \underline{P}_{uy}(f) & \underline{P}_{uu}(f) & \underline{P}_{uw}(f) \\ \underline{P}_{wy}(f) & \underline{P}_{wu}(f) & \underline{P}_{ww}(f) \end{bmatrix} .$$

We define $\underline{P}(f)^* = \underline{P}(-f)^\tau$. Then it follows that $\underline{P}(f)^* = \underline{P}(f)$, also. Because $\underline{C}_{wu}(s) \equiv \underline{0}$, $\underline{P}_{wu}(f) \equiv \underline{0}$. We define the transfer function matrix

$$\underline{H}(f) = \int_{-\infty}^{\infty} \underline{h}(s) e^{2\pi i s f} \, ds, \quad -\infty < f < \infty .$$

Because of (2), the elements of $\underline{H}(f)$ are bounded. Notice that our $\underline{H}(f)$ is the complex conjugate of the usual definition; our definition is the more convenient in the context of our multiple system.

We then transform (4), (5) to

$$\underline{P}_{yu}(f) = \underline{H}(f) \, \underline{P}_{uu}(f) , \tag{6}$$

$$\underline{P}_{yy}(f) = \underline{H}(f) \, \underline{P}_{uu}(f) \, \underline{H}(f)^* + \underline{P}_{ww}(f) . \tag{7}$$

For every f for which $\underline{P}_{uu}(f)$ is non-singular, $\underline{H}(f)$ and $\underline{P}_{ww}(f)$ are then identifiable as

$$\underline{H}(f) = \underline{P}_{yu}(f) \, \underline{P}_{uu}(f)^{-1} , \tag{8}$$

$$\underline{P}_{ww}(f) = \underline{P}_{yy}(f) - \underline{P}_{yu}(f) \, \underline{P}_{uu}(f)^{-1} \, \underline{P}_{uy}(f) . \tag{9}$$

From continuous records of $\underline{u}(t)$ and $\underline{y}(t)$, we can estimate $\underline{P}_{yy}(f)$, $\underline{P}_{yu}(f)$, $\underline{P}_{uu}(f)$, and thence $\underline{H}(f)$ and $\underline{P}_{ww}(f)$.

When the data are of the form

$$\{\underline{u}(n\Delta), \;\; \underline{y}(n\Delta); \;\; n = 0,1,\ldots,N-1\} ,$$

a natural way to proceed is as follows. One forms the discrete Fourier transforms

$$\underline{d}_u(f) = \frac{\Delta}{N} \sum_{n=0}^{N-1} \underline{u}(n\Delta) e^{-2\pi i f n\Delta}, \tag{10}$$

$$\underline{d}_y(f) = \frac{\Delta}{N} \sum_{n=0}^{N-1} \underline{y}(n\Delta) e^{-2\pi i f n\Delta}, \tag{11}$$

for f of the form $j/N\Delta$, j integral. Of course, such $\underline{d}_u$ and $\underline{d}_y$ have period $1/\Delta$, so we define them only for $-1/2\Delta < f \leq 1/2\Delta$. When N is large the computation of these quantities is still feasible, at least if N is "highly composite", particularly when it is a power of 2 (see Cooley, Lewis and Welch, 16). For other N the time series can be extended to this length, which we call N', by the addition of zeroes at either end, when, probably, a sequence $\{a_N(n)\}$, such as a cosine bell, will be introduced to "fade out" the original sequence near its extremities (see Hannan, 15, p. 265) and we would compute, instead of (10), (11)

$$\underline{d}_u(f) = \frac{\Delta}{N'} \sum_{n=0}^{N'-1} a_N(n)\underline{u}(n\Delta) e^{-2\pi i f n\Delta},$$

$$\underline{d}_y(f) = \frac{\Delta}{N'} \sum_{n=0}^{N'-1} a_N(n)\underline{y}(n\Delta) e^{-2\pi i f n\Delta},$$

for f of the form $j/N'\Delta$, j integral. The computation of the periodogram matrices

$$\underline{I}_{uu}\left(\frac{j}{N\Delta}\right) = \frac{N}{\Delta^2} \underline{d}_u\left(\frac{-j}{N\Delta}\right) \underline{d}_u\left(\frac{j}{N\Delta}\right)^{\tau}, \quad \underline{I}_{yy}\left(\frac{j}{N\Delta}\right) = \frac{N}{\Delta^2} \underline{d}_y\left(\frac{-j}{N\Delta}\right)\underline{d}_y\left(\frac{j}{N\Delta}\right)^{\tau},$$

$$\underline{I}_{yu}\left(\frac{j}{N\Delta}\right) = \frac{N}{\Delta^2} \underline{d}_y\left(\frac{-j}{N\Delta}\right) \underline{d}_u\left(\frac{j}{N\Delta}\right)^{\tau},$$

is the next step, followed by

$$\hat{\underline{P}}_{uu}(f) = \frac{1}{M} \sum_j \underline{I}_{uu}\left(\frac{j}{N\Delta}\right), \quad \hat{\underline{P}}_{yy}(f) = \frac{1}{M} \sum_j \underline{I}_{yy}\left(\frac{j}{N\Delta}\right),$$

$$\hat{\underline{P}}_{yu}(f) = \frac{1}{M} \sum_j \underline{I}_{yu}\left(\frac{j}{N\Delta}\right),$$

where the $M << N$ consecutive values of $j/N\Delta$ included in the sums are centered, roughly, at f. Then we have the estimators

$$\hat{\underline{H}}(f) = \hat{\underline{P}}_{yu}(f)\,\hat{\underline{P}}_{uu}(f)^{-1} ,$$

$$\hat{\underline{P}}_{ww}(f) = \hat{\underline{P}}_{yy}(f) - \hat{\underline{P}}_{yu}(f)\,\hat{\underline{P}}_{uu}(f)^{-1}\,\hat{\underline{P}}_{uy}(f) ,$$

where $\hat{\underline{P}}_{uy}(f) = \hat{\underline{P}}_{yu}(f)^*$. The programming of these computations thus begins from the discrete cosine and sine transforms, the real and imaginary parts of the $\underline{d}_u(f)$ and $\underline{d}_y(f)$. The periodogram and estimated spectral density matrices are thus linear combinations of products of these quantities. For a formula for the inverse of a complex matrix in terms of the inverse of its real part, see Hannan [15, p. 445].

Because of the periodicity of $\underline{d}_u$ and $\underline{d}_f$ we can regard $\hat{\underline{P}}_{uu}(f)$, $\hat{\underline{P}}_{yu}(f)$, $\hat{\underline{P}}_{yy}(f)$ as converging as $N \to \infty$, in a suitable mode, under suitable ergodicity conditions and for a suitable rate of increase of M with N, to the aliased spectra

$$\underline{P}^A_{uu}(f) = \sum_{j=-\infty}^{\infty} \underline{P}_{uu}(f+j/\Delta) , \tag{12}$$

$$\underline{P}^A_{yy}(f) = \sum_{j=-\infty}^{\infty} \underline{P}_{yy}(f+j/\Delta) , \tag{13}$$

$$\underline{P}^A_{yu}(f) = \sum_{j=-\infty}^{\infty} \underline{P}_{yu}(f+j/\Delta) , \tag{14}$$

respectively. However, as is well known, extreme care is needed in the choice of M, for given N. On the one hand, M must be chosen sufficiently large to damp down the variances of the periodograms (which are themselves inconsistent estimators of the spectral matrices). However, M must not be so big that $\underline{P}^A_{uu}$, $\underline{P}^A_{yy}$, $\underline{P}^A_{yu}$ vary considerably over the $j/N\Delta$ surrounding f, or serious biases may result. For example, if the influence of long lags in (1) is relatively great, $\underline{H}(f)$ may oscillate rapidly over some intervals. It may well be desirable, therefore, to begin by filtering the data in a way that seems likely to flatten the spectra in the neighborhood of f.

Now if

$$\underline{P}_{uu}(f+j/\Delta) = \underline{0}, \qquad j \neq 0, \tag{15}$$

then $\underline{P}^A_{uu}(f) = \underline{P}_{uu}(f)$, $\underline{P}^A_{yu}(f) = \underline{P}_{yu}(f)$. If also

$$\underline{P}_{ww}(f+j/\Delta) = \underline{0}, \qquad j \neq 0, \tag{16}$$

then $\underline{P}^A_{yy}(f) = \underline{P}_{yy}(f)$. Now define

$$\underline{\bar{H}}(f) = \underline{P}^A_{yu}(f)\,\underline{P}^A_{uu}(f)^{-1}, \tag{17}$$

$$\underline{\bar{P}}_{ww}(f) = \underline{P}^A_{yy}(f) - \underline{P}^A_{yu}(f)\underline{P}^A_{uu}(f)^{-1}\underline{P}^A_{uy}(f), \tag{18}$$

for those f, $-1/2\Delta < f < 1/2\Delta$, for which $\underline{P}^A_{uu}(f)$ is non-singular. We thus regard $\underline{\bar{H}}(f)$ and $\underline{\bar{P}}_{ww}(f)$ as the limits, as $N \to \infty$, of $\underline{\hat{H}}(f)$ and $\underline{\hat{P}}_{ww}(f)$. We can regard $\underline{\bar{H}}(f)$ also as the transfer function of a discrete time system, and $\underline{\bar{P}}_{ww}(f)$ as the noise spectrum for that system. Now from (8), (14), (17),

$$\underline{\bar{H}}(f) = \sum_{j=-\infty}^{\infty} \underline{H}(f+j/\Delta)\,\underline{P}_{uu}(f+j/\Delta)\,\underline{P}^A_{uu}(f)^{-1}.$$

If $\underline{H}(f)$ is periodic of period $1/\Delta$, so that $\underline{h}(s)$ gives delta function weights to points $s = n$, n integral, and zero weights to other points, $\underline{\bar{H}}(f) = \underline{H}(f)$, $\underline{\bar{P}}_{ww}(f) = \underline{P}^A_{ww}(f)$, where

$$\underline{P}^A_{ww}(f) = \sum_{j=-\infty}^{\infty} \underline{P}_{ww}(f+j/\Delta),$$

irrespective of whether (15) or (16) hold. If $\underline{H}(f)$ does not have period $1/\Delta$, a sufficient condition for $\underline{\bar{H}}(f) = \underline{H}(f)$, $\underline{P}_{ww}(f) = \underline{P}^A_{ww}(f)$ is (15). A sufficient condition for $\underline{\bar{P}}_{ww}(f) = \underline{P}_{ww}(f)$ also, is (16). If (15) holds but (16) does not, $\underline{\bar{P}}_{ww}(f) > \underline{P}_{ww}(f)$, because $\underline{P}_{ww}(f)$ is everywhere nonnegative definite.

Some further, physically meaningful, properties can be directly estimated from the basic spectral estimates described above, but, again, the interpretation of the estimates will be affected by aliasing. In particular, the coherence and phase properties are of interest. We define

$$\hat{\underline{\rho}}(f)^2 = \hat{\underline{P}}_{yy}(f)^{-1}\hat{\underline{P}}_{yu}(f)\hat{\underline{P}}_{uu}(f)^{-1}\hat{\underline{P}}_{uy}(f) ,$$

which, for $q = 1$, can be regarded as an estimator of the squared multiple coherence between the sequences $\underline{y}(n\Delta)$ and $\underline{u}(n\Delta)$. If $r = 1$ also, the estimated phase angle, $\hat{\phi}(f)$, between $\{\underline{y}(n\Delta)\}$ and $\{\underline{u}(n\Delta)\}$ is the arc tangent of the ratio between the imaginary and real parts of $\hat{\underline{P}}_{yu}(f)$. See Hannan [15, Chapter 5], Jenkins and Watts [17]. If $\underline{H}(f)$ is heavily damped outside Nyquist frequency, so that the coherence is small there, $\underline{P}^{A}_{yu}(f)$ could be close to $\underline{P}_{yu}(f)$ without $\underline{P}^{A}_{uu}(f)$ necessarily being close to $\underline{P}_{uu}(f)$. In that case we would expect aliasing effects to be small for the phase, and possibly for the coherence also, although underestimation of the coherence is likely because $\hat{\underline{P}}_{uu}(f)$, $\hat{\underline{P}}_{yy}(f)$ overestimate $\underline{P}_{uu}(f)$, $\underline{P}_{yy}(f)$.

3. IDENTIFICATION AND ESTIMATION OF TIME DOMAIN CHARACTERISTICS

Often the relationship between $\underline{u}(t)$ and $\underline{y}(t)$ is best thought of in the frequency domain, particularly when the form of relationship seems likely to change with frequency, when the signal-to-noise ratio is believed to be very low at some frequencies, and when no prior information on the nature of the relationship is available. However, the impulse response function, $\underline{h}(s)$, is also of interest because it represents the "density" of the regression of $\underline{y}(t)$ on $\underline{u}(t)$ at various time lags, the relative importance of long and short reaction times and the extent to which $\underline{y}(t)$ can be regarded as depending only on previous $\underline{u}(t)$. Moreover, the cross-spectral method described in the previous section provides no scope for the use of prior information, apart from the implicit assumptions of time-invariance and lack of feedback. Quite often, there seems to be justification for modelling $\underline{H}(f)$, and even $\underline{P}_{ww}(f)$, as explicit functions of a finite number of parameters, that possibly have a direct scientific interpretation, and when the record length is short the incentives for doing so are strong. Systems of this type are often constructed in control and econometrics. (See Mehra [18] for an interesting comparison of these two situations.) In some other cases, one might begin by proceeding as in Section 2, but conjecture *a posteriori* a functional relationship in terms of a few parameters.

The first situation, of reconstructing $\underline{h}(s)$ directly from $\underline{H}(f)$, is straightforward. The Fourier inverses of $\underline{H}(f)$ and $\underline{P}_{ww}(f)$ are

$$\left.\begin{aligned} \underline{h}(s) &= \int_{-\infty}^{\infty} \underline{H}(f)e^{-2\pi isf}\,df\,, \\ \underline{C}_{ww}(s) &= \int_{-\infty}^{\infty} \underline{P}_{ww}(f)e^{2\pi isf}\,df\,, \end{aligned}\right\} \quad -\infty < s < \infty\,,$$

when they exist. We shall suppose that $\underline{H}(f)$ and $\underline{P}_{ww}(f)$ have been estimated by $\underline{H}(f)$, $\underline{P}_{ww}(f)$ at each of the equidistant frequencies $f_\ell = \ell/\Delta K$, $\ell = 0, \pm 1, \ldots, \pm L$, where the integer L is such that $K = 2L+1$ is the closest odd integer to N/M, so that each spectral estimate is the average of roughly M periodogram matrices. Since $\hat{\underline{H}}(f)$ and $\hat{\underline{P}}_{ww}(f)$ are based on (15) and (16), we can consider the estimators

$$\hat{\underline{h}}(s) = \frac{1}{K} \sum_{|\ell| \leq L} \hat{\underline{H}}(f_\ell) \exp(-2\pi isf_\ell), \tag{19}$$

$$\hat{\underline{C}}_{ww}(s) = \frac{1}{K} \sum_{|\ell| \leq L} \hat{\underline{P}}_{ww}(f_\ell) \exp(2\pi isf_\ell)\,, \tag{20}$$

for $-\infty < s < \infty$. However, because these estimators are periodic of period $K\Delta$, they should be regarded as being defined over an interval of that length only, such as $[0,K\Delta)$, $(-K\Delta/2,K\Delta/2]$.

Notice that for those s that are multiples of Δ, $\underline{h}(s)$ and $\underline{C}_{ww}(s)$ are valid estimators of the coefficients of the delta function weights in the discrete time system (i.e., the Fourier coefficients of $\underline{H}(f)$), and of $\underline{C}_{ww}(s)$, respectively, irrespective of (15) and (16). An alternative way of estimating $\underline{h}(s)$ was proposed by Sims [19].

We introduce next a vector of unknown parameters, $\underline{\theta}_0$, such that the elements of $\underline{H}(f) = \underline{H}(f;\underline{\theta}_0)$ are given continuous functions of both f and $\underline{\theta}_0$. Then we consider as an estimator the value of $\underline{\theta}$, $\hat{\underline{\theta}}$, that minimizes, over the portion of the parameter space of interest, the function

$$S_N(\underline{\theta}) = \frac{1}{N}\sum_{\mathcal{B}} \operatorname{tr}\Big\{[\underline{H}(f_\ell;\underline{\theta})\hat{\underline{P}}_{uu}(f_\ell)\underline{H}(f_\ell;\underline{\theta})^* - 2\underline{H}(f_\ell;\underline{\theta})\hat{\underline{P}}_{uy}(f_\ell)]\hat{\underline{P}}_{ww}(f_\ell)^{-1}\Big\} , \tag{21}$$

where the summation is over those ℓ such that $f_\ell \in \mathcal{B}$, a prescribed, symmetric subset of $(-1/2\Delta, 1/2\Delta)$. (See Hannan and Robinson, 12). We can motivate $S_N(\underline{\theta})$ in the following way. A sequence of quantities

$$\sqrt{N}\,\underline{d}_w\left(\frac{-j}{N\Delta};\underline{\theta}\right) = \sqrt{N}\left[\underline{d}_y\left(\frac{-j}{N\Delta}\right) - \underline{H}\left(\frac{-j}{N\Delta};\underline{\theta}\right)\underline{d}_u\left(\frac{-j}{N\Delta}\right)\right]$$

for a finite number of consecutive values of j, such that the $j/N\Delta$ are close to a frequency f, can be regarded, for $\underline{\theta} = \underline{\theta}_0$, as being approximately independent and complex multivariate Gaussian with covariance matrix proportional to $\underline{P}^A_{ww}(f)$, under wide conditions on $\{\underline{w}(n\Delta);\ n = 0, \pm 1, \ldots\}$ (see Hannan, 15, Chapter 4), for those f for which (15) holds. Thus,

$$C\sum_{\mathcal{B}} \operatorname{tr}\left\{\underline{d}_w\left(\frac{-j}{N\Delta};\underline{\theta}_0\right)\underline{d}_w\left(\frac{j}{N\Delta};\underline{\theta}_0\right)^\tau \underline{P}^A_{ww}\left(\frac{j}{N\Delta}\right)^{-1}\right\}$$

can, for suitable C, be regarded approximately as the exponent in a complex-normal likelihood function. This is approximated by replacing $j/N\Delta$ in $\underline{H}(j/N\Delta;\underline{\theta})$ by its central value, f_ℓ, throughout each of the bands in $\mathcal{B}$, employing the fact that

$$\sum_{\mathcal{B}} \operatorname{tr}\{\underline{H}(f_\ell;\underline{\theta})\hat{\underline{P}}_{uy}(f_\ell)\underline{P}^A_{ww}(f_\ell)^{-1}\} =$$

$$= \sum_{\mathcal{B}} \operatorname{tr}\{\hat{\underline{P}}_{yu}(f_\ell)\underline{H}(f_\ell;\underline{\theta})^*\underline{P}^A_{ww}(f_\ell)^{-1}\} \quad ,$$

because of symmetry, and replacing $\underline{P}^A_{ww}(f)$ by $\hat{\underline{P}}_{ww}(f)$. On substituting $\underline{\theta}$ for $\underline{\theta}_0$, the part of this involving $\underline{\theta}$ is then proportional to $S_N(\underline{\theta})$. An alternative criterion to $S_N(\underline{\theta})$, which arises in a more obvious way and is considered in Robinson [13], is

$$T_N(\underline{\theta}) = \sum_{\mathcal{B}} \operatorname{tr}\left\{\underline{d}_w\left(\frac{-j}{N\Delta};\underline{\theta}\right)\underline{d}_w\left(\frac{j}{N\Delta};\underline{\theta}\right)^{\tau}\hat{\underline{P}}_{ww}\left(\frac{j}{N\Delta}\right)^{-1}\right\}, \tag{22}$$

where the sum is now over those $j/N\Delta$ in $\mathcal{B}$ and $\hat{\underline{P}}_{ww}(j/N\Delta) = \hat{\underline{P}}_{ww}(f_\ell)$, for that f_ℓ closest to $j/N\Delta$. It appears that $T_N(\underline{\theta})$ represents a closer approximation to the log likelihood than $S_N(\underline{\theta})$. However, there seem to be computational advantages in using $S_N(\underline{\theta})$ rather than $T_N(\underline{\theta})$, because in the former case $\underline{H}(f;\underline{\theta})$ is formed at fewer frequencies, and in some cases the formation of $\underline{H}(f;\underline{\theta})$ for given f, $\underline{\theta}$ may be rather expensive, possibly involving matrix inversion. Note that the $\hat{\underline{P}}_{uu}(f_\ell)$ and $\hat{\underline{P}}_{yu}(f_\ell)$ must be formed in any case, to estimate $\underline{P}_{ww}(f_\ell)$. Asymptotically (for M increasing at a suitable rate with N) the estimators minimizing (21) and (22) will have similar asymptotic properties, and will indeed be consistent and asymptotically efficient estimators of $\underline{\theta}_0$ under suitable ergodicity conditions, and (15). For given $\hat{\underline{\theta}}$, a more efficient estimator of $\underline{P}^A_{ww}(f_\ell)$ is

$$\tilde{\underline{P}}_{ww}(f_\ell) = \frac{N}{M\Delta^2}\sum_j \underline{d}_w\left(\frac{-j}{N\Delta};\hat{\underline{\theta}}\right)\underline{d}_w\left(\frac{j}{N\Delta};\hat{\underline{\theta}}\right)^{\tau} .$$

Sometimes, a system of the form (3) is parameterized, and the transforms

$$\underline{H}_1(f;\underline{\theta}_0) = \underline{H}_1(f) = \int_{-\infty}^{\infty} \underline{h}_1(s)e^{2\pi ifs}\,ds \quad ,$$

$$\underline{H}_2(f;\underline{\theta}_0) = \underline{H}_2(f) = \int_{-\infty}^{\infty} \underline{h}_2(s)e^{2\pi ifs} \, ds \quad ,$$

are linear in $\underline{\theta}_0$. In such cases a computationally preferable approach to that of using the just described method on $\underline{H}(f;\underline{\theta}) = \underline{H}_1(f;\underline{\theta})^{-1}\underline{H}_2(f;\underline{\theta})$, uses a method which can be regarded as a frequency domain version of instrumental variables. We have approximately,

$$\underline{d}_y(-f) \doteq [\underline{I}-\underline{H}_1(f;\underline{\theta}_0)] \, \underline{d}_y(-f) + \underline{H}_2(f;\underline{\theta}_0) \, \underline{d}_u(-f), \tag{23}$$

where $\underline{I}$ is the $q \times q$ identity matrix. We write the matrix $[\underline{I}-\underline{H}_1(f;\underline{\theta}),\underline{H}_2(f;\underline{\theta})]$ as a $q(q+r) \times 1$ column vector, such that the j-th column is placed directly below the $(j-1)$-th. This vector can be written in the form $\underline{G}(f)\underline{\theta}$, where $\underline{G}(f)$ is a $q(q+r) \times p$ known matrix function of f. By elementary operations, (23) can then be rewritten as

$$\underline{d}_y(-f) \doteq \{[\underline{d}_y(f)^*, \, \underline{d}_u(f)^*] \otimes \underline{I}\} \, \underline{G}(f) \, \underline{\theta}_0 \quad ,$$

where $\otimes$ is the tensor product operator. Then we premultiply the latter expression by the "instrumental" matrix

$$\underline{J}(f) \, \{\underline{d}_u(f) \otimes \underline{I}\} \quad ,$$

where $\underline{J}(f)$ is a given complex $p \times q^2$ matrix function of f, whose real components are even functions and whose imaginary components are odd functions. If we now proceed by summing over $f = j/N\Delta \in \mathcal{B}$ and replacing (for computational ease) $\underline{G}(j/N\Delta)$ and $\underline{J}(j/N\Delta)$ by their central values $\underline{G}(f_\ell)$ and $\underline{J}(f_\ell)$, within each of the aforementioned bands in $\mathcal{B}$, we get the estimator

$$\hat{\underline{\theta}} = \hat{\underline{A}}^{-1}\hat{\underline{a}} \quad , \tag{24}$$

where

$$\hat{\underline{A}} = \frac{1}{N} \sum_{\mathcal{B}} \underline{J}(f_\ell)\{[\hat{\underline{P}}_{yu}(f_\ell)^\tau,\hat{\underline{P}}_{uu}(f_\ell)^\tau] \otimes \underline{I}\} \, \underline{G}(f_\ell) \quad ,$$

$$\hat{\underline{a}} = \frac{1}{N} \sum_{\mathcal{B}} \underline{J}(f_\ell)\, \hat{\underline{p}}_{yu}(f_\ell) ,$$

where $\underline{p}_{yu}(f)$ is a $qr \times 1$ column vector constructed from $\hat{\underline{P}}_{yu}(f)$ in the way $\underline{G}(f)\underline{\theta}$ was constructed. A related estimator in which $\hat{\underline{P}}_{yu}$, $\hat{\underline{P}}_{uu}$ are replaced by $\underline{I}_{yu}$, $\underline{I}_{uu}$, and f_ℓ is replaced by $j/N\Delta$ and the summations are over j, is considered in Robinson [20]. Although such estimators can be justified as being consistent under conditions similar to those required for the estimators minimizing $S_N(\underline{\theta})$ and $T_N(\underline{\theta})$ (see [20]), a slight problem exists as far as finding an efficient estimator is concerned, for an optimum choice of $\underline{J}(f)$ depends upon the unknown $\underline{\theta}_0$. It is likely, then, that a consistent but inefficient $\hat{\underline{\theta}}$ will first be computed, and used in $\underline{J}(f)$ to generate a more efficient estimate. We refer the reader to [20], where the details of such a procedure are described.

4. BOUNDS FOR BIASES CAUSED BY ALIASING

It is of some interest to analyze the biases that are liable to accrue from invalidity of (15) and (16), and so to motivate the consideration of alternative, improved, estimators. We shall express bounds for the biases in terms of natural measures of the departures from (15) and (16), and of the departure from periodicity of $\underline{H}(f)$.

A. THE TRANSFER FUNCTION

We consider first, for some f, $-1/2\Delta < f < 1/2\Delta$,

$$\begin{aligned}\bar{\underline{H}}(f) - \underline{H}(f) &= [\underline{P}^A_{yu}(f) - \underline{H}(f)\underline{P}^A_{uu}(f)]\underline{P}^A_{uu}(f)^{-1} \\ &= \sum_{|j|\geq 1} [\underline{H}(f+j/\Delta) - \underline{H}(f)]\underline{P}_{uu}(f+j/\Delta)\underline{P}^A_{uu}(f)^{-1}.\end{aligned}$$

It follows immediately that

$$\|\bar{\underline{H}}(f) - \underline{H}(f)\| \leq \sup_j \|\underline{H}(f+j/\Delta)-\underline{H}(f)\|\, \|\underline{P}^A_{uu}(f)^{-1}\| \sum_{|j|\geq 1} \|\underline{P}_{uu}(f+j/\Delta)\| \tag{25}$$

where by $\|\underline{x}\|^2$ we now mean the greatest eigenvalue of $\underline{x}\underline{x}^*$. Notice that the first factor is a measure of the departure from periodicity of $\underline{H}(f)$ and the final factor is a measure of the departure from (15), the right hand side vanishing if either $\underline{H}(f)$ has period $1/\Delta$ or (15) holds.

B. THE NOISE SPECTRUM

We consider next

$$\bar{\underline{P}}_{ww}(f) - \underline{P}^A_{ww}(f) = \underline{P}^A_{yy}(f) - \underline{P}^A_{yu}(f)\underline{P}^A_{uu}(f)^{-1}\underline{P}^A_{uy}(f) - \underline{P}^A_{ww}(f) . \tag{26}$$

Because

$$\underline{P}^A_{yy}(f) = \sum_{j=-\infty}^{\infty} \underline{H}(f+j/\Delta)\underline{P}_{uu}(f+j/\Delta)\underline{H}(f+j/\Delta)^* + \underline{P}^A_{ww}(f) ,$$

(26) is

$$\sum_{j=-\infty}^{\infty} \underline{H}(f+j/\Delta)\underline{P}_{uu}(f+j/\Delta)\underline{H}(f+j/\Delta)^* - \underline{P}^A_{yu}(f)\underline{P}^A_{uu}(f)^{-1}\underline{P}^A_{uy}(f)$$
$$= \sum_{j=-\infty}^{\infty} [\underline{H}(f+j/\Delta) - \bar{\underline{H}}(f)]\underline{P}_{uu}(f+j/\Delta)\underline{H}(f+j/\Delta)^* . \tag{27}$$

Writing

$$\underline{H}(f+j/\Delta) - \bar{\underline{H}}(f) = [\underline{H}(f+j/\Delta) - \underline{H}(f)] + [\underline{H}(f) - \bar{\underline{H}}(f)] ,$$

(27) is

$$\sum_{|j|\geq 1} [\underline{H}(f+j/\Delta) - \underline{H}(f)]\underline{P}_{uu}(f+j/\Delta)\underline{H}(f+j/\Delta)^* + [\underline{H}(f)-\bar{\underline{H}}(f)]\underline{P}^A_{uy}(f) .$$

Thus,

$$\|\bar{\underline{P}}_{ww}(f) - \underline{P}^A_{ww}(f)\| \leq \sup_j \|\underline{H}(f+j/\Delta) - \underline{H}(f)\| \sum_{|j|\geq 1} \|\underline{P}_{uu}(f+j/\Delta)$$
$$\underline{H}(f+j/\Delta)^*\| + \|\bar{\underline{H}}(f) - \underline{H}(f)\| \, \|\underline{P}^A_{uy}(f)\| . \tag{28}$$

The second factor of the first term on the right is bounded by

$$\sup_j \|\underline{H}(f+j/\Delta)\| \sum_{|j|\geq 1} \|\underline{P}_{uu}(f+j/\Delta)\|$$

but this bound is conservative when $\underline{H}(f)$ is not periodic, particularly in cases where it vanishes as $f \to \pm\infty$. A bound for the second term on the right is given by (25). Notice that (28) is independent of $\underline{P}_{ww}(f)$. However, we also have

$$\|\bar{\underline{P}}_{ww}(f)-\underline{P}_{ww}(f)\| \leq \|\bar{\underline{P}}_{ww}(f)-\underline{P}^A_{ww}(f)\| + \sum_{|j|\geq 1} \|\underline{P}_{ww}(f+j/\Delta)\| \quad , \qquad (29)$$

by the triangle inequality.

C. THE COHERENCE

We consider the difference between

$$\underline{\rho}(f)^2 = \underline{P}_{yy}(f)^{-1}\underline{P}_{yu}(f)\underline{P}_{uu}(f)^{-1}\underline{P}_{uy}(f)$$

and

$$\bar{\underline{\rho}}(f)^2 = \underline{P}^A_{yy}(f)^{-1}\underline{P}^A_{yu}(f)\underline{P}^A_{uu}(f)^{-1}\underline{P}^A_{uy}(f)$$

which are true and aliased multiple squared coherences, when $q=1$. Then we have

$$\bar{\underline{\rho}}(f)^2 - \underline{\rho}(f)^2 = \underline{P}_{yy}(f)^{-1}\underline{P}_{ww}(f) - \underline{P}^A_{yy}(f)^{-1}\bar{\underline{P}}_{ww}(f) \ ,$$

from (9), (18). This is

$$\underline{P}_{yy}(f)^{-1}[\underline{P}^A_{yy}(f) - \underline{P}_{yy}(f)]\underline{P}^A_{yy}(f)^{-1}\underline{P}_{ww}(f) + \underline{P}^A_{yy}(f)^{-1}[\underline{P}_{ww}(f) - \bar{\underline{P}}_{ww}(f)] \quad .$$

Thus,

$$\|\bar{\underline{\rho}}(f)^2 - \underline{\rho}(f)^2\| \leq \|\underline{P}_{yy}(f)^{-1}\| \ \|\underline{P}^A_{yy}(f)^{-1}\underline{P}_{ww}(f)\| \ \|\underline{P}^A_{yy}(f) - \underline{P}_{yy}(f)\| + \|\underline{P}^A_{yy}(f)^{-1}\| \, \|\bar{\underline{P}}_{ww}(f) - \underline{P}_{ww}(f)\| \quad . \qquad (30)$$

Of course, as far the first term on the right is concerned,

$$\|\underline{P}^A_{yy}(f) - \underline{P}_{yy}(f)\| \leq \sum_{|j|\geq 1} \Big\{\|\underline{H}(f+j/\Delta)\underline{P}_{uu}(f+j/\Delta)\underline{H}(f+j/\Delta)^*\| + \|\underline{P}_{ww}(f+j/\Delta)\|\Big\} \quad ,$$

and a bound for the second term can be obtained by using (28), (29).

D. THE PHASE

We now consider the phase angle between a single element of $\underline{y}(t)$ and a single element of $\underline{u}(t)$. We translate this into the case $q = r = 1$, and consider

$$\phi(f) = \arctan\left\{\frac{H_I(f)P_{uu}(f)}{H_R(f)P_{uu}(f)}\right\} = \arctan\left\{\frac{H_I(f)}{H_R(f)}\right\} ,$$

$$\bar{\phi}(f) = \arctan\left\{\frac{\sum_{j=-\infty}^{\infty} H_I(f+j/\Delta)P_{uu}(f+j/\Delta)}{\sum_{j=-\infty}^{\infty} H_R(f+j/\Delta)P_{uu}(f+j/\Delta)}\right\} ,$$

where $H_R(f)$ and $H_I(f)$ are respectively the real and imaginary parts of $H(f)$, and since all quantities here are of necessity scalars, bold-face is not used. It follows that

$$\tan\bar{\phi}(f)-\tan\phi(f) = \frac{\sum_{|j|>1} [H_R(f)H_I(f+j/\Delta)-H_R(f+j/\Delta)H_I(f)]P_{uu}(f+j/\Delta)}{H_R(f)\sum_{j=-\infty}^{\infty} H_R(f+j/\Delta)P_{uu}(f+j/\Delta)}$$

Then by the mean value theorem

$$|\bar{\phi}(f) - \phi(f)| \leq |\tan\bar{\phi}(f) - \tan\phi(f)| ,$$

and the right hand side is bounded by

$$\frac{\sup_j |H_R(f)H_I(f+j/\Delta) - H_I(f)H_R(f+j/\Delta)| \sum_{|j|>1} P_{uu}(f+j/\Delta)}{|H_R(f)\sum_{j=-\infty}^{\infty} H_R(f+j/\Delta)P_{uu}(f+j/\Delta)|} . \tag{31}$$

E. THE IMPULSE RESPONSE FUNCTION

As $N \to \infty$, and M increases suitably,

$$\underline{\hat{h}}(s) \rightarrow \underline{\bar{h}}(s) = \int_{-1/2\Delta}^{1/2\Delta} \underline{\bar{H}}(f) e^{-2\pi isf} \, df .$$

Then

$$\underline{\bar{h}}(s) - \underline{h}(s) = \int_{-1/2\Delta}^{1/2\Delta} [\underline{\bar{H}}(f) - \underline{H}(f)] \, e^{-2\pi isf} \, df$$

$$- \int_{|f| \geq 1/2\Delta} \underline{H}(f) e^{-2\pi isf} \, df .$$

Thus, $\|\underline{\bar{h}}(s) - \underline{h}(s)\|$ is bounded by

$$\sup_f \left\{ \|\underline{P}^A_{uu}(f)^{-1}\| \sup_j \|\underline{H}(f+j/\Delta) - \underline{H}(f)\| \right\} \cdot \int_{-1/2\Delta}^{1/2\Delta} \sum_{|j| \geq 1} \|\underline{P}_{uu}(f+j/\Delta)\| \, df \qquad (32a)$$

$$+ \left\| \int_{|f| \geq 1/2\Delta} \underline{H}(f) e^{-2\pi isf} \, df \right\| , \qquad (32b)$$

where the supremum over f is taken over $(-1/2\Delta, 1/2\Delta)$, and we use (25). An alternative bound to (32a) is $\Delta^{-1} \sup_f \|\underline{\bar{H}}(f) - \underline{H}(f)\|$. Of course, when $\underline{H}(f)$ is periodic or almost periodic, so that $\underline{h}(s)$ is a linear combination of delta functions, applying atoms of mass to denumerably many t and zero weight to other t, or when $\underline{H}(f)$ diverges as $f \rightarrow \pm \infty$, as when $\underline{h}(s)$ is a linear combination of derivatives of delta functions, (32b) does not exist. In such cases, $\underline{h}(s)$ should be evaluated at only integral multiples of Δ, and given a different interpretation (see Section 3). In other cases the bias contributed by (32b) depends on the rate of decrease of $\underline{H}(f)$, as well as the discrepancy between $\underline{\bar{H}}(f)$ and $\underline{H}(f)$.

F. *THE NOISE AUTOCOVARIANCE*

We have also

$$\underline{C}_{ww}(s) \rightarrow \underline{\bar{C}}_{ww}(s) = \int_{-1/2\Delta}^{1/2\Delta} \underline{\bar{P}}_{ww}(f) e^{2\pi isf} \, df .$$

Then for s that are multiples of Δ,

$$\bar{\underline{C}}_{ww}(s) - \underline{C}_{ww}(s) = \int_{-1/2\Delta}^{1/2\Delta} [\bar{\underline{P}}_{ww}(f) - \underline{P}_{ww}^{A}(f)]e^{2\pi isf}\,df$$

so that

$$\|\bar{\underline{C}}_{ww}(s) - \underline{C}_{ww}(s)\| \leq \Delta^{-1} \sup_f \|\bar{\underline{P}}_{ww}(f) - \underline{P}_{ww}^{A}(f)\| \quad . \tag{33}$$

For s that are not multiples of Δ, we have

$$\bar{\underline{C}}_{ww}(s) - \underline{C}_{ww}(s) = \int_{-1/2\Delta}^{1/2\Delta} [\bar{\underline{P}}_{ww}(f) - \underline{P}_{ww}(f)]e^{2\pi isf}\,df - \int_{|f|\geq 1/2\Delta} \underline{P}_{ww}(f)e^{2\pi isf}\,df,$$

so that

$$\|\bar{\underline{C}}_{ww}(s) - \underline{C}_{ww}(s)\| \leq \Delta^{-1} \sup_f \|\bar{\underline{P}}_{ww}(f) - \underline{P}_{ww}(f)\| + \int_{|f|\geq 1/2\Delta} \|\underline{P}_{ww}(f)\|\,df \quad . \tag{34}$$

G. *THE ESTIMATORS OF* θ_0

Next we consider the estimators minimizing $S_N(\underline{\theta})$ or $T_N(\underline{\theta})$. As $N \to \infty$, with M increasing at a suitable rate and under suitable ergodicity conditions, $S_N(\underline{\theta})$, and the part of $T_N(\underline{\theta})$ involving $\underline{\theta}$, converge with probability one to

$$S(\underline{\theta}) = \int_{\mathcal{B}} \text{tr} \left\{[\underline{H}(f;\underline{\theta})\underline{P}_{uu}^{A}(f)\underline{H}(f;\underline{\theta})^{*} - 2\underline{H}(f;\underline{\theta})\underline{P}_{uy}^{A}(f)]\underline{P}_{ww}^{A}(f)^{-1}\right\} df \quad .$$

Then we can use the argument of Jennrich [21] to show that $\hat{\underline{\theta}}$ converges with probability one to $\bar{\underline{\theta}}$, the value that absolutely and uniquely minimizes $S(\underline{\theta})$ over the compact set over which the optimization is undertaken. A similar argument shows that $\bar{\underline{\theta}} \to \underline{\theta}_0$ as $\Delta \to 0$ if $\lim_{\Delta\to 0} S(\underline{\theta})$ is minimized uniquely by

$\underline{\theta} = \underline{\theta}_0$. Then by the mean-value theorem

$$\underline{S}'(\bar{\underline{\theta}}) = \underline{S}'(\underline{\theta}_0) + \underline{\underline{Q}}(\bar{\underline{\theta}} - \underline{\theta}_0) \quad ,$$

where, under suitable regularity conditions, $\underline{S}'(\underline{\theta}) = \partial S(\underline{\theta})/\partial\underline{\theta}$, and $\underline{\underline{Q}}$ is $\underline{S}''(\bar{\underline{\theta}}) = \partial\underline{S}'(\underline{\theta})/\partial\underline{\theta}^{\tau}$, with each of the rows being evaluated at a different value of $\underline{\theta}$ "in between" $\bar{\underline{\theta}}$ and $\underline{\theta}_0$. However, $\underline{S}'(\underline{\theta}) = \underline{0}$ by definition and so

$$\|\bar{\underline{\theta}} - \underline{\theta}_0\| \leq \|\underline{\underline{Q}}^{-1}\| \, \|\underline{S}'(\underline{\theta}_0)\| \quad . \tag{35}$$

When $\underline{S}'(\underline{\theta}_0)$ is evaluated it will be seen that it depends principally on the discrepancy between $\underline{H}(f;\underline{\theta}_0)\underline{P}^A_{uu}(f)$ and $\underline{P}^A_{yu}(f)$, and the right hand side of (35) is bounded by

$$2\,\|\underline{\underline{Q}}^{-1}\| \sup_{f\in\mathcal{B}} \left\{\|\underline{D}(f)\| \, \|\underline{P}_{ww}(f)^{-1}\| \sup_{j} \|\underline{H}(f+j/\Delta;\theta_0) - \underline{H}(f;\theta_0)\|\right\} \times \int_{\mathcal{B}} \sum_{|j|\geq 1} \|\underline{P}_{uu}(f+j/\Delta)\| \quad df \quad , \tag{36}$$

where $\underline{\underline{Q}}$ will certainly be nonsingular for Δ sufficiently small if $\lim_{\Delta\to 0} \underline{S}''(\underline{\theta})$ is nonsingular within a neighborhood of $\underline{\theta}_0$, and where $\underline{D}(f)$ is the derivative, with respect to $\underline{\theta}^{\tau}$ and evaluated at $\underline{\theta} = \underline{\theta}_0$, of the column vector formed by stacking the columns of $\underline{H}(f;\underline{\theta})$ beneath one another in sequence.

The estimators based on instrumental variates can be similarly studied. We suppose that, as $N \to \infty$,

$$\hat{\underline{A}} \to \bar{\underline{A}} = \int_{\mathcal{B}} \underline{J}(f) \left\{[\underline{P}^A_{yu}(f)^{\tau}, \underline{P}^A_{uu}(f)^{\tau}] \otimes \underline{I}\right\} \underline{G}(f) \quad df \quad ,$$

$$\hat{\underline{a}} \to \bar{\underline{a}} = \int_{\mathcal{B}} \underline{J}(f)\underline{p}^A_{yu}(f) \quad df \quad ,$$

where $\underline{p}^A_{yu}(f)$ is the vector form of $\underline{P}^A_{yu}(f)$. But from (25),

$$\underline{P}^A_{yu}(f) = \sum_{j=-\infty}^{\infty} \left\{[\underline{I}-\underline{H}_1(f+j/\Delta;\underline{\theta}_0)] \, \underline{P}_{yu}(f+j/\Delta) + \underline{H}_2(f+j/\Delta;\underline{\theta}_0) \, \underline{P}_{uu}(f+j/\Delta)\right\} \quad ,$$

so that

$$\underline{p}^{A}_{yu}(f) = \left\{[\underline{P}^{A}_{yu}(f)^{\tau}, \underline{P}^{A}_{uu}(f)^{\tau}] \otimes \underline{I}\right\} \underline{G}(f)\underline{\theta}_0$$

$$+ \sum_{j=-\infty}^{\infty} \left\{[\underline{P}_{yu}(f+j/\Delta)^{\tau}, \underline{P}_{uu}(f+j/\Delta)^{\tau}] \otimes \underline{I}\right\}$$

$$[\underline{G}(f+j/\Delta) - \underline{G}(f)]\ \underline{\theta}_0 \quad .$$

But the second term on the right is the vector form of

$$\sum_{j=-\infty}^{\infty} [\underline{H}_1(f;\underline{\theta}_0)\underline{H}_1(f+j/\Delta;\underline{\theta}_0)^{-1}\underline{H}_2(f+j/\Delta;\underline{\theta}_0)-\underline{H}_2(f;\underline{\theta}_0)]\underline{P}_{uu}(f+j/\Delta),$$

on using the relationship between $\underline{H}_1(f;\underline{\theta})$, $\underline{H}_2(f;\underline{\theta})$ and $\underline{G}(f)$. Thus, after substituting for $\underline{p}^{A}_{yu}(f)$ in $\bar{\underline{\theta}} = \bar{\underline{A}}^{-1}\bar{\underline{a}}$ an upper bound for $\|\bar{\underline{\theta}} - \underline{\theta}_0\|$ is

$$\|\underline{A}^{-1}\| \sup_{f\epsilon\mathcal{B}} \left\{\|\underline{J}(f)\| \sup_{j} \|\underline{H}_1(f;\underline{\theta}_0)\underline{H}_1(f+j/\Delta;\underline{\theta}_0)^{-1}\underline{H}_2(f+j/\Delta;\underline{\theta}_0)\right.$$

$$\left. -\underline{H}_2(f;\theta_0)\|\right\} \times \int_{\mathcal{B}} \sum_{|j|\geq 1} \|\underline{P}_{uu}(f+j/\Delta)\| \ df \quad . \qquad (37)$$

Notice that each of the bounds (25), (28), (29), (30), (31), (32a), (33), (34), (36), (37) vanishes if either the transfer function is periodic or if no aliasing is present.

If further assumptions are made, more perspicuous expressions for the biases can be obtained. The assumptions we choose are motivated by the fact that they are ones on which we can base relatively simple modifications of the cross-spectral method in Section 5. It is not a strong assumption to require that there exist some real $\alpha > 1$, $C > 0$ such that

$$\operatorname{tr}\{\underline{P}_{uu}(f)\} \leq C|f|^{-\alpha} \qquad (38)$$

for $|f| > 1/2\Delta$, since $\underline{u}(t)$ is in any case wide-sense stationary. (A familiar type of spectral density that satisfies (38) is the rational spectral density

$$\underline{P}_{uu}(f) = \left[\sum_{j=1}^{r} \underline{A}_j (if)^j\right]^{-1} \left[\sum_{j=1}^{s} \underline{B}'_j (if)^j\right] \left[\sum_{j=1}^{s} \underline{B}_j^T (-if)^j\right] \left[\sum_{j=1}^{r} \underline{A}_j^T (-if)^j\right]^{-1}, \quad -\infty < f < \infty.)$$

Since $\|\underline{P}_{uu}(f)\| \leq \text{tr}\,\{\underline{P}_{uu}(f)\}$,

$$\sum_{|j|\geq 1} \|\underline{P}_{uu}(f+j/\Delta)\| \leq c \sum_{|j|\geq 1} |f+j/\Delta|^{-\alpha} = c\Delta^{\alpha} \sum_{|j|\geq 1} |\Delta f+j|^{-\alpha} \tag{39}$$

Now for $f > 0$

$$\sum_{j\geq 1} |\Delta f+j|^{-\alpha} \leq \zeta(\alpha) \quad ,$$

where we define the Riemann zeta function

$$\zeta(\alpha) = \sum_{j\geq 1} j^{-\alpha} \quad ,$$

which exists for every $\alpha > 1$. Also

$$\sum_{j\leq -1} |\Delta f+j|^{-\alpha} = \sum_{j\geq 1} (j-\Delta f)^{-\alpha} \leq (1-\Delta f)^{-\alpha} + \sum_{j\geq 1} (j+\Delta f)^{-\alpha},$$

since

$$(j+1-\Delta f)^{-\alpha} \leq (j+\Delta f)^{-\alpha}, \quad 0 < f < 1/2\Delta \quad .$$

Then (39) is not greater than

$$c\Delta^{\alpha} [(1-\Delta f)^{-\alpha} + 2\zeta(\alpha)] \leq c\Delta^{\alpha} [2^{\alpha} + 2\zeta(\alpha)]$$

for $f > 0$, and, by a corresponding argument, not greater than

$$c\Delta^{\alpha} [(1+\Delta f)^{-\alpha} + 2\zeta(\alpha)] \leq c\Delta^{\alpha} [2^{\alpha} + 2\zeta(\alpha)]$$

for $f < 0$. Note that (38) implies that if j is that non-negative integer such that $2j+1 < \alpha \leq 2j+3$, $\underline{u}(t)$ is $j-1$ times sample function differentiable and j times mean square differentiable, and if α is itself an even integer we can use the result

$$\sum_{|j|\geq 1} (\beta+2j)^{-\alpha} = \frac{(-1)^{\alpha-1}}{(\alpha-1)!} \frac{d^{\alpha}}{d\beta^{\alpha}} \log \frac{\sin\pi\beta/2}{\pi\beta/2} \tag{40}$$

(Zorn, 22, p. 327). A stronger requirement than (38),

$$\mathrm{tr}\,\{\underline{P}_{uu}(f)\} \leq C \exp(-|f|) \quad , \tag{41}$$

ensures that $\underline{u}(t)$ is infinitely differentiable, and that

$$\int_{-\infty}^{\infty} \frac{\log \mathrm{tr}\,\{\underline{P}_{uu}(f)\}\, df}{1+f^2} = \infty \quad ,$$

(see Yaglom, 23, p. 189), so that $\underline{u}(t)$ is "singular" rather than "regular". Then

$$\sum_{|j|\geq 1} \|\underline{P}_{uu}(f+j/\Delta)\| \leq C \sum_{|j|\geq 1} \exp(-|f+j/\Delta|) \quad . \tag{42}$$

But

$$\sum_{j\geq 1} \exp[-(f+j/\Delta)] = \frac{\exp[-(f+1/\Delta)]}{1-\exp(-1/\Delta)} \quad ,$$

$$\sum_{j\leq -1} \exp[-|f+j/\Delta|] = \sum_{j\geq 1} \exp(f-j/\Delta) = \frac{\exp(f-1/\Delta)}{1-\exp(-1/\Delta)} \quad ,$$

so that (42) is

$$\frac{C \exp(-1/2\Delta) \cosh f}{\sinh(1/2\Delta)}$$

A still stricter requirement is that $\mathrm{tr}\,\{\underline{P}_{uu}(f)\}$ be band-limited, so that it vanishes everywhere outside some finite interval. We suppose that

$$\begin{aligned} \mathrm{tr}\,\{\underline{P}_{uu}(f)\} &\leq C-D|f|, \quad C/D \geq |f| \geq 1/2\Delta, \\ &= 0, \quad |f| > C/D \quad , \end{aligned}$$

for some positive C, D. Because of symmetry we need consider only positive f. Then

$$\sum_{|j|\geq 1} \|\underline{P}_{uu}(f+j/\Delta)\| \leq \sum_{j=1}^{c} [C-D(f+j/\Delta)] + \sum_{j=1}^{d} [C+D(f-j/\Delta)], \quad (43)$$

where c is the integer part of $\Delta(C/D-f)$ and d is the integer part of $\Delta(C/D+f)$. (Clearly $d = c$ or $d = c+1$.) Then the right hand side of (43) is

$$(c+d)C-(c-d)\,Df - \frac{D}{2\Delta}\,[c(c+1) + d(d+1)] .$$

This is

$$c\left[2C - \frac{D(c+1)}{\Delta}\right], \quad (44)$$

when $d = c$ and

$$(2c+1)C + Df - \frac{Dc}{2\Delta}\,(2c+3) ,$$

when $d = c + 1$. Each of these are approximately

$$\frac{Dc^2}{\Delta} \simeq \frac{\Delta C^2}{D} .$$

Using similar bounds to (38), (41), (43) for $\mathrm{tr}\,\{\underline{P}_{ww}(f)\}$ and possibly for $\|\underline{H}(f)\|$, bounds for

$$\sum_{|j|\geq 1} \|\underline{P}_{ww}(f+j/\Delta)\| \quad \text{and} \quad \sum_{|j|\geq 1} \|\underline{P}_{uu}(f+j/\Delta)\|\,\|\underline{H}(f+j/\Delta)\|$$

may be obtained. As far as (32a), (36), (37), are concerned, we have

$$\int_{\mathcal{B}} \sum_{|j|\geq 1} \|\underline{P}_{uu}(f+j/\Delta)\|\,df \leq \int_{-1/2\Delta}^{1/2\Delta} \sum_{|j|\geq 1} \|\underline{P}_{uu}(f+j/\Delta)\|\,df$$

$$\leq C\int_{|f|\geq 1/2\Delta} |f|^{-\alpha}\,df$$

$$= \frac{2C\,(2\Delta)^{\alpha-1}}{\alpha-1} ,$$

under (38). Similarly, under (41),

$$\int_{B} \sum_{|j|\geq 1} \|\underline{P}_{uu}(f+j/\Delta)\| \, df \leq 2C \exp(-1/2\Delta) \quad ,$$

and under (43),

$$\int_{B} \sum_{|j|\geq 1} \|\underline{P}_{uu}(f+j/\Delta)\| \, df \leq D\left(\frac{C}{D} - \frac{1}{2\Delta}\right)^2, \quad C/D \geq 1/2\Delta \ ,$$
$$= 0 \quad , \qquad C/D < 1/2\Delta \ .$$

5. MODIFICATIONS TO REDUCE THE EFFECTS OF ALIASING

When the sampling schedule is easily and inexpensively controllable, one is often content, in the estimation of power spectra, with a Δ such that the estimated spectrum around $f = 1/2\Delta$ is much smaller than the apparent maximum value achieved within $[-1/2\Delta, 1/2\Delta]$. In that event the raw spectral estimate for the sampled process can be taken fairly seriously as an estimate of the spectrum of the continuous process, at least for values of f that are well within $(-1/2\Delta, 1/2\Delta)$. Moreover, if the power spectra of two processes are small so also must be their co- and quadrature spectra, and indeed, because the latter, unlike the power spectra, can change in sign, there may in some cases be a tendency for departures from the off-diagonal part of (15) to cancel themselves out.

However, unless the cutoff point is rather sharp, the bias--in relative, if not absolute, terms--would tend to become substantial as f approaches $1/2\Delta$. Moreover, even if the spectra are smaller everywhere outside $(-1/2\Delta, 1/2\Delta)$ than inside, the tail may be long and the aggregate contribution of the aliased values may be significant. Added to this is the fact, already mentioned, that the sampling interval is often almost or entirely uncontrollable, in which case there would usually be no *a priori* reason for believing that $\underline{P}_{uu}(f-1/2\Delta)$, say, does not significantly contribute to $\underline{P}^{A}_{uu}(f)$. It seems that in most cases consideration should be given to the possibility of forming modified estimates for use in the estimates of transfer function, noise spectrum, coherence, phase, impulse response function and parametric models, discussed in Sections 2 and 3.

Because there are various spectral matrices--$\underline{P}_{uu}(f)$, $\underline{P}_{yu}(f)$, $\underline{P}_{yy}(f)$--that we wish to correct, it is simpler if we speak merely of determining the scalar $p(f)$ from the scalar $p^A(f)$, where $p(f)$ and $p^A(f)$ are either each cosine transforms or each sine transforms, and

$$p^A(f) = \sum_{j=-\infty}^{\infty} p(f+j/\Delta) \quad .$$

Power spectra and cospectra are cosine transforms; quadrature spectra are sine transforms. It is understood that each component of $\underline{P}_{uu}(f)$, $\underline{P}_{yu}(f)$, $\underline{P}_{yy}(f)$ is to be individually corrected, and the corrected matrices are then simply to be used in place of $\underline{P}_{uu}(f)$, $\underline{P}_{yu}(f)$, $\underline{P}_{yy}(f)$ in all the estimates of Sections 2 and 3. A natural way to proceed would involve specifying a functional form for $p(f)$ outside $(-1/2\Delta,\ 1/2\Delta)$, possibly in terms of a few unknown parameters for which an attempt at estimation is possible, and thence determine

$$\sum_{|j|\geq 1} p(f+j/\Delta) \quad ,$$

followed by

$$p(f) = p^A(f) - \sum_{|j|\geq 1} p(f+j/\Delta) \quad .$$

Zorn [22] takes, for integral α,

$$p(f) = Cf^{-\alpha}, \quad |f| > 1/2\Delta \quad , \tag{45}$$

an approximation (that improves as $f \to \pm\infty$), to a $p(f)$ that is a rational function of f, the difference between the orders of the denominator and numerator being α. Then

$$\sum_{|j|\geq 1} p(f+j/\Delta) = C\sum_{|j|\geq 1} (f+j/\Delta)^{-\alpha} = -C(2\Delta)^{\alpha}V_{\alpha}(\beta) \quad ,$$

where $-V_\alpha(\beta)$ is (40), and $\beta = 2f\Delta$. Zorn [22] tabulates $V_\alpha(\beta)$ for $\alpha = 1,2,3,4$ and for $\beta = 0.05(0.05)1.00$, and discusses the choice of α and C.

The larger α is, the more conservative the modification of $p^A(f)$ of course. Instead of calculating $V_\alpha(\beta)$ for larger α we can proceed to the limiting case

$$p(f) = Ce^{-f}, \quad 1/2\Delta < f < \infty \quad . \tag{46}$$

Then

$$\sum_{|j|\geq 1} p(f+j/\Delta) = \frac{C\exp(-1/2\Delta)\cosh f}{\sinh(1/2\Delta)}, \quad (p(f) \text{ even}), \tag{47}$$

$$= \frac{C\exp(-1/2\Delta)\sinh f}{\sinh(1/2\Delta)}, \quad (p(f) \text{ odd}) \; . \tag{48}$$

Now if (46) holds outside $(-1/2\Delta, 1/2\Delta)$ it is reasonable to assume that it holds also for some f in between $(0,1/2\Delta)$, but close to $1/2\Delta$. Then for such an f we have

$$p^A(f) = \frac{C\cosh(1/2\Delta - f)}{\sinh(1/2\Delta)}, \quad (p(f) \text{ even}),$$

$$= \frac{C\sinh(1/2\Delta - f)}{\sinh(1/2\Delta)}, \quad (p(f) \text{ odd}),$$

which may be solved to produce an estimate of C, which is then incorporated in (47), (48).

A still more conservative approach would assume that beyond $f = 1/2\Delta$, $p(f)$ changes linearly until it disappears altogether. Thus we have

$$\begin{aligned} p(f) &= C - Df, \quad 1/2\Delta \leq f \leq C/D \\ &= 0 \quad , \quad f > C/D \; . \end{aligned} \tag{49}$$

Using (44) and replacing c by $\Delta(C/D-f)$ we have

$$\sum_{|j|\geq 1} p(f+j/\Delta) \simeq \frac{\Delta}{D}(C^2 - D^2f^2) - (C - Df), \tag{50}$$

if $p(f)$ is even. If $p(f)$ is odd,

$$\sum_{|j|\geq 1} p(f+j/\Delta) \simeq -2Dfc \simeq -2\Delta f(C-Df) \quad . \tag{51}$$

Now if (49) holds also for some f close to, but to the left of $1/2\Delta$, for those f we have

$$p^A(f) \simeq \frac{\Delta}{D}(C^2 - D^2f^2), \quad (p(f) \text{ even}),$$

$$p^A(f) \simeq (C - Df)(1 - 2\Delta f), \quad (p(f) \text{ odd}).$$

For two such frequencies f_1, f_2 we would therefore estimate C and D by

$$\hat{C} = \frac{[p^A(f_2)-p^A(f_1)][f_1p^A(f_2)-f_2p^A(f_1)]}{\Delta(f_1-f_2)^2}$$

$$\hat{D} = \frac{p^A(f_2)-p^A(f_1)}{\Delta(f_1-f_2)} \quad ,$$

if $p(f)$ is even, and

$$\hat{C} = \frac{[f_1p^A(f_2)-f_2p^A(f_1)] - 2\,[f_1^2p^A(f_2)-f_2^2p^A(f_1)]}{(f_1-f_2)(1-2\Delta f_1)(1-2\Delta f_2)}$$

$$\hat{D} = \frac{[p^A(f_2)-p^A(f_1)] - 2\,[f_1p^A(f_2)-f_2p^A(f_1)]}{(f_1-f_2)(1-2\Delta f_1)(1-2\Delta f_2)}$$

if $p(f)$ is odd, and then insert these estimates in (50) and (51).

Notice that the possibility of cospectra or quadrature spectra changing sign over $(1/2\Delta,\infty)$ or $(-\infty, -1/2\Delta)$ is not allowed for by (45), (46), (49).

Partial support of NSF Grant GS-32327X2 is acknowledged

REFERENCES

1. Akaike, H., Some Problems in the application of the cross-spectral method, in *Spectral Analysis of Time Series* (B. Harris, Ed.), John Wiley, New York, 1967, pp. 81-107.

2. Parzen, E., "Regression Analysis of Continuous Parameter Time Series," Proc. Fourth Berkeley Symp. Math. Stat. and Prob., Vol. 1, University of California Press, 1961, pp. 469-489.

3. Hannan, E. J., "Linear Regression in Continuous Time," mimeographed, 1970.

4. Kholevo, A. S., "On Estimates of Regression Coefficients," *Th. Prob. Appl., 14,* 79 (1969).

5. Heble, M. P., "A Regression Problem Concerning Stationary Processes," *Trans. Amer. Math. Soc., 99,* 350 (1961).

6. Bartlett, M. S., *An Introduction to Stochastic Processes,* Cambridge University Press, Cambridge, 1955.

7. Ibragimov, I. A., "On Maximum Likelihood Estimation of Parameters of the Spectral Density of Stationary Time Series," *Th. Prob. Appl., 12,* 115 (1967).

8. Dzhaparidze, K. O., "On the Estimation of the Spectral Parameters of a Gaussian Stationary Process with Rational Spectral Density," *Th. Prob. Appl., 15,* 531 (1970).

9. Dzhaparidze, K. O., "On Methods for Obtaining Asymptotically Efficient Spectral Parameter Estimates for a Stationary Gaussian Process with Rational Spectral Density," *Th. Prob. Appl., 16,* 550 (1971).

10. Akaike H. and Y. Yamanouchi, "On the Statistical Estimation of Frequency Response Function," *Ann. Inst. Statist. Math., 14, 23* (1963).

11. Jenkins, G. M., "Cross-Spectral Analysis and the Estimation of Linear Open Loop Transfer Functions," in *Time Series Analysis* (M. Rosenblatt, Ed.), John Wiley, New York, 1963, pp. 267-276.

12. Hannan, E. J. and P. M. Robinson, "Lagged Regression with Unknown Lags," *J. R. Statist. Soc., 35,* 252 (1973).

13. Robinson, P. M., "The Estimation of Continuous Time Systems from Discrete Data," Ph.D. Thesis, Australian National University, 1972.

14. Granger, C. W. J. and M. Hatanaka, *Spectral Analysis of Economic Time Series,* Princeton University Press, Princeton, 1964.

15. Hannan, E. J., *Multiple Time Series*, John Wiley, New York, 1970.

16. Cooley, J. W., P. A. W. Lewis and P. D. Welch, "The Fast Fourier Transform Algorithm: Programming Considerations in the Calculation of sine, cosine and Laplace Transforms," *J. Sound Vib., 12,* 315 (1970).

17. Jenkins, G. M. and D. F. Watts, *Spectral Analysis*, Holden-Day, San Francisco, 1968.

18. Mehra, R. K., "Identification in Control and Econometrics, Similarities and Differences," *Ann. Econ. and Soc. Measurement, 3,* 21 (1974).

19. Sims, C. A., "Discrete Approximations to Continuous Time Distributed Lag Models in Econometrics," *Econometrica, 39,* 545 (1971).

20. Robinson, P. M., "Instrumental Variables Estimation of Differential Equations," forthcoming in *Econometrica*.

21. Jennrich, R. I., "Asymptotic Properties of Nonlinear Least Squares Estimators," *Ann. Math. Statist., 40,* 633 (1969).

22. Zorn, M., "The Evaluation of Fourier Transforms by the Sampling Method," *Automatica, 4,* 323 (1969).

23. Yaglom, A. M., *Stationary Random Functions*, Prentice-Hall, Englewood Cliffs, N.J., 1962.

FOUR CASES OF IDENTIFICATION OF CHANGING SYSTEMS

T. Bohlin

Royal Institute of Technology
Stockholm, Sweden

1. INTRODUCTION

A general way to approach the problem of identifying systems, whose dynamics or other characteristics change with time, is to attribute the variation to those of a number of "parameters". It will then be possible to treat these parameters, describing dynamics, as any other process variables and, in particular, to estimate them in real time. The current parameter estimates will at all times provide an updated model, that can be used for prediction, control, or other purposes. The point is that techniques for the estimation of process variables are well developed [1].

This holds in particular, when the model structure can be made linear in the parameters, for instance of the form

$$\begin{cases} \underline{\theta}(t+1) = \underline{\Phi}(t)\ \underline{\theta}(t) + \underline{w}_1(t) \\ \underline{y}(t) = \underline{H}(t)\ \underline{\theta}(t) + \underline{w}_2(t) \end{cases} \tag{1}$$

where $\underline{y}(t)$ is the output vector, $\underline{\theta}(t)$ is the parameter vector, and $\underline{w}_1(t)$ and $\underline{w}_2(t)$ are sequences of normal random vectors such that

$$E\underline{w}_1(t)\ \underline{w}_1(s)^T = \underline{R}_1(t)\ \delta(t-s)\ ,$$
$$E\underline{w}_2(t)\ \underline{w}_2(s)^T = \underline{R}_2(t)\ \delta(t-s)\ .$$

When $\theta(t)$ is interpreted as a state vector and $\underline{\Phi}$, $\underline{H}$, $\underline{R}_1$, $\underline{R}_2$ are given functions of time, the problem is known as "state estimation" and solved by the well-known "Kalman filter", see e.g. [2]. However, less well-known is the fact that the coefficient matrices may also be functions, known but not necessarily linear, of past input and output variables $\underline{u}(t-1)$, $\underline{u}(t-2)$,..., $\underline{y}(t-1)$, $\underline{y}(t-2)$,... . The same estimator is still valid. This is a consequence of the fact that past inputs and outputs are known at estimation time.

This fact has little importance in state estimation, but is crucial for parameter estimation [3-5]. It makes it feasible to use the second equation of (1), the "output equation", for describing the process dynamics (instead of the usual first), which makes the first equation available for describing the dynamics of parameter variations. It is clear that in the time-varying case one must always provide some description of how the parameters vary.

Thus, the solution algorithm is the same in the state- and parameter-estimation cases, but the ways coefficient matrices are interpreted in theory and calculated in practice are entirely different.

Now, the interpretation and calculation of $\underline{\Phi}$, $\underline{H}$, $\underline{R}_1$, and $\underline{R}_2$ are far from immediate in a given case and will be a main subject in this chapter. The dynamics of the process are unknown to start with, and therefore, and *a fortiori*, the coefficient matrices $\underline{\Phi}$, $\underline{H}$, $\underline{R}_1$, $\underline{R}_2$ are unknown. It may seem that introducing the unfamiliar structure (1) with its unknown coefficient matrices just adds to the ignorance. However, knowing $\underline{\Phi}$, $\underline{H}$, $\underline{R}_1$, $\underline{R}_2$ does make it feasible to estimate the dynamics, via $\underline{\theta}$, and the point is that the end result is normally less sensitive to coefficient values than to the original unknown θ. Thus it turns out that a very crude description $\underline{\Phi}$, $\underline{R}_1$ of the parameter variations will normally suffice, which requires only a minimum amount of *a priori* information, like how fast parameters vary on the average.

This facilitates the problem of specifying $\underline{\Phi}$ and $\underline{R}_1$; the first equation of (1) is not very critical. A reasonable assumption used throughout this chapter, is that the parameters change with independent, random increments and independently of each other. That is,

$$\underline{\Phi} = \underline{I}, \qquad \underline{R}_1 = \underline{\text{Diag}}\,\{\mu_i^2\}\,, \tag{2}$$

where μ_i is the assumed average rate of change of the parameter $\theta_i(t)$. If μ_i are not known either, there are ways to estimate them.

A philosophically minded reader may argue, that in order to estimate one unknown one must always assume another, *ad infin*. In order to estimate the rate of change of the process parameters one must assume the rate of change (= zero) of the rate of change of process parameters. This is certainly so, and the rationale for still trying to estimate characteristics, on which the primary estimation is founded, is not that one hopes ultimately to find truth, but to decrease the sensitivity to unavoidable assumptions, so that a wrong guess matters less.

In contrast to $\underline{\Phi}$ and $\underline{R}_1$, the forms of $\underline{H}$ and $\underline{R}_2$, as functions of inputs and outputs, depend very much on the actual problem and on the available *a priori* information. The original model structure, containing the unknown parameters, generally does not have the form (1), and so must be rewritten.

Generally, this is not possible to do directly, either because the unknown parameters do not enter linearly, or the resulting function H is not known, or both. This can sometimes be amended by series expansion:

Let, for instance, the original model structure be linear,

$$\underline{y}(t) = \underline{G}(q^{-1})\ \underline{u}(t) + \underline{F}(q^{-1})\ \underline{w}(t) , \tag{3}$$

where $\underline{w}$ is white and $\underline{G}$ and $\underline{F}$ are unknown input- and noise-transfer functions in the backward-shift operator q^{-1}, i.e. $q^{-1}\underline{x}(t) = \underline{x}(t-1)$. They are identical to the corresponding z-transforms, but must be interpreted here as operators, since $\underline{u}$ and $\underline{w}$ are time series.

Now, if $\underline{G}$ and $\underline{F}$ are expanded into [6,7]

$$\underline{G}(q^{-1}) \approx \underline{A}_1(q^{-1})^{-1}\ \underline{B}_1(q^{-1})$$

$$\underline{F}(q^{-1}) \approx \underline{A}_1(q^{-1})^{-1}\ \underline{D}_1(q^{-1})^{-1} ,$$

where $\underline{A}_1$, $\underline{B}_1$, $\underline{D}_1$ are polynomials in q^{-1}, then (3) becomes

$$\underline{A}_2(q^{-1})\ \underline{y}(t) = \underline{B}_2(q^{-1})\ \underline{u}(t) + \underline{w}(t) \quad , \tag{4}$$

where $\underline{A}_2 = \underline{D}_1\ \underline{A}_1$ and $B_2 = \underline{D}_1\ \underline{B}_1$. If all coefficients in $\underline{A}_2$ and $\underline{B}_2$ are taken as unknown parameters $\underline{\theta}$, then (4) has the form of the second equation in (1), where $\underline{R}_2 = \underline{I}$ and $\underline{H}(t)$ is a matrix composed of elements in $\underline{y}(t-1)$, $\underline{y}(t-2),\ldots,\underline{u}(t-1)$, $\underline{u}(t-2)$, ... The exact form depends on how one arranges the coefficients in $\underline{\theta}$ and does not matter in this context.

This gives a model with constant parameters, which can then be made to vary by augmenting the first equation in (1). If, further, the fact is disregarded that parameters in $\underline{\theta}$ are dependent through the common factor $\underline{D}_1$ in $\underline{A}_2$ and $\underline{B}_2$, it is possible to make the convenient choice (2) for the coefficient matrices $\underline{\Phi}$ and $\underline{R}_1$.

If the unknown model is nonlinear, one can sometimes expand it in a functional series, taking the unknown coefficients as parameters, and so obtain the crucial linearity in the parameters. Also more general structures than (1) can be identified easily, as long as they retain the linearity in the parameters. For instance, the first equation in (1) may be replaced by a higher-order stochastic matrix difference equation [8].

It is apparent that the form (1), of models possible to identify conveniently in real time, is quite general, in the sense that a large class of time-varying dynamic relationships can be described, or at least approximated by (1). However, if a given original model structure has to be rewritten to conform to (1), this means that the physical significance of the structure and parameters is lost. This is sometimes a disadvantage *per se*, but not necessarily so, since after finding the parameters $\underline{\theta}$, one can usually also recover the original ones. Also, if rewriting involves series expansion, it increases the number of parameters, which, again, may or may not be a significant disadvantage. More serious, though, is the fact that embodied in the original form and parameters often lies *a priori* information, which may be hard to carry over to the new form.

In the case of series expansion outlined above, rewriting introduces dependences between the coefficients $\underline{\theta}$ in $\underline{A}_2$ and $\underline{B}_2$. These dependences carry *a priori* information, which is sacrificed for the advantages of a simpler model structure and easier computations. This is obviously permissible; there is nothing mathematically illegal in not using all available *a priori* information. It can even be an advantage, since it provides an opportunity to check the original assumptions on structure.

The main advantage of using the structure (1) is thus ease of computation. The estimator is akin to the usual Kalman filter; it is thus an exact solution to the estimation problem, it has minimum estimation error, it is a one-shot method, and it is a true real-time estimation in the sense that it is recursive in time and the storage space needed to evaluate the estimates remains constant, as time progresses and the amount of data increases. This means also that it is feasible to cope with a large number of parameters, which increases the number of relationships that can be approximated by (1).

These considerations form the background to doing real-time identification in the way it will be done in this chapter. In summary: The physical problem, i.e. the purpose and *a priori* information, is given in an unstructured way, the mathematical framework, i.e. the structure (1), is set by the problems one can solve conveniently, and the human identifier's task is to squeeze as much as possible of the former into the latter. He/she has to make the transformation from physical reality to mathematical assumptions, and this transformation should end somewhere in the class of easily solvable problems (1). His/her problem is not so much going from assumptions to solution; this road is well chartered, and computers can handle the task. The problem is, what should $\underline{\Phi}$, $\underline{H}$, $\underline{R}_1$, and $\underline{R}_2$ be like, and what physical variables should constitute $\underline{u}$ and $\underline{y}$?

Unfortunately, methods for finding structures and making mathematical assumptions are poorly developed and even poorly understood. It will probably remain an art for some time, and one has mainly to rely on case studies to acquire skill in the art.

In the four case studies reviewed in Sections 3-6 the starting points are a number of physical problems, and the main subject is the reasoning and the approximations that, in each case, make it possible to use the so convenient recursive filtering techniques associated with (1).

The narratives all follow a rather stereotyped pattern of "purpose", "physical background", "mathematical description", "analysis", "test/verification", and "interpretation of the results in physical terms". This is so on purpose, to emphasize that identification, as well as design based on identification, is a sequence of steps leading from reality, via theory, and back into reality, where the "analysis" step often is the easy one. As indicated, "description" and "interpretation" are the critical steps, since there are few rules to guide them.

An approach to the problem of finding a structure, which is followed in two of the cases, and which is general enough to be called "principle", is the following two-step procedure:

1. Set up a primary possibly crude, stochastic model for the behavior of the physical process, based on *a priori* information only. Write it in such a way that it is possible to compute a residual sequence $\{\underline{e}(t)\}$ from observations, that is uncorrelated in time and uncorrelated with inputs $\underline{u}$, when the *a priori* assumptions hold. Such a sequence is, for instance, the innovations sequence associated with a linear state-vector model.

2. Describe the residual sequence by a linear difference equation with stochastic coefficients of the form

$$\begin{aligned}\underline{e}(t) &+ \underline{a}_1(t)\ \underline{e}(t-1) + \ldots + \underline{a}_n(t)\ e(t-n)\\ &= \underline{b}_1(t)\ \underline{u}(t-1) + \ldots + \underline{b}_m(t)\ \underline{u}(t-m) + \underline{\kappa}(t) + \underline{w}_2(t).\end{aligned} \quad (5)$$

The result has the form (1), where $\underline{e}$ substitutes $\underline{y}$ and (for a scalar e)

$$\underline{H}(t) = \{-e(t-1) \dots -e(t-n)\ u(t-1) \dots u(t-m)\ 1\} \quad (6)$$

$$\underline{\theta}(t) = \underline{\mathrm{Col}}\{a_1(t) \dots a_n(t)\ b_1(t) \dots b_m(t)\ \kappa(t)\}. \quad (7)$$

The idea is, that as long as *a priori* assumptions hold, the parameters will be zero; when they do not hold, the residuals will not be uncorrelated, and real-time identification will modify the total model accordingly.

The scheme is a two-line defense against "malignant nature". If the first line breaks down, because reality contradicts assumptions, the second line is brought into action. Also a third line is feasible--by the fact that the innovations of (5) can be tested for whiteness, uniformity, etc. It is needed in two of the case studies. In this way one can obviously have as many defense lines as one cares.

2. THEORY AND METHODS

Two fundamental analysis problems associated with (1) are i) estimating $\underline{\theta}(t)$ when $\underline{\Phi}$, $\underline{H}$, $\underline{R}_1$, $\underline{R}_2$, $\underline{y}_t$, $\underline{u}_t$ are given and ii) predicting $\underline{y}$ when $\underline{\Phi}$, $\underline{H}$, $\underline{R}_1$, $\underline{R}_2$, $\underline{y}_t$, $\underline{u}$ and $\underline{\theta}$ are given Here $\underline{y}(t)$ is the value at time t, while $\underline{y}_t = \{\underline{y}(t), \underline{y}(t-1), \dots\}$, and $\underline{y} = \{\dots, \underline{y}(t+1), \underline{y}(t), \underline{y}(t-1), \dots\}$. Also, introduce the notation $\hat{\underline{x}}(t+j|t)$ for the expected value of $\underline{x}(t+j)$ conditional on everything known at time t. It is well known that this makes $\hat{\underline{x}}$ the minimum-variance estimate of $\underline{x}$. Repeatedly estimating $\underline{\theta}$ for increasing t will be referred to as "tracking", since $\underline{\theta}$ changes with time.

A. *PREDICTING OUTPUT*

The prediction problem is very easy to solve: The one-step-ahead prediction is

$$\hat{\underline{y}}(t+1|t) = \underline{H}(t+1)\ \underline{\theta}(t+1) \quad (8)$$

with the error

$$\underline{\tilde{y}}(t+1|t) = \underline{w}_2(t+1) \quad . \tag{9}$$

Also, if $\underline{H}(t+1)$ is linear in $\underline{y}_t$ (and possibly nonlinear in $\underline{u}_t$), then the multi-step predictor is obtained by repeated use of (8):

$$\underline{\hat{y}}(t+\tau|t) = \underline{H}[\underline{\hat{y}}(t+\tau-1|t),\ldots,\underline{\hat{y}}(t+1|t),\underline{y}(t),\underline{y}(t-1),\ldots]\underline{\theta}(t+\tau), \tag{10}$$

where $\underline{\hat{y}}(t+\tau|t) = E\{\underline{y}(t+\tau)|\underline{y}_t,\underline{u}_{t+\tau},\underline{\theta}_{t+\tau}\}$, that is, parameters are assumed to be known.

In practice they are not known, and one has to use the estimates $\underline{\hat{\theta}}(t+\tau|t)$. The latter are also easy to find, if $\underline{\hat{\theta}}(t|t)$ is known: For $\underline{\Phi} = \underline{I}$,

$$\underline{\hat{\theta}}(t+\tau|t) = \underline{\hat{\theta}}(t|t), \qquad \tau = 1,2,\ldots \tag{11}$$

If the prediction errors caused by estimation errors in $\underline{\hat{\theta}}$ are much smaller than those caused by the process noise $\underline{w}_2$, which is reasonable to demand from a satisfactory estimation, then it is also reasonable to substitute (11) for $\underline{\theta}(t+\tau)$ in (10). Substitution is not correct generally.

B. TRACKING PARAMETERS

The estimation problem has a solution in the form of a set of recursive equations. In particular, for $\underline{\Phi} = \underline{I}$ and constant $\underline{R}_1$:

$$\left\{\begin{array}{lcl} \underline{R}_{yy}(t|t-1) &=& \underline{H}(t)\,\underline{R}_{\theta\theta}(t|t-1)\,\underline{H}(t)^T + \underline{R}_2 \\ \underline{K}(t) &=& \underline{R}_{\theta\theta}(t|t-1)\,\underline{H}(t)^T\,\underline{R}_{yy}^{-1}(t|t-1) \\ \underline{\tilde{y}}(t|t-1) &=& \underline{y}(t) - \underline{H}(t)\,\underline{\hat{\theta}}(t|t-1) \\ \underline{\hat{\theta}}(t+1|t) &=& \underline{\hat{\theta}}(t|t-1) + \underline{K}(t)\,\underline{\tilde{y}}(t|t-1) \\ \underline{R}_{\theta\theta}(t+1|t) &=& \underline{R}_1 + \underline{R}_{\theta\theta}(t|t-1) - \underline{K}(t)\,\underline{R}_{yy}(t|t-1)\,\underline{K}(t)^T , \end{array}\right. \tag{12}$$

where

$$\underline{\hat{\theta}}(t+1|t) = E\{\underline{\theta}(t+1)|\underline{y}_t,\underline{u}_t\} ,$$

$$\underline{R}_{\theta\theta}(t+1|t) = \text{Var}\{\underline{\theta}(t+1)|\underline{y}_t,\underline{u}_t\}$$

are the mean and the covariance matrix of $\underline{\theta}(t+1)$ conditional on everything known at time t.

The algorithm has the familiar structure of the Kalman filter, see e.g. [2], where $\tilde{y}$ are the innovations or "model errors", and the first, second, and fifth equations together form the Riccati equation. A difference of great importance is that the Riccati equation depends on the input and output data, via $\underline{H}$, and the gain matrix $\underline{K}$ therefore cannot be precomputed. It has to be evaluated in real time. In fact, this constitutes the bulk of computations using (12).

However, it also introduces a nonlinear action into the estimator, which is very important for its performance. It works mainly by the fact that large values of $\underline{H}(t)$ increase the estimator gain $\underline{K}(t)$. The latter is in this way "adapted" to the signal level. If there appears temporary, large changes in the input signal affecting $\underline{H}(t)$, with or without a corresponding, large response in the output signal, this causes the algorithm to adjust $\underline{\hat{\theta}}$ more (for a given model error $\underline{\tilde{y}}$) than otherwise. If the signal is temporarily weak, there will be only small changes in $\underline{\hat{\theta}}$ even for comparable model errors, because the latter are interpreted as due to noise in the signal. In this way the estimator is able to utilize effectively any substantial variation there happen to be in the data and still to suppress spurious variation in the estimate due to noise. The result is vastly superior to linear estimation, where the gain must be a compromise between tracking ability and in sensitivity to noise.

The effect of $\underline{R}_1$ is easy to see from the equations: It keeps $\underline{R}_{\theta\theta}(t|t-1)$ from becoming infinitely small, since $\underline{R}_{\theta\theta}(t|t-1)$ is always $\geq \underline{R}_1$. This has the secondary effect that the gain $\underline{K}(t)$ is prevented from going to zero, which would otherwise happen, meaning that the updating of the estimates would eventually cease. A large gain means that new innovations are weighted more at the expense of old ones, and so increases the rate at which old data are forgotten. Thus, besides having the meaning of average rate of change of parameter values, $\underline{R}_1$ also affects the speed with which the estimator is able to follow changing parameters. This is of course what would be expected, but the point is that the tracking speed of the estimator is something tangible, on which to base the choice of $\underline{R}_1$, while the rates of changes of the actual parameters often are not.

The estimator yields the highest possible accuracy--but only if the coefficient matrices have the right values. It is clear that $\underline{R}_1$ affects the covariance matrix $\underline{R}_{\theta\theta}$ of the estimates directly. This means that the accuracy of the estimation is a matter of choice to some extent; if $\underline{R}_1$ is large, the estimates become erratic, and $\underline{R}_{\theta\theta}$, indicating this, becomes large. If $\underline{R}_1$ decreases, so will both the variations in $\underline{\hat{\theta}}$ and the value of $\underline{R}_{\theta\theta}$, but the latter will be a reasonable measure of estimation accuracy only as long as $\underline{R}_1 \geq$ the rate of change of the true parameters (if it exists). If $\underline{R}_1$ becomes still less, then estimates will be biased, since $\underline{\hat{\theta}}$ will be lagging behind $\underline{\theta}$, and the bias is not included in $\underline{R}_{\theta\theta}$ (which is computed on the assumption that $\underline{R}_1$ is right).

Thus, $\underline{R}_1$ has three interpretations: It i) specifies the average rate of change of the parameters $\underline{\theta}$, it is a design parameter that ii) controls the tracking ability, by controlling indirectly the rate at which old data are forgotten, and iii) affects the estimation accuracy. These intuitive interpretations are important as a background for choosing $\underline{R}_1$.

The effect of $\underline{R}_2$ is the opposite of that of $\underline{R}_1$. In the case of a scalar output, which is the only case appearing in the applications in this chapter, it is only the ratio $\underline{R}_1/R_2$ that affects the estimates $\underline{\hat{\theta}}$. However, the value of R_2 is not immaterial; it does affect the covariance estimate $\underline{R}_{\theta\theta}$ proportionally. From (9) follows that R_2 can be interpreted as the variance of the one-step-ahead prediction error of y when $\underline{\theta}$ is known. It is therefore roughly equal to the high-frequency noise level, which sometimes gives a clue to what value should be assigned to R_2.

C. *ESTIMATING RATE OF CHANGE*

In cases where the choice of $\underline{R}_1$ and $\underline{R}_2$ cannot be based on *a priori* information or on the resulting performance of the estimator (12), there is a further possibility: Given a long data record $(\underline{y}_N, \underline{u}_N)$, the values of $\underline{R}_1$ and $\underline{R}_2$ can be estimated using the Maximum-Likelihood principle, on the assumption that they are known functions of time and a few unknown constants. The method reviewed here will involve a search procedure and will not be a real-time method, since the complete data record is needed several times.

In the case of a scalar output, the logarithm of the likelihood function is [8]

$$L[\underline{R}_1, R_2 | y_N, \underline{u}_N] = -\frac{1}{2} \sum_{t=1}^{N} \log R_{yy}(t|t-1) - \frac{1}{2} \sum_{t=1}^{N} \tilde{y}^2(t|t-1)/R_{yy}(t|t-1) , \qquad (13)$$

where R_{yy} and $\tilde{y}$ are implicit functions of $\underline{R}_1$ and R_2, computed from (12).

The Maximum-Likelihood estimates are the values $\hat{\underline{R}}_1$ and $\hat{R}_2$ that maximize L with respect to the unknown constants in $\underline{R}_1$ and R_2. One constant, λ say, can be estimated explicitly. If R_2 is assumed constant, it is convenient to choose $\lambda^2 = R_2$. If R_2 is not constant, but its variation is known, except for a constant factor, it is reasonable to choose the latter for λ^2. This possibility will be utilized in one of the case studies.

To find the estimate of λ, rewrite (12) and (13) in such a way that the dependence on λ becomes explicit. For this purpose introduce

$$\begin{aligned} \underline{R}_1'(t) &= \underline{R}_1(t)/\lambda^2 \\ R_2'(t) &= R_2(t)/\lambda^2 \\ R'_{yy}(t|t-1) &= R_{yy}(t|t-1)/\lambda^2 \\ \underline{R}'_{\theta\theta}(t|t-1) &= \underline{R}_{\theta\theta}(t|t-1)/\lambda^2 \quad . \end{aligned} \tag{14}$$

Then it is easy to see that (12) remains unchanged if primed variables are substituted for unprimed ones, while (13) becomes

$$\begin{aligned} L[\lambda^2\underline{R}_1',\lambda^2 R_2'|y_N,\underline{u}_N] = & -\frac{N}{2}\log\lambda^2 - \frac{1}{2}\sum_{t=1}^{N}\log R'_{yy}(t|t-1) \\ & -\frac{1}{2\lambda^2}\sum_{t=1}^{N}\tilde{y}^2(t|t-1)/R'_{yy}(t|t-1) \ . \end{aligned} \tag{15}$$

Neither $\tilde{y}$ nor R'_{yy} depends on λ, so that the maximum with respect to λ is attained for

$$\hat{\lambda}^2 = \frac{1}{N}\sum_{t=1}^{N}\tilde{y}^2(t|t-1)/R'_{yy}(t|t-1) \tag{16}$$

and has the value

$$\hat{L}[\underline{R}_1',R_2'|y_N,\underline{u}_N] = -\frac{N}{2}[U(N) + \log V(N)] + \text{constant}, \tag{17}$$

where $U(N)$ and $V(N)$ can be computed recursively from

$$\left\{\begin{aligned} U(t) &= U(t-1) + \frac{1}{t}[\log R'_{yy}(t|t-1) - U(t-1)] \\ V(t) &= V(t-1) + \frac{1}{t}[\tilde{y}^2(t|t-1)/R'_{yy}(t|t-1) - V(t-1)] \\ U(0) &= V(0) = 0 \quad . \end{aligned}\right. \tag{18}$$

By augmenting (18) with the estimator of $\underline{\theta}(t)$ it is easy to evaluate the log-likelihood (17) recursively in time for given values of $\underline{R}_1'$ and R_2'. It is then possible to search for the maximum with respect to $\underline{R}_1'$ and R_2' by processing the data record several times.

This may be cumbersome in case of many parameters. Fortunately, a very crude maximization will often suffice--the method has been used without too much trouble in two of the four case studies.

It follows from (16) and (18) that λ can also be estimated in real time from

$$\hat{\lambda}^2(t) = V(t) \quad . \tag{19}$$

The variable $\hat{\lambda}(t)$ will tend to a constant as $t \to \infty$.

It is conceivable that also λ may change slowly, in particular, this is so when the value of R_2 has been taken for λ^2. Theory does not allow for changing λ. However, it is easy to modify the algorithm for $\hat{\lambda}^2(t)$ heuristically as follows:

$$V(t) = V(t-1) + \rho(t)[\tilde{y}^2(t|t-1)/R'_{yy}(t|t-1) - V(t-1)], \tag{20}$$

where $\rho(t) = \max(1/t, \rho)$. A positive ρ has the same general effect on $\hat{\lambda}^2(t)$ as has a positive $\underline{R}_1$ on $\hat{\theta}(t)$, viz. old data are weighted less, and $\hat{\lambda}^2(t)$ does not tend to a constant. In this case the weights are decreasing exponentially as $(1-\rho)^t$.

It is possible to modify also the likelihood function in the same way, by substituting $\rho(t)$ for $1/t$ in (18). An approximate real-time estimation of $\underline{R}_1$ is then feasible by means of a peak-holding algorithm applied to the function $-U(t) - \log V(t)$, which is the log-likelihood divided by the effective sample length.

Other methods to estimate covariance matrices in connection with recursive filtering have been devised by Bélanger [9], Sage and Husa [10], and Mehra [11].

D. *TESTING VALIDITY OF THE STRUCTURE*

The *a priori* choice of structure, i.e. the forms or values of $\underline{\Phi}$, $\underline{H}$, $\underline{R}_1$, $\underline{R}_2$, and the structure (1) itself may be in doubt. This can be tested as follows:

If all assumptions hold, then the innovations sequences $\underline{\tilde{y}}(t|t-1)$ generated by (12) will be Gaussian, uncorrelated, and have covariance matrix $\underline{R}_{yy}(t|t-1)$. Conversely, from the computed innovations one can observe whether this is actually so, in which case one infers that observations do not contradict assumptions.

A simple test for the scalar case is checking whether the magnitudes of the innovations are in accordance with what would be expected from a normal variable with a given variance, e.g. less than the 3-sigma limit

$$\tilde{y}^2(t|t-1) < 9\ R_{yy}(t|t-1)\ . \qquad (21)$$

This is a crude test, and also improper, if, as would be reasonable, it is repeated for every t. Even if assumptions hold, the innovations will surpass the test limit about once in 400. However, in practice the test can still be used, provided one takes care that the consequences of occasional wrong test results do not become fatal. For instance, it would be unwise to abort the identification just because the test (21) fails.

Other properties than the magnitude can be tested, e.g. the correlation. If estimated values of $\underline{R}_1$ and R_2 are substituted for postulated ones, the test will still be valid approximately for large t.

A case where the test (21) indicates "error" is when the data point $y(t)$ is an "outlier", i.e. is inconsistent with previous data points, while the latter do agree with assumptions. The test will therefore be referred to as a "test of consistency". Outliers occur in practice, when measurement equipment malfunctions, and ever so often for no apparent reason. In such cases it is reasonable to wait and see if outliers will disappear spontaneously, and meanwhile indicate the fact that $y(t)$ is an outlier by setting $R_2(t) = \infty$. This gives $\underline{K}(t) = \underline{0}$, which simply means not updating $\hat{\underline{\theta}}(t+1|t)$. If outliers persist, this is reason for alarm and for activating diagnostic routines to find out the possible cause of the inconsistency.

How to design such procedures and what to do in general, if test results show that *a priori* assumptions cannot hold, is a new problem. Generally, test results cannot indicate conclusively *what* is wrong. They can possibly discriminate between alternative hypotheses for what might have gone wrong. A simple example of such an alternative hypothesis, which was used above, is that the error level $R_2(t)$ has increased temporarily (due to malfunctioning equipment).

E. SUMMARY OF ALGORITHMS

The algorithms for estimating parameters, predicting output, evaluating likelihood, and testing consistency are collected here for easy reference. They are valid for a scalar output:

Start values:

$$\hat{\underline{\theta}}(1|0) = 0, \quad \underline{R}'_{\theta\theta}(1|0) = \underline{R}_0, \quad U(0) = 0, \quad V(0) = 0 \ .$$

For all t:

Relative variance of innovations,

$$R'_{yy}(t|t-1) = R'_2(t) + \underline{H}(t)\ \underline{R}'_{\theta\theta}(t|t-1)\ \underline{H}(t)^T \ . \tag{22a}$$

Estimator gain,

$$\underline{K}(t) = \underline{R}'_{\theta\theta}(t|t-1)\ \underline{H}(t)^T / R'_{yy}(t|t-1) \ . \tag{22b}$$

Innovation,

$$\tilde{y}(t|t-1) = y(t) - \underline{H}(t)\ \hat{\underline{\theta}}(t|t-1) \ . \tag{22c}$$

Consistency test,

$$\text{If} \quad \tilde{y}^2(t|t-1) > 9\ V(t-1) R'_{yy}(t|t-1), \quad \text{then} \quad \underline{K}(t) = 0 \ . \tag{22d}$$

Parameter estimate,

$$\underline{\hat{\theta}}(t+1|t) = \underline{\hat{\theta}}(t|t-1) + \underline{K}(t)\ \tilde{y}(t|t-1) \quad . \tag{22e}$$

Relative covariance of estimate,

$$\underline{R}'_{\theta\theta}(t+1|t) = \underline{R}'_1(t) + \underline{R}'_{\theta\theta}(t|t-1) - \underline{K}(t)R'_{yy}(t|t-1)\underline{K}(t)^T \quad . \tag{22f}$$

Auxiliary variables for likelihood and residual variance,

$$\rho(t) = \max(1/t,\rho)$$

$$U(t) = U(t-1) + \rho(t)[\log R'_{yy}(t|t-1) - U(t-1)], \tag{22g}$$

$$V(t) = V(t-1) + \rho(t)[\tilde{y}^2(t|t-1)/R'_{yy}(t|t-1) - V(t-1)] \quad . \tag{22h}$$

For arbitrary $t = k$:

Predicted output,

For $j = 1,\ldots,\tau$ do

$$\hat{y}(k+j|k) = \underline{H}[\hat{y}(k+j-1|k),\ldots,\hat{y}(k+1|k),y(k),y(k-1),\ldots]\ \underline{\hat{\theta}}(k+1|k) \quad . \tag{23a}$$

Estimated covariance of parameter estimates,

$$\underline{R}_{\theta\theta}(k+1|k) = V(k)\ \underline{R}'_{\theta\theta}(k+1|k) \quad . \tag{23b}$$

Estimated residual variance,

$$\hat{\lambda}^2(k) = V(k) \quad . \tag{23c}$$

Time-average of log-likelihood,

$$\ell[\underline{R}'_1,R'_2|y_k,u_k] = -U(k) - \log V(k) \quad . \tag{23d}$$

Estimated rate of change of parameters,

Find $\underline{\hat{R}}'_1(k)$ and $\hat{R}'_2(k)$ that maximize $\ell[\underline{R}'_1,R'_2|y_k,\underline{u}_k]$

$$\hat{\underline{R}}_1(k) = \hat{\underline{R}}'_1(k)\ V(k) \quad , \tag{23e}$$

$$\hat{R}_2(k) = \hat{R}'_2(k)\ V(k) \quad . \tag{23f}$$

(If R_2 is constant, it is convenient to choose $\lambda^2 = R_2$, so that $R'_2 = 1$).

3. ANALYSIS OF DRYER CONTROL

The first of the four case studies was carried out in 1967 [12] in connection with the computer-automating of a paper mill at Billerud AB, Sweden [13]. The purpose was to find out why, initially, control of the dryers proved unexpectedly difficult. It appeared that feedback control that was well tuned and working satisfactorily for long periods, suddenly failed to do so, even under apparently unchanged operating conditions.

A hypothetical reason would be that dryer dynamics had changed, possibly spontaneously, but most likely as a consequence of unmeasurable external disturbances. Two available input/output records were therefore analyzed in order to confirm or reject the hypothesis.

A. TECHNICAL BACKGROUND

A multi-cylinder dryer unit of a paper machine dries the paper by running it over a number of rotating, steam-heated cast-iron cylinders. The heat evaporates the water in the paper, and this brings the moisture content of the finished paper down to less than 10%. The residual moisture content is an important quality variable and the subject of control. This is exerted by manipulation of the steam pressures in the cylinders, which affects the rate of evaporation of water from the paper.

In the paper machine under study the dryer unit was divided into four sections, comprising a number of cylinders each; the cylinders in each section shared steam-feed system. This means that the entire unit was controlled via four steam valves. However only that of the fourth section, nearest to the end of the dryer unit, was controlled directly by feedback from the moisture gauge at the end of the dryer. Thus, the dynamics of interest for the feedback control were those of the fourth dryer section only, but including the steam generator and feed-control systems.

Drying is, physically, a complicated and nonlinear process. It is still reasonable to believe that it can be linearized around an operating point and thus, for control purposes, can be described by linear dynamics at each such point. However, dynamics may then depend on the operating point, and it is possible that it will vary with external variables (i.e. external from the point of view of the fourth section). There are a great many candidates for such external variables, e.g. steam-generator load, condensor load, and the operating points of the first three dryer sections.

B. MATHEMATICAL DESCRIPTION

Earlier investigations [14] indicated that the last dryer section could be described (around an operating point) by a model structure of the form

$$\begin{cases} x(t) = ax(t-1) + b_0u(t-k) + b_1u(t-k-1) \\ v(t) = v(t-1) + \lambda w(t) \\ y(t) = x(t) + v(t) \end{cases} \tag{24}$$

where $y(t)$ is the measured moisture content (after smoothing), $u(t)$ is the steam-pressure regulator set point (controlling the steam valve), and $\{w(t)\}$ is a sequence of uncorrelated random variables with zero means and unit variances. The parameters a, b_0, b_1, and λ depend on the operating point. They have simple relations to commonly used, practical process characteristics:

$$\begin{aligned} a &= \exp(-h/T) , \\ b_0 &= g[1 - \exp(-(kh+h-T_0))] , \\ b_1 &= g[-a + \exp(-(kh+h-T_0))] , \end{aligned} \tag{25}$$

where T = time constant, T_0 = dead time, $kh \le T_0 < kh + h$, g = gain, h = sampling interval. This is valuable, since it facilitates a comparison with what is known about the process from experience.

Smoothing of the output serves to decrease the high-frequency noise, which is necessary for the simple form of the disturbance model $v(t)$ in (24) to hold. Thus, the model covers plant plus smoothing filter, which is the process sensed by the feedback control law. If smoothing is not effective, then a term $-\lambda c w(t-1)$ must be added to $v(t)$. This is equivalent to adding an independent white-noise sequence to $y(t)$, which makes the output signal more high-frequent.

A natural way to generalize (24) to changing dynamics is to assume that the process' time constants, gains, etc., and hence a, b_0, b_1, vary slowly with time. They are then no longer "constants", and the theory on which those concepts are founded may get into jeopardy, but the modification has intuitive appeal for slow changes, and is as valid a generalization of the original model structure as anything else.

However, with a, b_0, and b_1 as unknown parameters $\underline{\theta}$, and varying with time, the model (24) is not possible to write on the form (1) for which the convenient estimator (12) applies. For constant parameters it can be written as

$$y(t) = \frac{b_0 + b_1 q^{-1}}{1 - aq^{-1}} q^{-k} u(t) + \frac{\lambda}{1 - q^{-1}} w(t) , \qquad (26)$$

which is of the form (3), but it cannot be rewritten in the form (4) needed to generalize to variable coefficients. In essence, it is necessary, for the linear least-squares method to apply, that the driving noise sequences are white. Rewriting (26), to bring out $w(t)$, yields

$$(1 - aq^{-1})y'(t) = (b_0 + b_1 q^{-1})q^{-k}u'(t) + \lambda w(t) , \qquad (27)$$

where

$$y'(t) = P(q^{-1})y(t), \quad u'(t) = P(q^{-1})\,u(t),$$

with

$$P(q^{-1}) = (1 - q^{-1})(1 - aq^{-1})^{-1} \quad .$$

The result is, formally, of the desired form (4), where the input and output sequences are "prefiltered" by the common operator $P(q^{-1})$. However, a appears in the prefilter. Prefiltering is the idea behind the "Generalized Least-Squares method." For constant parameters this increases the applicability of least-squares identification to other processes than those driven by white noise, but a difficulty is that the prefilter must be specified. Often it is not known; in the present case a is unknown. Estimating also the prefilter from data leads to the "Modified Generalized Least-Squares method" [7], or to the "Maximum-Likelihood method" [15], depending on the algorithm used for finding the solution. However, none of these are readily extended to variable parameters [16].

Now, if one writes

$$y'(t) = a(t)y'(t-1) + b_0(t)u'(t-k) + b_1(t)u'(t-k-1) + \lambda w(t) \quad , \tag{28}$$

where y' and u' are prefiltered using

$$P(q^{-1}) = \frac{1-q^{-1}}{1-a_0q^{-1}} \quad , \tag{29}$$

this represents a generalization of (27). Notice that this also makes it a generalization of (24) or (26), since all forms are equivalent for constant parameters. It is possibly not the most immediate generalization from an intuitive point of view, since it has been necessary to employ the rather unphysical trick of splitting the a-parameter in (24) into two, and then letting only one vary. However, it is certainly allowed mathematically, and it does make the identification problem easy. Intuitively, the generalization should be valid as long as $a(t)$ stays close to a_0. Also, as soon as it is not valid, this can be detected, see Section 2.D.

It remains to describe the way in which parameters change, if at all. Little is known about this *a priori*, except that it is reasonable to imagine that the effects of small changes accumulate. The simplest way to express this is by a "random-walk" model

$$\theta_i(t) = \theta_i(t-1) + \mu_i w_i(t) \quad , \tag{30}$$

where $\theta_i(t)$ is $a(t)$, $b_0(t)$, or $b_1(t)$, and $w_i(t)$ are uncorrelated and have zero means and unit variances. The constants $\mu_i \ll 1$ are unknown and probably different.

This does not exhaust the *a priori* knowledge about the parameter variations. Mainly, b_0 and b_1 should remain negative (more steam means dryer paper), and a must remain in the interval $(0,1)$. The model (30) yields unlimited variation. In the case of a this suggests a stationary model, preferably centered around a_0, for instance

$$a(t) - a_0 = (1-\beta)[a(t-1)-a_0] + \mu_1 w_1(t) \quad , \tag{31}$$

where β is small but positive. However, even if β affects the *a priori* model (31) very much, it affects the end result, the *a posteriori* variation, only little. Thus, one gains little by complicating the model (30).

In summary, the following model structure has been conceived for the dynamic relation of steam-pressure set point to smoothed moisture-gauge readings:

$$\begin{cases} y'(t) = a(t)y'(t-1) + b_0(t)u'(t-k) + b_1(t)u'(t-k-1) + \lambda w_0(t) \\ a(t) = a(t-1) + \mu_1 w_1(t) \\ b_0(t) = b_0(t-1) + \mu_2 w_2(t) \\ b_1(t) = b_1(t-1) + \mu_3 w_3(t) \\ u'(t) = a_0 u'(t-1) + \Delta u(t) \\ y'(t) = a_0 y'(t-1) + \Delta y(t) \end{cases} \qquad (32)$$

where $w_0(t),\ldots,w_3(t)$ are Gaussian and orthonormal, and a_0 is known.

C. *ANALYSIS AND INTERPRETATION*

The analysis was carried out on historical data, collected in open loop and for other purposes. The set point of the analog steam-pressure regulator had been varied stepwise by means of one of the control computer's actuators, and the moisture readings at the dry end had been recorded simultaneously. The two samples analyzed had the following specifications:

Sample 1: $N = 331$ points, interval $h = 18$ sec, length = 1.66 hours. $y(t)$ is an average of five measurements, taken with 3.6 sec interval.

Sample 2: $N = 484$ points, interval $h = 36$ sec, length = 4.84 hours. $y(t)$ is an exponential average updated every 7.2 sec. The weight on raw measurements is 0.25. During the recording the set point of the pulp flow to the paper machine had also been varied. Pulp flow affects the moisture. Before analyzing the effect of steam pressure, the computed effect of pulp flow changes was therefore subtracted from the moisture recording.

With

$$\underline{H}(t) = \{-y'(t-1)\ u'(t-k)\ u'(t-k-1)\}$$

$$\underline{\theta}(t) = \underline{\text{Col}}\{a(t)\ b_0(t)\ b_1(t)\} \tag{33}$$

$$\underline{\Phi} = \underline{I},\quad \underline{R}_1 = \underline{\text{Diag}}\{\mu_i^2\},\quad R_2 = \lambda^2,$$

the model (32) takes the form (1), and the tracking algorithm (22a-f) is applicable for estimating the values of coefficients as they change.

The specificiations needed to do this are the values of a_0, h, k, λ, μ_1, μ_2, μ_3 and start values $\hat{\underline{\theta}}(1|0)$ and $\underline{R}_{\theta\theta}(1|0)$. Of these the sampling interval was given and k and a_0 were postulated. The product kh represents an assumed (nominal) dead time, in both cases set to 36 sec. The actual dead time may vary according to eq. (25) in addition to the nominal value. For prefiltering was used $a_0 = 0.6$, interpreted as a nominal value, around which a(t) was assumed to vary. The start value $\hat{\underline{\theta}}(1|0)$ was unknown. To simulate this $\hat{\underline{\theta}}(1|0)$ was set to zero and $\underline{R}_{\theta\theta}(1|0)$ was set to a high value, = 100 $\underline{I}$.

The values of μ_1, μ_2, μ_3, and λ were estimated using (22g,h) and (23b-e). No systematic search strategy was used for maximizing the likelihood, but trial values were entered manually into the computer. With three parameters to search for, and using a small computer (IBM 1130), the search took approximately one hour per sample. Using automatic search this time would be reduced drastically. However, since the analysis was to be limited to the two cases, the programming of a search procedure would have to be added, and this would have taken more time.

Postulated and estimated constants are given in Table 1.

Results of the second part of the analysis (tracking) are illustrated by Figures 1 and 2. Standard deviations D of the estimation errors have been computed as the square roots of the diagonal elements in $\hat{\underline{R}}_{\theta\theta}(t+1|t) - \hat{\underline{R}}_1$.

TABLE 1

Specifications for Coefficient Tracking

	Sample 1	Sample 2
Postulated		
N	331	484
h	18 sec	36 sec
k	2	1
Estimated		
$\hat{\lambda}$	0.258	0.179
$\hat{\mu}_1$	0	0.003
$\hat{\mu}_2$	0	0.057
$\hat{\mu}_3$	0.041	0.013

Comments on the results:

- The data contain additive disturbances of approximately the same magnitude as the effect of variations of the input variable.
- One of the coefficients has varied significantly in each sample, viz. $\hat{b}_1$ in Sample 1 and $\hat{b}_0$ in Sample 2. Other coefficients have varied little or not at all. This is also seen from the values of $\hat{\mu}_i$, interpreted as estimated rates of change. The range of variation is substantial and may well explain the observed poor performance of a fixed-parameter linear regulator. Thus, the results support the initial hypothesis.
- The coefficients in the two samples are not comparable directly, partly because the different smoothings of measurements change the apparent process dynamics differently, and partly because the different sampling intervals mean different model structures (different k). Possibly, the estimated drift rates $\hat{\mu}_i$ can be compared. Those of Sample 1 must first be multiplied by $\sqrt{2}$, since the sampling interval is half that of Sample 2.

This gives the value 0.58 for the highest drift rate in Sample 1, to be compared with 0.57 for Sample 2. This is obviously too good an agreement to be anything more than a coincidence. The fact that the a-coefficients agree, in spite of the fact that structures are different, is possibly a consequence of the common prefilter parameter $a_0 = 0.6$.

●The beginning of each sample shows large variations in estimates and large standard deviations. Variations are here mainly a manifestation of uncertainty. This is a consequence of the large initial uncertainty adopted. As information is collected, uncertainty decreases. In cases were $\hat{\mu}_i$ are much different from zero, the variances reach approximately stationary values, and the variations in the coefficient estimates then indicate real variations in plant parameters.

●The estimates often change stepwise. This cannot be interpreted as meaning that the plant parameters necessarily have changed their values in the same way. Large changes in the estimates may occur at large changes in the input u, if, before that, u has varied only little. Then the system is suddenly excited, and the information contents in y(t) increase, which the estimator utilizes for updating $\hat{\underline{\theta}}(t)$. If the deviation between model and plant has become large during the "quiescent" interval, the adjustment will be large. The effect is seen clearly in Figure 2 for $\hat{b}_0$ at time 1.3 hours. Also, compare with the situation at time 2.8 hours, where the change in estimate should indicate a real change, since, before that, the plant has been well excited.

●Since the analysis was carried out off line, and a possible failure therefore would not have any serious consequences (as long as it would be detected), it was not considered necessary to use the consistency test (22d). Instead, the residuals (the innovations) $\tilde{y}(t|t-1)$ were plotted (bottom curves in Figures 1 and 2) for the purpose of assessing the validity of *a priori* assumptions after the analysis. Usually, serious deficiencies in

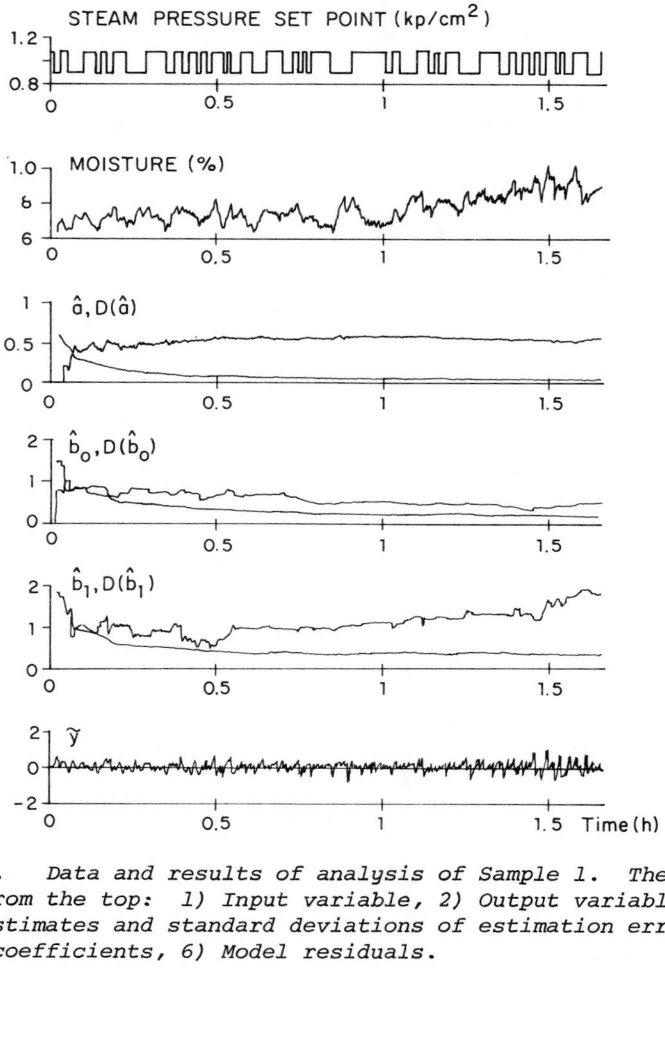

Fig. 1. Data and results of analysis of Sample 1. The curves show from the top: 1) Input variable, 2) Output variable, 3-5) Estimates and standard deviations of estimation errors for model coefficients, 6) Model residuals.

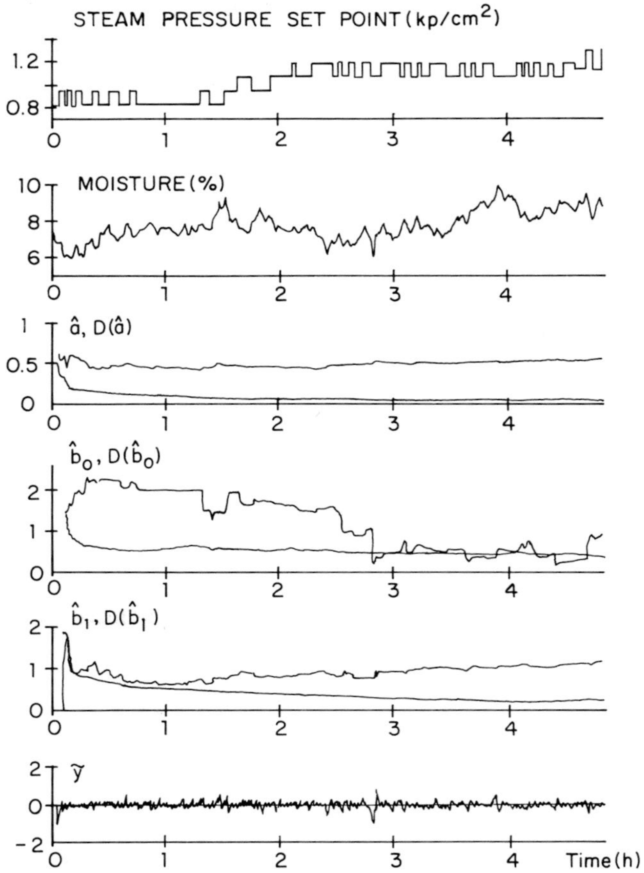

Fig. 2. Data and results of analysis of Sample 2. The meanings of the curves are the same as in Fig. 1.

the model structure reveal themselves as irregularities, patterns, or correlation in the ideally white-noise appearance of the residuals. In this way it is feasible to support the thorough but cumbersome numerical tests for whiteness by the quicker and more versatile diagnostic ability of the eye.

Thus, judging from the residuals, nothing is seriously wrong with the assumptions. Possibly, the variance of the residuals of Sample 1 is greater towards the end of the sample, indicating that the assumption of a constant disturbance level λ might not hold. This is probably a consequence of the fact that the average moisture content has also risen towards the end. The change of level is of such a magnitude that nonlinear effects are likely. Also, there is a short burst of large residuals in Sample 2 at 2.8 hours, associated with a sudden change in $\hat{b}_0$ and a temporary drop in moisture. It is reasonable to infer that the cause of these simultaneous events, even if unknown, is one and the same, that it is an external disturbance, and, therefore, that it does not necessarily indicate that the model is wrong. An investigation of the normalized residuals $\xi(t) = \tilde{y}(t|t-1)/\sqrt{R_{yy}(t|t-1)}$ was carried out for Sample 1. The sample distribution of $\xi(t)$ agreed well with the normal distribution.

D. *EPILOGUE*

The control problem was resolved by Wittenmark and Borisson [17] by applying a self-tuning regulator, and after the dryer unit and the steam-generating plant had been rebuilt to a larger capacity. The results in the reference indicate less variation in dryer dynamics than was observed before the rebuilding.

4. ANALYSIS OF EEG-SIGNALS WITH CHANGING SPECTRA

This case study [18] was carried out in collaboration with Dr. Wennberg of the Karolinska hospital in Stockholm, Sweden. Its purpose was to find a new description of such electroencephalographic signals that would, intuitively, be classified as nonstationary. The general idea of having such a description is that it may relate to the patient's medical status and eventually have diagnostic value.

A. PHYSIOLOGICAL AND METHODOLOGICAL BACKGROUND

Electroencephalographic signals, or EEG-signals for short, are induced by the electrical activity in the brain and are the results of measurements by electrodes placed on the patient's scalp. The recordings of such measurements resemble random signals, occasionally containing a dominating frequency called "rhythm" and/or pulse-like activity called "spikes". EEG-recordings are generally believed to hold information about the brain and are used by the physician as a means for diagnosis.

Several attempts have been made to describe EEG-signals mathematically [19-21]. The purpose of this is to supplement the diagnostician's visual inspection of EEG-recordings with mathematical analysis, and in general to provide means for accessing the information in the EEG. Essentially two kinds of description have been tried:

1. Stationary stochastic processes. The information obtained comprises various time-invariant statistical characteristics of the EEG-signal, such as power spectrum or amplitude distribution or other characteristics derived from these. Methods using correlation analysis [22], frequency analysis [23], reversed correlation analysis [24], and the fitting of time-series models [25-27] belong to this category.

2. Wave patterns. This yields essentially the frequency of occurrence and the form of particular, characteristic patterns [28,29].

The essential prerequisite for applying the first alternative is that characteristics be constant in time. Long samples (of the order of tens of seconds) are used for the analysis. In the second alternative any wave form is allowed in principle; however, to render the analysis practical, the wave forms must belong to only few and restricted classes. This means that the attention is focused on transient properties, and short samples are used.

In some cases neither description would be adequate. Such a case is when a pure wave-pattern analysis would describe only parts of the sample--essentially those parts showing a dominating pattern, such as spikes--and, further, the remaining parts, the so-called "ongoing activity", are not stationary but still believed to hold information [30]. It is then customary, for analysis, to divide time into segments, and base the analysis on the assumption that characteristics are constant during each segment [31]. In other cases an algorithm based on the stationary theory is modified heuristically to meet changing conditions, in essence by applying decreasing weights to older parts of the signal. Frequency analyzers, in effect, do this [32]. However, the basic theory still assumes stationarity, and this means that changing characteristics distort the results of analysis.

That nonstationary EEG-signals exist is generally recognized--even if the concept of "stationarity" itself may be debated in this context--and is intuitively clear from inspecting some recordings (see e.g. Figure 4). EEG-signals are generally "stationary" only for a limited time, e.g. 20 sec [33]. Even shorter samples contain "nonstationarities", which occur spontaneously, and particularly, it seems, in pathological cases. If this is indeed so, then the property of "nonstationarity" itself (if it can be measured) would have physiological relevance.

Also, changes in signal character appear as a consequence of the very severe measurement problems. Even the slightest physical activity, like for instance eye movements, causes so-called "artefacts", which appear as large transient disturbances of low frequency. Before analysis using a "stationary" method such disturbances must be located and removed, which is cumbersome, or suppressed by high-pass filtering, which also suppresses some of the information in the EEG at low frequencies.

For these and other reasons a description that obviates the assumption of stationarity, even in theory, is desirable. The present case study reviews an attempt to establish and verify such a description [18]. It generalizes a model previously used for stationary signals--the "autoregressive series" [25,27,34]--by allowing characteristics to vary in time. It is believed that this will increase the applicability of mathematical analysis of EEG-signals, in particular:

1. when it is difficult to attain the ideal measurement conditions needed for obtaining a stationary signal,

2. when the character of the EEG changes spontaneously, as has been observed in some cases,

3. when a change in character is induced by stimuli, and the transient course of the change is interesting, as is conceivable in physiological experiments.

B. *MATHEMATICAL DESCRIPTION*

The model used for a description was conceived in the following way: Zetterberg proposed and applied successfully the following class of models to stationary EEG-signals sampled with the constant interval h [26,35,36]:

$$y(t) + \dots + a_n y(t-n) = \lambda[w(t) + \dots + c_m w(t-m)] + \kappa \tag{34}$$

where a_i and c_i are used to characterize individual signals. This model has a power spectrum of the form

$$S(f) = \lambda^2 h \left| \frac{C(\exp 2\pi i f h)}{A(\exp 2\pi i f h)} \right|^2 , \tag{35}$$

where A and C are the polynomials

$$\begin{aligned} A(z) &= 1 + a_1 z + \dots + a_n z^n, \\ C(z) &= 1 + c_1 z + \dots + a_m z^m \end{aligned} \quad . \tag{36}$$

Thus, identifying a signal using a model of the class (34) means approximating its spectrum by an expression of the form (35). This form is always possible to write as a ratio of polynomials in $\cos 2\pi fh$. It follows that the form (35) can approximate a continuous spectrum in the domain $f \epsilon (0, 1/2h)$, (A and C exist due to the factorization theorem). However, the same is true for a single polynomial. Therefore, it should be possible to characterize the EEG-spectrum by the A-polynomial only and let the C-polynomial be common to all signals. That this is feasible has been demonstrated by Gersch [25] and Bohlin [34]. The order n of the A-polynomial and thus the number of a-coefficients needed for a satisfactory description increases, but not too much.

Now, by letting the a-coefficients and the constant κ vary randomly with time in the manner discussed in 3.B, eq. (30), one obtains the following model

$$\begin{cases} e(t) + a_1(t)e(t-1) + \ldots + a_n(t)e(t-n) = \lambda w_0(t) + \kappa(t) \\ a_i(t) = a_i(t-1) + \mu w_i(t) \\ \kappa(t) = \kappa(t-1) + \mu_1 w_{n+1}(t) \\ y(t) = e(t) + c_1 e(t-1) + \ldots + c_m e(t-m) \end{cases} \quad (37)$$

where $\{w_i(t)\}$ are independent sequences of uncorrelated, normal variables with zero means and unit variances.

This class of models is clearly a generalization that has the properties one would intuitively attribute to a "nonstationary" random process. The a-coefficients, as functions of time, characterize the particular EEG described, as distinguished from other EEG's describable by the same class of models. The implicit assumption is that the characteristics and the κ-term vary less than the signal itself on the average. The term $\kappa(t)$ has been introduced to account for various additive low-frequency disturbances, caused by random variation in the measurement conditions; it is not to be regarded as a characteristic of the EEG.

It would be conceivable to let also the c-coefficients vary, and possibly advantageous, since it would undoubtedly reduce the orders n and m. However, an exact solution of the corresponding analysis problem is not known. There are approximate solutions, see for instance [16], but they are not yet sufficiently well developed for the present purposes.

The rate at which characteristics change is specified by μ. Thus, μ is a property of the EEG, while μ_1 is not. The generalization in (37) means using nonzero values of μ and μ_1. This, of course, implies that one either has to specify μ, μ_1 *a priori* to select a model from the class (37), or else estimate them from an EEG-sample, and so assign an individual model for each sample. The former is clearly more convenient and may be adequate in practice, the latter may be feasible in some applications.

It would be feasible to use different μ-values for different coefficients, but this would make the problem of finding their values much more difficult, without an obvious return in increased applicability.

The a-coefficients are difficult to interpret physically. Other characteristics will therefore have to be computed from the numbers $n, m, c_1, \ldots, c_m, a_1(t), \ldots, a_n(t)$, describing the behavior of the EEG-signal around time t. Such a characteristic is the power spectrum. Since the model is not stationary, it does not have a power spectrum in the usual sense. However, it is possible to define a "time-variable power spectrum" by

$$S(f|t) = \lambda^2 h \left| \frac{C(\exp 2\pi i f h)}{A(\exp 2\pi i f h|t)} \right|^2 , \qquad (38)$$

where the denominator is the A-polynomial with variable coefficients. It coincides with the ordinary spectrum, when coefficients become constants, and has the properties one would intuitively require from a "time-variable spectrum". For instance, it is defined as an ensemble average, i.e. as an instantaneous property of the signal, and not as the usual time average. Its estimate depends only on the recent and not on the older history of the signal. In this way the model (37) provides a description that has intuitive appeal.

The power spectrum is not the only, nor necessarily the best physical characterization of the model, but the point is that the bulk of the analysis of the EEG-signal--finding the a-coefficients--will need the form (37) only, and will be independent of what form one eventually wants the result in.

C. *THE ANALYSIS ROUTINE*

Given an EEG-sample, the analysis problems associated with the model (37) are several and are solved as follows:

1. Tracking

Given $n, m, c_1, \dots, c_m, \lambda, \mu, \mu_1$,
estimate $a_1(t), \dots, a_n(t), \kappa(t)$.

This is solved using the equations (22a-f) applied to the sequence $\{e(t)\}$, where, from (37)

$$e(t) = y(t) - c_1 e(t-1) - \dots - c_m e(t-m) \quad . \qquad (39)$$

The consistency test (22d) is included, since it is not reasonable to expect that a member of the class of models (37) will apply for all t-values. Therefore, a "reject" result of the test activates a spike-recognition scheme [18], not detailed here. The rationale for this is that since the test result has indicated that data contradict the hypothesis (37) at the particular point t_i, alternative hypotheses must be employed instead and tested for that point. One such alternative is the appearance of a spike at time t_i.

2. Interpretation

Given $n, m, c_1, \dots, c_m, a_1(t), \dots, a_n(t), \lambda$, find characteristics that have physical meanings.

The power spectrum is computed from (38).

3. Estimation of rate of change

Given n, m, $c_1,\ldots,c_m$,
estimate λ, μ, μ_1.

It follows from the model structure (37) that μ is a measure of how fast coefficients change, or, loosely, how non-stationary the EGG is. Also, the constant μ_1 is the rate of change of the term $\kappa(t)$. Since neither is known *a priori*, it is reasonable to use the method discussed in Section 2.C and leading to the equations (22g,h) and (23c-f).

Now, $\kappa(t)$ was introduced to account for various additive low-frequency disturbances. Hence, if $\mu_1 \neq 0$, so that $\kappa(t)$ is allowed to vary, this means in effect that a certain, low-frequency part of the measured signal will be absorbed in $\kappa(t)$ and thus excluded from what is *a priori* regarded as the interesting part of the signal, the part to be characterized by $a_1,\ldots,a_n$. It is clear that this separation into "interesting" and "uninteresting" properties of the measured signal is arbitrary and determined by choosing the value of μ_1. In particular, if $\mu_1 = 0$, then all the signal frequencies are subject to modeling, including any disturbances. It follows that choosing $\mu_1 > 0$, in effect, high-pass filters the signal. Also, it is reasonable to expect that an estimated value of μ_1 would be influenced mainly by the low-frequency disturbances, the "artefacts", and therefore would say little about the EEG.

For these reasons it would not seem worth the effort to try and estimate μ_1, and would seem better to choose μ_1 *a priori*; for instance, to set $\mu_1 > 0$ only in cases where prefiltering is deemed necessary for special reasons; otherwise set $\mu_1 = 0$ and keep in mind that part of the low frequencies in the model spectrum may correspond to artefacts.

To see the relation between μ_1 and the characteristic of the equivalent prefilter rewrite (37):

$$A(q^{-1}|t)C(q^{-1})^{-1}y(t) = \lambda w_0(t) + \mu_1(1-q^{-1})^{-1}w_{n+1}(t) . \tag{40}$$

The right-hand side may be written as $\lambda F'(q^{-1})w_0(t)$, where the equivalent filter F' has the power transmission

$$|F'(\exp 2\pi ifh)|^2 = 1 + \lambda^{-2}\mu_1^2|1 - \exp 2\pi ifh|^{-2} . \tag{41}$$

Now, if, as assumed, $A(q^{-1}|t)$ does not change fast with t on the average, then (40) becomes approximately

$$A(q^{-1}|t)C(q^{-1})^{-1}F'(q^{-1})^{-1}y(t) = \lambda w_0(t) , \tag{42}$$

so that the signal described by A and C is $F'(q^{-1})^{-1}y(t)$. Hence, the equivalent prefilter is F'^{-1} with the power transmission

$$[1 + \lambda^{-2}\mu_1^2|1 - \exp 2\pi ifh|^{-2}]^{-1} , \tag{43}$$

which defines a high-pass filter.

The 3 dB cut-off frequency f_1 is obtained by equating (43) to 1/2, which gives for low frequencies

$$f_1 \approx \mu_1/2\pi\lambda h = \mu_1'/2\pi h . \tag{44}$$

This provides a basis for choosing μ_1'.

With fixed f_1 tracking depends on $\mu' = \mu/\lambda$. This means in particular, that for a constant μ tracking is not invariant under a change of scale of the signal amplitude, since this changes λ. In order to define a measure of nonstationarity that i) is a characteristic of the signal as well as ii) makes the equations invariant under scale changes, introduce

$$d = y_e \, \mu/\lambda , \tag{45}$$

where y_e is the rms value of the signals. Then d is clearly independent of scale. Secondly, the tracking equations, in essence, depend only on d. Thirdly, d is a characteristic of the signal; it is zero for stationary signals and has higher values the faster coefficients vary.

Thus, it is proper to call d an "index of nonstationarity" and consider classifications of EEG's according to the values of d.

Also the likelihood is a function only of d and is evaluated by (22g,h) and (23d). Since d is scalar, it is not unduly difficult to determine the nonstationarity index by maximizing the likelihood. This has been done in a number of test cases.

In summary, the design parameters μ', μ_1' for the tracking are set by the following rules:

$\mu' = d/y_e$, where d is the nonstationarity index of the EEG,

$\mu_1' = 2\pi h f_1$, where f_1 is the lowest frequency to be modeled, normally zero.

d can be estimated by searching for the maximum of

$$\hat{L} = \frac{N}{2}\left[U(N) + \log V(N)\right], \qquad (46)$$

where U and V are evaluated using (22g,h). Finally, $\hat{\lambda}^2 = V(N)$.

It remains to determine values of orders n, m and to determine coefficients in the C-polynomial. Orders $m = 9$ and $n = 14$ have been used, since prior experimentation indicated that this should be sufficient [34]. The values may be unnecessarily high in some cases. However, this caused no serious problems, and no attempt was therefore made to find least possible values.

The C-polynomial affects only the equation (39), which prefilters the data sequence. Physically, the purpose of prefiltering is normally to reduce disturbances by suppressing the parts of the sample spectrum where little signal power is expected. For EEG-signals this is the case for frequencies higher than about 25 Hz. Also, experiments indicated that the values of c-coefficients did not affect the end result appreciably, the spectrum or other relevant characteristics of the signal, provided C was chosen in such a way that higher frequencies were filtered out. This means that C can be chosen quite arbitrarily, as long as $|C(\exp 2\pi fh)|^{-2}$ is small for $f >$ 25 Hz. Thus, low-pass filtering using C pays off, while high-pass filtering using μ_1 does not. Details are found in [18].

The C-polynomial does affect the $\hat{a}$-coefficients substantially. However, this is of little concern, since the $\hat{a}$-coefficients play the role of auxiliary variables and are not physically relevant except in combination with the c-coefficients.

D. *TEST RESULTS*

A number of test cases were evaluated. This served to investigate the feasibility of the new method, assess its performance, and generally to demonstrate what kind of results to expect.

The test cases were mainly samples of physiological EEG-signals, selected as being more or less "difficult" for stationary analysis. In addition, a few simple, artificially generated test samples were used, as well defined references, for appraising accuracy experimentally. The tests were confined to off-line analysis, i.e. segments of recorded EEG's were sampled and the data stored in a disk memory prior to the analysis. However, real-time analysis was simulated by the fact that data were retrieved from the disk memory and analyzed one sample point at a time.

The physiological test samples were selected from a collection of EEG's recorded by Wennberg, who also investigated some of them by other methods [35]. No systematic way was used to select samples, but the intention was to include samples that looked different to the eye, and samples where stationary analysis had caused problems.

Ten samples were analyzed. They have been numbered from #1 to #11, since sample #2 was excluded as too short (20 sec). The length of the remaining samples ranges from 50 to 160 sec. The sampling frequency was, alternatively, 62.5 Hz and 100 Hz.

In view of the discussion in Section 4.C.3, the constant $\mu_1 = 0$ was used throughout the analysis, with one exception (see below). A full account of the analysis of all test samples is found in [18]. The discussion is here confined to the determination of the nonstationarity index, and to the spectral analysis of one of the test cases, which the value of the nonstationarity index indicated to be particularly fast-changing. In addition [18] reports the results of tests of integrated spike detection and recognition, and an assessment of estimation accuracy.

1. The Nonstationarity Index

When the analysis of an EEG-signal is not carried out in real time, the nonstationarity index d can be estimated from a sample, while in real-time analysis the EEG must be preclassified with respect to its degree of nonstationarity--for instance on the basis of prior analysis. In order to illustrate what values of $\hat{d}$ can be expected and what ranges of d are reasonable for preclassification, the log-likelihood function (16) has been computed as a function of d for the ten test cases. The results are plotted in Figure 3, after change of sign and subtraction by a constant.

The likelihood maxima correspond to the estimated nonstationarity index $\hat{d}$ in each case. The statistical variation of $\hat{d}$ is asymptotically for long samples

$$D^2(\hat{d}) \to |\partial^2 L/\partial d^2|^{-1} \quad , \tag{47}$$

which corresponds to half a unit variation in the log-likelihood function. The formula is actually conjectured from the properties of other Maximum-Likelihood estimators, viz. asymptotic efficiency [15], and has not been proven in the present case. Still, it is reasonable to say that variations in the likelihood up to about 9/2 (corresponding to a 3-sigma or 99.7% confidence interval of d) are not significant, and that, for practical purposes, d-values within the band $|L(d) - L(\hat{d})| \leq 9/2$ are equivalent. The confidence bands are indicated in Figure 3.

It appears from Figure 3 that all likelihood functions fall into one of two distinct categories with respect to the range of likely d-values, viz. those ranging from approximately 2^{-15} to 2^{-11}, and those from 2^{-11} to 2^{-9}. For numerical reasons (a 16-bit machine) d-values below 2^{-15} are equivalent to zero. Thus, the value $d = 0$ would be adequate for samples #1, #4, #5, #10, #11, and those samples are--as a result of the analysis--classified as "stationary" or "unchanging". Among the remaining samples the value $d = 2^{-11}$ is adequate for #3, #7, #8, #9, while sample #6 would seem to require $d = 2^{-9}$. Label the last two classes "slow-changing" and "fast-changing" respectively.

Now, sample #6 is special. It was recorded for the sole purpose of testing the new method. The character of the EEG was varied intentionally by having the patient open and close his eyes during the recording. In samples #3, #7, #8, #9 changes were spontaneous.

Samples #8 and #9 were recorded simultaneously from different positions on the scalp. Similarly, samples #10 and #11 were recorded simultaneously and on another occasion. It is therefore a point in favor of the method of estimating non-stationarity that, on both occasions, it did result in a common classification for signals that were recorded simultaneously.

An essential difference between the recordings #8 and #9 is that #8 contains a large artefact. If it is removed (by equivalent high-pass filtering using $f_1 = 0.99$ Hz), the corresponding likelihood curve becomes more similar to the one for #9, and the minimum is not changed appreciably. Notice, however, that even *with* the artefact the $\hat{d}$-estimates are the same within estimation accuracy. This gives some further experimental

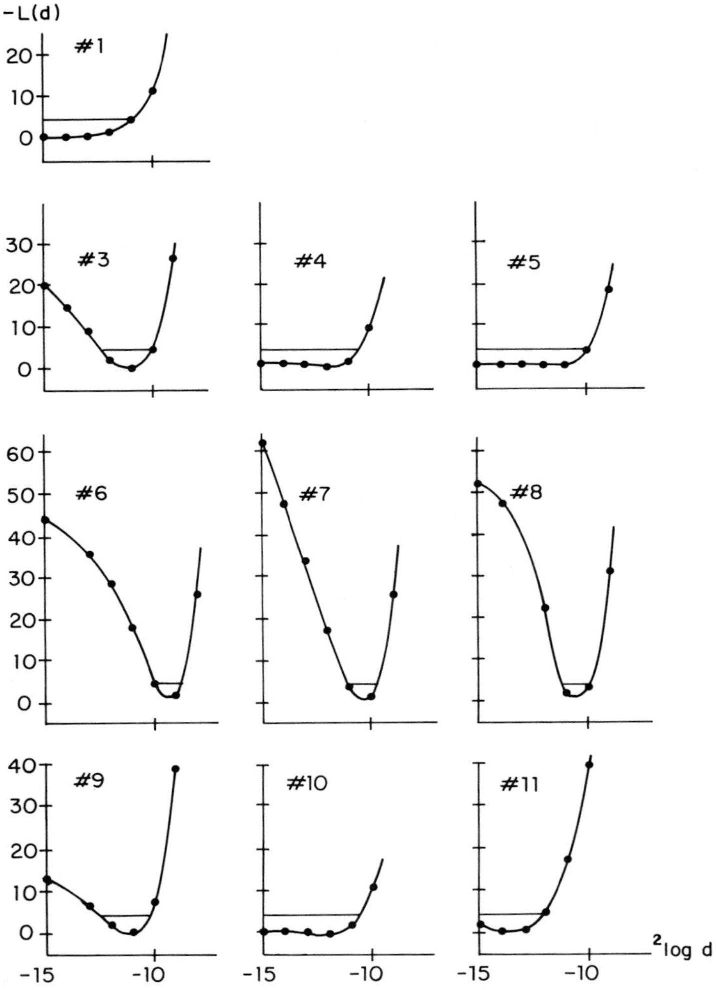

Fig. 3. Negative log-likelihood functions for the ten test samples, as functions of the nonstationarity index d.

support to the idea that the nonstationarity index is indeed a property of the EEG and only to a lesser degree of the measurement conditions.

Generally, none of the test cases disagreed with the intuitive ideas behind the analysis method, and the results discussed gave positive support.

2. The Power Spectrum

With known classifications, according to the index of nonstationarity, first the $\hat{a}$-coefficients were computed according to Section C.1, and then the time-variable spectra were computed using (38). Again, details are given in [18]. The most nonstationary case, #6, is plotted in Figure 4. The character of the signal clearly changes. Three intervals are dominated by a frequency of approximately 10 Hz, the so-called "α-rhythm". They correspond to periods of closed eyes.

Figure 5 shows the computed spectrum. The estimated instantaneous power-density distribution $\hat{S}(f|t)$ has been plotted for every 100th sample point, the curves displaced somewhat to give a three-dimensional impression. Thus, the time interval between the curves is one second.

Comments on the results:

- The loss of "α-activity", caused by open eyes, shows clearly. Also, some "β-activity" (frequencies around 20 Hz) is visible. The very low-frequent components appearing occasionally, are probably due to disturbances. If necessary, they can be suppressed by taking $f_1 > 0$. This changes only frequencies below f_1 appreciably.
- The accuracy of the estimated α-rhythm is lower than it would be in an analysis of a stationary signal $(d = 0)$. By having $d > 0$ the tracking algorithm is more alerted to changes and therefore also more susceptible to random variation. The first few curves are more uncertain, since initial uncertainty has been assumed high.

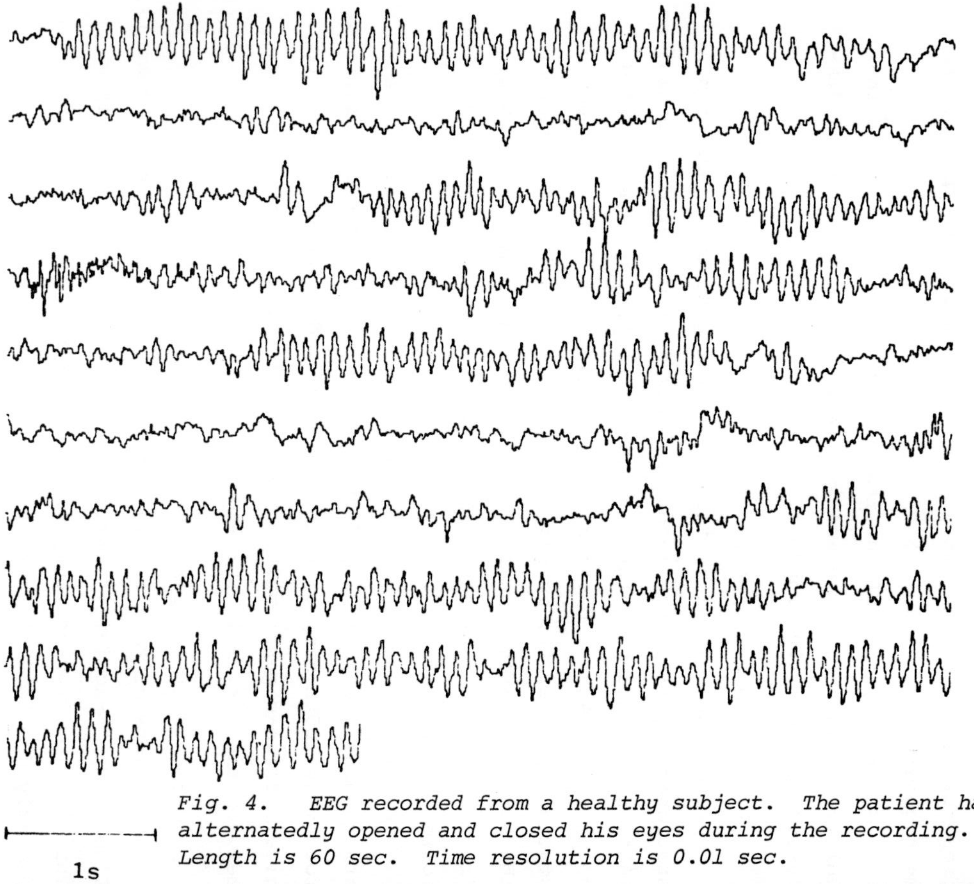

Fig. 4. EEG recorded from a healthy subject. The patient has alternatedly opened and closed his eyes during the recording. Length is 60 sec. Time resolution is 0.01 sec.

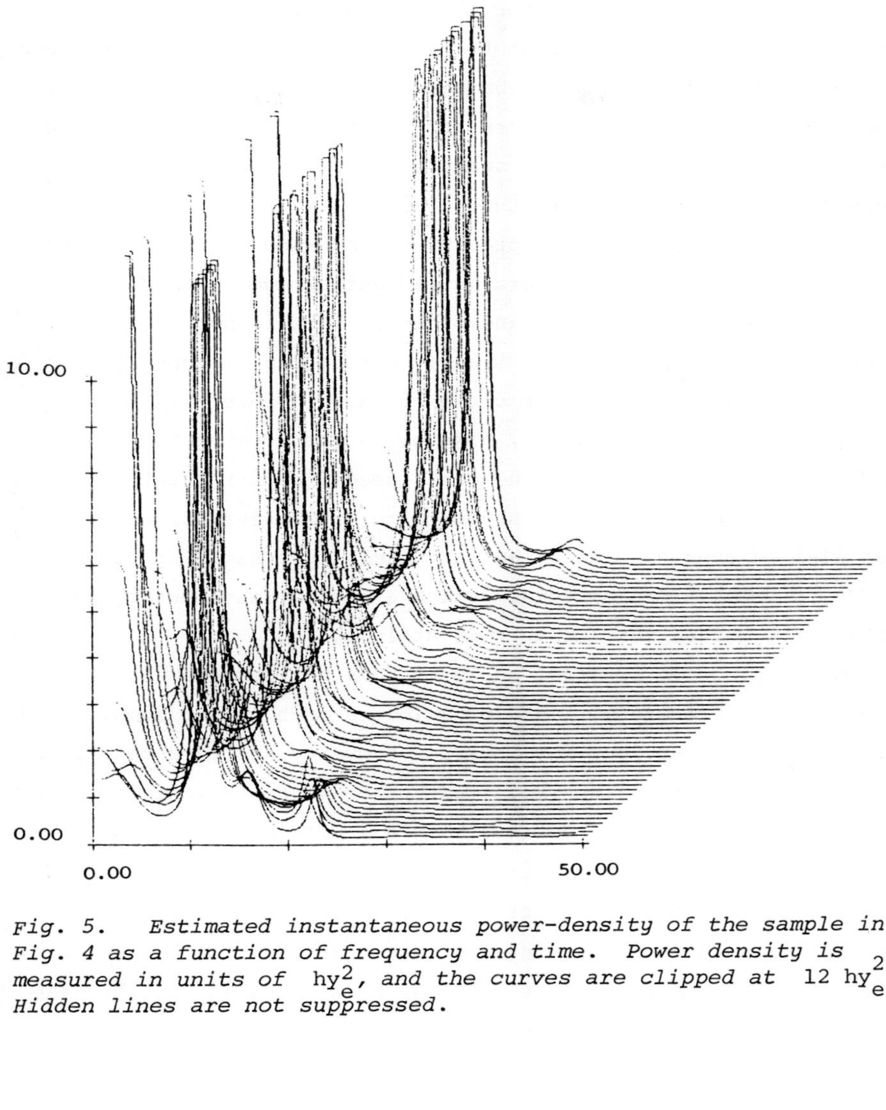

Fig. 5. Estimated instantaneous power-density of the sample in Fig. 4 as a function of frequency and time. Power density is measured in units of hy_e^2*, and the curves are clipped at* $12\ hy_e^2$*. Hidden lines are not suppressed.*

●A general observation for all test samples is the fact that the analysis is very tolerant to artefacts and other low-frequency disturbances. If such disturbances are present, they do not ruin the spectrum, except, of course, at the times they occur and in the frequency intervals they occupy. It is therefore not necessary to remove such disturbances.

●The way used here to represent the surface $\hat{S}(f|t)$ graphically has the drawback that graphs may get clogged with details, some of which are hidden behind others. This is not so in real-time analysis, since the spectral functions can be generated in real time and displayed as a changing curve, preferably on a screen. To do this requires, however, a fast computer. The IBM 1130 computer used for the tests needed approximately five times the sample length (in seconds) to track coefficients. This can also be expressed as a maximum permissible sampling rate of 20 Hz for real-time analysis. Using a newer minicomputer, the PDP-11/45, pushed this to approximately 80 Hz.

●The method of spectral analysis does not compete with the Fast Fourier Transform. The FFT allows a much higher sampling rate, about 10 kHz, but has inferior accuracy and time resolution. The latter is set by the facts that piece-wise stationarity is assumed, and that the effective sample length must be a substantial number (256, say) times the sampling interval to yield a reasonable frequency resolution. For the parametric spectral estimation used here tracking response times down to 0.1 - 0.2 sec, or 10-20 sampling intervals have been measured for the case of a suddenly appearing 10 Hz frequency, without this ruining the frequency resolution [18]. Since this corresponds to only one or two periods of the 10 Hz α-rhythm, it is reasonable to infer that the method has a time/frequency resolution close to the theoretical limit--it is obviously impossible to distinguish a frequency among other possible wave forms, until at least one period has evolved.

5. FORECASTING MACHINE FAILURE

The third case study was carried out in 1973 for the IBM Nordic Laboratory in Sweden. Permission to publish a general account is gratefully acknowledged.

A general characteristic of the case is that design is made on much less *a priori* information than in the other cases reviewed in this chapter. Also, the application requires that the result have a higher degree of automation. The solution is to be applied in a routine fashion, and it is not possible to count on the human inventiveness to amend cases where the routine fails because certain effects in the data were not foreseen--at least not too often.

A. TECHNICAL BACKGROUND

The purpose of the study was to design a method, in the form of a general computer algorithm, for forecasting the performances of a number of different units, such as pumps, turbines, or boilers in a steam-turbine plant. The result was to be used for scheduling maintenance, repair, or replacement of each unit on the basis of its individual performance.

A basic understanding was that, for each unit, an efficiency factor be computable from measurements, taken at regular intervals. This factor would be 100% for a new unit and would deteriorate with time, e.g. due to wear, corrosion, or clogging. The rate of deterioration was expected to be more or less irregular and to vary between individual units in an unforeseeable way, e.g. depending on hidden initial defects, fuel condition, and how hard the unit would be driven.

Also, a unit was to be considered operating satisfactorily, as long as its efficiency stays above a given limit, say 80%. Efficiencies reaching below this limit will call for maintenance. The problem was to forecast this event from a series of observed efficiency factors, and to do so sufficiently far ahead in time to be able to schedule maintenance.

Apart from this, no *a priori* information on individual units or records of efficiency factors were available for the design.

This summarizes the prerequisites for the design problem. The following narrates the deductions and the reasoning leading to a solution:

Efficient forecasting generally requires that the predictor be tuned to the particular characteristics of each individual unit. If these are known, a well-established theory can be applied for designing optimal predictors [37]. It seems reasonable to expect that pure speculation on how a typical unit would behave (which is in fact the only "*a priori* information" available) would suffice for finding the structure of the predictor, but would leave a number of parameters undetermined. The latter would specify, in essence, the level of high-frequency errors and the average rate and persistance of deterioration, once it has started; does the unit go down fast, or does it die hard?

The obvious plan of action would be to log data from each unit, analyze the records to find the characteristic parameters, and use the theory to find the individual predictors for each unit. However, the units are many, and it is also likely that many deterioration processes will be slow, so that the times needed for satisfactory records would be prohibitive. In any case, costs and development time-schedules would make the plan unfeasible.

A way out is real-time identification, that is to automate the analysis and tuning *procedure*, include this procedure with the predictor in the form of auxiliary programs, to be put to use at run time. Thus, at start-up the predictors will be untuned, but their performances will improve, as data become available, and the identification and tuning routines do their jobs. The predictor will be self-tuning.

An additional advantage will be that the tuning is always updated, even if a unit, for one of many possible reasons, would change its characteristics.

In this way the predictor, identifier, and tuner will be routines common to all units, but operating with the aid of a data bank, containing the characteristics and the status of each individual unit. At start-up *a priori* information (which is not individual) is embodied in the routines, while the data bank is empty.

The main disadvantage of the scheme is that it is slightly more hazardous than one including prior analysis, since the design of the predictor structure and the analysis and tuning programs must be based on speculations on what the physical situation might be, and not on data from all units. There is always a risk that a particular, odd unit might behave radically different from what was assumed or observed as typical. The proper way to counteract this in the design is to make the scheme insensitive to *a priori* assumptions. In fact, the collecting of *a posteriori* information (which is facts) carried out by the real-time identification, does make the predictor performance less sensitive to *a priori* information (which is here speculation and guesswork). However, the risk always remains. It must be watched in the field test.

The additional disadvantages of a more complex program and the fact that predictors will not be optimal from start-up are believed to be acceptable.

B. MATHEMATICAL DESCRIPTION

Even if the predictor is to modify its behavior to suit each individual unit, it is worthwhile to include into the structure of the predictor as much as possible of the ideas--or speculation--of what might be a reasonable behavior of a unit. The self-tuning capacity is limited, and its task is facilitated if the predictor has a reasonable performance to start with. The ideal case is when the self-tuning feature will modify the original tuning only little. The two-step approach outlined in the introduction will therefore be followed.

A crude model of the deterioration process is conceived as follows: It is assumed that the sequence $\{y(t)\}$ of computed efficiency factors is the sum of four random components, viz.

- H.f. noise $\sigma_0 e_0(t)$
- Drift $\sigma_1 \Delta^{-1} e_1(t)$
- Trend $\sigma_2 \Delta^{-2} e_2(t)$
- Acceleration $\sigma_3 \Delta^{-3} e_3(t)$,

where $e_0(t),\ldots,e_3(t)$ are orthonormal sequences of Gaussian, random variables, Δ^{-1} is the summation operator (Δ = backward-difference operator), and $\sigma_0,\ldots,\sigma_3$ are constants.

The "trend" component is meant to describe a steady deterioration of the unit, i.e. a deterioration that, once started, continues at nearly the same rate. Notice that this does not mean a linearly decreasing component, but a component whose increments in consecutive time intervals differ by a small random quantity of standard deviation σ_2. The "acceleration" component is introduced to take into account the presumption that deterioration, that has started, may proceed at an increasing rate. The "drift" component models slow changes that cannot be described as "trend" or "acceleration", i.e. as also having a slow-changing increment. It has a more erratic appearance. The "noise" component includes observation errors as well as any other effects with frequencies higher than the sampling rate.

Observation errors are thus included in the sequence to be predicted. Predicting an observed performance value is obviously equivalent to predicting a hypothetical, "true" performance value without observation error, provided measurements are uncorrelated and unbiased.

The intuitive meanings of the various components can also be derived from their associated predictors: The h.f. errors are predicted by zero, the drift by the last observed value, the trend by a linear extrapolation, and the acceleration by a parabolic extrapolation. Any sequence that is reasonably extrapolated in any of those ways can also be described as above. The predictor for the sum is a well-defined compromise between the four extrapolations depending on the weights $\sigma_0, \ldots, \sigma_3$ to the components.

Thus, the model

$$y(t) = \sigma_0 e_0(t) + \ldots + \sigma_3 \Delta^{-3} e_z(t) \tag{48}$$

summarizes the *a priori* assumptions made about the physical signal and appeals to intuition. However, it is neither suited for forecasting, nor is it immediately clear how a residual sequence is to be computed. To amend this write

$$y(t) = \Delta^{-m} C(q^{-1})\, e(t) \quad , \tag{49}$$

where $e(t)$ are Gaussian and orthonormal, and C is a polynomial of order m, where m is the highest negative power in (48) having a nonzero coefficient. This is a so-called "Integrated Moving Average Model" in the terminology of Box and Jenkins [37]. The two structures (48) and (49) are equivalent in the sense that they generate sequences with the same statistical properties. Also, there is a one-to-one correspondence between $\sigma_0, \ldots, \sigma_3$ and the coefficients in C. A difference is that a residual sequence can obviously be computed from (49) by solving for $\{e(t)\}$. It would therefore be advantageous to replace the original model (48) by the more convenient (49).

However, the weights σ_i have an intuitive meaning, and would therefore be easier to specify *a priori*, than would the coefficients in C. It is therefore worth the trouble to compute the coefficients that correspond to given weights, i.e. to find

$$C(z) = c_0 + \ldots + c_m z^m \quad , \tag{50}$$

such that the models are equivalent.

Multiply both models by Δ^m, and set spectra equal:

$$|\sigma_0 \Delta^m(z)|^2 + \dots + \sigma_m^2 \equiv |C(z)|^2 \quad , \tag{51}$$

where $\Delta(z) \equiv 1-z$ and $z = \exp(2\pi i f h)$. This is the well-known spectral factorization problem. Setting

$$|C(z)|^2 \equiv r_0 + \dots + r_m(z^m + z^{-m}) \tag{52}$$

yields an equation for $c_0,\dots,c_m$:

$$\begin{cases} c_0^2 + \dots + c_m^2 = r_0 \\ c_0 c_1 + \dots + c_{m-1} c_m = r_1 \\ \dots\dots\dots\dots\dots\dots\dots\dots \\ c_0 c_m = r_m \end{cases} . \tag{53}$$

$r_0,\dots,r_m$ are computed from (51) and (52):

$$m = 3: \quad \begin{cases} r_0 = 20\sigma_0^2 + 6\sigma_1^2 + 2\sigma_2^2 + \sigma_3^2 \\ r_1 = -15\sigma_0^2 - 4\sigma_1^2 - \sigma_2^2 \\ r_2 = 6\sigma_0^2 + \sigma_1^2 \\ r_3 = -\sigma_0^2 \end{cases} \tag{54}$$

$$m = 2: \quad \begin{cases} r_0 = 6\sigma_0^2 + 2\sigma_1^2 + \sigma_2^2 \\ r_1 = -4\sigma_0^2 - \sigma_1^2 \\ r_2 = \sigma_0^2 \end{cases} \tag{55}$$

Eq. (53) is solved by iteration [38]. A fast algorithm for the factorization problem is given in [39]. However, there is no requirement on speed, since C has to be computed only once. This will be referred to as "initial tuning".

When the C-polynomial has been found, it is an easy matter to compute the residual sequence

$$e(t) = C^{-1}(q^{-1})\ \Delta^m y(t) \tag{56}$$

from

$$e(t) = [-c_1 e(t-1) - \ldots - c_m e(t-m) + \Delta^m y(t)]/c_0 \ . \tag{57}$$

These computations are done in real time and termed "prewhitening", since they will result in an uncorrelated sequence, if *a priori* assumptions are correct.

The second step is to model $\{e(t)\}$ in case assumptions are not correct. From (1), (6), and (7), and since there is no input:

$$e(t) = \underline{H}(t)\ \underline{\theta}(t) + w_2(t)\ , \tag{58}$$

where

$$\underline{H}(t) = \{-e(t-1)\ldots -e(t-n)\},\quad \underline{\theta}(t) = \underline{\mathrm{Col}}\{a_1(t)\ldots a_n(t)\}. \tag{59}$$

The additive term $\kappa(t)$ has been excluded, since it is reasonable to expect that the m-th order difference operation Δ^m in (56) will eliminate persistent deviations of $e(t)$ from zero. It is true that if m is assigned a too small value, then there could still remain a low-frequency component in $e(t)$, which a κ-term would take care of. However, as discussed in the case of EEG-analysis, also a-coefficients can model low-frequency variations.

It remains to describe the way in which coefficients $\theta(t)$ vary. As in the other case studies little is known about this *a priori*, and the "random-walk" model is adopted:

$$a_k(t+1) = a_k(t) + \mu_k(t)\ w_k(t)\ , \tag{60}$$

where $w_k(t)$ are uncorrelated, $Ew_k(t) = 0$, $Ew_k(t)^2 = 1$. Also, it is simplest to assume no difference in the average rate at which coefficients change, so that $\mu_k(t) = \mu(t)$, independent of k. Further, one must assume that $\mu(t)$ is independent of t, since the deterioration process is certainly autonomous. The constant μ is unknown.

Finally, $a_k(0)$ are unknown. It is therefore reasonable to assume $a_k(0)$ normally distributed, independent, and with large standard deviations.

The assumptions on parameters $\underline{\theta}$ imply the model

$$\begin{cases} \underline{\theta}(t+1) = \underline{\theta}(t) + \underline{w}_1(t) \\ \underline{R}_1 = \mu^2 \underline{I}, \quad E\underline{\theta}(0)\underline{\theta}(0)^T = \underline{R}_0 = \mu_0^2 \underline{I} \end{cases} , \tag{61}$$

where $\mu << 1$ and $\mu_0 >> \mu$.

In summary, the following structure has been conceived for the deterioration process:

$$\begin{cases} \Delta^m y(t) = c_0 e(t) + \ldots + c_m e(t-m) \\ e(t) + a_1(t)e(t-1) + \ldots + a_n(t)e(t-n) = \lambda w_0(t) \\ a_k(t+1) = a_k(t) + \mu w_k(t) \end{cases} , \tag{62}$$

where $w_0(t), \ldots, w_n(t)$ are Gaussian and orthonormal, and n, m, $c_0, \ldots, c_m$, and μ are known.

If, in particular, $\mu = 0$, so that $a_k(t)$ are constant, the model can be written

$$y(t) = \Delta^{-m} C(q^{-1}) A(q^{-1})^{-1} \lambda w(t) , \tag{63}$$

where $C(q^{-1}) = c_0 + \ldots + c_m q^{-m}$ and $A(q^{-1}) = 1 + \ldots + a_n q^{-n}$. C is known and A is unknown. The form (63) reveals that the case can also be regarded as one of estimating an unknown spectrum $|\Delta^{-m} C A^{-1}|^2$, as in the case of EEG-analysis. However, the process has a different behavior, and C is derived and computed in a different way.

C. *THE FORECASTING ROUTINE*

The real-time identification algorithm (22) is used for the estimation of $\underline{\theta}(t)$, with $e(t)$ substituted for $y(t)$ and taking

$$\underline{H}(t) = \{-e(t-1) \ \ldots \ -e(t-n)\}$$

$$\underline{R}_1' = \mu^2/\lambda^2 \ \underline{I} \ , \quad R_2' = 1 \ .$$

Since only one parameter, viz. μ, has been used to describe $\underline{R}_1$, it is not necessary to estimate it. Instead, μ will be regarded as a design parameter, to be specified *a priori* and common to all units. Hence, the eqs. (22g) and (23d,e) are excluded. However, the parameter λ, specifying R_2, is estimated via (22h) and (23c).

The consistency test (22d) is included. Failure to satisfy the test causes the routine first to inhibit the updating of $\hat{\underline{\theta}}$. Secondly, a diagnostic routine is called to discriminate between the hypotheses of i) a short burst of outliers and ii) a lasting change. How this is done in detail will not be described here. The inclusion of an automatic diagnostic routine is, however, important for the following reasons:

The proper way to recover from an error will depend on what caused the error. However, the error indications will be frequent, when there are many units, and because the false-alarm rate of the test is not negligible. With 100 units, a false error indication is to be expected every four sampling intervals. Therefore, it is not reasonable to rely on a human operator to discriminate between all causes of error indications. The short-burst case, which is believed to constitute the overwhelming majority of error indications, must be acted upon automatically, and the action is to ignore the corresponding data points.

In the remaining cases it is necessary to call the operator's attention, since such an error is significant, and its cause may be one of many. For instance, the measuring may have failed. A second possibility is that it has not, but the deterioration process has made a sudden change of character of such a magnitude, that the identification is unable to follow. Also in this case the operator should be called upon, since there is probably a technical reason for the sudden change; maybe the unit is really going down. In both cases it is reasonable to let the changes affect the prediction; they are both "real" in the sense that they call for action, in contrast to the case of temporary outliers.

1. *Predictor and Prediction Error Variance*

For a discussion of the predictor it is convenient to regard present time t as fixed and consider the values of $y(t+\tau)$, where τ is variable.

The applied predictor has been derived under the assumption that the coefficients $a_k(t+\tau)$ and λ are known for $\tau > 0$ and equal to the estimated values $\hat{a}_k(t+\tau|t)$ and $\hat{\lambda}(t)$. This is an approximation. It follows from (11) that $a_k(t+\tau)$ must also be assumed independent of τ for $\tau > 0$, although not necessarily for $\tau < 0$. Hence, it is possible to write the model (62) for the process to be predicted as (63), or, equivalently, as

$$\begin{cases} \Delta^m(a^{-1})y(t+\tau) = C(q^{-1})e(t+\tau) \\ A(q^{-1})e(t+\tau) = \lambda w(t+\tau) \end{cases} \tag{64}$$

for $\tau > 0$, where Δ, A, C operate with respect to τ with t fixed.

Taking expectations conditional on y_t yields

$$\begin{cases} \Delta^m(q^{-1})\hat{y}(t+\tau|t) = C(q^{-1})\hat{e}(t+\tau|t) \\ A(a^{-1})\hat{e}(t+\tau|t) = \lambda\hat{w}(t+\tau|t) = 0 \end{cases} \tag{65}$$

for $\tau > 0$, since $w(t)$ is white. Start values are

$$\begin{cases} \hat{e}(t+\tau|t) = e(t+\tau) \\ \hat{y}(t+\tau|t) = y(t+\tau) \end{cases} \tag{66}$$

for $\tau \leq 0$. The algorithm for computing $\hat{y}(t+\tau|t)$ recursively from (65) and (66) for $\tau = 1,2,\ldots$ constitutes the multistep predictor used in the forecasting routine.

Derivation of the prediction error variance $R_{yy}(t+\tau|t)$ is more involved. Let the error be

$$\tilde{y}(t+\tau|t) = y(t+\tau) - \hat{y}(t+\tau|t) \quad . \tag{67}$$

It follows from (64) and (65) that $\tilde{y}$ satisfies

$$\begin{cases} \Delta^m(q^{-1})\tilde{y}(t+\tau|t) = C(q^{-1})\tilde{e}(t+\tau|t) \\ A(q^{-1})\tilde{e}(t+\tau|t) = \lambda w(t+\tau) \end{cases} \quad . \tag{68}$$

For convenience, drop the explicit reference to the time t prediction takes place, and introduce the following representation of (68):

$$\begin{cases} \underline{x}(\tau) = \underline{\Phi}_x \, \underline{x}(\tau-1) + \underline{\Gamma}_x \, \tilde{e}(\tau) \\ y(\tau) = \underline{H}_x \, \underline{x}(\tau) \end{cases} \tag{69}$$

$$\begin{cases} \underline{z}(\tau) = \underline{\Phi}_z \, \underline{z}(\tau-1) + \underline{\Gamma}_z w(\tau) \\ \tilde{e}(\tau) = \underline{H}_z \, \underline{z}(\tau) \end{cases} \quad , \tag{70}$$

where

$$\begin{cases} \underline{\Phi}_x = \begin{pmatrix} -d_1 & -d_2 & \ldots & -d_m & 0 \\ 1 & 0 & \ldots & 0 & 0 \\ 0 & 1 & \ldots & 0 & 0 \\ \ldots & \ldots & \ldots & \ldots & \ldots \\ 0 & 0 & \ldots & 1 & 0 \end{pmatrix}, \quad \underline{\Gamma}_x = \begin{pmatrix} 1 \\ 0 \\ 0 \\ \ldots \\ 0 \end{pmatrix} \\ \underline{H}_x = \begin{pmatrix} c_0 & c_1 & \ldots & c_m \end{pmatrix}, \quad d_k = (-1)^k \begin{pmatrix} m \\ k \end{pmatrix} \end{cases} \tag{71}$$

$$\begin{cases} \underline{\Phi}_z = \begin{pmatrix} -a_1 & -a_2 & \cdots & -a_n & 0 \\ 1 & 0 & \cdots & 0 & 0 \\ 0 & 1 & \cdots & 0 & 0 \\ \cdots & \cdots & \cdots & \cdots & \cdots \\ 0 & 0 & \cdots & 1 & 0 \end{pmatrix}, \quad \underline{\Gamma}_z = \begin{pmatrix} \lambda \\ 0 \\ 0 \\ \cdots \\ 0 \end{pmatrix} \\ \underline{H}_z = \begin{pmatrix} 1 & 0 & \cdots & 0 \end{pmatrix} . \end{cases} \tag{72}$$

Define

$$\begin{cases} \underline{P}_{xx}(\tau) = E\underline{x}(\tau)\underline{x}(\tau)^T \\ \underline{P}_{xz}(\tau) = \underline{P}_{zx}(\tau)^T = E\underline{x}(\tau)\underline{z}(\tau)^T \\ \underline{P}_{zz}(\tau) = E\underline{z}(\tau)\underline{z}(\tau)^T . \end{cases} \tag{73}$$

Inserting (69) and (70) yields after some calculation

$$\begin{cases} \underline{P}_{zz}(\tau) = \underline{\Phi}_z \underline{P}_{zz}(\tau-1)\underline{\Phi}_z^T + \underline{\Gamma}_z \underline{\Gamma}_z^T \\ \underline{P}_{xz}(\tau) = \underline{\Phi}_x \underline{P}_{xz}(\tau-1)\underline{\Phi}_z^T + \underline{\Gamma}_x \underline{H}_x \underline{P}_{zz}(\tau) \\ \underline{P}_{xx}(\tau) = \underline{\Phi}_x \underline{P}_{xx}(\tau-1)\underline{\Phi}_x^T + \underline{\Gamma}_x \underline{H}_z \underline{\Phi}_z \underline{P}_{zx}(\tau-1)\underline{\Phi}_x^T \\ \qquad + \underline{\Phi}_x \underline{P}_{xz}(\tau-1)\underline{\Phi}_z^T \underline{H}_z^T \underline{\Gamma}_x^T + \underline{\Gamma}_x \underline{H}_z \underline{P}_{zz}(\tau)\underline{H}_z \underline{\Gamma}_x^T . \end{cases} \tag{74}$$

The variance of the prediction error is from (69) and (71)

$$R_{yy}(t+\tau|t) = \underline{H}_x \underline{P}_{xx}(\tau)\underline{H}_x^T . \tag{75}$$

The equations (74) can be used for computing covariances recursively. However, since the coefficient matrices are sparse, it is better to write the equations out in components before programming. This saves much computing.

2. *Forecasting Fault Conditions*

A fault condition occurs, by definition, when the efficiency factor of a unit falls below a given value y_f. Several measures are conceivable for forecasting this event. The ideal information would be a time t_f, such that the probability of unit failure prior to this time has a given, small value. Even if theoretically feasible, this measure would, however, lead to an unreasonably complicated algorithm. The following measure has therefore been used instead: Compute and display a curve of probabilities of fault at each future time $t+\tau$. This curve can be used to schedule maintenance; the unit should be serviced when the probability of fault becomes significant.

Notice that the time t_f defined above is not possible to compute as a level intercept of the probability curve. The relation is involved, and t_f is not even uniquely determined by the curve.

Assuming a Gaussian probability distribution of the efficiency factors, the probability of fault at time $t+\tau$ is

$$p_f(t+\tau|t) = \text{Errf}\ [(y_f - \hat{y}(t+\tau|t))/\sqrt{R_{yy}(t+\tau|t)}], \tag{76}$$

where Errf is the error function (=cumulative Gaussian distribution).

D. PERFORMANCE TEST

When, as in this case, process data cannot be made available for tests prior to the installation of the computer, the remaining alternative is test by simulation. Generally, a risk present in simulation is that the assumptions behind the generation of the data be unrealistic. Simplest, and unsatisfactory, would be to generate the data on the same assumptions only, as those for which the solution has been constructed. Such a test shows only that the derivations are correct, and, if approximations have been made, that they are allowed. What has also to be tested is that the solution is robust, i.e. that it works also for reasonable changes in the *a priori* assumptions.

In the present case there are three features to be tested, viz. prediction, self-tuning, and error detection. The way of predicting is not new [37]; linear extrapolation is known to work for all kinds of data that are not Gaussian, Markovian, etc.--provided its coefficients are reasonable.

The self-tuning is designed to find those coefficients, and is therefore the important feature to test. However, in this case data can in fact be generated by the same structure (48) as that for which the solution was designed. Provided the weights used for generating the data are different from those assumed for initial tuning, the predictor is in fact applied to a different structure from that for which it was designed, viz. $y = \Delta^{-m}C^{0}w$ instead of $y = \Delta^{-m}CA^{-1}w$. By varying the parameters used in the initial tuning, it is possible to investigate experimentally whether the self-tuning is able to compensate satisfactorily a wrong *a priori* guess of the character of the deterioration process.

How far one should reasonably go in testing different structures, before one is satisfied, is very difficult to say. Of course one can never test all relevant cases, but this is true also if one has got real data. Quite few, substantial variations may be sufficient, since they are sufficient to investigate the sensitivity to *a priori* assumptions, which is really all that is needed. A minimum demand is that there be some variation.

The robustness of the error detection is difficult to check by simulation, since real errors can be of a large number of unexpected kinds. However, the error detection has also been used before and tested on industrial data with very severe and irregular disturbances and was found to be sufficiently robust [40].

A test case was generated from

$$y(t) = 0.1w_0(1) + 0.1\Delta^{-1}w_1(t) + 0.02\Delta^{-2}w_2(t) , \qquad (77)$$

where the weights have been chosen *ad hoc*, although so that the result would simulate the behavior of a deterioration process. Mainly, the curve should be somewhat irregular and drift in one direction. It is plotted in Figure 6 and labelled "medium noise".

Since much irregularities is generally detremental to prediction, a second, "high-noise" case was also generated according to

$$y(t) = 0.5w_0(t) + 0.1\Delta^{-1}w_1(t) + 0.02\Delta^{-2}w_2(t) \quad . \tag{78}$$

It is believed to be an extreme case, although this belief is, admittedly, based on little real evidence. The sequence is plotted in Figure 6.

Results of test runs under different initial mistuning are compiled in Table 2. The average observed and estimated prediction errors are computed from

$$\bar{y}^2(\tau) = \frac{1}{100} \sum_{t=101-\tau}^{200-\tau} \tilde{y}^2(t+\tau|t)$$

$$\bar{R}_{yy}(\tau) = \frac{1}{100} \sum_{t=101-\tau}^{200-\tau} . \; R_{yy}(t+\tau|t)$$

Values for $n = 0$ are obtained without self-tuning. The order of the *a priori* model (62) has been $m = 2$. Thus, the effect of an "acceleration" component, $m = 3$, has not been tested. Obviously, it would not be sensible to try and estimate a random second derivative from any of the test cases in Figure 6. In order to test the third component in (48) one would have to generate a much smoother test case, and it is doubtful, whether this would be realistic. It is very likely that an acceleration term is not needed at all. Actually, it was introduced as a safety measure, in case the presumption on order would be wrong. It might help, and it does not cost much to have it.

The order of the residuals model has been $n = 2$. Experiments showed that higher orders did not improve much on the prediction.

TABLE 2

Average Observed and Estimated Prediction Errors $\bar{y}(\tau)$, $\sqrt{\bar{R}_{yy}(\tau)}$

For Ten- and Twenty-Step Prediction

#	σ_0	σ_1	σ_2	$\bar{y}(\tau)$, $\sqrt{\bar{R}_{yy}(\tau)}$							
				$\tau = 10$				$\tau = 20$			
				$n = 2$		$n = 0$		$n = 2$		$n = 0$	
"Medium noise" $\sigma = \{0.1\ \ 0.1\ \ 0.02\}$											
1.0	0.1	0.1	0.02	0.53	0.68	0.52	0.65	0.96	1.48	0.96	1.37
1.1	0.1	0.1	0.002	0.78	0.63	0.81	0.39	1.53	1.01	1.57	0.58
1.2	0.1	0.1	0.2	0.73	1.26	1.10	2.65	1.34	3.21	2.05	7.08
1.3	1	0.1	0.02	0.52	0.67	0.55	0.67	0.96	1.43	0.98	0.65
1.4	0.01	0.1	0.02	0.53	0.68	0.61	1.02	0.97	1.46	1.06	2.23
1.5	0.1	1	0.02	0.81	0.52	0.79	0.56	1.57	0.80	1.55	0.87
1.6	0.1	0.01	0.02	0.59	0.75	0.57	0.63	1.07	1.82	1.04	1.48
1.7	1	0.01	0.001	0.53	0.49	2.31	0.52	1.00	0.78	3.00	0.49
1.8	0.01	0.01	1	1.07	2.03	1.68	5.05	1.89	5.35	3.32	13.8
"High noise" $\sigma = \{0.5\ \ 0.1\ \ 0.02\}$											
2.0	0.5	0.1	0.02	0.75	0.90	0.75	0.95	1.14	1.58	1.14	1.67
2.1	0.5	0.1	0.002	1.11	0.82	1.16	0.70	1.84	1.05	1.89	0.84
2.2	0.5	0.1	0.2	1.32	2.20	1.71	3.86	2.27	5.16	3.03	9.60
2.3	5	0.1	0.02	0.96	0.82	1.03	0.74	1.48	0.99	1.56	0.81
2.4	0.05	0.1	0.02	0.95	1.58	1.28	3.43	1.49	3.21	2.04	7.43
2.5	0.5	1	0.02	0.98	0.97	1.01	1.84	1.69	1.39	1.70	2.82
2.6	0.5	0.01	0.02	0.76	0.92	0.76	0.91	1.16	1.67	1.15	1.62
2.7	5	0.01	0.001	1.20	1.50	6.22	2.81	2.24	2.10	6.99	2.63
2.8	0.05	0.01	1	2.64	6.51	6.59	21.9	5.01	16.8	13.6	59.7

Comments on the test results:

- Self-tuning does not impair prediction significantly in any case. It improves prediction substantially in some cases, e.g. #1.7 and #2.7. In these cases the guessed high-frequency noise is much higher than the actual one.
- Self-tuning also improves on the estimation of the prediction errors. These estimates are to be used for estimating the

probability of a fault condition according to (76).

●The predictor is reasonably insensitive to initial tuning. The most difficult cases are #1.8, #2.2, and #2.8, which all correspond to guessing much less h.f.-noise than there actually is.

A conclusion drawn from these observations, is that in order to make the self-tuning predictor applicable to a wide set of units, the initial tuning should be conservative, i.e. *a priori* assumptions should be "much noise and little trend". If the initial tuning is wrong, the self-tuning has an easier task correcting this, when the assumed variations are more high-frequency, than if they are more low-frequency than the actual variations. Thus, having bias in the initial tuning (with respect to the average unit) increases the robustness of the self-tuning predictor.

At the time this was written, tests under operating conditions were not concluded, and no results were available.

6. FORECASTING LOAD ON POWER NETWORKS

This case study was carried out in 1974 for Stockholm Energiverk, which company is the distributor of electrical energy to the Stockholm region in Sweden. The purpose was to demonstrate the feasibility of a new approach to forecasting variations in the total power demand from consumers.

Forecasting is desired for two purposes:

1. For planning about one day ahead. The business of the company is mainly distributing energy, purchased from a number of independent power plants. Energy is ordered currently, and the business is subject to a rather involved set of trade rules. Mainly, it is economically advantageous not to order more energy than will be consumed, and there are heavy penalties on using more energy than ordered.

2. For short-term control of the energy supply. The distributing company is charged by the suppliers according to the

energy consumed during each hour. Thus, the energy figure at the end of each hour interval determines the amount paid and any penalties incurred. It is therefore necessary to forecast the energy consumption figure at the end of each whole hour and balance it with the supply. This can be done by ordering additional energy, in practice as late as thirty minutes before check-off. At this stage the price rises sharply when energy is ordered on short notice.

An optimal buyer's strategy is likely to be complicated and is not to be considered here. But whatever it will be, it is apparent that reduced forecasting errors, both in the long and in the short range, can be utilized for reducing expenditure. Also the forecasting error variance is desired, since it will affect the margins used in any rational planning.

In contrast to the previous case of forecasting machine failure, much is here known about the typical behavior of power demand variations, and ample data were available for the design of a forecasting routine.

A. GENERAL BACKGROUND

The power demand over 24 hours normally varies around one of three characteristic profiles, depending on whether the day is a work day, a holiday, or a labor-free day before a holiday. Call the latter "Saturday" for short, even if it need not always be. The difference between profiles is significant during day hours, about 6 to 17 o'clock, and can be attributed to different kinds of activity in the general community on these days. Peak power on work days is about twice that on holidays.

Occasionally, other profiles are observed, namely on single days occurring between two holidays, or on Midsummer Eve, Walpurgis Night, and similar special days. Call them "odd" days for short. Such days are usually labor-free for part of the community, yielding an energy consumption profile between those of a full work day and a holiday, and less predictable.

The profiles change considerably with the seasons. This variation is particularly pronounced in Sweden, since the duration of daylight varies from about 6 to about 20 hours. An even slower variation is the increase of energy consumption following the rise in GNP.

Superimposed on the profiles and the seasonal and long-term variations one can distinguish at least three types of random variation [41]:

● One component changing with weather conditions, attributed to the extra energy needed for light during cloudy days and heat during cold and windy days. This means variations with a time constant of the order of days. Weather forecasts can be made available.

● Short peaks in demand, that can be associated with particular events, such as unusually popular television programs. They can, possibly, be foreseen.

● Variations in demand that cannot be foreseen. Time constant is of the order of hours.

Thus, components of the total demand can be classified roughly by how fast they vary and how predictable they are. These properties are not necessarily dependent. For instance, the basic profiles have a period of 24 hours, and are predictable a long time ahead, while weather effects usually have slower variation but are less predictable.

B. MATHEMATICAL DESCRIPTION

With this background in mind it seems reasonable to treat the different components of the energy consumption in the following different ways:

● Estimate the profile (indexed by "work day", "holiday", "Saturday", or "odd day") under the assumption that it varies slowly (seasonal and long-term variations).

● Predict energy consumption, given the estimated profile and observed deviations. This will take into account weather

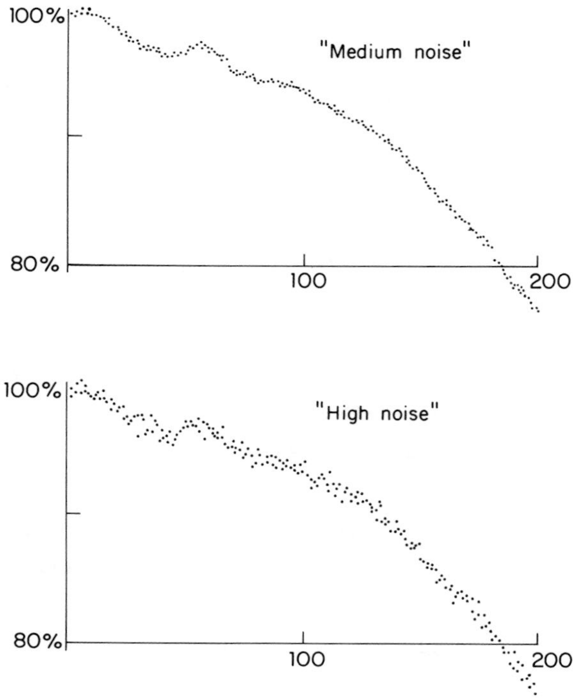

Fig. 6. Test cases: Simulated deterioration of efficiency factor.

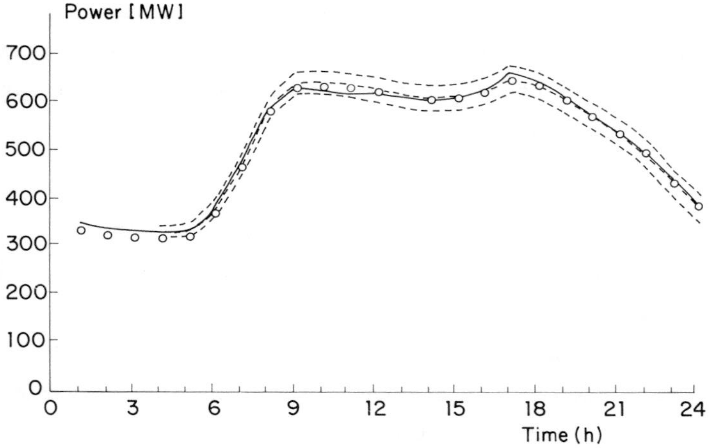

Fig. 7. Results of long-range forecasting on a winter work day. Full line: actual power consumption. Dashed lines: forecast and 95% confidence band. Circles: estimated current consumption profile. Forecasting took place at 3 hours.

conditions. In the short-range forecasting this will also take into account effects of other unforeseen causes of comparable duration.

●Include as inputs to the model such known, exogenous variables that can be assumed to affect the consumption appreciably. Although weather forecasts are not, strictly, exogenous variables (it ought to be the true weather and not the forecast that affects energy consumption), it is reasonable to conjecture that if used as such, weather forecasts will help.

The last feature is not indispensable and was not included in the feasibility study. However, weather forecasts are available, and in the following the possibility of input (exogenous) variables will be taken into account.

As before, the two-step procedure will be applied for deriving the model used for forecasting. The first step--setting up a crude model from *a priori* assumptions--is here even more strongly motivated than in the case of forecasting machine failure, since much more is known *a priori*.

Let $\{x_i(t)\}$, $i = 1,2,3,4$ be the four alternative energy consumption profiles, and let $i = 1$ correspond to holidays. Further, since profiles are different only during daytime, let $i = 2$ and $i = 3$ correspond to the additional energy consumption on work days and Saturdays respectively, and $i = 4$ to that on "odd" days. It turns out to be feasible to divide the latter class into subclasses, but this has not been done. Pooling of "odd" days is sufficient, partly because such days are few, partly because one is not very interested in forecasts on such days.

Thus, assume for a crude, primary model that the observed energy consumption between sampling intervals is

$$y(t) = x_1(t) + x_2(t)\delta_2(t) + x_3(t)\delta_3(t) + x_4(t)\delta_4(t) + e_0(t), \tag{79}$$

where $\delta_2(t) = 1$ on work days, $\delta_3(t) = 1$ on Saturdays, $\delta_4(t)=1$

on "odd" days, and $\delta_i(t) = 0$ otherwise. $\{e_0(t)\}$ is a sequence of independent, normal random variables with variances $R_2 = \sigma_2^2$ and zero means.

For the slow variation of the profiles assume

$$x_i(t+T) = x_i(t) + e_i(t) \quad , \tag{80}$$

where $Ee_i(t) = 0$, $Ee_i(t)e_j(s) = \sigma_1^2\delta(t-s)\delta(i-j)$, and T corresponds to 24 hours. For long-range forecasting measurements are taken every hour and $T = 24$. For short-range forecasting energy is measured every 15 minutes, so that $T = 96$. Thus, the assumption is that each point on each profile has changed from that of the previous day by a random quantity $e_i(t)$. It is reasonable that σ_1 be small but positive.

Notice that nothing in the *a priori* assumptions indicates the reasonable belief that the energy consumption profile is continuous in time. It would be feasible to take this into account, for instance by adding a term $\beta_i x_i(t+T-1)$. However, this turns out to complicate the computations considerably. Also, it turns out not to be needed; since the observations sample a smooth function, also the estimated profiles will be smooth without further ado.

The assumptions define a state-vector model of the order 4T in the state variables $\{x_i(j) | i=1,2,3,4; \ j=t,\ldots,t-T+1\}$. Applying Kalman's filter yields a residual sequence (the innovations) for the model. Thanks to the assumption that a point on a profile is coupled only to that of the previous day, the model separates into T independent fourth-order models, all defined by (79) and (80), and so does the Kalman filter. Since all models are equal, so are the solutions of the Riccati equations (σ_1 and σ_2 have been assumed independent of time) and a common equation will do for all points on the profile.

"Prewhitening" (see Section 5.B) is therefore carried out as follows: Let $t = kT + \tau$, where k indicates the day and τ the time of the day, counted in units of the sampling interval. For $\tau = 1$ and all k compute

$$\begin{cases} R_{yy}(k|k-1) = \underline{H}(t)\underline{R}_{xx}(k|k-1)\underline{H}(t)^T + R_2 \\ \underline{K}(t) = \underline{R}_{xx}(k|k-1)\underline{H}(t)^T/R_{yy}(k|k-1) \\ \underline{R}_{xx}(k+1|k) = \underline{R}_1 + \underline{R}_{xx}(k|k-1) - \underline{K}(k)R_{yy}(k|k-1)\underline{K}(k)^T \end{cases} \tag{81}$$

where

$$\begin{cases} \underline{H}(t) = \left(1 \ \delta_2(t) \ \delta_3(t) \ \delta_4(t)\right) \\ \underline{R}_1 = \sigma_1^2\underline{I} \\ R_2 = \sigma_2^2 \end{cases} \tag{82}$$

For $\tau = 1,\ldots,T$ and all k compute

$$\begin{cases} \tilde{y}(t|t-T) = y(t) - \underline{H}(t)\hat{\underline{x}}(t|t-T) \\ \hat{\underline{x}}(t+T|t) = \hat{\underline{x}}(t|t-T) + \underline{K}(k)\tilde{y}(t|t-T) \end{cases} \tag{83}$$

The residual sequence is taken as $e(t) = \tilde{y}(t|t-T)$. It is uncorrelated and has variance $R_{yy}(k|k-1)$, if *a priori* assumptions hold.

Since they may not hold, and as a second step in the modeling, assume for $e(t)$ a model of the same form as before, viz. (5-7). Since the variance of $e(t)$ changes from holiday to workday in a known way, it is reasonable to set $Ew_2(t)^2 = \lambda^2 R_{yy}(k|k-1)$.

In summary, the following structure has been conceived for the variations in the energy consumption:

$$\begin{cases} \hat{\underline{x}}(t+T|t) = \hat{\underline{x}}(t|t-T) + \underline{K}(k)e(t) \\ y(t) = \underline{H}(t)\hat{\underline{x}}(t|t-T) + e(t) \\ e(t) + a_1(t)e(t-1) + \ldots + a_n(t)e(t-n) = \underline{b}(t)\underline{u}(t) + \\ \qquad + \lambda w_0(t)\sqrt{R_{yy}(k|k-1)} \\ a_i(t+1) = a_i(t) + \mu w_i(t) \\ \underline{b}(t+1) = \underline{b}(t) + \underline{\mu}_1 \underline{w}_{n+1}(t) \end{cases} \tag{84}$$

where $w_0(t),\ldots,w_n(t)$, $\underline{w}_{n+1}(t)$ are Gaussian and orthonormal, k is the integer part of t/T, and $\underline{K}(k)$ is given by (81, 82). $\underline{u}(t)$ is an exogenous variable, such as weather forecast data; it may have more than one component. Influence of lagged variables are conceivable and may be introduced into the model. The first two equations in (84) follow from (83) and form a realization equivalent to (79, 80), see [2].

Parameters assumed known are n, σ_1, σ_2, μ, and $\underline{\mu}_1$. Since they must be postulated, their interpretations are important: σ_1 and σ_2 may be interpreted as "rms lasting change per day of a profile" and "rms high-frequency variation". Similarly, it would be feasible to construct interpretations in words for μ and μ_1, but this would clarify little.

C. *THE FORECASTING ROUTINE*

As before, the forecasting routine comprises three separate routines, executed each time a new measurement $y(t)$ arrives:

1. *Prewhitening*

The purpose is twofold: i) to estimate energy consumption profiles and ii) to compute residuals $e(t)$. Equations (81) and (83) are used; (81) needs to be evaluated only once a day.

2. *Identification*

The purpose is to model any behavior of the observations not in agreement with the assumptions of the primary model, as reflected in nonwhiteness of the residuals sequence. The same routines as before are used, viz. (22a-c,e,f,h) and (23c,f). The consistency test (22d) is not used, since energy measurements are believed to be reliable. Also, equations (22g) and (23b,d,e) are not used for estimating μ and μ_1, but the latter are to be regarded as design parameters. This has not caused any problems in the feasibility study, since no exogenous variable has been used and only μ remains to be set, but may do so otherwise. For $R_2'(t)$ is used $R_{yy}(k|k-1)$ for the following reason: If

σ_1 and σ_2 are known, then $R_{yy}(k|k-1)$ is known. However, since σ_1 and σ_2 must be guessed *a priori*, they may be wrong and R_{yy} may be wrong by a constant factor λ, so that only relative variation is actually known. Therefore, the unknown factor λ must be estimated, which is done using (23c).

3. Prediction

Prediction is here different from that in Section 5.C.1, since the model structure (84) is different from (62). It follows from (84) that

$$\hat{y}(t+\tau|t) = \underline{H}(t+\tau)\hat{\underline{x}}(t+\tau|t+\tau-T) + \hat{e}(t+\tau|t) \quad , \qquad (85)$$

i.e. the predicted value of the residuals sequence is added to the current estimate of the energy consumption profile. If e is non-white, then $\hat{e}$ is nonzero and takes into account faster variations (e.g. due to weather) than those of the profiles (e.g. due to seasonal variations).

A predictor for e is found by arguments similar to those used in 5.C.1 and is given by

$$\hat{e}(t+\tau|t) = -\hat{a}_1(t)\hat{e}(t+\tau-1|t) -\ldots- \hat{a}_n(t)e(t+\tau-n|t). \quad (86)$$

Start values are $\hat{e}(t+\tau|t) = e(t+\tau)$ for $\tau \leq 0$.

Computing the predictor variance is simpler than in 5.C.1. It follows from (84) and (85) that $\tilde{y}(t+\tau|t) = \tilde{e}(t+\tau|t)$, i.e. the sole contribution to the prediction error is that of the residuals. This is a consequence of the definition of the residuals as the deviation of $y(t)$ from the *estimated* profile and not from a "true" one. But the variance of $\tilde{e}(t+\tau|t)$ is computed from (74) and (75), where $\Phi_x = 0$, $H_x = 0$, and λ^2 is replaced by $\lambda^2 R_{yy}(k|k-1)$.

D. PERFORMANCE TEST

In this application it was possible to base the test on actual measurements. The company provided historical records of hourly measurements of energy consumption for a period of three years. About half the record, viz. from 1970-01-01 to 1971-05-28, was used for simulating long-range forecasting. A summary of the results is given here.

For programming convenience it was assumed that planning takes place at 3 o'clock in the morning (it does not) and results in an estimated profile for the remainder of the day, i.e. the range of forecasting is from 1 to 21 hours.

The estimation routine was initiated in the simulated time of one o'clock on 1970-01-01 with zero estimates $\hat{\theta}(1|0)$ and large variances. After about one month (simulated time) predicted error variances settled down to approximately stationary values, indicating the approximate amount of data needed to model the process to within the accuracy allowed by the variations in demand.

The following constants were assumed:

$n = 2$, $\sigma_1 = 10$ MW, $\sigma_2 = 10$ MW, $\mu = 0$. These values were found suitable by running the test record a few times and observing the average forecasting errors. Due to the estimation of λ the absolute values of σ_1 and σ_2 are not critical. Positive values of μ did not improve the forecasting, which indicates that the test data did not contradict the assumption that the unknown coefficients in the residuals model $\theta_i = a_i$ are constant.

The predicted and actual energy consumption and the predicted forecasting errors reached on a winter work day are shown in Figure 7; the forecasting errors are illustrated by a 95% confidence band (i.e. the difference between the upper and lower limits is four standard deviations). The graph represents a typical behavior of the forecasting routine on a work day, i.e. the result is neither unusually good nor unusually bad.

Figure 8 shows the average actual and predicted forecasting errors for work days as functions of the lead used in forecasting. Values are evaluated over the period 1970-01-31 to 1971-05-28.

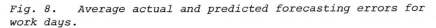

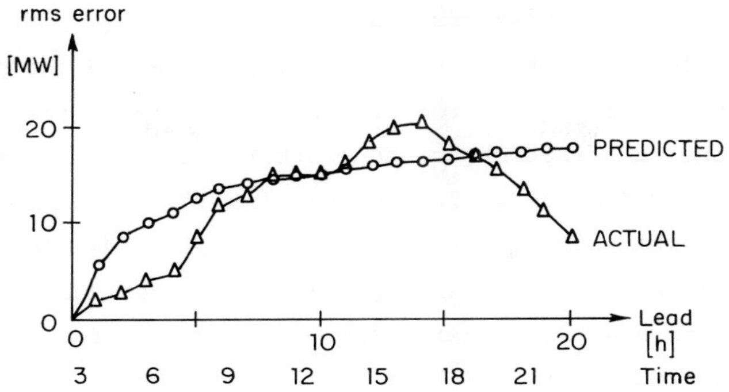

Fig. 8. Average actual and predicted forecasting errors for work days.

Comments on the test results:

- The difference between the current profile and the predicted value is the contribution from predicting the residuals, interpreted as the contribution due to weather and other short-range variations. It decreases with increased lead. It improves the forecasting substantially, but only for small leads, less than 6 hours, say. For larger leads prediction of residuals does not improve the forecasting significantly over that offered by the estimated current profile.

- The forecasting accuracy is comparable to or better than earlier manual practice, in spite of the fact that no external information, such as weather forecasts has been used. It is about equal to that reached by Farmer and Potton [41, 42], also using past data only, for 24 hours and 1 hour forecasting, viz. 3.4% and .7% respectively during peak load. The company decided to go ahead with the project.

- Observed and predicted forecasting errors agree well only in daytime. Actual errors are smaller for small leads and decrease for large leads, which the model is unable to predict. It is reasonable to attribute this to the assumption that the unpredictable variation in demand σ_1 and σ_2 are independent of the clock. This saves computing by yielding one Riccati equation instead of twenty-four. However, it is apparently an oversimplification. Actually, variations are smaller during hours of low power demand--for about 21 to 7 o'clock--and Figure 8 indicates that this must be taken into account, if a better prediction of forecasting errors is needed during these hours. The curve strongly suggests a constant *relative* variation in energy demand. However, for planning purposes those during daytime are most important, and there agreement is good.

- Simulated short-range forecasting brought the errors down to the vicinity of measurement errors (which are approximately 2 MW) for a lead time $\leq$ 45 minutes.

Acknowledgments

The first case study was made within the generous framework of a joint development project by Billerud AB and IBM Nordic Laboratory. Project managers were O. Alsholm for BAB and Å. Ekström for IBM. Access to full-scale production units for experimentation has been an asset, the value of which cannot be overestimated. S. Wensmark carried out the experiments used for the analysis.

The second case study was part of a research program on EEG-analysis, headed by L. H. Zetterberg of the Royal Institute of Technology in Stockholm. Test data were provided by A. Wennberg. During the first two case studies the author was employed by IBM.

The third case was carried out under a development contract between IBM and the author. B. Ek provided the problem, and S. Wensmark did part of the simulations.

The fourth case was an informal feasibility study made for Stockholm's Energiverk. F. Sandin provided the problem and supplied the data records.

To the people mentioned and to a great many others, who have been indirectly involved I would like to express my gratitude, in particular to K. J. Åström for much exchange of thoughts on the many facets of identification.

REFERENCES

1. Lainiotis, D. G., *Inform. Sci.*, *7*, 191 (1974).

2. Åström, K. J., *Introduction to Stochastic Control Theory*, Academic Press, New York, 1970.

3. Åström, K. J. and P. Eykhoff, *Automatica*, *7*, 123 (1971).

4. Eykhoff, P., *System Identification*, Wiley, London, 1974, p. 253.

5. Lee, R. C. K., *Optimal Estimation, Identification, and Control*, MIT Press, Cambridge, Mass., 1964, p. 113.

6. Åström, K. J., Preprints First IFAC Symposium on Identification in Automatic Control Systems,Prague, 1967, paper 1.8.

7. Clarke, D. W., Preprints First IFAC Symposium on Identification in Automatic Control Systems, Prague, 1967, paper 3.17.

8. Bohlin, T., *IEEE Trans. Automatic Control, AC-15,* 104 (1970).

9. Bélanger, P. R., Proc. 5th IFAC World Congress, Paris, 1972, paper 38.3.

10. Sage, A. O. and G. W. Husa, Preprints JACC, Boulder, 1969.

11. Mehra, R. K., *IEEE Trans. Automatic Control, AC-13,* No. 2 (1970).

12. Bohlin, T., IBM SDD Nordic Laboratory, Sweden, Technical Paper 18.190, 1968.

13. Integrated Control of Paper Production (Editorial), *Control, 10,* 583 (1966).

14. Bohlin, T., Proc. PRP-Automation Congress, Antwerp, 1966, paper C7.

15. Åström, K. J. and T. Bohlin, Proc. Theory of Self-Adaptive Control Systems, Teddington, 1965 (P. Hammond, editor).

16. Söderström, T., Lund Institute of Technology, Division of Automatic Control, Report 7308, 1973.

17. Wittenmark, B. and U. Borisson, Preprints IFAC Symposium on Digital Computer Applications to Process Control, Zurich, 1974.

18. Bohlin, T., IBM SDD Nordic Laboratory, Sweden, Technical Paper 18.212, 1971.

19. Brazier, M., *J. EEG and Clin. Neurophys., Supplement No. 20,* 1961.

20. Livanov, M. N. and V. S. Rusinov, *Mathematical Analysis of the Electrical Activity of the Brain,* Harvard University Press, Cambridge, Mass., 1968.

21. Walter, D. O. and A. B. Brazier, *J. EEG and Clin. Neurophys. Supplement No. 27,* 1968.

22. Grindel, O. M., Proc. Symposium on the Mathematical Analysis of the Electrical Activity of the Brain, Erivan, USSR, 1964, Harvard University Press, Cambridge, Mass., 1968.

23. Dumermuth, G. and H. Flühler, *Med. and Biol. Eng., 5,* 319 (1967).

24. Kaiser, E. and I. Petersén, *Acta Neurologica Scandinavica, 42,* Suppl. 22 (1966).

25. Gersch, W., *Math. Biosciences, 7,* 205 (1970).

26. Zetterberg, L. H., *Math. Biosciences, 5,* 227 (1969).

27. Fenwick, P. B. C., P. Michie, J. Dollimore, and G. W. Fenton, *Biomedical Computing, 2,* 281 (1971).

28. Farley, B. G., *J. EEG and Clin. Neurophys., Supplement No. 20,* 1961 (M. Brazier, editor).

29. Peimer, I. A., Proc. Symposium on the Mathematical Analysis of the Electrical Activity of the Brain, Erivan, USSR, 1964, Harvard University Press, Cambridge, Mass., 1968.

30. Dumermuth, G., P. J. Huber, B. Kleiner, and T. Gasser, *IEEE Trans. Audio and Electroacoustics, AU-18,* No. 4, 404 (1970).

31. Meshalkin, L. D. and T. M. Efremova, Proc. Symposium on the Mathematical Analysis of the Electrical Activity of the Brain, Erivan, USSR, 1964, Harvard University Press, Cambridge, Mass., 1968.

32. Van Leeuwen, W. S., *J. EEG and Clin. Neurophys., Supplement No. 20,* 1961 (M. Brazier, editor).

33. Dumermuth, G., P. J. Huber, B. Kleiner, and T. Gasser, *J. EEG and Clin. Neurophys., 31,* 137 (1971).

34. Bohlin, T., *IBM J. Res. Development, 17,* 194 (1973).

35. Wennberg, A. and L. H. Zetterberg, *J. EEG and Clin. Neurophys., 31,* 457 (1971).

36. Zetterberg, L. H., *Automation of Clinical Electroencephalography* (P. Kellaway and I. Petersén, editors), Raven Press, New York, 1973.

37. Box, G. E. P. and G. M. Jenkins, *Time Series Analysis Forecasting and Control,* Holden-Day, San Francisco, 1970.

38. Rissanen, J. and L. Barbosa, *Inform. Sci.*, *1*, 221 (1968).

39. Rissanen, J., *Mathematics of Computation*, *27*, 147 (1973).

40. Pehrson, B., *IBM J. Res. Development*, *6*, 703 (1969).

41. Matthewman, P. D. and H. Nicholson, *Proc. IEE*, *115*, 1451 (1968).

42. Farmer, E. D. and M. J. Potton, Proc. 3rd IFAC Congress, London, 1966.

MODELING AND IDENTIFICATION OF A NUCLEAR REACTOR

Gustaf Olsson
Department of Automatic Control
Lund Institute of Technology
S-22007 Lund, Sweden

1. INTRODUCTION

Some representative results from modeling and identification experiments on the Halden Boiling Water Reactor, (HBWR), Norway, are presented in this paper. Linear input-output models as well as time invariant and time variable linear state models have been used as model structures. Some of the results are presented previously in [1-3], while others are new.

The purpose of the paper is to describe the different phases of identification and modeling of a complex dynamical system. Different identification methods have been used to demonstrate the applicability of identification techniques as a tool to explore the dynamics of a nuclear reactor.

A nuclear reactor is an example of a very complex dynamical system and offers some special features. There is a wide span of time constants in the system. The neutron kinetics is very fast, and the dominating kinetics time constant is about 0.1 second. The typical time constants for actuators and instrument dynamics vary between fractions of a second and about one second. The fuel element heat dynamics are of the order of a few seconds. The heat transfer in moderator and coolant channels as well as the hydraulics is of the order of some seconds up to some minute. The heat transfer through the heat removal circuits will take one to several minutes. Xenon oscillations have a time period of the order of days. On an even longer time scale there are the burn out phenomena due to fuel consumption.

Several nonlinear phenomena are important in a nuclear reactor. The dynamics of the coolant channels are very complex. The relation between boiling boundary, void contents and reactivity is generally highly nonlinear and very difficult to model. The heat exchanger dynamics and steam generation are also significantly nonlinear.

Many phenomena are spatially dependent. Power distribution oscillations due to xenon are not negligible in a large reactor. The spatial variations of void content and temperature in the coolant channels are essential, dynamical phenomena. The neutron distribution is not homogeneous since the fuel elements are burnt out at different rates in different parts of the core.

A model used for controller design cannot include all the mentioned phenomena in detail. A large number of compromises must be made in order to make the model not too large and still accurate. The purpose of this paper has primarily been to find linear models for steady state control. The nuclear power and the primary pressure then are the most important outputs to be controlled. This limits the interesting span of time constants to be smaller than some minutes. The results of the investigation show that the dynamics of the reactor generally can be described by quite low order models. It will be demonstrated that identification is a useful tool to find simpler descriptions of such a complex process.

Modeling and identification problems for nuclear power reactors have been considered extensively. The Maximum Likelihood (ML) method is compared with other methods for a reactivity-nuclear power model by Gustavsson [4]. Sage *et al.* [5] use a least squares approach to identify parameters in a reactor model. Ciechanowicz *et al.* [6] use spectral analysis to identify parameters in a simple linear model. Recursive identification or parameter tracking has been reported by different authors. Habegger *et al.* [7] apply Extended Kalman techniques to track parameters in a nuclear system. Moore *et al.* [8] use a combination of least squares and ML approach to get an adaptive control scheme of a model of a pressurized water reactor.

The dynamics of the Halden reactor has been studied extensively before. Single input experiments have been performed, e.g. step response analysis by Brouwers [9], frequency analysis by Tosi *et al.* [10], pseudo random reactivity perturbation experiments by Fishman [11] and noise experiments by Eurola [12]. Bjørlo *et al.* [13] have reported a linear multivariable model of the HBWR. The vessel pressure dynamics and core dynamics have been studied with recursive least squares techniques by Roggenbauer [14].

Four different approaches to the model building techniques are investigated in this report:

- multiple-input--single-output models with no *á priori* assumption about physical behavior,
- multivariable (vector difference) models without physical *á priori* knowledge,
- estimation of parameters in linear time invariant state models with known structure and *á priori* noise structure assumptions,
- estimation of time variable parameters in linear stochastic state models.

It is natural that a model with no *á priori* assumption about the physics does not demand physical insight into the process, at least not to get parameter values. In general there is no physical interpretation of the parameters, and it is therefore sometimes difficult to verify the models in more general terms. On the other hand, such a model can give a good insight into the required complexity of a more structured model. The validity of the model is limited to the same operational conditions for the plant as those during the identification experiment.

As a nuclear reactor is a multivariable system, the second approach is an attempt to take the couplings of the system into consideration without too many *á priori* assumptions. Compromises about the noise have to be made. The approach gives a better idea of the couplings in the system, and it is then easier to derive reasonable structures for more advanced models.

A state model with some of the parameters unknown naturally requires more insight into the process. In such a model the parameters have physical interpretations. If the assumptions on the structure are perfect, the model accuracy can be high. On the other hand, if the assumptions are imperfect, the model can be more inaccurate than an input-output model without *á priori* assumptions. The identification would then be constrained into too few degrees of freedom, either because of too few free parameters or of a wrong *á priori* structure.

In order to be valid for varying operating conditions the plant model should be nonlinear. Alternatively it has here been assumed a time varying linear state model. Some of the variable parameters then have been tracked by recursive identification techniques.

The paper is organized as follows. In Section 2, the reactor plant is described and its dynamics are studied qualitatively. A summary of the experiments selected is made in Section 3. Experimental design is also considered, as well as instrumentation and actuator characteristics. The identification methods used are briefly presented in Section 4. Maximum likelihood identification technique has been applied predominantly. The multiple-input--single-output models are discussed in Section 5. Although accurate models were found, the linearity of the models is a limitation, and it is doubtful if they are valid in a large operational range. Improvements of the accuracy were obtained by introducing other couplings by a vector difference equation approach in Section 6. In Chapter 7 a linear state vector model structure is presented. Parameters of this structure are identified. The recursive parameter tracking is finally described briefly in Section 8.

2. DESCRIPTION OF THE NUCLEAR REACTOR

A short description of the reactor is given to provide a physical background. In the first paragraph the different parts of the plant are briefly described. In paragraph B the most important dynamical reactivity feedbacks are considered. Finally it is discussed how changes in the three actual inputs propagate through the system.

A. *PLANT DESCRIPTION*

The reactor plant has been described elsewhere in great detail, e.g. in Jamne *et al.* [15] and several other reports from the Halden Reactor Project, e.g. [9-13]. For easy reference some main features of the plant are described here.

A simplified sketch of the plant with its heat removal circuits is shown in Fig. 1. The HBWR is a natural circulation, boiling heavy water reactor. It can be operated at power levels up to 25 MW and at 240°C.

1. *Core and Primary Circuit*

In the primary circuit heavy water is circulated in a closed loop. This circuit consists of the reactor vessel, steam transformers and a subcooler A. The latter ones are heat exchangers for the steam and water circulation loops respectively.

The core consists of enriched uranium fuel moderated by heavy water. There are 100 fuel assemblies in the core arranged in a hexagonal pattern each element being 88 cm in length. The core diameter is 167 cm and is surrounded by a radial reflector with a thickness 51 cm. The bottom reflector is 38 cm thick.

The fuel elements in the core have shrouds into which heavy water from the moderator enters through the holes in the bottom section. The shrouds create a defined flow pattern, and can separate the upstreaming mixture of steam and water from the downstreaming water between the elements. The lower part of the core is not boiling while the upper part is. The water in the system is close to the saturation temperature.

The mixture of steam and water leaves the shrouds through holes at the top and separation of water and steam takes place. The steam passes from the reactor vessel through the primary side of a heat exchanger called the steam transformer. As the steam is condensed it is pumped together with water from the bulk of the moderator through the primary side of the subcooler A (Figs. 1, 2). The water is cooled a few degrees below the

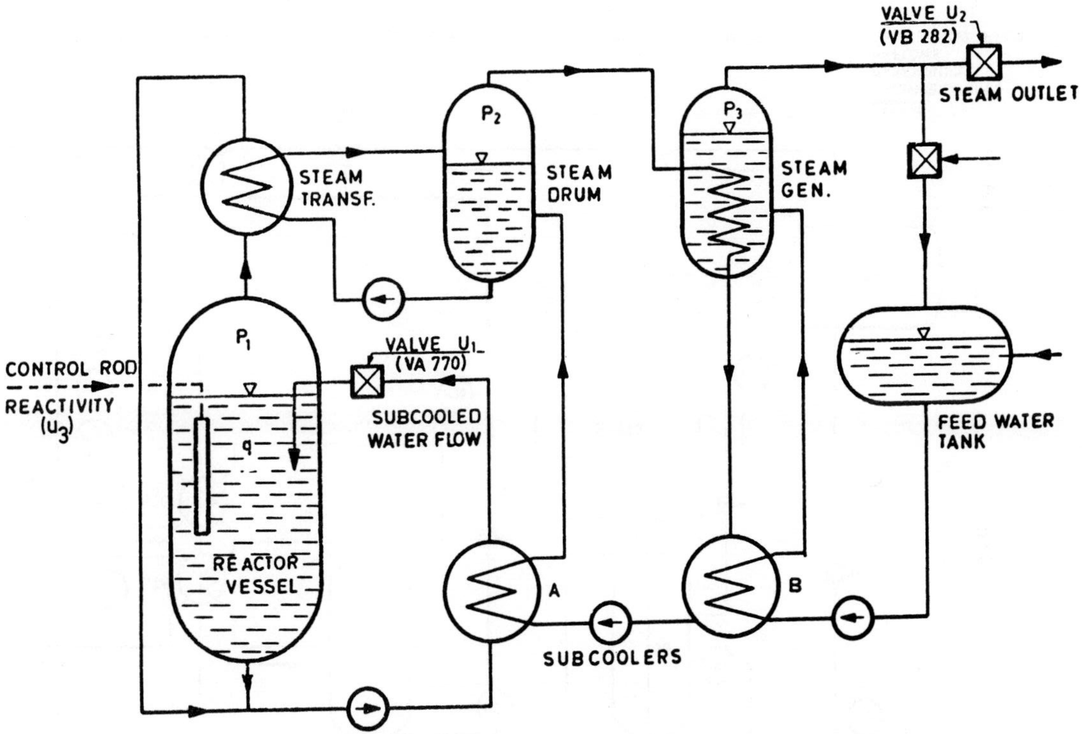

Fig. 1. Simplified flow sheet of the HBWR plant. Reprinted by courtesy of the Halden Reactor Project.

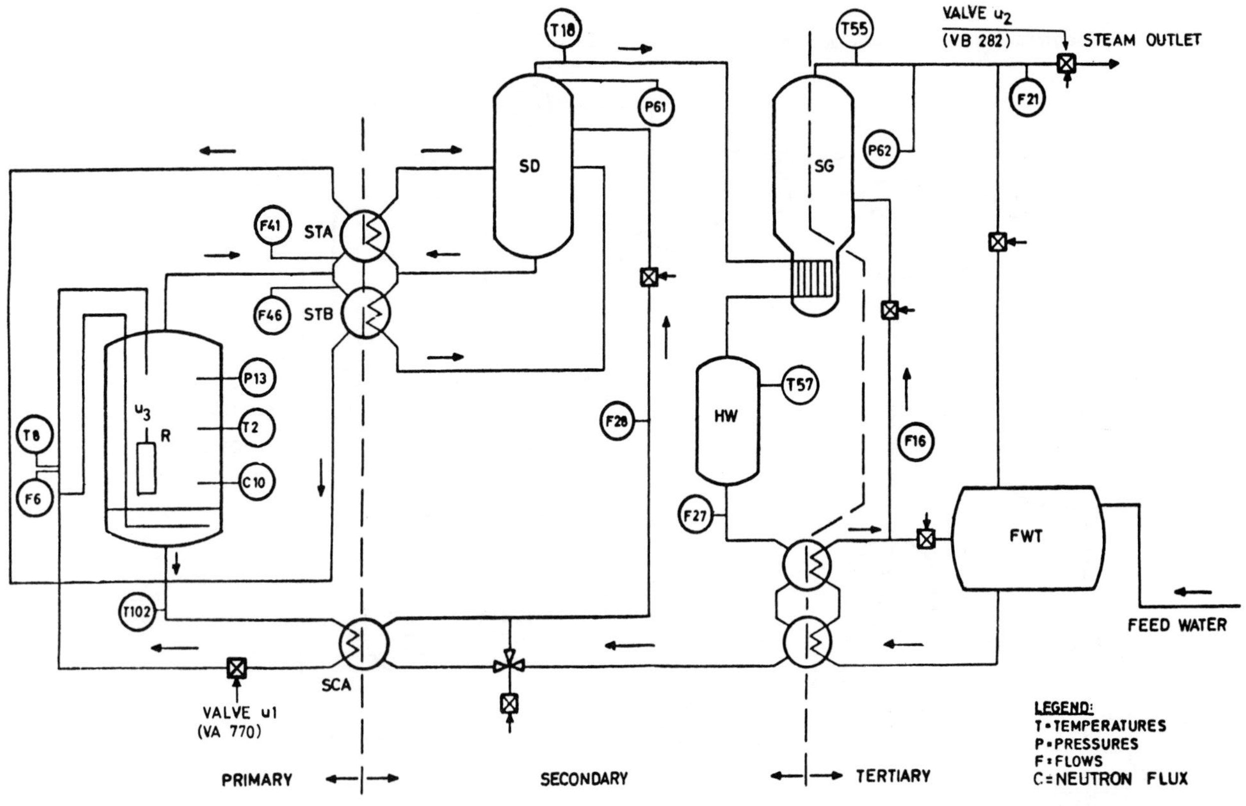

Fig. 2. Variables registered for the identifications. Reprinted by courtesy of the Halden Reactor Project.

saturation temperature and then recirculated into the vessel.

The reactivity is controlled by 30 absorption rods which can be inserted into the core.

2. Subcooling Circuit

The mass flow of subcooled water is controlled by a valve u_1 (VA 770). As the subcooled water enters the moderator it mainly affects the moderator temperature. The main purpose of the subcooling circuit is to suppress boiling of the moderator. To a lower extent it controls the reactivity of the core.

The water loop is to some extent similar to the coolant flow circulation system in a light water BWR, even if there are major differences. In the HBWR this system is not primarily designed for control purposes, and thus the flow and also the reactivity feedback are much smaller than in a light water BWR. The void reactivity feedback is about 20 pcm/% void (1 pcm = 10^{-5}) in the HBWR compared to about 125 pcm/% void in a BWR.

Because of the limited control authority of the valve u_1 it cannot alone control the nuclear power in the HBWR over a wide range as compared to a light water BWR. It must be complemented by the absorption rods.

3. Secondary and Tertiary Circuits

The secondary circuit is closed and filled with light water (see Figs. 1, 2). Water coming from the stream drum is circulated through the secondary side of the steam transformer and back to the steam drum. The secondary circuit also includes a steam flow from the steam drum to the primary side of a steam generator where it is condensed.

The condensed water returns via the hot well to the subcooler B, where primarily feedwater is preheated. The water is further heated up in the subcooler A before it returns to the steam drum. This steam drum mainly serves as a separator for steam and water.

The tertiary circuit is an open loop circuit of light water. The water is heated up to form steam in the secondary side of the steam generator. The steam can be used by consumers through a valve u_2 (VB 282). The plant has no turbine, but u_2 should normally be the turbine controller. The steam can be recirculated via the feedwater tank and the subcooler B to the steam generator.

B. *REACTIVITY FEEDBACKS*

The essential part of the dynamics has to do with the reactivity feedbacks. For the discussion we refer to Fig. 3. The net reactivity determines the nuclear power which is produced in the core. This net reactivity is a sum of several feedback effects. The nuclear power is created through the fission, which can be described by the kinetic equations, including delayed neutrons. This power generates heat which is transferred through the fuel elements. A change in fuel temperature causes a negative reactivity feedback. The heat flux transfers heat via the fuel elements and the moderator into the coolant. The moderator dynamics describes the temperature and void distribution in the moderator. It is related to the steam pressure, and water and steam velocities.

It should be remarked that there are some important differences between light water and heavy water boiling reactors. In H_2O systems almost all the moderator is boiling. In D_2O systems the boiling takes place only in a fraction of the moderator space, because the moderator-to-fuel ratio is relatively large. Therefore models of light water boiling reactors, which are described in the literature, such as Fleck [16], differ from the HBWR in basic assumptions.

The heat flux consists of several components. Except the nuclear power it is determined by gamma and neutron heating as well as the subcooling power. The coolant channel dynamics (the void and temperature distributions in the coolant channels) is primarily determined by the heat flux, but also by the vessel pressure, the steam and water velocities as well as the channel inlet temperature. This one in turn depends on the moderator

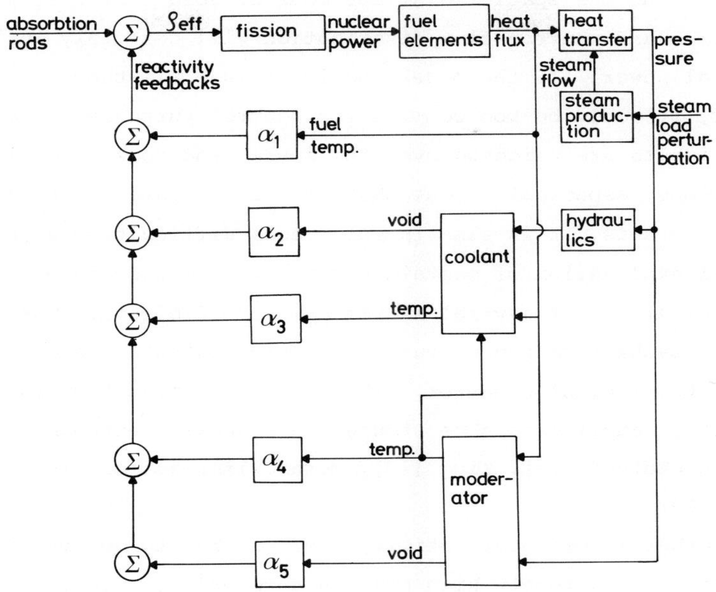

Fig. 3. Reactivity feedbacks in the reactor dynamics.

temperature. Naturally those phenomena are spatially dependent. Therefore it should be emphasized, that not only the total heat flux but also the spatial distributions of void contents, water velocities and neutron flux distribution will certainly influence the total power. If the model should include all those phenomena, however, it would be too complex for control purposes. Therefore the variables are weighted over the space, and some crucial assumptions, especially about the hydraulics, have to be made. Because of this, it is also in some cases difficult to give a physical explanation of certain parameters, as they in essence are combinations of several microscopic coefficients. The reactivity feedbacks from temperatures and void contents are crucial for the total plant behavior. The physical explanation for reactivity couplings can be studied in standard textbooks, like Glasstone-Edlund [17], King [18], Meghreblian-Holmes [19], Weaver [20].

Another important reactivity feedback has to do with fission products with extremely high neutron absorption, such as xenon. Transients due to xenon can appear in two ways. One type of xenon transients appears at high neutron flux levels and is enforced due to power changes. This varies the average concentration of xenon, and consequently the neutron level. As all the experiments have been performed at almost constant power, no such power transients are actual.

The other type of xenon feedback occurs in reactors with large geometrical dimensions. There the xenon concentration can oscillate spatially between different parts of the core, thus creating hot spots of power, while the average power is constant. Such phenomena have been analyzed by several authors, e.g. Wiberg [21] and Olsson [22] and will not be considered here because of two reasons. First, the oscillations are too slow to be of interest here, as the primary purpose is to keep nuclear power and primary pressure constant. Second, the Halden reactor has small geometrical dimensions so that the spatial oscillations are too much damped to be of any interest.

The essential disturbances to the system consist of reactivity perturbations from the absorption rods or changes in the steam consumption.

A quite comprehensive description of the details of the HBWR dynamics can be found in Vollmer *et al.* [23] and Eurola [24].

C. *STEP RESPONSES*

For the following discussion it is useful to have an overview of the major physical phenomena of the plant. The purpose is to provide this by qualitative discussion of step responses and the major physical phenomena that are involved. The results are based on both theoretical considerations and practical experiences.

1. *Subcooling Valve* u_1

Assume that the valve u_1 (VA 770) is closed stepwise. As only small changes are discussed linear relations are assumed. The downcomer subcooled flow F6 (see Figs. 2, 4) decreases rapidly as the valve closes. The water temperature T102 just before the subcooler is not affected, but the subcooled water flow temperature T8 is decreased with a few seconds time constant (Fig. 4).

The heat flow delivered to the subcooler A is called the subcooling power Q. This power is calculated from energy balances over the heat exchanger (subcooler A) and is a function of the product of the temperature change of T8 and the flow change of F6.

It is possible to empirically relate the subcooling power in a simple fashion, to F6, T8 and u_1, as can be visualized by Fig. 4. As the flow F6 is closely related to the valve opening u_1, the subcooling power change can be written

$$\delta Q(s) = a_1 \delta(T8(s)) + a_2 \frac{1}{1+Ts} \delta(U_1(s))$$

where s is the Laplace operator.

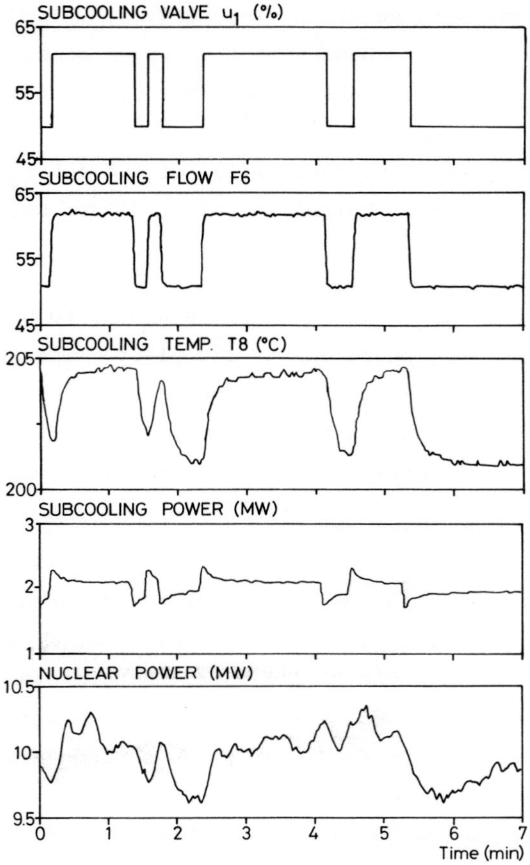

Fig. 4. *Typical responses to changes in* u_1.

The subcooling power can also empirically be written as a function of T8:

$$\delta Q(t) = \alpha_1 \delta(T8) + \alpha_2 \frac{d(T8)}{dt}$$

where $\alpha_1 > 0$, $\alpha_2 > 0$.

The effect of closing the value is thus, that more heat energy is returned to the core. The bubble formation in the moderator is amplified, and this phenomenon directly causes a negative reactivity feedback. Because of this the nuclear power decreases quite rapidly.

In a longer time scale several secondary effects take place, which is illustrated by the step response in Fig. 5. As the nuclear power decreases, the vessel pressure and the temperatures also decrease. Other reactivity feedbacks now are beginning to act and the nuclear power is slowly returned to a more positive value.

The vessel pressure naturally is coupled through the steam transformer to the secondary and the tertiary circuits. Those pressures therefore slowly follow the pressure decrease in the vessel. The steam production in the primary circuit is, however, influenced to a lesser degree.

When u_1 is closed only a slight decrease of the steam inlet flow F41 (Fig. 2) can be observed. The same is true for the flow F28 in the secondary circuit.

2. Consumers Steam Valve u_2

A sudden increase of the valve opening u_2 (VB 282) for the tertiary steam flow directly increases the tertiary steam flow F21 (see Figs. 6 and 2). Consequently the tertiary pressure (P62) will be decreased with a dominating time constant of about one minute.

The temperature T55 is strongly coupled to the pressure variations and it follows the pressure P62 closely. Also the flow F16 is increased, but delayed a few seconds after the flow

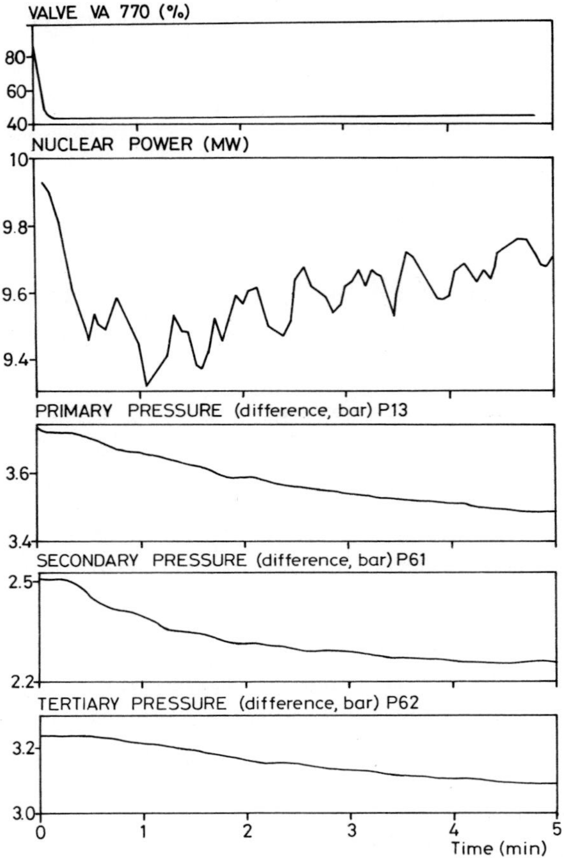

Fig. 5. Typical responses to a step change in u_1.

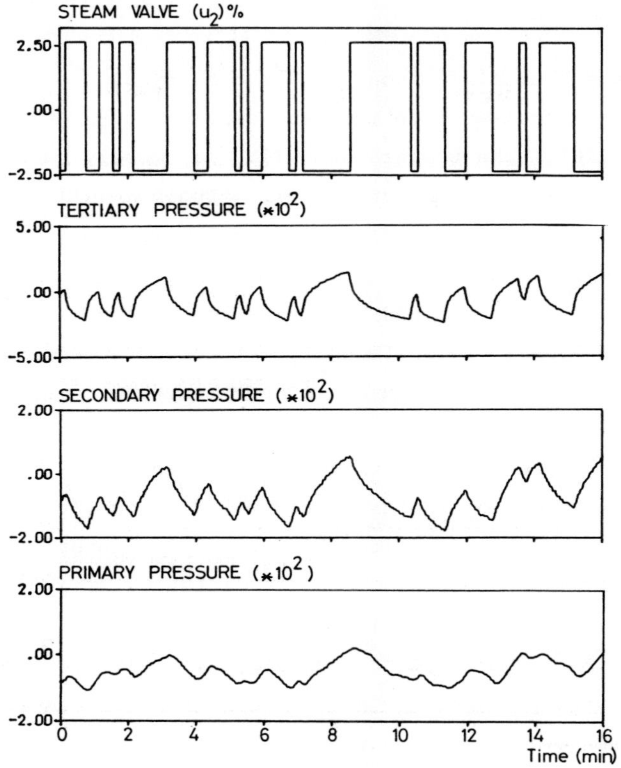

Fig. 6. Typical responses to changes in u_2.

F21. The feedwater temperatures T60 and T61 are quite unaffected by u_2.

When the heat flow through the secondary side of the steam generator is decreased also the secondary pressure P61 will decrease (Fig. 6). The temperature T18 is closely coupled to the pressure and follows P61 quite well. The hot well temperature T57 and the secondary water temperature T81 are relatively constant despite changes in u_2.

The flows F27 and F28 are varying quite noticeably. The dynamics is, however, significantly influenced by an internal controller. The hot well level is kept constant in all the experiments by a valve controlling the flow F28. The flows F27 and F28 increase when the valve u_2 is opened.

The pressure drop in the tertiary and secondary circuits is propagated to the primary circuit with a 2-3 minutes' time delay, and thus the vessel pressure P13 is decreased (Fig. 6). A pressure drop in the core will cause the void to increase in the first moment, and the boiling boundary will fall. The reactivity feedback from void therefore has the effect to decrease the nuclear power in the first moment. When the power decreases, however, the steam production also decreases, thus creating a smaller void content and a higher boiling boundary again. This causes the nuclear power to increase. As indicated by the experimental step response in Fig. 7 the nuclear power shows a nonminimum phase behavior.

The control power from u_2 is significantly larger than that of u_1, a fact which is illustrated by the step responses in Figs. 5 and 7. On the other hand, the valve u_1 can change the nuclear power much more rapidly than the valve u_2, so they complement each other dynamically. In a light water BWR, as mentioned before, there is not such a great difference in control authority between u_1 and u_2.

From an identification point of view the valve u_2 is certainly the best input for studies of the heat removal circuit dynamics. The valve u_1 naturally has the strongest influence on the subcooling circuit.

3. Control Rod Reactivity u_3

By inserting or withdrawing the absorption rods the nuclear power can be rapidly and significantly changed. This dynamic is very rapid and is governed mainly by the delayed neutrons. If the rods are properly positioned, so that the reactivity change per step is large enough, the rods can control the nuclear power very well. It should be observed, however, that also the flux distribution generally is affected by the rods. Moreover, wearing out problems should be considered, which means that the rods should not be used for frequent control movements.

The nuclear power transfers heat to the fuel elements quite rapidly with a time constant of the order 5-10 seconds. The temperatures of the moderator and coolant increase more slowly. The pressure changes are quite slow, of the order half a minute for the vessel pressure to about a few minutes for the tertiary pressure.

3. EXPERIMENTS

In this section we will consider experimental design problems, such as choice of input signals and measurements. The selected experiments are summarized and the data handling problems are mentioned. In all the experiments the input disturbances were generated in the IBM 1800 computer, connected to the plant. All measurements were also registered using the computer.

A. *SUMMARY OF THE EXPERIMENTS*

In Table 1 the main features of the operating conditions are shown for the selected experiments.

The valve amplitudes are defined in % opening. The reactivity is defined in "steps", where one step reactivity is defined as the reactivity corresponding to the movement of the rod step motors one step. It corresponds to 7-10 pcm reactivity, depending on the position of the rods. The figures in brackets under u_3 in Table 1 define the rod numbers. The rods are moved in parallel one step up and down.

TABLE 1

Summary of the Identification Experiments

Exp.	u_1 (%) VA770	u_2 (%) VB282	u_3 (steps) Rods	Nuclear power (MW)	Subcooling power (MW)
1	-	-	3(13,15,17)	9.7	1.35
2	-	-	3(13,17,18)	9.95	1.85
3	-	±2.5	3(13,17,18)	10	1.95
4	-	±3	2(20,21)	10	1.95
5	±7	±3	-	10	2.0
6	-	±2	1(20)	8.0	1.1
7	-	±2.5	2(20,21)	10.0	1.95→1.35

Most of the experiments contain more than 3000 samples, i.e. 6000 seconds. For identification purposes not more than 2000 samples have been used at a time.

For safety reasons it was sometimes necessary to move some control rod manually in order to keep the nuclear power and vessel pressure within permitted limits.

B. *INTERNAL CONTROLLERS*

It was important to study the plant in open loop operation, and therefore some controllers were removed, primarily the nuclear power controller, which keeps the nuclear power within desired limits by adjusting the absorption rods.

The primary (vessel) pressure is controlled by a PID controller acting on the valve u_2 (VB 282). For safety reasons it was not allowed to remove this control in the first experiment series, here represented by experiment 1. As this control loop has a time constant of more than one minute, the fast time constants still could be determined. When more experiences had been gained, it was allowed to remove also the pressure controller, experiments 2-7.

Other local controllers were acting as before, i.e. control of the hot well level and steam generator level as well as return flow to the feedwater tank. These controllers, however, do not influence the determination of the overall dynamics.

C. *EXPERIMENTAL DESIGN CONSIDERATIONS*

In the design of input signals and operating levels a large number of conditions have to be considered. A general survey of such problems have been described in Gustavsson [25].

In order to gain a good signal-to-noise ratio a large input amplitude is desired. Through preliminary experiments it was found, that three rods moved one step in parallel could disturb the nuclear power about 0.5 MW from the operating level of about 10 MW. The upper limit of the changes in u_1, u_2 and u_3 were determined by nonlinear effects.

One experiment was done in order to cover a wider range of operational conditions, experiment 7. The subcooling power was changed along a desired ramp. The parameters of a time variable model then were identified recursively (see 8).

The major time constants were discussed in 1 and 2. They will determine the desired frequency content of the input signals. The upper limit of the frequency was determined by practical reasons, as the computer sampling time was fixed to 2 seconds. By experience we also know, that in one identification experiment it is difficult to accurately determine time constants spanning more than about 2 decades, i.e. here from some second to a few minutes.

In all the reported experiments pseudo random binary sequences (PRBS) have been applied as inputs as it was desirable to get persistently exciting signals. In the case of several inputs, the signals have been chosen so as to be independent. As the pressure control was in action in experiment 1, the input signal was chosen to excite time constants essentially smaller than one minute. In experiments 2-7 the sequence was chosen with longer pulses in order to get better estimation of the long time constants.

There are different rules of thumb in the literature how to choose a suitable PRBS sequence, and those rules can give quite different results, as demonstrated here. Briggs *et al.* [26] have made a detailed analysis of the PRBS sequence. According to their rules the period time of the sequence should be at least 5 times the longest time constant T_m of the process. Another rule of thumb says, that the longest pulse of the sequence should be at least $3*T_m$. Then the process is allowed to reach a new steady state during the pulse, and the estimation of the gain and largest time constant will be improved. The PRBS sequence for experiment 1 was chosen with the shortest pulse length of 2 seconds, a period time of 991 samples (almost 2000 seconds) and a longest pulse length of only 18 seconds. With the cited rules applied to this sequence it limits the longest time constant either to 400 seconds or to 6 seconds, a significant difference. Thus it is found that the PRBS sequence can be too fast for the low frequencies. This fact has been observed also e.g. by Gustavsson [25] and Cumming [27], [28].

For the second PRBS sequence the period time is still about 2000 seconds but the longest pulse is 196 seconds. According to referred rules the longest time constant then could be 60-400 seconds. The shortest pulse was chosen 12 seconds, but still the sampling time is 2 seconds. It is shown in Section 5, that the sampling time--and not only the input sequence--is important for the accuracy of the long time constants.

D. INSTRUMENTS AND ACTUATORS

The variables recorded during the experiments are indicated in Fig. 2. The meaning of the letters are

P - pressure

F - steam or water flow

T - temperature

C - nuclear power

The HBWR instrumentation is described in detail elsewhere, see [29]. Here only the main features are summarized. The pressures are registered as differential pressures in the three circuits (P13, P61, P62) with conventional DP cells with a range of about ±0.3 bar.

The flows are generally measured with venturi meters plus differential pressure cells. The temperatures are measured by thermocouples. The nuclear power is measured by an ion chamber C10.

The pressure cells and flow meters in the primary circuit have time constants around one second. The temperatures, however, are registered much faster, at about 0.1 second. The instrumentation does not generally cause any problem, as the important dynamics generally are much slower. The actuator time constants are not negligible. To move a valve through its whole range takes about 6 seconds. A typical time delay for the valve u_1 in the experiments was therefore about 1 second. For the valve u_2 the corresponding delay was about half a second.

The instrument noise of the pressure meters and the nuclear channel are well known from previous experiments, see e.g. [9-12]. For the nuclear power the measurement noise is about ±0.03 MW. The standard deviation for the differential pressure meters has also been experimentally determined. Typical values are $0.5 \cdot 10^{-4}$ units. The pressure unit is expressed as pressure variation divided by total pressure. During the experiments the total variation of e.g. the vessel pressure was about $\pm 0.5 \cdot 10^{-2}$ units. This means, that the noise to signal ratio was about 1%.

The A/D converter has 11 bit resolution, and conversion errors must be considered. For the nuclear power measurements the total power is measured and converted. As the power variations are most about 5% of the total power the conversion errors are not negligible, especially for long input pulses, when the variations of the signal are small. The error is estimated to be about $5 \cdot 10^{-3}$ MW. The quantization errors must also be considered for the pressure meters, see 5.D.

During the experiments 35 variables were recorded, some of them only for checking up purposes. The data were logged on the IBM 1800 computer and were measured with 2 seconds sampling interval by a 100 Hz relay multiplexer. Because of the multiplexer the measurements could be up to 0.3 seconds separated in time for the same sampling interval. The sample and hold circuit also introduced a time constant, about 0.35 sec.

4. IDENTIFICATION METHODS

For the preliminary analysis of the experimental data and for the first model approaches simple methods were used to find rough estimates of the input-output relationships. Step response analysis and correlation analysis were used to verify preliminary models and to design new experiments. For the parameter estimation the Maximum Likelihood method has been used except for the recursive estimation, where an Extended Kalman filter is applied.

In this section the methods are summarized. For detailed descriptions a large number of papers are available, see e.g. Åström-Eykhoff [30], Eykhoff [31], and Mehra *et al.* [32].

A. *MULTIPLE-INPUT--SINGLE-OUTPUT (MISO) STRUCTURE*

The plant dynamics is represented by the canonical form, introduced by Åström *et al.* [33]

$$(1 + a_1 q^{-1} + \dots + a_n q^{-n}) y(t) = \sum_{i=1}^{p} (b_{i1} q^{-1} + \dots + b_{in} q^{-n}) \cdot u_i(t) + \lambda (1 + c_1 q^{-1} + \dots + c_n q^{-n}) \, e(t)$$

or

$$A^*(q^{-1})y(t) = \sum_{i=1}^{p} B_i^*(q^{-1})u_i(t) + \lambda C^*(q^{-1})e(t) \qquad (1)$$

where q is the shift operator and p the number of inputs. A^*, B^* and C^* are defined as corresponding polynomials in q^{-1}. It is trivial to extend the model to include both time delays and direct input terms, corresponding to a coefficient b_{i0} in (1). Moreover initial conditions can be estimated.

If $e(t)$ is assumed to be a sequence of independent Gaussian random variables the parameters a_i, b_i, c_i and λ can be determined using the method of Maximum Likelihood (ML). The method is described in detail elsewhere, e.g. [30-33], and only some remarks will be made here.

The likelihood function $L(\theta;\lambda)$ for the unknown parameters

$$\theta^T = (a_1 a_2 \dots a_n b_{11} \dots b_{pn} c_1 \dots c_n) \qquad (2)$$

is given by

$$\ell n\, L(\theta;\lambda) = -\frac{1}{2} \sum_{1}^{N} \frac{1}{\lambda^2} \varepsilon^2 - N\, \ell n\, \lambda + \text{const.} \qquad (3)$$

where the residuals $\varepsilon(t)$ are defined by

$$\varepsilon(t) = [\hat{C}^*(q^{-1})]^{-1} \left[\hat{A}^*(q^{-1})y(t) - \sum_{i=1}^{p} \hat{B}_i^*(q^{-1})u_i(t) \right]$$

and $\hat{A}^*$, $\hat{B}_i^*$ and $\hat{C}^*$ are estimates of the polynomials A^*, B_i^* and C^*. N is the number of samples and λ^2 is the covariance of the residuals.

The maximization problem reduces to the problem of minimizing the loss function

$$V = \frac{1}{2} \sum_{t=1}^{N} \varepsilon^2(t)$$

with respect to the unknown parameters. When the estimate $\hat{\theta}$ is calculated the parameter λ can be solved from the minimum value of the loss function

$$\lambda^2 = \frac{2}{N} V(\hat{\theta}) \tag{5}$$

In [33] it is shown that the estimates are consistent, asymptotically normal and efficient under quite mild conditions. The parameter λ can be interpreted as the standard deviation of the one step prediction error. The technique gives not only the estimates but also their standard deviations from the Cramér-Rao inequality.

As the number of parameters in the model or the system order is not given *á priori* a statistical test can be done in order to find the proper model. The loss function should not decrease significantly if the right order has been reached and more parameters are added. It is shown in [33] that the quantity

$$F_{n_1,n_2} = \frac{V_{n_1} - V_{n_2}}{V_{n_2}} \cdot \frac{N - n_2}{n_2 - n_1} ; \quad n_2 > n_1 \tag{6}$$

asymptotically has an F-distribution, where n_i is the number of parameters and V_{n_i} the corresponding loss functions. The residuals should also be tested for independence in time and in relation to the inputs.

An alternative test function due to Akaike [34] has also been used besides the F test. An Information Criterion is defined,

$$J = \frac{N + k}{N - k} \ln|\lambda^2| \tag{7}$$

where N is the number of samples, k the number of parameters and λ^2 the measurement noise covariance. Typically J as a function of k has a minimum for the right number of parameters.

The ML identification method has been extensively used in a large number of applications. Surveys are given in [25], [30] and [32].

B. *MULTIVARIABLE STRUCTURES*

The ML method has been generalized to the multivariable case. It is desirable to estimate a parameter vector θ of a linear continuous model

$$dx = Axdt + Budt + dv \tag{8}$$

$$dy = Cxdt + Dudt + de \tag{9}$$

The model is written in discrete time and in innovations form in order to simplify the noise estimation, according to Åström [35] or Mehra [36],

$$\begin{aligned} \hat{x}(t+1) &= \phi\hat{x}(t) + \Gamma u(t) + K\varepsilon(t) \\ y(t) &= C\hat{x}(t) + Du(t) + \varepsilon(t) \end{aligned} \tag{10}$$

where

$$\phi = e^{A(\theta)}$$

$$\Gamma = \left\{ \int_0^1 e^{A(\theta)s}\, ds \right\} B(\theta)$$

and $\hat{x}(t)$ denotes the conditional mean of $x(t)$, given previous measurement values $y(t-1), y(t-2), \ldots$.

The nose $\varepsilon(t)$ is now a sequence of independent Gaussian random vector variables. The likelihood function (3) is generalized to the form

$$\ell n\, L(\theta,R) = -\frac{1}{2} \sum_{t=1}^{N} \varepsilon^T(t) R^{-1} \varepsilon(t) - \frac{N}{2}\, \ell n \det R + \text{const.} \tag{11}$$

where R is the covariance of $\varepsilon(t)$ and is assumed to be constant. The loss function (cf. (4)) is

$$V = \det \left| \sum_{t=1}^{N} \varepsilon(t)\varepsilon^T(t) \right| \tag{12}$$

Eaton [37] has shown that the loss function can be minimized independently of R. As soon as the minimum of V is found an estimate of R can be achieved,

$$\hat{R} = \frac{1}{N} \sum_{t=1}^{N} \varepsilon(t)\varepsilon^{T}(t) \tag{13}$$

which is a generalization of (5).

Several strong theorems have also been stated about the multivariable case, e.g. see Åström *et al.* [30], Mehra [36], Woo [38], Caines [39], Ljung [40] and Mehra *et al.* [41].

C. A VECTOR DIFFERENCE EQUATION APPROACH

In order to find alternative models for the reactor also a vector difference approach was tried. Simplifying assumptions of the noise are made in order to identify the vector difference equation row by row. The noise assumptions are only adequate if there are weak couplings between the outputs considered.

The structure of the system is generalized from (1) to

$$\begin{aligned}[I + A_1q^{-1} + \ldots + A_nq^{-1}]y(t) &= [B_1q^{-1} + \ldots + B_nq^{-1}]u(t) \\ &\quad + [I + C_1q^{-1} + \ldots + C_nq^{-n}]e(t)\end{aligned} \tag{14}$$

where the capital letters assign constant matrices, while y, u and e are vectors. It is clear that there is no one-to-one correspondence between (14) and (10). In e.g. Guidorzi [42] this relation is further discussed. The likelihood function is still (11) where the residuals are defined by

$$\begin{aligned}\varepsilon(t) &= [I + C_1q^{-1} + \cdots + C_nq^{-n}]^{-1} * \\ &\quad * \left\{[I + A_1q^{-1} + \ldots + A_nq^{-n}]y(t) - [B_1q^{-1} + \ldots + B_nq^{-n}]u(t)\right\}\end{aligned} \tag{15}$$

If it is desired to identify the model row by row, then the loss function has to be written as a sum of n functions. This is possible if R is diagonal,

$$R = \text{diag}(\lambda_1^2,\ldots,\lambda_n^2) \tag{16}$$

and each matrix C_i is diagonal as well. The assumption means, that every output of the model is disturbed by a separate noise source, independent of other noise sources. With such assumptions all the parameters of A_i, B_i and C_i are identifiable.

The parameter estimates are not unbiased, consistent or with minimum variance as for the single output case. Still these multivariable models might indicate interesting couplings which will be shown in Section 6.

D. RECURSIVE PARAMETER ESTIMATION

If the unknown parameters θ in the system (10) are time variable there is no computationally simple optimal method to track the parameters recursively. A large number of suboptimal methods therefore have been proposed, and the Extended Kalman filter is one of the simplest ones to find the parameters. The unknown parameter vector is estimated as part of an extended state vector. The algorithm used here is described in detail in Olsson-Holst [43], where a literature survey of the application of suboptimal filters has been done as well.

The parameter vector θ is assumed to be constant but driven by independent noise w,

$$\theta(t+1) = \theta(t) + w(t) \tag{17}$$

The artificial noise covariance determines how fast the parameter can be tracked. In the use of Extended Kalman filter there is no simple way to choose the value of $\text{cov}(w)$. It has to be found by trial and error, and depends on the system noise as well as the variability of the parameters. It may, however, be found off-line using the ML method [32], [41] and then kept fixed in the Extended Kalman filter.

The sample covariance matrix of the residuals

$$\varepsilon(t) = y(t) - C\hat{x}(t|t-1) - Du(t) \tag{18}$$

can be used as a test quantity to judge the quality of the results. The residuals should be a sequence of zero mean independent stochastic variables.

E. MODEL VERIFICATION

Generally the problem of verifying a model is still an art. Many different types of tests have to be performed in order to check the model behavior. Here only the open loop behavior of different models has been compared. It should, however, be emphasized that the final test of a model should be performed in closed loop. Then the real process should be controlled by a controller based on the achieved model. The model has also to be tested if it is really predictive. Then a model achieved from one experiment should be compared with the real output from another experiment.

Even if the parameters of two models are close to each other, their step responses might be quite different. If two models have similar Bode diagrams they could reveal quite different time behavior. Even if the residuals are zero mean and white it does not mean that a better model cannot be found. These examples indicate, that the model verification is most important and also difficult.

As the ML method is based heavily on the residual properties, the residuals should primarily be tested for independence and normality and independence to the inputs.

The loss function changes are tested against the F-test quantity (6) in the MISO case complemented with the Akaike test (7). The model error, defined as the difference between the real output and the output of the deterministic part of the model, is computed.

The standard deviation of the parameters has been checked. If the model order is too high, then the Fisher Information matrix becomes singular, which means that corresponding parameter estimates are linearly correlated and the parameter covariances will be very high.

The discrete models have often been transformed to continuous models in order to compare time constants and zeroes with physical knowledge. Bode plots have been calculated and simulations have been performed.

Single-input--single-output models then have been written in the transfer function form

$$G(s) = k_0 + \sum_{i=1}^{n_1} \frac{k_i}{T_i s+1} + \sum_{j=1}^{n_2} \frac{\kappa_j\left(\frac{1}{z_j} s + 1\right)}{\left(\frac{s}{\omega_0}\right)^2 + 2\zeta\left(\frac{s}{\omega_0}\right) + 1} \tag{19}$$

where $n_1 + 2*n_2$ is the order of the system.

F. COMPUTATIONAL ASPECTS

Some practical considerations on the computations are given in this paragraph.

1. Data Analysis

Before the measurement data is used for parameter estimation, several stages of preliminary data analysis are executed. The variables are plotted in order to detect outliers, trends and abnormal behavior. The relation between inputs and outputs can be inspected and the signal to noise ratios could be visualized. Mean values are subtracted and trend corrections are made in some cases. Cross correlation analysis has also been performed in order to verify relations between the different variables.

The data preparation and analysis part of the identification work should not be underestimated. Data must be in suitable form, programs must be stream-lined and be supplied with adequate inputs and outputs.

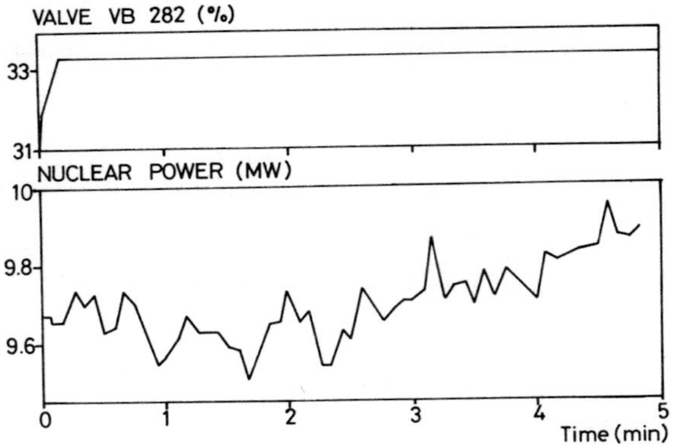

Fig. 7. Nuclear power response to step change in u_2.

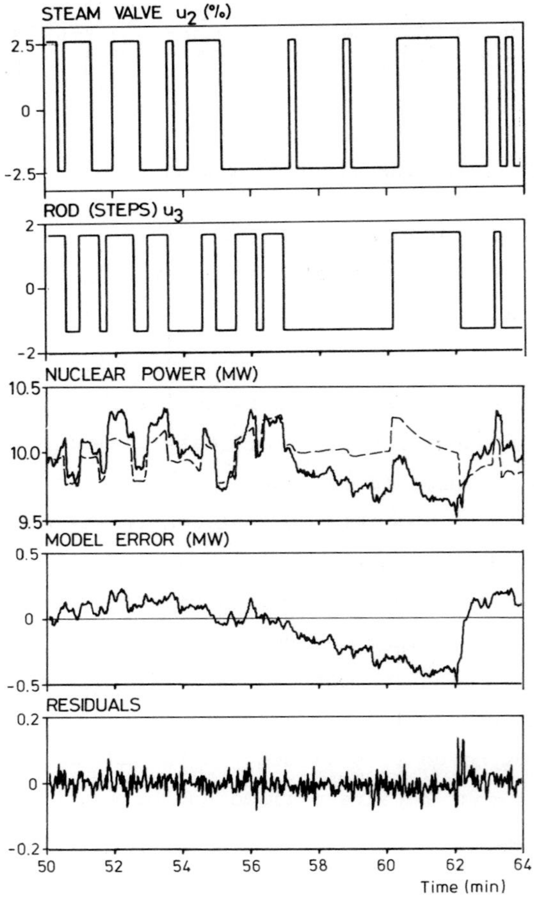

Fig. 8. Model of the nuclear power (broken line) related to u_2 *and* u_3. *The observed values are from a part of expt. 3.*

2. *Identification Programs*

Most of the data analysis and identifications have been performed on the Univac 1108 computer at the Lund University Data Center. The program package for MISO identification was written by Gustavsson [4]. The ML identification program for multivariable systems has been written by Källström, see [4]. The Extended Kalman program is described in Olsson-Holst [4].

In data analysis, parameter estimation or model verification the control engineer must often check intermediate results before he can proceed to the next step of the modeling phase. It is therefore virtually impossible and not even desirable to automate all the different partial decisions and create one general model building program.

The need for interactive programs was realized a long time ago at the Department of Automatic Control at Lund Institute of Technology, and such a program system IDPAC has now been constructed to solve MISO identification and data analysis problems on an interactive basis, see Gustavsson [25], Wieslander [46]. However, most of the identifications discussed in the present paper were performed before the interactive program was completed.

5. MULTIPLE-INPUT--SINGLE-OUTPUT MODELS

In this section we will consider models for four important variables of the plant, *viz.* the nuclear power and the primary (vessel), secondary and tertiary pressures, called C10, P13, P61 and P62 respectively in Fig. 2.

Correlation analysis between the actual inputs and outputs has been applied in order to get a more substantial information about the couplings in the plant, than was presented in Section 2. The actual cross correlations are drawn in Table 2. Some correlations are quite clear (e.g., $u_2 \rightarrow$ P62) while some others are obscure ($u_3 \rightarrow$ P61). The ML identification gave, however, a significant relation in the latter case.

TABLE 2

Qualitative Correlations Between the Examined Inputs and Outputs (max. time lag 10 min.)

Output / Input	Nuclear power C10	Vessel pressure P13	Secondary pressure P61	Tertiary pressure P62
u_1 (VA770)	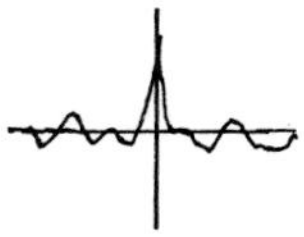	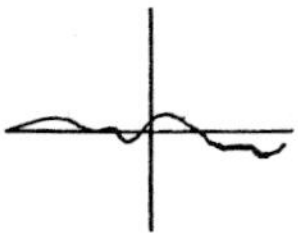	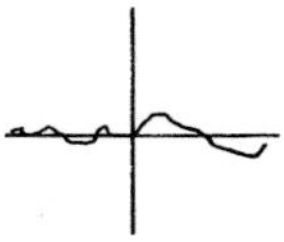	
u_2 (VB282)	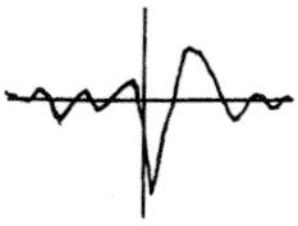	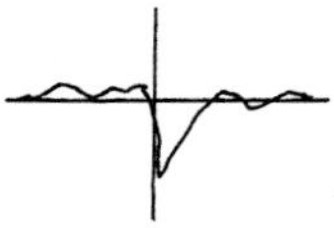	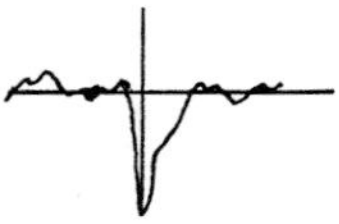	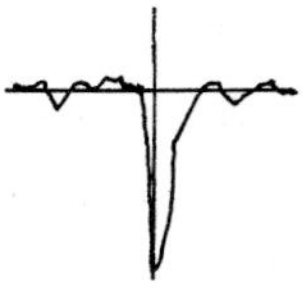
u_3 (rods)	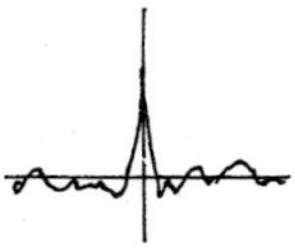	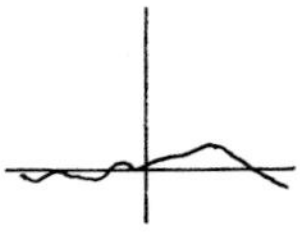	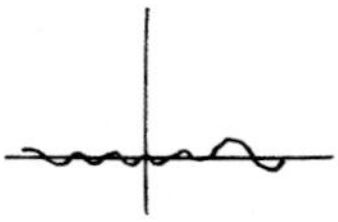	

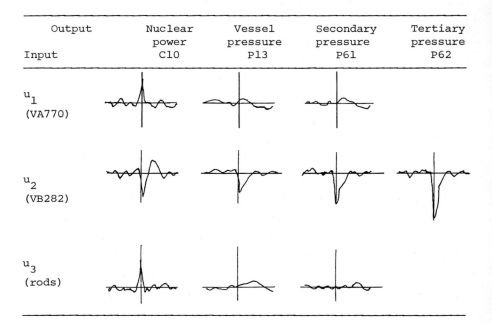

The interaction between the actual inputs and outputs could be qualitatively understood if Fig. 2 is considered. The influence of the different inputs were discussed in 2.C. The valve u_1 (VA 770) has a limited control authority but influences the nuclear power significantly. The influence on the pressures is, however, quite small. It is natural that a disturbance from u_1 is successively damped out from the subcooling circuit to the core and further to the secondary and tertiary heat removal circuits. The valve u_2 (VB 282) has a much higher control authority than u_1 and therefore the relations to all the actual outputs are quite clear. Naturally the valve has the fastest and greatest response in the tertiary circuit but the response is damped into the secondary and primary circuits. In an analog way it is understood, that the rod (u_3) influence on the nuclear power is significant while the influence on the primary, secondary and tertiary pressures is getting successively smaller.

A. NUCLEAR POWER

In 2.B it was demonstrated that the nuclear power response on reactivity disturbances is very fast. Compared to the sampling time of 2 seconds it is prompt, which corresponds to a direct term b_0 in the model (1). The valves will disturb the nuclear power through the reactivity feedbacks and consequently the dominating time constants for these loops will be longer.

1. Reactivity Input (u_3)

In preliminary experiments, see [1], it was found, that the reactivity input--nuclear power output loop could be described by third or fourth order dynamics. The time constants in experiment 1 were found to be 0.7, 8.9 and about 500 seconds respectively. Typically the input PRBS sequence was very fast (see 3.C) and the slow time constant has consequently been determined poorly.

Now, we consider experiments 2 and 3 where the rod reactivity input is used. In experiment 2 there is only this input, but in experiment 3 the valve u_2 is also perturbed independently. Now, if the system is linear, the superposition principle should be valid. As the experimental conditions are essentially the same for experiments 2 and 3 the model parameters should be similar. Table 3 shows the parameters for model (1) with corresponding standard deviations from the Cramér-Rao inequality. The results show that at least the a_i and b_i parameters are close to each other with the differences well within one standard deviation. The c_i parameters, however, show a larger discrepancy This is quite reasonable, as different modes have been excited in the two experiments.

It is noticed that the c_2 coefficients are quite small in both cases. A model with only $c_2 = 0$, however, should have no clear physical interpretation. If instead c_3 is neglected no better model could be obtained.

TABLE 3

Identification Results Relating the Nuclear Power to the Steam Valve (u_2) *and to Reactivity* (u_3)

Experiment	2	3		2	3
	N=2000	N=1900			
a_1	-1.662±.041	-1.626±.078	c_1	-.726±.047	-.579±.084
a_2	.713±.045	.683±.082	c_2	-.049±.043	.025±.034
a_3	- .044±.016	- .044±.017	c_3	-.063±.031	-.054±.025
		$u_2(*10^2)$			
b_1		-.151±.057	λ	$.253*10^{-1}$	$.282*10^{-1}$
b_2		.134±.098			
b_3		-.098±.060	Poles	.981;.607; .074	.966;.581 .078
	$u_3(*10)$	$u_3(*10)$			
b_0	.236±.009	.233±.010			
b_1	.221±.020	.232±.026			
b_2	-.853±.024	-.837±.034			
b_3	.402±.025	.386±.045			

Now consider the continuous transfer functions corresponding to the parameters in Table 3. Their coefficients (see eqn. (19)), are listed in Table 4. The term k_0 corresponds to the prompt input b_0 in model (1). No standard deviations are derived from the results in Table 3.

TABLE 4

Continuous Transfer Functions of Nuclear Power

Exp.	2	3	3
Input	u_3	u_3	u_2
T_1 (sec.)	0.8	0.8	
T_2 (sec.)	4.0	3.7	
T_3 (sec.)	104	59	
k_0*10	0.24	0.23	0
k_1*10	0.76	0.78	-0.021
k_2*10	-0.23	-0.29	0.088
k_3*10	0.096	0.35	-0.95

There is found a very fast time constant of 0.8 seconds. It is clearly significant despite the sampling interval of 2 seconds. It can be explained by the actuator dynamics. Due to the sampling theorem it is still possible to detect the fast time constant. Similar experiences are reported by Gustavsson [25].

The next time constant is determined to 3.7 or 4 seconds. The fuel dynamics should have a time constant of about 8-10 seconds and the result from experiment 1 seems to be reasonable. There are, of course, other dynamical effects added to the computed time constant, such as pressure and flow variations, which explain the smaller value. The longest time constant is determined quite poorly, especially in experiment 1. It comes from the heat removal circuit dynamics and it should be of the order one or two minutes.

As remarked before the poor accuracy is partly due to the

input sequence. The longest pulse of 196 seconds is apparently not long enough, see 3.C. The short sampling interval is also important. The actual discrete pole is situated close to the unit circle, see Table 3. Therefore a small numerical error in the computations can create a significant change of the time constant. For example, if the pole 0.981 is changed ±0.001 the corresponding time constant would be moved from 104 to 110 or 99 seconds respectively.

Now consider the coefficients k_i of Table 4, which indicate how the different modes are amplified. First compare the rod influence on different modes. The reactivity input is most significant in the fast modes. Thus both k_0 and k_1 are significant and quite similar in the two experiments. Especially k_3 is much larger in experiment 3. This might indicate, that the low frequencies have been more excited in experiment 3 due to the extra input from u_2. We also notice the negative sign of k_2. It shows a clear negative reactivity feedback from the fuel temperature.

2. Steam Valve Input (u_2)

Table 3 shows clearly, that the b_i parameters corresponding to u_2 (VB 282) are less accurate than those corresponding to u_3 (rods). This is natural, as the nuclear power is perturbed more by the rods than by the valve u_2.

An attempt was made to get better model accuracy by introducing different time delays for u_2, but no improvement was obtained. The time constants are, of course, the same as for the rod input in experiment 3, but the mode amplifications are different. Table 4 shows that the low frequencies are more amplified by u_2 than the high ones. The relative influence of u_3 and u_2 is also shown by Table 4. The rod input u_3 dominates in the fast modes $(k_1$ and $k_2)$, while the valve dominates in the low frequency range (k_3). The static amplification from the valve u_2 to the nuclear power should be positive (see 2.C). In Tables 3 and 4 it is negative, and the model has no non-minimum phase behavior. Experiment 5 gives similar results. The explanation for this discrepancy has to do with the sampling time, experiment length and input sequence. Previous step responses showed a slow non-minimum phase response (Fig. 7). It takes about two minutes for the step response to get positive after the negative undershoot. This behavior is too slow to be detected in the experiments. Therefore the model has a negative numerator (k_3 in Table 4) for the slow time constant. Observe, however, that the signs of k_1 and k_2 are reasonable in accordance with the discussion of 2.C.

The standard deviation λ of the one step prediction error in Table 3 is 0.025 and 0.028 MW respectively, which is close to the instrument noise level, see 3.D.

A section of experiment 3 has been plotted in Fig. 8. The plots can demonstrate some features of the identification method. The nuclear power has a negative trend between 56 and 62 minutes. At about $t = 62$ it suddenly increases again. The model, however, does not follow the slow trend and the positive change. The residuals ε are large at time 62. The reason is, that an absorption rod was moved manually during the experiment to keep the power within permitted limits. This input could, of course, have been added to the other inputs. It was not included here in order to show, how the ML method can detect abnormal behavior during an experiment.

3. *Subcooling Valve Input* (u_1)

In experiment 5 the valves u_1 and u_2 were moved independently of each other and a corresponding model of the nuclear power was obtained. This model is also of third order. In contrast to previous models there are complex poles. The continuous transfer function is written in one real and one complex mode, according to (19). The coefficients are shown in Table 5.

TABLE 5

Continuous Transfer Function of the Nuclear Power Experiment 5

Input	u_1		u_2
T_1 (sec.)		68	
z_1	-0.58		0.18
ω_0		0.23	
ζ		0.27	
k_1*10^3	1.62		-103
K_1*10^3	9.8		5.2

The complex poles are lightly damped. The period time is about 28 seconds. Similar oscillations have been observed earlier when the subcooling valve has been moved, see Bjørlo *et al.* [47]. A significant amplification of the nuclear power was achieved when the valve was exciting the system at a period of about 25 seconds.

The fast time constants which were excited by the absorption rod have not been detected here by the valves. A slow time constant of 68 seconds is found and is not too far away from what was obtained in experiment 3, where valve u_2 was also perturbed.

The negative value of z_1 indicates that the system is non-minimum phase.

B. *PRIMARY PRESSURE*

Primary pressure input-output models have been studied in a similar way to those for the nuclear power. The steam valve u_2 is the dominating input, and generally the pressure dynamics is much slower than the nuclear power dynamics, as the pressure has to be influenced through the heat flux (see Fig. 3). Most of the identified models are of order three or four. In most cases the fourth order models have large parameter covariances, even if the loss function is acceptable, indicating that the third order models may be adequate.

1. Reactivity Input (u_3)

The influence from u_3 is much less than for the nuclear power. From experiments 1 and 2 the models obtained were quite poor, though third order models were accepted when parameter accuracy, loss function, and residual tests were considered. The parameters of experiment 2 are shown in Table 6 and its continuous transform (19) in Table 7.

TABLE 6

Models From Different Experiments Relating Primary (Vessel) Pressure to the Different Input Signals

Exp. *N*	2 1000	3 1900	4 1000	5 1000
a_1	-2.304±.006	-2.077±.007	-2.121±.031	-2.155±.017
a_2	1.665±.011	1.349±.012	1.414±.058	1.478±.031
a_3	- .361±.006	- .269±.006	- .291±.027	- .321±.015
		$u_2(*10^4)$	$u_2(*10^4)$	$u_1(*10^5)$
b_1		- .060±.014	-	- .159±.080
b_2		- .221±.025	- .252±.010	- .075±.140
b_3		.079±.016	-	.390±.083
	$u_3(*10^4)$	$u_3(*10^4)$	$u_3(*10^4)$	$u_2(*10^4)$
b_1	.497±.017	.453±.025	.092±.055	.013±.015
b_2	- .490±.017	- .181±.045	.467±.098	- .229±.017
b_3	0	- .180±.025	- .498±.057	0

Table 6 continued

Table 6 continued

c_1	-1.176±.029	-.893±.027	-.790±.047	-.751±.036
c_2	.417±.047	.395±.036	.328±.058	.405±.037
c_3	- .081±.029	-.024±.027	.032±.046	-.031±.033
λ	$.663*10^{-4}$	$.724*10^{-4}$	$.749*10^{-4}$	$.753*10^{-4}$
Poles	.997;.909; .398	.983;.706; .387	.984;.736; .401	.987;.708; .459

TABLE 7

Continuous Transfer Functions Relating Primary Pressure to the Different Inputs

Exp.	2	3		4		5	
Input	u_3	u_2	u_3	u_2	u_3	u_1	u_2
T_1 sec	2.2	2.1		2.2		2.6	
T_2 sec	20	5.8		6.5		5.8	
T_3 sec	665	119		123		157	
$k_1 \cdot 10^4$	-0.64	-0.13	-1.6	-0.87	-2.5	0.45	-1.4
$k_2 \cdot 10^4$	8.0	4.1	3.2	8.5	4.8	-1.3	7.7
$k_3 \cdot 10^4$	39	-71	29	-106	22	8.6	-114

The fast time constant related to the actuator dynamics is still statistically significant. A combination of actuator dynamics and the fuel dynamics might explain the 2 second time constant. The longest time constant is again related to the heat removal circuit dynamics.

2. *Steam Valve Input* (u_2)

Different results from experiments 3, 4 and 5 will now be compared. In all the fourth order models a negative discrete pole was found. As such a model has no continuous corresponding model it is difficult to make any physical interpretations. Therefore the third order models are discussed. The problem with negative discrete poles is considered further in paragraph D. In all models the parameter c_3 is poorly determined and may be set to zero.

There is a long time constant corresponding to a pole very close to the unit circle in the discrete model. As before, this causes a poor accuracy of the long time constant, and the static amplification is also inaccurate. The following points should be noted:

(i) The a_i parameters in the three experiments are quite close to each other.

(ii) Consider the b_i parameters corresponding to u_2 in Table 6. Experiments 3 and 4 are compared. In experiment 4 b_1 and b_3 were cancelled in order to get better parameter covariances. No significant change of the loss function was observed. Corresponding parameter b_1 in experiment 5 could also have been eliminated. Now look at the b_i parameters for the reactivity input u_3. In experiment 4, b_1 is much smaller than in experiments 2 and 3. There is no obvious explanation available. The elimination of b_1 and b_3 in experiment 4 for the input u_2 changed the actual parameter a little amount. Probably the difference between the experiments has to do with the fac⁺, that different rods were used in experiment 4 than in previous experiments.

(iii) The parameter standard deviation depends asymptotically on $\sqrt{N}$, where N is the number of samples. The results in experiments 3 and 4 can be compared, and the parameter covariances roughly follow such a law.

(iv) The time constants of about 2 and 6 seconds probably represent combinations of actuator dynamics and fuel dynamics.

(v) Fig. 9 shows a plot of the primary pressure related to the steam valve and reactivity inputs in experiment 3. The model is based on data from 40 to 72 min. and the simulation of the model is made for the time after 72 min. Observe, that the model error makes a positive jump at about $t = 85$. The reason is, that a control rod was moved manually. As the manual change is not included in the simulation a model error results. At the same time there is a large value in the residuals ε which can be observed as a pulse in the plot.

(vi) The model error varies slowly with a period of several minutes. This indicates that there are slow time constants which are not accurately found in the model. In closed loop, however, such slow variations can be taken care of easily by the controller.

3. Subcooling Valve Input (u_3)

The subcooling valve (u_1) has been used as an input in experiment 5, and the model is shown in the Tables 6 and 7. The time constants were discussed in previous section. In order to compare the influence from the different valves u_1 and u_2 the coefficients k_i from experiments 4 and 5 are compared in Table 7. The following points should be noted:

(i) Even though the static amplification has a poor accuracy in the identification it is clear from experiment 5, that the steam valve amplification is about 10 times larger and of different sign than that of the subcooling valve. A better determination of the static amplification must be made with larger sampling intervals and longer input pulses. The reason to use a longer sampling interval is, that the poles then are not situated so close to the unit circle. Numerical inaccuracies do not become so critical.

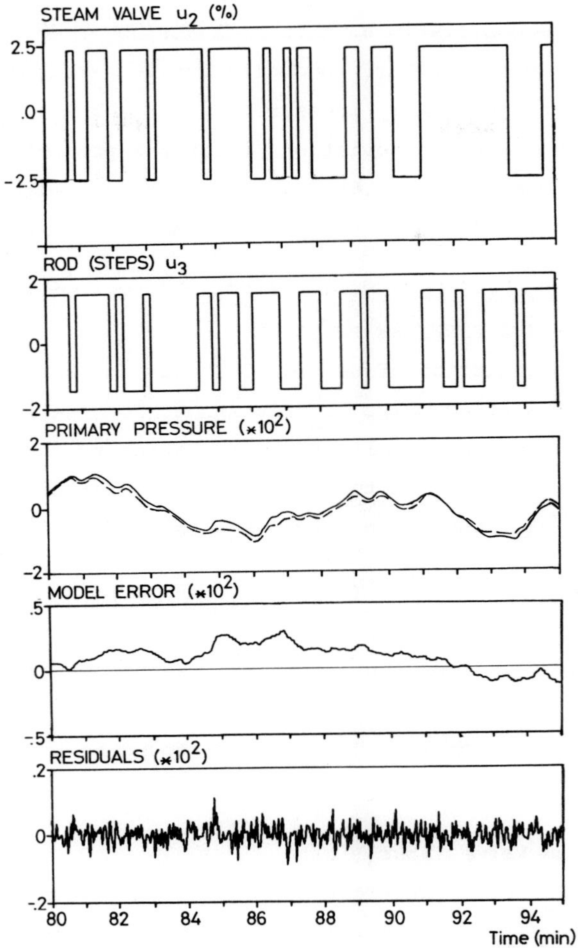

Fig. 9. Model of the primary pressure (broken line) related to u_2 and u_3. The observed values are from a part of expt. 3.

(ii) The oscillations which could be observed in the nuclear power as a result of subcooling valve perturbations are not observed in the primary pressure.

(iii) The standard deviation λ of the prediction error (see Table 6) varies from 0.66×10^{-4} to 0.75×10^{-4}. It is considered satisfactory compared to the instrumentation noise level, discussed in 3.D.

C. *SECONDARY AND TERTIARY PRESSURES*

In the introduction of Section 5 it was emphasized that the influence of the steam valve u_2 is strong for the secondary and tertiary circuits. Especially the correlation to the tertiary pressure is very good. On the other hand the influences from the reactivity or the subcooling valve changes are poor or negligible.

1. Reactivity Input (u_3)

Because of the poor correlation only a first order significant model was found for the secondary pressure. A time delay of 6 seconds was estimated (cf. Table 2). No relation at all between u_3 and the tertiary pressure was found by the ML identifications.

2. Steam Valve Input (u_2)

In Table 8 the identification results are shown. Consider the first column, where the secondary pressure is related to the steam valve and the reactivity. When the present model is compared with a second order model, a high test quantity (6) is achieved ($F = 124$). Therefore the third order model is accepted over the second order model. The table shows that the b_i estimates corresponding to reactivity input are much more inaccurate than those for the steam valve input.

The continuous model (19) time constants are shown in Table 9. The longest time constant is not very precise. The other two can be compared to corresponding results for the primary pressure, Table 7. Instead of accepting the fast time constant

TABLE 8

Models from Different Experiments Relating Heat Removal Circuit Pressures to Different Inputs

Output Exp. N	Sec. press. 4 1000	Sec. press. 5 1000	Tert. press. 3 1900	Tert. press. 3 1900
a_1	-1.877±.023	-1.898±.015	-1.559±.012	-1.534±.003
a_2	.918±.044	.938±.031	.568±.011	.543±.003
a_3	- .040±.023	- .038±.017	-	-
	u_2(*10)	u_1(*10^4)	u_2(*100)	u_2(*100)
b_0	-	-	-	- .039±.0003
b_1	- .115±.005	.574±.168	- .129±.003	- .077±.0006
b_2	- .087±.012	-	.120±.003	.103±.0004
b_3	.173±.009	-	-	-
	u_3(*100)	u_2(*10)		
b_1	- .170±.131	- .109±.004		
b_2	.388±.252	- .075±.010		
b_3	- .183±.135	.163±.006		
c_1	-1.323±.043	-1.256±.021	- .385±.073	.434±.028
c_2	.575±.082	.450±.028	- .030±.047	.089±.026
c_3	- .076±.054	- .023±.023	-	-
λ	$.194 \cdot 10^{-3}$	$.188 \cdot 10^{-3}$	$.478 \cdot 10^{-3}$	$.174 \cdot 10^{-3}$
V			$2.171 \cdot 10^{-4}$	$.2873 \cdot 10^{-4}$
Poles	.986;.843; .048	.989;.865; .044	.979;.580	.977;.556

0.66 seconds a direct input term ($b_0 \neq 0$ in (1)) was tried out, and a significantly better result was achieved. However, only models with negative discrete poles were found.

TABLE 9

Continuous Transfer Functions Relating Secondary and Tertiary Pressures to the Different Inputs

Output Expt.	Sec. press. 4	Sec. press. 5	Tert. press. 3	Tert. press. 3
T_1 (sec)	0.7	0.6	3.7	3.4
T_2 (sec)	11.7	13.8	126	98
T_3 (sec)	138	178	-	-
	$u_2(*10^3)$	$u_2(*10^3)$	$u_2(*10^4)$	$u_2(*10^4)$
k_0	-	-	-	-0.039
k_1	0.24	0.22	-0.27	-0.25
k_2	-1.02	-1.21	-1.37	-1.15
k_3	-12.6	-13.6	-	-
	$u_3(*10^3)$	$u_1(*10^6)$		
k_1	-0.023	0.0015		
k_2	-0.131	-31.6		
k_3	1.79	430		

For the tertiary pressure it is natural to expect the fastest time constant to be even smaller. In fact, this time constant is too small to be estimated with the actual sampling time and a second order model is found with the shortest time constant 3.7 seconds (Tables 8 and 9, third columns). A closer examination of Fig. 6 will also reveal one long and one short time constant for the tertiary pressure. By adding a direct input term b_0 an exceptional improvement of the loss function is found, corresponding to a very high F test quantity (column 4 in Tables 8 and 9). Also the parameter accuracy is improved. A significant improvement of the loss function can be achieved for third order models, but negative discrete poles or pole--zero cancellation appears. The time constants for the tertiary pressure are smaller than for the secondary pressure, which is natural (see Table 9).

Fig. 10 shows a plot of the secondary pressure in experiment 3, related to steam valve and reactivity. The model is based on an observation record from 40-72 min. in the experiment and is used to predict from 80 to 94 min. The residuals have a distinct spike at about 84 min. and the model error makes a positive change. The reason is the same as for the primary pressure, Fig. 9. The tertiary pressure from experiment 3 is plotted in Fig. 11. It is based on 1900 data and simulated on the the same data set. The plot shows the same part of the experiment as Fig. 8. The manual movement of a rod is revealed also here by the model error change at about 62 min.

3. *Subcooling Valve Input* (u_1)

The correlation between u_1 and the secondary and tertiary pressures is poor, which has been discussed before. A significant ML model was, however, found for the secondary pressure, and the parameters are shown in Tables 8 and 9 column 2.

The standard deviation of the one step prediction error is larger than for the primary pressure (cf. Tables 6 and 8) but is still considered satisfactory with respect to the instrumentation noise.

D. *THE PROBLEM OF NEGATIVE REAL DISCRETE POLES*

In several models, especially those of high order (third or fourth) negative real poles of the discrete model have appeared. Since these models have no continuous analog they cannot be given physical interpretations. Still they may be useful for

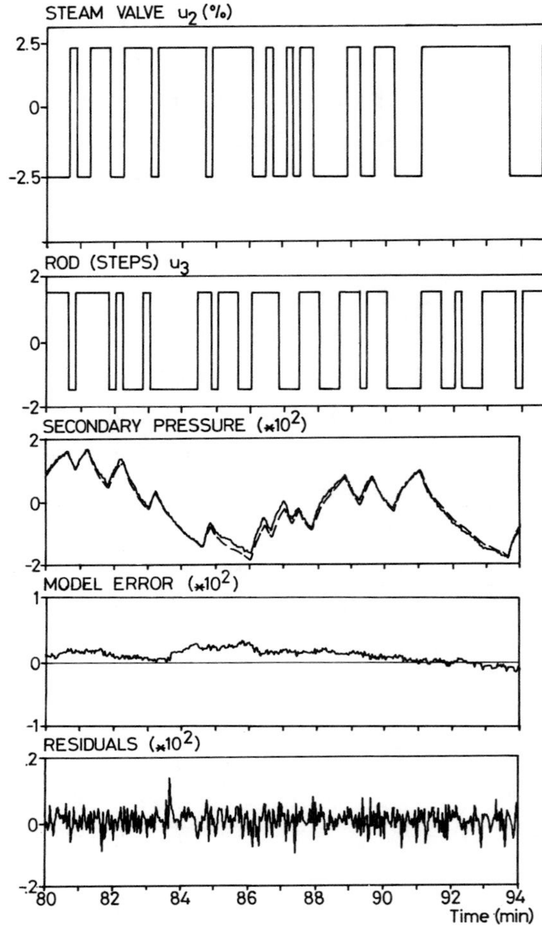

Fig. 10. Model of the secondary pressure (broken line) related to u_2 *and* u_3. *The observed values are from a part of expt. 3.*

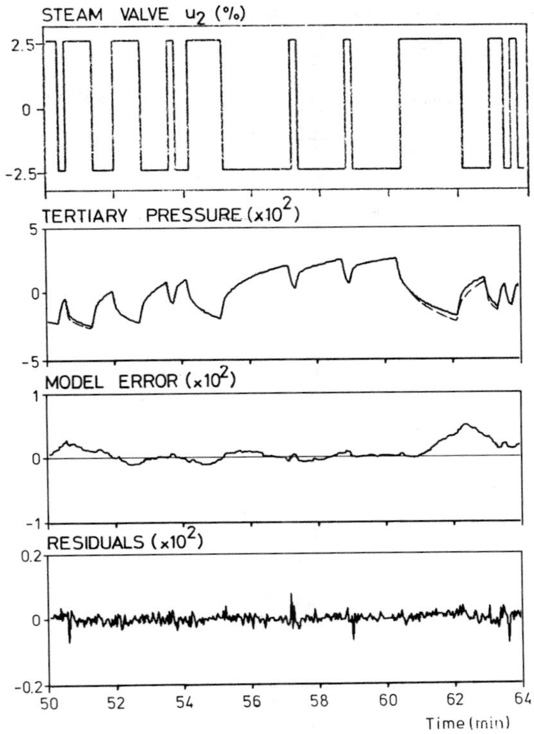

Fig. 11. *Model of the tertiary pressure (broken line) related to* u_2. *The observed values are from a part of expt. 3.*

time discrete regulators. The following reasons may be given for negative discrete poles:

(i) The negative pole may reflect that the order is too high. Generally there is a corresponding zero close to the pole in the C^* or in the B^* polynomial, but not always in both. Cancellation may be possible. In the reactor models cancellation between the A^* and C^* have been the most common case. The noise thus can be represented by a lower order transfer function, a fact which has been observed in many practical situations by e.g. Bohlin [48]. Söderström [49] has also analyzed cancellation problems.

(ii) Quantization error may cause negative discrete poles as pointed out by Åström [50]. For the secondary pressure and to a lesser extent for the primary pressure, negative poles were quite common. The quantization error of the 11 bit converter for the pressures is at least $0.8*10^{-4}$ normalized units. The standard deviation of the one step predictor error is $0.7*10^{-4}$ for the primary and $1.9*10^{-4}$ for the secondary pressure. Thus the quantization error cannot be neglected in comparison with the residuals. As the nuclear power one step prediction errors have been about 0.025 MW in comparison with the quantization error 0.005 MW (see 3.D) this quantization error is not so serious, even though only the total power is measured.

(iii) For the secondary pressure models of second order two minima of the loss function appeared in experiments 3 and 4. The models have about the same loss function. In one model there is one negative real pole, in the other both the poles are positive real. This problem of non-uniqueness of the ML estimates has been analyzed by Söderström [51]. Similar results can also be found for the nuclear power related to the reactivity input.

6. VECTOR DIFFERENCE EQUATIONS

In preceding MISO models the couplings between the outputs or state variables of the plant have been neglected. In order to take the couplings between the inputs and outputs into account the vector difference approach, described in 4.C, was tried out. The results then are compared with the MISO models.

From a computational point of view this approach is also a MISO identification, as one row at a time of the vector difference equation is identified. Then the other outputs are used as auxiliary variables. Apart from the noise approximation there is also another error source as the different "inputs" are not independent of each other. This will also be discussed.

A. *CORRELATION ANALYSIS*

In Table 2 the correlation between the "real" inputs and the actual outputs is shown. Here also other pairs of inputs and outputs have been studied to find out the significant causality relations. The input has been whitened and corresponding impulse response has been estimated using Fast Fourier Transform technique. Generally 2000 data points were used.

TABLE 10

Qualitative Correlations Between Some Selected Variables

	P13	P61	P62
Nuclear power C10	-	-?	0
Vessel pressure P13		++	+
Secondary pressure P61			++
Tertiary pressure P62			

The correlation results are shown in Table 10. The signs indicate a positive or negative correlation between the variables.

Two signs means a clear cross correlation, one sign a low signal to noise ratio, a question mark a poor correlation while a zero means insignificant correlation.

B. MAXIMUM LIKELIHOOD (ML) IDENTIFICATIONS

Some specific results from experiment 3 will now be discussed in order to demonstrate the model characteristics when couplings are taken into account.

Table 3 shows, that the best possible model with nuclear power (C10) as function of u_2 and u_3 is characterized by 13 parameters (plus 3 initial conditions), i.e. 16 parameters which give $\lambda = 0.0282$. If the vessel pressure (P13) is added to the model, it can be improved significantly. A second order model--now with three inputs--corresponds to 13 parameters and $\lambda = 0.0282$. Significant improvements of the loss function is obtained for a third order model (19 parameters) with $\lambda = 0.0277$ ($F = 6$). Without the primary pressure as an auxiliary input no improvement is found when the number of parameters is increased to more than 13. The correlation analysis indicated that the secondary and tertiary pressures are not coupled to the nuclear power. It is verified by the ML identification, as no improvement is obtained by adding those variables as auxiliary inputs.

The plot of the model output in Fig. 12 shows an interesting behavior compared to the previous model output, Fig. 8. The new model can follow the drift of the nuclear power between 56 and 62 minutes much better. During this time the input u_2 is negative most of the time, and consequently the pressures are forced to rise, which in turn decreases the nuclear power. Thus the drift of the nuclear power is noticed through the vessel pressure P13, and the model error change at $t \approx 62$ is consequently not so distinct. Even if the loss function is significantly smaller for the new model, the residuals in the two models look similar to each other. In principle there is only a slight scaling of the residuals. The autocovariance does not change much.

The smallest loss function related to u_2 and u_3 in experiment 3 was obtained for order four and 20 parameters (cf. Table 6). It corresponds to $\lambda = 0.714 * 10^{-4}$. If the nuclear power and secondary pressure are added, a third order model with 21 parameters gives $\lambda = 0.633 * 10^{-4}$. This corresponds to an F test quantity (6) of 270 (1000 data). The model improvement is quite understandable. The gains from both the rods and the steam valve are quite small. A reactivity change is first noticed in the nuclear power (Fig. 3) before it propagates to the vessel pressure. A change in the steam flow valve causes pressure changes in the heat removal circuits which propagate towards the reactor vessel.

For the tertiary pressure the ML identification gave significant models with both the steam valve u_2 and the secondary pressure as inputs. According to Table 8 the best model with only u_2 as input has 9 parameters (including initial conditions) and $\lambda = 0.174 * 10^{-3}$. With P61 added to the model a second order model with 12 parameters is the best one with $\lambda = 0.156 * 10^{-3}$. The correlation analysis showed that P62 should be related also to the vessel pressure, but the ML identifications did not reveal this. The reason is, that the primary and secondary pressures are strongly correlated, so all causality relations from primary to tertiary pressures can be explained by the secondary pressure alone.

Now compare the plots of the model errors in Figs. 11 and 13. As for the nuclear power, the residual amplitudes are decreased but the two realizations and their covariances are quite similar.

C. *SIMULATIONS*

The whole vector difference equation (VDE) with the two inputs u_2 and u_3 and the four outputs can now be written in the form (14). The model contains three A_i matrices, four B_i matrices and three C_i matrices. The deterministic part of the model contains 47 parameters, 28 in the A_i matrices and 19 in the B_i matrices. The diagonal C_i matrices contain 10 parameters.

The different assumptions of the noise (see 3.C) and the inputs are tested by simulation of the VDE. When each row of the VDE was simulated separately, as in Figs. 12 and 13, then all the auxiliary variables had their observed values. When the whole VDE is simulated, then only the true inputs u_2, u_3 have their observed values given. It is natural, that the output error then is larger. Because of the new relations found, the model error is, however, still smaller than for the MISO models, as in Figs. 8-11. In Figs. 14 and 15 the nuclear power and the tertiary pressure are plotted from the VDE simulation made for inputs from experiment 3. Figs. 14 and 15 should be compared to the Figs. 8, 12 and 11, 13 respectively.

It is demonstrated that the VDE model output error is (in mean square) between the results of MISO identification and row-by-row VDE identification. Observe, however, that slow variations occur quite obviously also here.

7. A STATE MODEL

From a control point of view it is interesting to get an accurate and still reasonably small model of the plant. An attempt is made to formulate a state model in order to achieve better physical interpretation of the model parameters. In this section a model is identified from experiment 4 using only two inputs u_2 and u_3. A structure of the plant model is derived first and the essential approximations are accounted for. Then the identification results are presented and discussed.

A. DERIVATION OF A MODEL STRUCTURE

In Section 2 the qualitative behavior of the plant was discussed, and by identification some of the most essential relations were confirmed. Here an attempt is made to quantify the assumptions of physical couplings between the different process

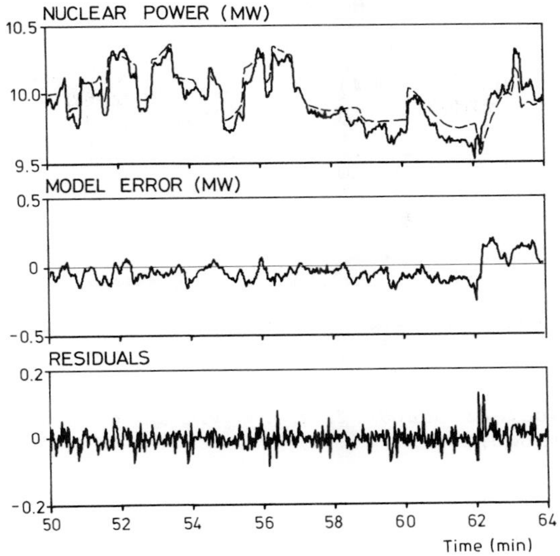

Fig. 12. Model of the nuclear power (broken line) related to u_2, u_3 and the primary pressure. The observed values are from part of expt. 3.

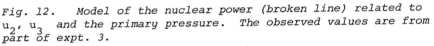

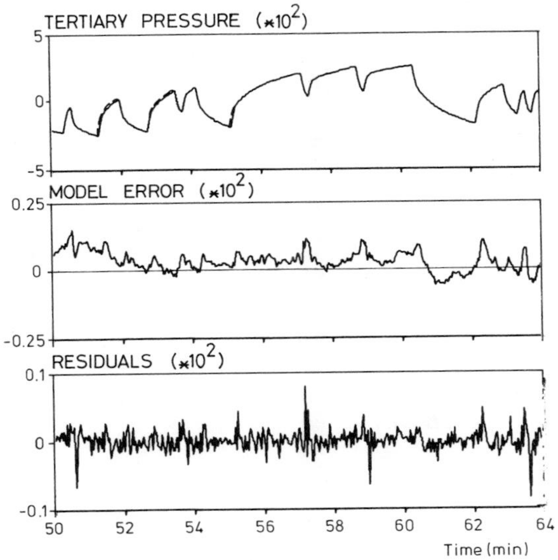

Fig. 13. Model of the tertiary pressure (broken line) related to u_2 and the secondary pressure. The observed values are from a part of expt. 3.

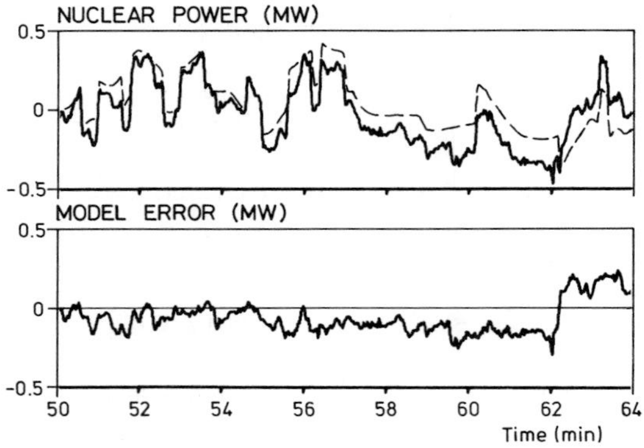

Fig. 14. Nuclear output from the VDE simulation. A part of expt. 3 is shown.

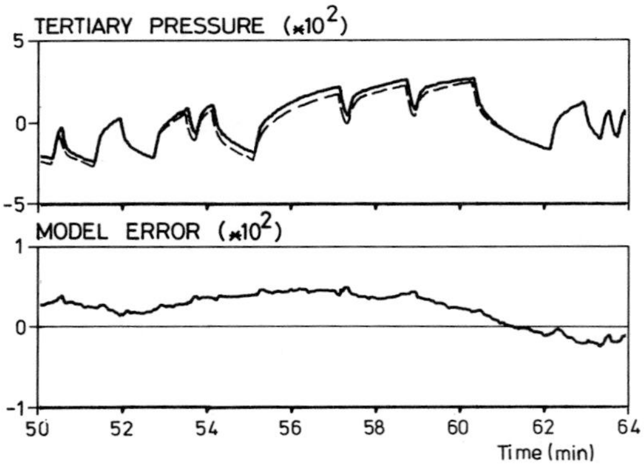

Fig. 15. Tertiary pressure output from the VDE simulation. A part of expt. 3 is shown.

variables. The goal is to find a linear state variable description.

It is assumed, that the variations are small, so that the nonlinear effects are negligible. The state variables are defined as deviations from stationary values.

1. Kinetics

The neutron level n^* is proportional to the nuclear power C10. If one group of delayed neutrons is assumed the neutron density equation is

$$\frac{dn^*}{dt} = \frac{\delta k-\beta}{\ell} \cdot n^* + \lambda c \qquad (20)$$

where n^* is the neutron density, c the concentration of delayed neutrons, β the delayed neutron fraction, λ a weighted average value of the decay constants of the precursors of the six groups of delayed neutrons, ℓ the neutron generation time and $\delta k = k_{eff} - 1 \approx$ reactivity. The last term is discussed in paragraph 6.

The one group description of delayed neutrons is

$$\frac{dc}{dt} = \frac{\beta}{\ell} n^* - \lambda c \qquad (21)$$

As the neutron kinetics is very fast compared to other phenomena in the plant a prompt jump approximation is made, i.e. dn^*/dt is put to zero. This makes the nuclear power an algebraic equation of the other state variables, according to (20).

2. Fuel Temperature Dynamics

The heat content of the fuel elements is represented by the average fuel temperature θ_f. As it is influenced by heat transfer through the fission and is decreased by the coolant, the following dynamics

$$T_f \frac{d\theta_f}{dt} = - \theta_f + \gamma_1 n^* + \gamma_2 \theta_c \qquad (22)$$

is assumed, where γ_i are constants and T_f is an average time constant for the fuel elements determined by their total heat capacity. It is initially assumed to be 8 seconds. The coolant temperature θ_c will later be represented by an average water temperature θ_w and the coefficient γ_2 is found to be close to zero.

3. Coolant and Moderator Dynamics

The hydraulics, coolant and moderator dynamics are probably the most complex features of the plant. Here several crucial approximations are made. All the water content in the core is represented by an average water temperature θ_w, which then (together with fuel temperature) represents the heat flux in the core. The void content is strongly related to both heat flux and vessel pressure, and therefore it is here involved in those state variables.

The vessel pressure p_1 gives, of course, no information about the void distribution along the coolant channels. The reactivity feedback from void depends not only on the average void but also on the spatial distribution of the void. Moreover the boiling boundary is not taken into account, and it is a critical variable.

It has been demonstrated in Section 2 that the subcooled flow temperature ($\Delta T8$) is related to both the water temperature θ_w and to the reactivity. This dynamic has not been included in the present state model, as the valve u_1 was not moved in the selected experiment. The temperature changes therefore were not significant.

The structure of the water temperature equation can now be formulated. Because of the large water mass the heat capacity is large, and corresponding time constant is of the order minutes. Initially it is assumed to be 100 seconds.

The heat flux which can change the water temperature can be represented by the three states, fuel temperature, subcooled water temperature and the heat transfer through the steam transformer. Part of the heat is also due to the fact that all the fission power is not captured in the fuel, but in the moderator. The coupling to the subcooled water has been neglected. The heat flux through the steam transformer is for the moment represented by the term q_1. Then the water equation is formulated as

$$T_w \frac{d\theta_w}{dt} = -\theta_w + \gamma_3\theta_f + \gamma_4 q_1 \tag{23}$$

4. *Vessel Pressure Dynamics*

According to the assumptions about the coolant the pressure must reflect many different features. This means, that the equation parameters are combinations of many physical phenomena and it is therefore very difficult to make any theoretical derivation of their numerical values.

The vessel pressure is certainly related to the heat flux from the fuel elements and the water temperature. To a very small extent it is related to the subcooled temperature. No identification has verified any significant relation. In any case the influence from the subcooled temperature is neglected in the present experiment.

The vessel pressure also depends on the steam removal through the steam transformer to the secondary circuit. If this energy flux is represented as before by q_1 the pressure equation structure is

$$\frac{dp_1}{dt} = \gamma_5 p_1 + \gamma_6\theta_f + \gamma_7\theta_w + \gamma_8 q_1 \tag{24}$$

5. *Heat Removal Circuit Dynamics*

The dynamical coupling between the reactor core and the steam circuits is through the vessel pressure and the primary steam flow. As remarked before there is also a weak coupling to the water circuits through the subcooler A. The subcooling temperature and flow then can represent the essential variables for this coupling.

The heat transfer in the steam transformers and in the subcoolers now is considered. The functional difference between the steam transformers and the subcoolers is, that the latter ones have one phase flow (water) both in the primary and in the secondary circuits. In order to simplify the model as much as possible only the steam phase is considered. It is known, that the water is only slightly subcooled in the circuits. Variations in the subcooling are considered as stochastic disturbances to the pressures.

The mass and energy balance equations for the heat exchangers have been formulated earlier by Eurola [24]. As the steam is close to saturation it is reasonable--as remarked in Section 2--that the temperature variations are assumed proportional to the pressure variations. Therefore the pressure is used to represent the enthalpy.

The primary steam flow variations (F41) are not negligible, as soon as the steam valve u_2 has been moved. Identifications have shown that it is also significantly related to the primary pressure and to some extent to the nuclear power. Therefore we assume here that the enthalpy on the primary side of the steam transformer is described only by the vessel pressure. In Section 2.C it is indicated that the temperature variations on the secondary side are small. Therefore the secondary side enthalpy is also represented just by the pressure. The consequence of these arguments is, that the energy term q_1 in eqs. (23) and (24) can be replaced by the secondary pressure p_2. With similar arguments the tertiary circuit dynamics is described by only one state variable, the tertiary pressure p_3.

The secondary pressure dynamics is consequently assumed to be

$$\tau_2 \frac{dp_2}{dt} = q_{12} - q_{23} \tag{25}$$

where q_{12} and q_{23} are the heat fluxes from the primary to secondary and from secondary to tertiary circuits respectively. The heat fluxes are assumed to be related to the pressures in the following way:

$$q_{12} = \nu_1 p_1 - \nu_2 p_2$$

$$q_{23} = \nu_3 p_2 - \nu_4 p_3$$

where ν_i are constants. This results in

$$\frac{dp_2}{dt} = \gamma_9 p_1 + \gamma_{10} p_2 + \gamma_{11} p_3 \tag{26}$$

For the tertiary system we have

$$\tau_3 \frac{dp_3}{dt} = q_{23} - q_3 \tag{27}$$

where q_3 is the heat removed from the tertiary system. We assume

$$q_3 = \nu_5 p_3 + \nu_6 u_2$$

The state equation then is

$$\frac{dp_3}{dt} = \gamma_{12} p_2 + \gamma_{13} p_3 + \gamma_{14} u_2 \tag{28}$$

6. *Reactivity Feedbacks*

The reactivity term δk in eq. (20) defines the coupling between the kinetic equations and the rest of the plant. The feedback effects have been indicated in Fig. 3. The void content has been represented by vessel pressure and by water temperature. As the steam removal influences the void content we also include the secondary pressure among the reactivity feedbacks. It is assumed that a linear relation holds,

$$\delta k = u_3 + \gamma_{15} \theta_f + \gamma_{16} \theta_w + \gamma_{17} p_1 + \gamma_{18} p_2 \tag{29}$$

where u_3 represents the net reactivity from the rods. The feedback from the subcooled water is neglected.

7. *Summary*

To summarize the structure, the state vector of the linear model is defined as

x_1	delayed neutrons c	(21)
x_2	fuel temperature θ_f	(22)
x_3	water temperature θ_w	(23)
x_4	vessel pressure p_1	(24)
x_5	secondary pressure p_2	(26)
x_6	tertiary pressure p_3	(28)

The input vector has only the two components steam valve and reactivity.

The model is described by

$$\frac{dx}{dt} = Ax + Bu \tag{30}$$

$$A = \begin{pmatrix} 0 & \underline{a_{12}} & \underline{a_{13}} & \underline{a_{14}} & \underline{a_{15}} & 0 \\ a_{21} & \underline{a_{22}} & a_{23} & \underline{a_{24}} & \underline{a_{25}} & 0 \\ 0 & a_{32} & a_{33} & a_{34} & a_{35} & 0 \\ 0 & a_{42} & a_{43} & a_{44} & a_{45} & 0 \\ 0 & 0 & 0 & a_{54} & a_{55} & a_{56} \\ 0 & 0 & 0 & 0 & a_{65} & a_{66} \end{pmatrix} \qquad B = \begin{pmatrix} 0 & b_{12} \\ 0 & b_{22} \\ 0 & 0 \\ 0 & 0 \\ 0 & 0 \\ b_{61} & 0 \end{pmatrix} \tag{31}$$

The underlined elements will be discussed in Section 8.

The three pressures p_1-p_3 are measured, but the nuclear power has not been used as an output. The general form of the nuclear power related to the other state variables is derived from (20) and (29) but the parameters are unknown. In order to limit the complexity of unknown parameters the nuclear power measurements therefore are not used.

The output equation then is

$$y = Cx$$

where

$$C = \begin{pmatrix} 0 & 0 & 0 & 1 & 0 & 0 \\ 0 & 0 & 0 & 0 & 1 & 0 \\ 0 & 0 & 0 & 0 & 0 & 1 \end{pmatrix} \qquad (32)$$

B. PARAMETER IDENTIFICATION

The identification of the state model is now presented. First the noise is discussed.

1. Noise Description

In 3.D the instrument noise is considered and is found to be quite small. Thus the major contribution to the residuals are due to process noise and model errors.

There are many noise sources in the plant, a fact which is demonstrated by the MISO identifications. The boiling is a large noise source term, which affects x_4. Temperature variations in coolant, subcooled water affect x_3. The saturation temperature is changed due to heat flux variations. Varying degrees of subcooling in the water phase in the heat removal circuits will disturb the pressures x_5 and x_6. Also the flow variations in the circuits create disturbances.

The process noise terms also can represent modeling errors to some extent.

2. Identification Results

The stochastic structure of the system is described by eqs. (8)-(9). From experiment 4 a sequence of 800 samples have been used.

In the first approach the matrices K and D of (10) were assumed to be zero. With 8 parameters assumed unknown in the A and B matrices a minimum point was found corresponding to

$$\mathrm{tr}(\hat{R}) = 0.123\ 10^{-3}$$

(see (13)). This corresponds to standard deviations of the prediction errors $0.50 * 10^{-2}$ (vessel pressure), $0.74 * 10^{-2}$ (secondary pressure) and $0.66 * 10^{-2}$ (tertiary pressure). These errors are very large compared to previous MISO results ($0.7 * 10^{-4}$, $0.19 * 10^{-3}$, $0.17 * 10^{-3}$ respectively). Moreover, the residuals were not accepted to be white noise.

It is clear, that process noise must be included. First only three non-zero elements of the K matrix were tried, k_{41}, k_{52} and k_{63}.

In order to limit the computations not more than 15 parameters at a time were assumed to be unknown in the A, B and K matrices. With K included a significant improvement was obtained. The loss function decreased noticeably. The standard deviations of the prediction errors for the three pressures were $0.88 * 10^{-3}$, $0.96 * 10^{-3}$ and $0.15 * 10^{-2}$ respectively. Those values are still too large compared to the MISO results.

It is demonstrated that it is not trivial to find a correct structure in state form. Several improvements can be made, and work is in progress to improve the model structure. It is clear, that the number of parameters in the A and B matrices (31) can be increased. In the VDE approach 34 significant parameters were found in the deterministic part of the model relating two inputs to the three selected output pressures. In the state equation identification the number of degrees of freedom (equal to the number of parameters to be identified) for fitting the observed data has been reduced. In A and B (31) there are only 25 parameters. Clearly the number of states should be increased. The assumptions about the core dynamics have to be more elaborate. One variable describing the void content and two different states for the coolant and moderator temperature would be a significant improvement. Moreover, previous identifications showed, that the primary steam flow probably should be considered a separate state variable. It is also clear, that one state for each heat removal circuit is too little. The present state model has no time constant smaller than 6 seconds, and the results in 5.C clearly demonstrated that fast modes are important. Thus additional states are needed to describe the secondary and tertiary pressures better.

It is difficult to find good initial values of the K matrix, as they do not have any intuitive physical interpretation. It is maybe easier to guess parameters in the process noise covariance matrix, and then transform to K by using a Riccati equation [45].

The computational work is by no means trivial. The likelihood function is minimized numerically. The gradients are computed numerically using finite differences, and a Fletcher-Powell algorithm is used for minimization. Manual interaction has to be done to a large extent during the minimization. The intermediate results have to be judged if they are reasonable. Otherwise it is easy to get unreasonable computational times, depending on too slow convergence, wrong step lengths etc.

In Fig. 16 the first part of experiment 4 is plotted. The best model hitherto is compared to the real output values.

8. RECURSIVE IDENTIFICATION

In experiment 7 the operating level is changed significantly by means of the subcooling. A time variable linear model could describe this phenomenon. Here an Extended Kalman filter has been applied in order to recursively track the varying parameters.

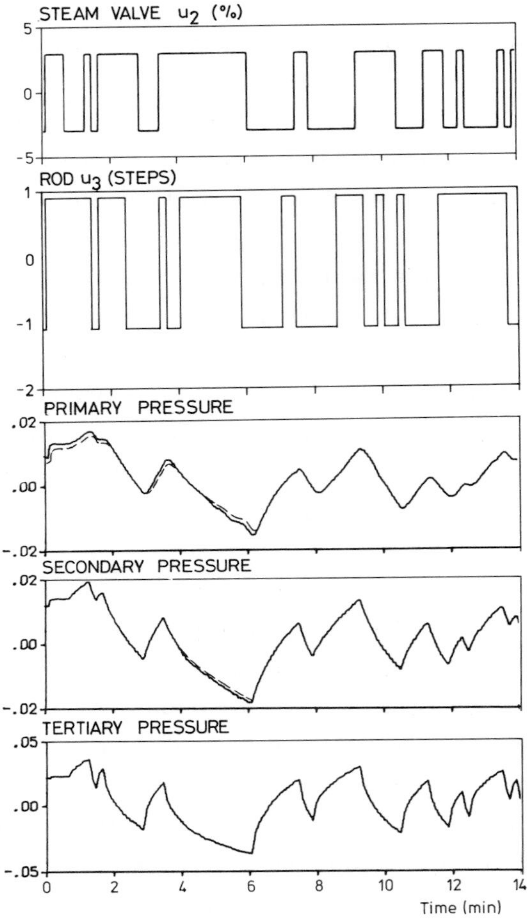

Fig. 16. Observed and predicted (broken lines) outputs of the state model from the first half of expt. 4.

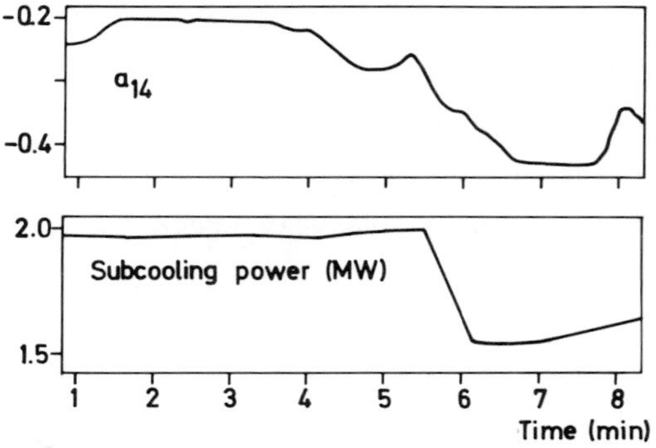

Fig. 17. Estimation of time-varying parameters with the Extended Kalman filter in expt. 7. Among the six estimated parameters a_{14} *is shown. The subcooling is changed manually.*

A. *INFLUENCE OF SUBCOOLING POWER*

The general nature of the subcooling effects were discussed in Sections 2.A and C. The quantitative influence of varying subcooling power has been studied by comparing experiments 6 (Table 1) and 4. It was found, that not all the parameters in the A matrix (31) changed, except mainly the underlined ones.

Generally a lower subcooling power means a lesser degree of stability. Mainly the reactivity feedback coefficients (Fig. 3) will be affected. As they are hidden in the system equation coefficients a couple of examples are given here.

As soon as the subcooling power decreases there is a higher probability for boiling in the moderator. The total void content increases. The sensitivity to pressure changes will then rise and the vessel pressure influence on reactivity will grow. With a prompt jump approximation this means that x_1 and x_2 are primarily influenced, i.e. the parameters a_{14} and a_{24} (31). As an example $a_{14} = -0.24$ from experiment 4 with 1.95 MW subcooling, and $a_{14} = -0.64$ at experiment 6 with 1.1 MW subcooling.

In experiment 7 the subcooling power was changed manually from 1.95 to 1.4 MW during 15 minutes, while u_2 and u_3 were disturbed (Table 1). The subcooling power is shown in Fig. 17 (upper fig.). It was not included in the model but considered as an external disturbance source. The initial condition for experiment 7 is the same as the operating level of experiment 4. Therefore the model described in 7.B is used as the starting model for the recursive parameter estimation.

B. *PARAMETER TRACKING*

The observed variables from experiment 7 were put into an Extended Kalman filter (see 4.D) and the six time-varying parameters were tracked. There is no way to find optimal estimates of time-variable parameters in a multivariable system. It is known, that the Extended Kalman filter most often gives unreliable confidence limits on the parameter estimates. Several compensations for this have been proposed [43]. Here, however, the main interest has been to test the simplest possible filter to track the parameters.

The six unknown parameters were described as eq. (17) with an artificial noise w. Initially the covariance matric of w was chosen diagonal, and only trial and error methods were used to find suitable values. It was found that the diagonal elements of cov(w) should lie between 10^{-6} and 10^{-7}, i.e. somewhat smaller than the process noise covariance elements. This is reasonable, as the parameters are assumed to vary slowly compared to the state variables. With too small values of cov(w) the tracking was too insensitive.

In Fig. 17 an example is shown. Six parameters were estimated simultaneously and a_{14}, discussed above, is displayed. The parameter is approaching -0.4 which seems to be plausible result, as the subcooling reaches 1.4 MW.

It is natural to try to minimize the number of time variable parameters, as the computing time grows very fast with the size of the extended state vector. Attempts with only two time-variable parameters were not successful, but three parameters could be reasonably accurate.

The computing time for the Extended Kalman filter may be a severe constraint on an on-line computer. Here the extended state vector consists of 12 states which means a considerable computational burden. Probably even more state variables should be included in order to improve the model. Therefore it is crucial to simplify the calculations as much as possible and a tailor made filter has to be defined.

Acknowledgements

This work has been partially supported by the Swedish Board for Technical Development. The research has been performed in cooperation with the OECD Halden Reactor Project. The author is especially indebted to Dr. R. Grumbach, Mr. H. Roggenbauer and Mr. R. Karlsson (now with Atomenergi AB, Sweden) with the Halden Reactor Project for their participation and interest. The permission of the Project to publish the results is also gratefully acknowledged.

The team work at the Department of Automatic Control has been most valuable. Professor K. J. Åström has contributed with constructive criticism, new ideas and never failing encouraging support. Dr. I. Gustavsson has throughout the work shared his knowledge of identification and has been of invaluable help. Mr. C. Källström wrote the state model identification program. Mr. J. Holst has cooperated with the author on suboptimal filtering problems. Dr. B. Wittenmark has contributed with valuable comments and corrections when reading the manuscript. Mrs. G. Christensen has typed the manuscript and Miss B. M. Carlsson has prepared the figures.

REFERENCES

1. Olsson, G., "Maximum Likelihood Identification of Some Loops of the Halden Boiling Water Reactor," Report 7207, Department of Automatic Control, Lund Institute of Technology, Lund, 1972, also OECD Halden Reactor Project, HPR-176, 1975.

2. Roggenbauer, H., W. Seifritz and G. Olsson, "Identification and Adjoint Problems of Process Computer Control, *Enlarged Halden Programme Group Meeting*, Loen, Norway, 1972.

3. Olsson, G., "Modeling and Identification of Nuclear Power Reactor Dynamics from Multivariable Experiments, Proc. 3rd IFAC Symp. on Identification and System Parameter Estimation, The Hague, the Netherlands, 1973.

4. Gustavsson, I., "Comparison of Different Methods for Identification of Linear Models for Industrial Processes, *Automatica, Vol. 8,* 1972, pp. 127-142.

5. Sage, A. P. and G. W. Masters, "Identification and Modeling of Nuclear Reactors, *IEEE Trans. Nucl. Sci., NS-14,* 1967, pp. 279-285.

6. Ciechanowicz, W. and S. Bogumil, "On the On-Line Statistical Identification of Nuclear Power Reactor Dynamics," *Nucl. Sci. Engr., Vol. 31,* 1968, pp. 474-483.

7. Habegger, L. J. and R. E. Bailey, "Minimum Variance Estimation of Parameters and States in Nuclear Power Systems," Proc. 4th IFAC Congress, Warsaw, Paper 12.2, 1969.

8. Moore, R. L. and F. Schweppe, "Model Identification for Adaptive Control of Nuclear Power Plants," *Automatica, Vol. 9,* 1973, pp. 309-318.

9. Brouwers, A., "Step Perturbation Experiments with the HBWR Second Fuel Charge," OECD Halden Reactor Project, HPR-51, 1964.

10. Tosi, V. and F. Åkerhielm, "Sinusoidal Reactivity Perturbation Experiments with the HBWR Second Fuel Charge," OECD Halden Reactor Project, HPR-49, 1964.

11. Fishman, Y., "Pseudorandom Reactivity Perturbation Experiments with the HBWR Second Fuel Charge," OECD Halden Reactor Project, HPR-50, 1964.

12. Eurola, T., "Noise Experiments with the HBWR Second Fuel Charge," OECD Halden Reactor Project, HPR-53, 1964.

13. Bjørlo, T. J. *et al.*, "Digital Plant Control of the Halden BWR by a Concept Based on Modern Control Theory," *Nucl. Sci. Engr., Vol. 39,* 1970, pp. 231-240.

14. Roggenbauer, H., "Real-Time Nuclear Power Plant Parameter Identification with a Process Computer," Proc. 3rd IFAC Symp. on Identification and System Parameter Estimation, The Hague, the Netherlands, 1973.

15. Jamne, E. and J. G. Siverts, "Description of the HBWR Plant," OECD Halden Reactor Project, HPR-95, 1967.

16. Fleck, J. A., Jr., "The Dynamic Behavior of Boiling Water Reactors," *J. Nucl. Energy, Part A., Vol. 11,* 1960, pp. 114-130.

17. Glasstone, S. and M. C. Edlund, *The Elements of Nuclear Reactor Theory,* Van Nostrand, Princeton, N.J., 1952.

18. King, C. D. G., *Nuclear Power Systems,* Macmillan, New York, 1964.

19. Meghreblian, R. V. and D. K. Holmes, *Reactor Analysis,* McGraw-Hill, New York, 1960.

20. Weaver, L. E., *Reactor Dynamics and Control,* American Elsevier, New York, 1968.

21. Wiberg, D., "Optimal Control of Nuclear Reactor Systems," *Advances in Control Systems* (C. Leondes, Ed.), Vol. 5, Academic Press, N.Y., 1967.

22. Olsson, G., "Simplified Models of Xenon Spatial Oscillations," *Atomkernenergie, Vol. 16, No. 2,* 1970, pp. 91-98.

23. Vollmer, H. and A. J. W. Anderson, "Development of a Dynamic Model for Heavy Water Boiling Reactors and Its Application to the HBWR," OECD Halden Reactor Project, HPR-54, 1964.

24. Eurola, T., "Dynamic Model of the HBWR Heat Removal Circuits," OECD Halden Reactor Project, HPR-62, 1964.

25. Gustavsson, I., "Survey of Applications of Identification in Chemical and Physical Processes, *Automatica, Vol. 11,* 1975, pp. 3-24.

26. Briggs, P. A. N., K. R. Godfrey, and P. H. Hammond, "Estimation of Process Dynamic Characteristics by Correlation Methods Using Pseudo Random Signals," Proc. 1st IFAC Symp. Identification in Automatic Control Systems, Prague, 1967.

27. Cumming, I. G., "Frequency of Input Signal in Identification," Proc. 2nd IFAC Symp. Identification and Process Parameter Estimation, Prague, 1970.

28. Cumming, I. G., "On-Line Identification for the Computer Control of a Cold Rolling Mill," *Automatica, Vol. 8,* 1972, pp. 531-541.

29. Pettersen, F., "Description of System Hardware for the Main Process Computer Installation at the HBWR," OECD Halden Reactor Project, HPR-123, 1971.

30. Åström, K. J. and P. Eykhoff, "System Identification, a Survey," *Automatica, Vol. 7,* 1971, pp. 123-162.

31. Eykhoff, P., *System Identification,* Wiley, 1974.

32. Mehra, R. K. and J. S. Tyler, "Case Studies in Aircraft Parameter Identification," Proc. 3rd IFAC Symp. on Identification and System Parameter Estimation, The Hague, the Netherlands, 1973.

33. Åström, K. J. and T. Bohlin, "Numerical Identification of Linear Dynamic Systems from Normal Operating Records," IFAC Symp. Theory on Self-Adaptive Control Systems (P. H. Hammond, ed.), Teddington, Engl., Plenum Press, N.Y., 1965.

34. Akaike, H., "Statistical Predictor Identification," *Ann. Inst. Statist. Math., Vol. 22, No. 2,* 1970, pp. 203-217.

35. Åström, K. J., *Introduction to Stochastic Control Theory,* Academic Press, N.Y., 1970.

36. Mehra, R. K., "Identification of Stochastic Linear Systems Using Kalman Filter Representation," *AIAA Journal, Vol. 9, No. 1,* 1971, pp. 28-31.

37. Eaton, J., "Identification for Control Purposes," *IEEE Winter Meeting,* N.Y., 1967.

38. Woo, K. T., "Maximum Likelihood Identification of Noisy Systems," Proc. 2nd IFAC Symp. on Identification and Process Parameter Estimation, Prague, 1970.

39. Caines, P. E., "The Parameter Estimation of State Variable Models of Multivariable Linear Systems," Control Systems Centre Report No. 146, The Univ. of Manchester, Inst. of Sci. and Techn., April, 1971.

40. Ljung, L., "On Consistency for Prediction Error Identification Methods," Report 7405, Dept. of Automatic Control, Lund Inst. of Technology, Lund, 1974; see also chapter in this volume.

41. Mehra, R. K. and P. S. Krishnaprasad, "A Unified Approach to the Structural Estimation of Distributed Lags and Stochastic Differential Equations," Third NBER Conference on Stochastic Control and Economic Systems, Washington, D. C., May, 1974.

42. Guidorzi, R., "Canonical Structures in the Identification of Multivariable Systems," *Automatica, Vol. 11,* 1975, pp. 361-374.

43. Olsson, G. and J. Holst, "A Comparative Study of Suboptimal Filters for Parameter Estimation," Report 7324, Dept. of Automatic Control, Lund Inst. of Techn., Lund, 1973.

44. Gustavsson, I., "Parametric Identification of Multiple Input, Single Output Linear Dynamical Systems," Report 6907, Dept. of Automatic Control, Lund Inst. of Techn., Lund, 1969.

45. Åström, K. J. and C. Källström, "Identification of Ship Steering Dynamics," *Automatica, Vol. 12,* 1976, pp. 9-22.

46. Wieslander, J., "IDPAC User's Guide," Report 76, Dept. of Automatic Control, Lund Inst. of Techn., Lund, 1976.

47. Bjørlo, T. J., *et al.*, "Application of Modern Control Theory for Regulation of the Nuclear Power and the Reactor Vessel Pressure of the HBWR," OECD Halden Reactor Project, HPR-131, Halden, 1971.

48. Bohlin, T., "On the Maximum Likelihood Method of Identification," *IBM J. Res. and Dev., Vol. 14,* 1970, pp. 41-51.

49. Söderström, T., "Test of Pole-Zero Cancellation in Estimated Models," *Automatica, Vol. 11,* 1975, pp. 537-539.

50. Åström, K. J., private communication, 1975.

51. Söderström, T., "On the Uniqueness of Maximum Likelihood Identification," *Automatica, Vol. 11,* 1975, pp. 193-197.

A 6
B 7
C 8
D 9
E 0
F 1
G 2
H 3
I 4
J 5